Mortalin Biology: Life, Stress and Death

Sunil C. Kaul • Renu Wadhwa
Editors

Mortalin Biology: Life, Stress and Death

Springer

Editors
Sunil C. Kaul
National Institute of Advanced Industrial Science and Technology
Tsukuba, Ibaraki
Japan

Renu Wadhwa
National Institute of Advanced Industrial Science and Technology (AIST)
Tsukuba, Ibaraki
Japan

ISBN 978-94-017-8279-1 ISBN 978-94-007-3027-4 (eBook)
DOI 10.1007/978-94-007-3027-4
Springer Dordrecht Heidelberg London New York

Printed on acid-free paper

Springer is part of Springer Science+Business Media (www.springer.com)

Foreword

The heat shock response was originally discovered over 40 years ago in the fruit fly and heat shock proteins were subsequently identified in all organisms. It was found that these protect organisms against exposure to suboptimal temperatures and numerous other stresses many of which are also inducers of the response. Research on the mechanisms of action of these stress proteins was aided by development of molecular technologies, identification and cloning of new genes encoding stress proteins in a variety of assay systems both at the cellular and organism levels and it became clear that these proteins have essential constitutive functions in normal unstressed conditions. While *in vivo* animal systems are necessary to make certain firm conclusions, culture cell systems still offer an excellent platform to dissect the molecular mechanisms of action of a protein and to uncover the signal transduction pathways involved in the response. In my view, mortalin is an excellent example of this. Besides its cloning in the normal and cancer cell hybrid screening assay, many of its characteristics such as multiple subcellular residences, impact on p53 protein activity and carcinogenesis, and involvement in neuro-degenerative pathologies have been found since its discovery. The present book offers a single volume reading on the discovery of mortalin biology by experts from different fields and different parts of the globe. It is a unique volume compiling structural, evolutionary and functional aspects of a single stress protein in a variety of model systems ranging from invertebrates to human cells in culture and clinical samples. Besides making an easily understandable reading, it will be very helpful in asking further questions and designing experiments to advance mortalin-based diagnostics and therapeutics.

IBIS, Pavillon C.E. Marchand
1030 Ave de la médecine
Université Laval
Québec, Qc, Canada G1V 0A6
Phone: 418 656–3339
Fax 418 656–7176
E-mail: robert.tanguay@ibis.ulaval.ca

Robert M. Tanguay, D.Sc.
Professor and Associate Head
Dept. Molecular Biology
Medical Biochemistry & Pathology
Lab Cell & Developmental Genetics

Preface

Since the discovery of heat shock response by Ferruccio Ritossa in 1962, the phenomenon has been well characterized in a variety of cells and organisms as induction of a family of proteins called "heat shock proteins" (HSP). Based on their molecular weight, these proteins are classified into, at least, 6 major subfamilies named as HSP100, HSP90, HSP70, HSP60, HSP40 and small HSPs. The fact that the heat shock protein synthesis can be triggered by a variety of other stress conditions such as, infection, inflammation, exercise, starvation, oxygen-, nitrogen- or water-deprivation and exposure to chemical and physical toxins, they are also classified as "stress proteins". Then came the surprise that the HSP also exist under non-stressful conditions and perform housekeeping functions, such as folding and assisting in the establishment of correct protein conformation, mediating protein-protein interactions, intra-cellular trafficking of other proteins, preventing unwanted protein aggregation and channelizing their degradation. A new term "chaperones" evolved to express such functionality of this highly conserved class of proteins.

A new member of HSP70 family of proteins was first cloned in 1993 in a cell hybrid protein-screening assay. Since it was identified to be associated with cellular mortal phenotype, it was named 'mortalin'. Endorsing its multiple functionality, mortalin made its manifestation in many independent experimental regimes, such as those aimed to identify molecules involved in antigen processing, stress-survival and mitochondrial functions. With nearly two decades of experimentation, mortalin has been recognized as an essential protein that not only acts as a chaperone and stress-survival factor but also plays a key role in mitochondrial import motor function, energy generation, ROS management, immune response, control of centrosome duplication and activities of tumor suppressor protein p53. Stemming from these multiple functions is its role in human cancers on one-hand and neurodegenerative diseases on the other. With an aim to introduce mortalin at the graduate and advanced undergraduate levels, this book is organized as a chapter-wise description of structure, evolution and functional role of mortalin in normal and diseased physiology. We hope that this sketch of mortalin biology by the team of experts will help in asking new questions, advancing knowledge and developing mortalin-based diagnostic and therapeutic reagents and technologies.

We are very grateful to all the authors for their interest, enthusiasm and devotion to mortalin research that made this book necessary and possible. Without their hard work to contribute chapters, it was not possible to accomplish this volume suitable for general and specialized reading.

Sunil C. Kaul
Renu Wadhwa

Contents

Contributors

Abdussalam Azem Department of Biochemistry and Molecular Biology, George S. Wise Faculty of Life Sciences, Tel Aviv University, Tel Aviv 69978, Israel

Walter A. Baseler West Virginia University School of Medicine, Division of Exercise Physiology, Center for Cardiovascular and Respiratory Sciences, 1 Medical Center Drive Morgantown, WV 26506, USA

Renata Batistoni Dipartimento di Biologia, Unità di Biologia Cellulare e dello Sviluppo, Università di Pisa. S.S.12 Abetone e Brennero 4, 56127 Pisa, Italy

S. Anne Böttger Department of Biology, West Chester University, West Chester, Pennsylvania 19383, USA

Lena F. Burbulla Laboratory of Functional Neurogenomics, Center of Neurology, Hertie-Institute for Clinical Brain Research and German Center for Neurodegenerative Diseases (DZNE), University of Tübingen, Hoppe-Seyler-Str. 3, 72076 Tübingen, Germany

Francesco Cappello IEMEST, Istituto Euro-Mediterraneo di Scienza e Tecnologia, Palermo, Italy

Travis J. Cook Department of Environmental & Occupational Health Sciences, University of Washington School of Public Health, Seattle, WA 98195, USA

James A. Cowan Evans Laboratory of Chemistry, The Ohio State University, 100 West 18th Avenue, Columbus, Ohio 43210, USA

Tara L. Croston West Virginia University School of Medicine, Division of Exercise Physiology, Center for Cardiovascular and Respiratory Sciences, 1 Medical Center Drive Morgantown, WV 26506, USA

Ritesh P. Daya McMaster University, 1200 Main St. West, HSC 4N81, Hamilton, Ontario, L8N 3Z5, Canada

Custer C. Deocaris Department of Food Science and Technology, College of Home Economics, University of the Philippines, Diliman, Quezon City, Philippines

Sinclair R. Dundas Department of Pathology, University Medical Buildings, Foresterhill, Aberdeen, AB25 2ZD, Scotland, United Kingdom

Zvi Fishelson Department of Cell and Developmental Biology, Sackler School of Medicine, Tel Aviv University, Tel Aviv 69978, Israel

Kenji Fukasawa Molecular Oncology Program, H. Lee Moffitt Cancer Center and Research Institute, Tampa, FL 33612, USA

Ran Gao National Institute of Advanced Industrial Science & Technology (AIST), Central 4, 1-1-1 Higashi, Tsukuba, Ibaraki 305-8562, Japan

Joseph P. Gabriele McMaster University, 1200 Main St. West, HSC 4N81, Hamilton, Ontario, L8N 3Z5, Canada

Rona G. Giffard Dept. of Anesthesia, Stanford University School of Medicine, 300 Pasteur Drive, S272 Grant Building, CA 94305-5117, USA

Sarah E. Groleau McMaster University, 1200 Main St. West, HSC 4N81, Hamilton, Ontario, L8N 3Z5, Canada

Jake G. Hoekstra Department of Pathology, University of Washington School of Medicine, Seattle, WA 98104, USA

John M. Hollander West Virginia University School of Medicine, Division of Exercise Physiology, Center for Cardiovascular and Respiratory Sciences, 1 Medical Center Drive Morgantown, WV 26506, USA

Ohad Iosefson Department of Biochemistry and Molecular Biology, George S. Wise Faculty of Life Sciences, Tel Aviv University, Tel Aviv 69978, Israel

Jinghua Jin Department of Neurobiology, Zhejiang University School of Medicine, Hangzhou, Zhejiang 310058, China

Masayuki Kanai Molecular Oncology Program, H. Lee Moffitt Cancer Center and Research Institute, Tampa, FL 33612, USA

Sunil C. Kaul National Institute of Advanced Industrial Science & Technology (AIST), Central 4, 1-1-1 Higashi, Tsukuba, Ibaraki 305-8562, Japan

Zeenia Kaul Center for Childhood Cancer, The Research Institute at Nationwide Children's Hospital, Columbus, Ohio, USA

Rejko Krüger Laboratory of Functional Neurogenomics, Center of Neurology, Hertie-Institute for Clinical Brain Research and German Center for Neurodegenerative Diseases (DZNE), University of Tübingen, Hoppe-Seyler-Str. 3, 72076 Tübingen, Germany

Nikki P. Lee Department of Surgery, Li Ka Shing Faculty of Medicine, The University of Hong Kong, Hong Kong, China

Ben Low The Jackson Laboratory, 600 Main Street, Bar Harbor, Maine 04609, USA

Wen-Jing Lu Department of Surgery, Li Ka Shing Faculty of Medicine, The University of Hong Kong, Hong Kong, China

John M. Luk Department of Oncology, Roche R&D Center (China) Ltd., 720 Cai Lun Road, Shanghai 201203, China

Wen-I Luo Ohio State Biochemistry Program, The Ohio State University, 100 West 18th Avenue, Columbus, OH 43210, USA

Alberto J. L. Macario Department of Microbiology and Immunology, School of Medicine, University of Maryland at Baltimore; and IMET, Columbus Center, 701 East Pratt Street, Baltimore, Maryland 21202, USA

Everly Conway de Macario Department of Microbiology and Immunology, School of Medicine, IMET, University of Maryland at Baltimore, Columbus Center, 701 East Pratt Street, Baltimore, Maryland 21202, USA

Ram K. Mishra McMaster University, 1200 Main St. West, HSC 4N81, Hamilton, Ontario, L8N 3Z5, Canada

Geneviève Morrow Laboratory of Cell and Developmental Genetics, Department of Molecular Biology, Medical Biochemistry and Pathology, Institut de Biologie Intégrative et des Systèmes and PROTEO, Université Laval, Québec, G1V 0A6, Canada

Oren Moskovich Department of Cell and Developmental Biology, Sackler School of Medicine, Tel Aviv University, Tel Aviv 69978, Israel

Graeme I. Murray Department of Pathology, University Medical Buildings, Foresterhill, Aberdeen, AB25 2ZD, Scotland, United Kingdom

Masanori Nishi National Institute of Advanced Industrial Science & Technology (AIST), 1-1-1 Higashi, Tsukuba, Ibaraki 305-8562, Japan

Angheliki Nomikos Department of Pathology, University Medical Buildings, Foresterhill, Aberdeen, AB25 2ZD, Scotland, United Kingdom

Yi-Bing Ouyang Dept. of Anesthesia, Stanford University School of Medicine, 300 Pasteur Drive, S272 Grant Building, CA 94305-5117, USA

Marie Le Pécheur Laboratory of Cell and Developmental Genetics, Department of Molecular Biology, Medical Biochemistry and Pathology, Institut de Biologie Intégrative et des Systèmes and PROTEO, Université Laval, Québec, G1V 0A6, Canada

Zdenek B. Pristupa McMaster University, 1200 Main St. West, HSC 4N81, Hamilton, Ontario, L8N 3Z5, Canada

Moran Saar Department of Cell and Developmental Biology, Sackler School of Medicine, Tel Aviv University, Tel Aviv 69978, Israel

Robert M. Tanguay Laboratory of Cell and Developmental Genetics, Department of Molecular Biology, Medical Biochemistry and Pathology, Institut de Biologie Intégrative et des Systèmes and PROTEO, Université Laval, Québec, G1V 0A6, Canada

Toshimasa Uemura National Institute of Advanced Industrial Science & Technology (AIST), 1-1-1 Higashi, Tsukuba, Ibaraki 305-8562, Japan

Renu Wadhwa National Institute of Advanced Industrial Science & Technology (AIST), Central 4, 1-1-1 Higashi, Tsukuba, Ibaraki 305-8562, Japan

Charles W. Walker Department of Molecular, Cellular and Biomedical Sciences, Center For Marine Biology and Marine Biomedical Research Group, The University of New Hampshire, Durham, NH 03824, USA

Robin E. White Dept. of Anesthesia, Stanford University School of Medicine, 300 Pasteur Drive, S272 Grant Building, CA 94305-5117, USA

Tomoko Yaguchi National Institute of Advanced Industrial Science & Technology (AIST), Central 4, 1-1-1 Higashi, Tsukuba, Ibaraki 305-8562, Japan

Chae-Ok Yun Department of Bioengineering, College of Engineering, Hanyang University, 17 Haengdang-Dong, Seongdong-Gu, Seoul 133-791, Korea

Jing Zhang Department of Pathology, University of Washington School of Medicine, Seattle, WA 98104, USA

Part I
Structure and Function of Mortalin

Chapter 1
Birth of Mortalin: Multiple Names, Niches and Functions Connecting Stress, Senescence and Cancer

Renu Wadhwa and Sunil C. Kaul

Abstract The mitochondrion, arising from the historical endosymbiosis during the stressful period of the Great Oxidation Event 2.4 billion years ago, marks the existence of all eukaryotes. Retaining only a handful of genes from its ancestral symbiont and yet performing life-essential tasks, it is heavily dependent on the nucleus and a consortium of stress chaperones that maintain its structural and functional integrity by regulation of transport of the nuclear-encoded proteins, their quality control by chaperoning and proteolysis, and energy-generation as a part of their housekeeping and stress-survival functions. Mortalin, first identified in 1993 from cell fusion studies as a marker of mortal cell phenotype, was characterized as an Hsp70 family stress chaperone based on its sequence homology. Nearly two decades of experimental data have revealed its residence beyond the mitochondrial boundaries, life essential functions in and outside the mitochondria and those that specifically promote carcinogenesis on one hand and neurodegeneration on the other. Aimed to portrait mortalin characteristics, both in structure and function and drive the mortalin biology to drug discovery, this chapter reviews the events leading to its identification and role in old age diseases including cancer along with its possibility of being a therapeutic target.

Keywords Mortalin · Hsp70 family · Stress protein · Identification · Functions

1.1 Mortalin as a Member of Hsp70 Family of Proteins

Origin of heat shock proteins (Hsp-s, often called stress chaperones) preceded the birth of the mitochondria that marked the Great Oxygenation Event and the origin of the first eukaryote (Margulis 1975). Central to the mitochondrial evolution was the transition from individualistic bacteria to host-dependent organelles. Phylogenetic studies suggest that the earliest bacterial symbiont may have probably carried a genome of 630 distinct genes (Gabaldon and Huynen 2003) that got gradually lost

R. Wadhwa (✉) · S. C. Kaul
National Institute of Advanced Industrial Science and Technology (AIST), Central 4, 1-1-1 Higashi, Tsukuba, Ibaraki 305-8562, Japan
e-mail: renu-wadhwa@aist.go.jp

S. C. Kaul, R. Wadhwa (eds.), *Mortalin Biology: Life, Stress and Death*,
DOI 10.1007/978-94-007-3027-4_1, © Springer Science+Business Media B.V. 2012

and transformed to the present day human mitochondrion that encodes for only 13 polypeptides, 22tRNAs and 2 rRNAs (Anderson et al. 1981), and functions to actively convert free atmospheric oxygen into water. For this gene-depleted present day organelle to function, its proteome will need to be synthesized in the cytosol; traverse the mitochondrial boundaries, the outer (OM) and inner (IM) membranes; refold back into their native forms; and finally, get sorted into various intra-mitochondrial locations (Elstner et al. 2008; Huynen et al. 2009). Intuitively, the development of machineries for import, post-import folding, maturation and segregation must have originated as 'adaptive' evolutionary phenomenon. Mitochondrial chaperones, being the stress proteins that arose from harsh planetary conditions, are likely to be the most competent guardians of the mitochondrial proteome: its import, protein quality control and stress protective functions. They act both in housekeeping and stress responses based on their ability to bind with unfolded (nascent) and misfolded (denatured) proteins (Ecroyd and Carver 2008; Tatsuta 2009).

It was more than half a century ago when Ferruccio Ritossa reported the unusual puffing patterns in the polytene chromosomes of *Drosophila* after 30-min exposure of its larvae to elevated temperatures (37 °C) and their return to ambient temperatures for recovery (Ritossa 1962). The term Hsp-s was later dubbed when follow-up experiments revealed the increased expression of 70- and 26-kDa proteins suggesting that these gene products may be indispensable molecules that assisted protein refolding to overcome heat stress (Ananthan et al. 1986; Tissieres et al. 1974). The concept of Hsp-s as molecular chaperones was built from the earlier ideas of Laskey and colleagues that described chaperonization, an activity associated with nucleoplasmin in *Xenopus* oocytes (Laskey et al. 1978). The term was expanded to include a diverse class of proteins that aid polypeptide folding, transit across cellular and organelle membranes, assist the disassembly of macromolecular complexes or aggregates, regulate their conformation and target them for proteolysis to assure protein quality control that widely affect bio-signaling and functions (Ellis 1987; Hartl 1991).

Hsp family of proteins is composed of at least 40 members in humans. They are grouped into at least 6 major subfamilies named as Hsp100, Hsp90, Hsp70, Hsp60, Hsp40 and small Hsp-s based on their molecular weights (Powers and Workman 2007). Hsp-s of the same family share similar domain structure, whereas members of each specific family associate with unique pattern of expression and cellular localization (Lindquist and Craig 1988). These are known for their multifunctional ability such as housekeeping functions in maintaining the protein structure, gene transcription, signal transduction and immunity (Helmbrecht et al. 2000), and induction in response to stress (like high temperature, chemical and physical stress resulting in augmentation of the biological functions for sustaining cell survival) (Sherman and Multhoff 2007). The present chapter portraits events on identification, cloning and functional role of mortalin (a member of Hsp70 family of proteins) in normal and abnormal, stressed and diseased, scenarios.

1.2 Mortalin-Multiple Births

1.2.1 Cell Fusion Studies for Markers of Mortality and Immortality

In order to track proteins involved in regulation of cellular mortality and immortality, normal and immortal mouse fibroblasts were fused to generate hybrid mortal cells and screened for proteins associated with either the mortal or immortal phenotype. An approximately 66-kDa cytoplasmic protein segregated with the loss of immortality in cybrids and was named 'mortalin' (Wadhwa et al. 1993a). An antibody raised against the protein revealed its pancytoplasmic distribution in normal mortal cells. Surprisingly, it also detected the protein in immortal mouse fibroblasts but was perinuclear in localization. Using the antibody, the cDNA for mortalin was cloned. It is 2850 bp in length, encoding a 74-kDa protein constituting 679 amino acids with a high degree of homology with members of the Hsp70 family, including *Escherichia coli* DnaK (51%), *Saccharomyces cerevisiae* SSC1p (65%), the constitutive cytosolic Hsp70 from rat, Hsc70 (46%) and the rat endoplasmic reticulum isoform, BiP (49%) (Wadhwa et al. 1993a). Although the complete crystal structure of mortalin has not yet been resolved, based on the conserved homology and bioinformatics, its three-dimensional structure was unraveled. Like most Hsp-s, mortalin has 2 principal domains, the amino-terminal ATPase region and carboxyl-terminal region, as illustrated by the kettle pot model (Kaul et al. 2007). A second cloning from mouse immortal cells revealed that the mouse mortalin exists in two isoforms of opposing phenotypes: the mortality-associated pancytoplasmic form (which was renamed as mortalin-1, mot-1) and the immortalization-associated perinuclear mortalin (mot-2) (Wadhwa et al. 1993b, 1996; Kaul et al. 1998). Mouse mot-1 and mot-2 cDNA differed by only two amino acids (V618M and R624G) in the carboxy-terminus and segregated in F1 and F2 progenies suggesting that these were encoded by two alleles on chromosome 18 (Kaul et al. 1995; Wadhwa et al. 1996). The two minutely different proteins were found to have different structural and functional characteristics to the extent that overexpression of mot-1 in immortal NIH3T3 cells induced senescence, while overexpression of mot-2 in the same cells mediated malignant transformation (Kaul et al. 1998; Wadhwa et al. 1993b). Recently, another variant of mouse mortalin, D626G, was identified and awaits functional characterization (Chardonnet et al. 2007). Studies on mortalin in humans revealed only one form of mortalin that possessed malignant transformation activity as mouse mot-2 (Kaul et al. 1998, 2007).

1.2.2 Mortalin as CSA (C3H Strain Specific Antigen)

Nearly at the same time when mortalin took its birth in mortality/immortality screen as described above, in a totally independent scenario, it was identified as a strain-specific antigen, found only in C3H mice and was named CSA (C3H-specific antigen). Mouse mortalin gene was sequenced and shown to contain 17 exons interrupted by 16 introns and has two dimeric repeats of the consensus sequence of the heat-shock element in the 5′-flanking region (Michikawa et al. 1993); surprising to this fact, the protein is heat un-inducible. Another feature that aligned well with its mitochondrial residence was described as its first intron interrupted within the amino-terminal leader sequence, a pattern found similar to that of cytochrome c1, a well-known mitochondrial protein (Michikawa et al. 1993). Interestingly, the presence of two isoforms of mortalin correlated with the immortalization tendencies of fibroblasts derived from specific mouse strains. Fibroblasts from C3H strain of mouse that contain mot-1 were difficult to immortalize as compared to the mot-2 harboring fibroblasts from Balb/c and C57BL/6 strains. Despite these differences, both types of mortalin are essential for cell survival (Domanico et al. 1993; Michikawa et al. 1993) and one out of these two residues, arginine at residue 578 of C3H mouse, contributed to the immunogenicity of the protein. Using anti-CSA monoclonal antibody, the subcellular localization of CSA was shown to be the mitochondria and the fact that new genetic marker in mice was located on a gene encoding for a mitochondrial protein caught lot of attention.

1.2.3 Mortalin as PBP74 (Peptide-Binding Protein) in Immune Regulation

Yet another lab looking for proteins involved in antigen-processing, identified peptide-binding proteins (PBP72/74) by their ability to bind to a model antigenic peptide from pigeon cytochrome C (Pc). PBP72/74 did not bind to the native Pc and thus were suspected to recognize some feature of peptides not found in the native antigens. Antisera raised against PBP72/74 blocked the presentation of native antigen and of the corresponding fragment. Although the role of PBP72/74 in antigen processing is still a matter of research, it was interesting that the investigators detected the protein on cell surface, endosomes, golgi, ER, plasma membrane and the vesicular cytoplasmic structures. Cloning of PBP74 cDNA revealed that it is identical to mortalin (Domanico et al. 1993). Recently, it was demonstrated that the surface-expressed mortalin plays important role in antigen presentation and in innate immunity (Pilzer and Fishelson 2005; Pilzer et al. 2005). It was shown to bind to complement C8 and C9, shed in vesicles containing C9 and complement membrane attack complex (MAC) and involved in MAC elimination. Anti-mortalin antibodies increased cell sensitivity to MAC-mediated lysis suggesting that mortalin promotes the shedding of membrane vesicles loaded with complement MAC and protects cells from complement-mediated lysis.

1.2.4 Mortalin as Grp75 (Glucose Regulated Protein 75) in Stress Response

To discover novel metabolic stress markers for the central nervous system, Massa et al. screened candidate genes with degenerate RT-PCR primers (Massa et al. 1995) and reported a rat-brain cDNA that encoded the glucose-regulated protein 75 (Grp75). It turned out to be a mitochondrial member of the Hsp70 family with sequence homology to mortalin and mortalin/PBP74/CSA. *In situ* analysis of normal brain revealed an abundance of Grp75 in neurons of the basal forebrain, reticular and subthalamic nuclei, globus pallidus and amygdala. With focal brain ischemia, Grp75 mRNA was upregulated in a peculiar fashion depending on the degree of injury. If ischemic focus was small, induction occurred only within the affected area, whereas, with a more extensive damage, Grp75 acquired a more global expression pattern. Consistent to this, five isoforms of Grp-s at 74–75 kDa mass were found from proteomic profiling of 2-deoxyglucose-treated murine and human fibroblasts. Cells treated with 2-deoxyglucose provide an *in vitro* model for glucose deprivation (Merrick et al. 1997). Stress from energy deprivation in the tibialis anterior muscle, a type II muscle, after 10 days of chronic contractile activity, stimulated mortalin/Grp75 protein, but not its mRNA (Ornatsky et al. 1995). Orsini et al. showed that a fraction of cytosolic p66Shc (regulates lifespan in mammals and is a critical component of the apoptotic response to oxidative stress) localizes within mitochondria where it forms a complex with mitochondrial Hsp70/mortalin (Orsini et al. 2004). Mortalin was shown to inhibit p66Shc function activated during oxidative stress (ultraviolet radiation) that induced the dissociation of p66Shc-mortalin complexes. Another study identified mortalin as one of the factors responsible for superior stress defense in murine embryonic stem cells (Saretzki et al. 2004) suggesting that it is an important component of the glucose and oxidative stress response of cells.

1.2.5 Mortalin as mtHsp70, a Mitochondrial Chaperone

Combination of immunological, biochemical and functional approaches both *in vitro* and *in vivo* revealed that human mortalin was imported into and stayed in the mitochondrial compartment (Dahlseid et al. 1994). By confocal immunofluorescence microscopy, mortalin was found inside the organelle and co-localized with the mitochondrial Hsp60. Deletion of the N-terminal 46-amino acid pre-sequence resulted in a cytosolic localization of the epitope-tagged protein (Dahlseid et al. 1994). Bhattacharyya et al. (1995) cloned human mtHsp70 gene by screening of an expression library with a monoclonal antibody and demonstrated that the 3A3-reactive protein co-fractionated with mitochondrial proteins. The nucleotide sequence of the respective cDNA clone matched with mortalin.

1.2.6 Mortalin as Tumor Necrosis Factor Receptor-Associated Protein 1

Tumor necrosis factor receptor-associated protein 1 (TRAP-1) was originally isolated from yeast two-hybrid system as a protein that interacted with the intracellular domain of the type 1 tumor necrosis factor receptor (TNFR-1). It was also identified as a mitochondrial heat shock protein in Saos-2 (human osteosarcoma) cells adapted to mild oxidative stress induced by diethylmaleate (DEM). A recent study mentioned mortalin to be TRAP-1/mtHsp70 stimulated by ischemia. To understand the role of TRAP-1 in brain injury, it was overexpressed in astrocytes and was found to drop ROS production with glucose-deprivation. In addition, TRAP-1 preserved mitochondrial membrane potential, maintained ATP levels and cell viability during stress. Such findings endorsed TRAP-1 to be mortalin/mtHsp70/Hsp75/Grp75/PBP74 and an interesting gene that provided protection against ischemia-like *in vitro* injury (Voloboueva et al. 2007). It conferred greater resistance to hydrogen peroxide and cisplatin, and inhibited release of apoptosis-inducing factor upon cisplatin treatment (Montesano et al. 2007). Just like mortalin in other studies, TRAP-1 was detected in the mitochondrial matrix and non-mitochondrial locations, including pancreatic zymogen granules, insulin secretory granules, cardiac sarcomeres, nuclei of pancreatic and heart cells, and on the cell surface of blood vessel endothelial cells (Cechetto and Gupta 2000).

1.3 Mortalin: Inside and Outside the Mitochondria

Representing approximately 1% of the total protein content, mortalin is one of the most abundant proteins in the mitochondrial matrix (Naylor et al. 1996). It fulfills two special needs of the mitochondria: (i) constitutes an essential component of the import machinery and (ii) protein quality control by assisting in functional folding and degradation of unfunctional proteins. It has been identified as the only ATPase component of the pre-protein mitochondrial import complex (Schneider et al. 1994; Brunner et al. 1995) and plays a crucial role in mitochondrial biogenesis: the translocation of cytosolic precursors, their partitioning within the matrix and across the two mitochondrial membranes (Rehling et al. 2004). Along with the second mitochondrial chaperone-Hsp60, mortalin has been shown to maintain mitochondrial homeodynamics by taking care of degradation of the misfolded nonfunctional proteins and ROS by mitochondrial stress response signaling mediated by the transcription factor CHOP (Zhao et al. 2002; Yoneda et al. 2004). Unlike the well-understood ER stress response signaling that is mediated ER-resident Hsp70 BiP/Grp78 and proximal signal transducers IRE1, PERK and ATF6, mechanism of mitochondrial stress response is yet to be resolved.

Despite the fact that mitochondrion was frequently assigned as mortalin's home, it was seen traveling to many other subcellular sites and the idea of mortalin being a mitochondrion's permanent resident, indeed, was ramified to include its 'adventuring' tendencies. Ran et al. (2000) by undertaking subcellular fractionation and immunoelectron microscopy in a variety of human cancer cell lines revealed that mortalin exists in mitochondria of all the tested cells and travels to other subcellular organelles, ER and Golgi, in a cell line specific way. Ma et al. (2006) detected it in the nucleus of dividing cells at the time of chromosome duplication. Most recently, mortalin is also found as a secreted protein and detected in the extra-cellular space similar to Hsp60 so besides its primary home as mitochondria, mortalin lives in many subcellular sites that is thought to be relevant to its multiple functionality. Some other examples of mitochondrial proteins found at unexpected locations both in the normal and pathogenic states include the mitochondrion proteins aspartate aminotransferase (mAsmAT) and Hsp60. mAsmAT is regarded as a transporter of free fatty acids into the mitochondria (Passarella et al. 1990), albeit, containing an N-terminal mitochondrial targeting sequence, was found on the cell surface and into the culture medium. Hsp60 mitochondrial chaperone was initially discovered in mammalian cells as a protein altered in Chinese Hamster Ovary (CHO) cells that were made resistant to the microtubule (MT)-inhibitor podophyllotoxin (Singh et al. 1997; Soltys and Gupta 2000). Found in various extra-mitochondrial locations, it has been shown to play role as an amino acid transporter, biosynthesis and packaging of insulin, modulating chromatin packing by histone 2B, and the regulation of cell cycle via the plasma membrane resident p21ras protein (Gupta et al. 2008; Gupta and Knowlton 2005; Knowlton and Gupta 2003; Soltys and Gupta 2000). How mortalin could arrive at specific extra-mitochondrial destinations is yet to be understood. Nevertheless, the phenomenon of multiple localizations and multiple functions may reflect a molecular-evolutionary protein economics that argues for a single protein to acquire distinct roles in more than one cellular compartment obviating the need to create a new gene. A comprehensive review of some of the likely mechanisms that could control the export of resident proteins from the mitochondrial matrix to other intra- and extra-cellular sites from evolutionary perspectives can be found elsewhere (Soltys and Gupta 1999, 2000).

Outside the mitochondria, mortalin is expected to collaborate with an expanding list of binding partners, described in several reviews, that endows it an assumption of wider cellular roles ranging from intracellular trafficking, control of centrosome duplication, regulation of p53 activity, calcium and ROS signaling, differentiation among many others (Deocaris et al. 2006, 2007a, b; Kaul et al. 2007; Takano et al. 2001; Wadhwa et al. 2003b, 2005).

1.4 Mortalin: Stress, Aging and Cancer

Sequence homology had placed mortalin in heat shock 70 family of proteins. Although remained unresponsive to the heat shock (Wadhwa et al. 1993a), several other stress conditions such as glucose deprivation, ionizing radiations, hypoxia and increase in the reactive oxygen species (ROS) were shown to induce mortalin that acts as a stress-survival factor (Carette et al. 2002; Hori et al. 2002; Liu et al. 2005; Merrick et al. 1997; Yang et al. 2011b). Suppression of mortalin by antisense mortalin oligonucleotide was found to sensitize cells to ionizing radiation and oxidative stress (Sadekova et al. 1997; Yang et al. 2011a, b). Another study proposed mortalin as a DNA-PK regulated protein that plays a protective role against drug-induced apoptosis and determines drug sensitivity (Um et al. 2003b).

In worms, overexpression of mortalin led to increased longevity (Yokoyama et al. 2002), and its knockdown led to accelerated aging syndrome (Kimura et al. 2007), associated with defects in mitochondrial import, a reduction in the levels of ATP-2, Hsp60 and CLK-1. Mortalin-compromised worms showed progeria like phenotypes including lower motility, defects in oogenesis, earlier accumulation of auto-fluorescent material, and a shorter life span (Kimura et al. 2007). Since a close correlation exists between stress resistance and longevity mechanisms, it may be true for even more complex models, such as mice. In general, when stress and impairment of the chaperone system are combined, the resulting gene-environment interaction may amount to causative impetus to premature aging (Macario and Conway de Macario 2002). Caloric restriction (CR), the only effective experimental manipulation known to retard aging in rodents and primates, restored age-impaired chaperone induction, while reversing the age-induced changes in constitutive level of Hsp (Berner and Stern 2004; Boxenbaum 1991; Kirkwood and Shanley 2000). CR rats were seen to have increased mortalin expression level in the testis (Um et al. 2003a, b). These examples support the hypothesis that mortalin endows a better adaptation capacity to various forms of stresses and increases the longevity in an organism. On the other hand, a sick version of mortalin (oxidized mortalin), in fact, was found in brain tissues of a rodent model for Alzheimer's disease (Berner and Stern 2004), and this may likely be actively involved in etiology of the disease rather than as mere molecular fossil of neurodegeneration. The oxidized form of mortalin was tested to act as an anti-chaperone, promotes protein aggregation, and overrides the chaperone activity of undamaged mortalin protein (Deocaris et al. unpublished data). Several studies have endorsed the involvement of mortalin to age-pathologies including cardiovascular diseases (Massa et al. 1995), diabetes (Matsuoka et al. 2005; Zhang et al. 2006) and neurodegenerative disorders, Alzheimer's and Parkinson's Diseases (PD) (Calabrese et al. 2001; Choi et al. 2004; Jin et al. 2006; Osorio et al. 2007; Seyb et al. 2006; Sirk et al. 2007). It was also shown to be one of the five proteins that form complex with alpha-synuclein and DJ-1 (an oncogene and causative gene for familial form of the PD) and is critically involved in the pathogenesis of PD (Jin et al. 2006, 2007; Shi et al. 2008). Li et al. showed that DJ-1 is associated with HSP70, CHIP and mortalin and the complex is involved in regulation of oxidative stress (Li et al. 2005). An association of wild type DJ-1, but not the mutants found in PD patients, was enhanced by treatment of cells with H_2O_2. Van Laar et al. also showed that the level of mortalin decrease during dopamine oxidation leading to selective dopaminergic terminal degeneration *in vivo* and altered mitochondrial function *in vitro* (Van Laar et al. 2008). In a quantitative proteomic study on comparison of the nigral mitochondrial proteins of Parkinson's disease (PD) patients with those from age-matched controls, mortalin was detected as downregulated in the PD patients (Jin et al. 2006). Indeed, manipulations of mortalin levels in dopaminergic neurons resulted in significant changes in sensitivity to PD phenotypes *via* pathways involving mitochondrial, proteasomal and oxidative stress response functions (Jin et al. 2006) revealing that mortalin is involved in the PD pathogenesis.

Within the ROS-bathed cellular environment, mutations stochastically accumulate with time contributing to genomic instability and cancers. Consistent with the major involvement of ROS-related mutational events, Bert Vogelstein's group found that the majority of mutations in ten human colorectal cancer cell lines were: (a) somatically acquired mtDNA mutations (b) the detected mutations however were not associated with major perturbations of mitochondrial functions, as oxygen consumption (Polyak et al. 1998). Given that cancer cells carry biologically risky and numerous mitochondrial gene mutations that could exacerbate mitochondrial dysfunction, it may be suggested that a strong chaperone buffering system within this organelle could be one plausible strategy how cancer cells are able to tolerate high mutational loads. Several studies have found that the level of mortalin was elevated in many human tumors, the tumor-derived and *in vitro* immortalized cells. Remarkably, overexpression of mortalin matched with the increase in aggressiveness of brain tumors from astrocytoma to glioblastoma (Takano et al. 1997) and was sufficient to increase the malignancy of breast carcinoma cells suggesting that an upregulation of mortalin contributes significantly to tumorigenesis (Wadhwa et al. 2006). Comparative proteomic analysis identified the correlation of mortalin overexpression with poor patient survival in colorectal adenocarcinomas (Dundas et al. 2005) and post-surgery hepatocarcinoma recurrence (Yi et al. 2008). In chronic myeloid leukemia (CML), a hematopoietic stem cell disease containing an aberrant Bcr-Abl protein tyrosine kinase activity, mortalin was identified as a major protein down-regulated during the progression of a benign chronic phase to a rapidly fatal blast crisis. The absence of correlation between mRNA and protein levels pointed at the possible post-translational events that modify protein content (Smith et al. 2002). Another proteomic study on bone marrow cells from CML patients also identified mortalin as one of the 31 proteins that described a chronic phase molecular phenotype (Pizzatti et al. 2006).

In a study on the disease models of hematopoiesis in which Zebrafish mutants with abnormalities at various stages in blood development were used, positional cloning of a developmental blood mutant (crimsonless (*crs*)- anemic) revealed that the mutated gene was mortalin/HSPA9B that shows 84.8% identity and 89.4% similarity with human mortalin (Craven et al. 2005). A single amino acid mutant (G492E) within the substrate-binding domain of HSPA9B was shown to be the cause of the *crs* phenotype. Interestingly, a near-identical mutation in the conserved glycine at position 443 in DnaK (53.5% identity and 63.3% similarity to Zebrafish mortalin) completely abolished pro-peptide binding and chaperone function (Burkholder et al. 1996). To verify that the mutation in HSPA9B is sufficient to cause the *crs* phenotype, investigators used both rescue and antisense morpholino knockdown strategies (Craven et al. 2005). Injection of capped RNA encoding wild-type HSPA9B rescued approximately 95% (53 of 56) injected mutant embryos. Conversely, inactivation of Zebrafish mortalin using antisense morpholino-modified oligonucleotides recapitulated the anemic phenotype demonstrating that a single amino acid change, G492E that abolishes chaperone function of mortalin is the cause of *crs* phenotype in Zebrafish, a model of human MDS (Craven et al. 2005). The role of mortalin in cancer is best explained by its interactions with tumor suppressor protein p53 (Deocaris et al. 2007b; Kaul et al. 2001, 2005; Lu et al. 2011a, b; Ma et al. 2006; Wadhwa et al. 1998, 2010; Walker et al. 2006). Mortalin-p53 interactions were abrogated both in mammalian and clam cells by a cationic inhibitor (MKT-077) that binds to mortalin and dissociates p53 from the complex resulting in the reactivation of wild type p53 function (Wadhwa et al. 2000; Walker et al. 2006; Deocaris et al. 2007c; Pilzer et al. 2009).

1.5 Mortalin: Therapeutic Target

Considering the wide diversity of mortalin functions, it is likely that mortalin-based therapy would be useful in dampening the impact of some of the chaperone-associated maladies involving both the chaperone-hyperfunction (such as in cancers) and chaperone-deficiency (such as in neurodegenerative disorders) (Macario and Conway de Macario 2007). A variety of reagents hold promises and await further studies on validation. Some worth-mentioning here are mortalin injectibles, mortalin peptides, small molecules and antibodies.

Given the growing interest on developing recombinant stress proteins for chaperonotherapy, the field is expected to expand and validate information on the use of individual stress chaperones as chaperonotherapeutic tools. Some of the initial evidence that an Hsp70 member might serve as a chaperonotherapeutic agent is from the purified bovine brain Hsc70. When administrated exogenously, the chaperone proved useful for repair of peripheral sensory nerve damage. In this particular experiment, axotomy induced death in 33% of dorsal root ganglion neurons and 50% of motoneurons, and damage-control by the recombinant Hsc70 was apparent in virtually all sensory neurons (Houenou et al. 1996). Gifondorwa et al. tested whether the recombinant human Hsp70 could be used for treating amyotrophic lateral sclerosis (ALS), a debilitating neurodegenerative disorder that results in the progressive loss of motor neurons in the central nervous system (Gifondorwa et al. 2007). Using the G93A mutant SOD1 mouse, the group intraperitoneally-injected Hsp70 (3 times/per week) from postnatal day 50 until the end-stage of the disease. Such regimen was observed to lead to increased lifespan, delayed symptom onset, and intact motor functions. Interestingly, it also resulted in a more robust innervations of the neuromuscular junctions compared with control tissues. It was thus suggested that an Hsp70-based chaperonotherapy might be used to delay the disease progression in an ALS model via an unknown peripheral mechanism (Gifondorwa et al. 2007). An alternative approach would involve the use of *protein inducers* (Macario and Conway de Macario 2007). Given the roles played by stress chaperones in the maintenance of proteome integrity during aging and stressed conditions, it would be of considerable benefit to discover new compounds that will induce HSPs without any toxic effects. One of the first Hsp70 inducing agents introduced is Bioclomol, a hydroxylamine derivative developed by the Hungarian biotech company, Biorex Research and Development Co. Originally marketed to prevent microangiopathy in diabetes patients, the drug amplifies induction of Hsp70 when cells are subjected to stressful conditions (Hargitai et al. 2003). Two molecules are found to boost cellular production of mortalin: 2-deoxyglucose (Merrick et al. 1997) and glycerol (Deocaris et al. 2008).

Interestingly, treatment of cells with a chemical chaperone, glycerol caused heat-shock and oxidative stress responses, induction of mortalin and stimulation of proteasomal system. In *C. elegans,* it resulted in decreased accumulation of old age-associated lipofuscin suggesting that mortalin inducers may serve as anti-aging reagents. Further studies are warranted to resolve the molecular effects and mechanisms of action.

Mortalin was shown to bind to tumor suppressor protein and inactivate its transcriptional activation function by cytoplasmic sequestration (Kaul et al. 2005). Finding that the carboxy-terminal 312–352 residues of tumor suppressor protein p53 bind to mortalin, shorter peptides were used as mortalin binding antagonists. These peptides were seen to activate endogenous p53 function by displacing mortalin from p53-mortalin complexes and relocating p53 to the nucleus. This was sufficient to activate p53 function causing growth arrest in human osteosarcoma and breast carcinoma cells (Kaul et al. 2005). Similar to the peptides, MKT-077, a cationic rhodacyanine dye analogue and withanone, a phytochemical binds to mortalin and abrogates its interactions with p53 resulting in the release of p53 from cytoplasmically sequestered p53-mortalin complexes and reactivation of its transcriptional activation and apoptotic functions (Deocaris et al. 2007c). Thus, MKT-077 and withanone are the anti-mortalin reagents and candidate anticancer drugs. Induction of senescence like growth arrest by bromodeoxyuridine (Michishita et al. 1999) and 5-aza-2′ deoxycytidine (Widodo et al. 2007) caused shift in subcellular distribution of mortalin from perinuclear to pancytoplasmic type. It was shown that mortalin was direct target of these drugs and undergoes structural changes that may affect its function as chaperone, importer or regulator (Widodo et al. 2007; Deocaris et al. 2008). Besides the peptides and the chemicals, mortalin-specific ribozymes and siRNA that caused suppression of mortalin expression resulted in the growth arrest/apoptosis of transformed human cells (Wadhwa et al. 2003a; Yoo et al. 2010). Anti-mortalin antibodies were seen to have antitumor activity in nude mice xenografts. Cell internalizing feature of these antibodies was also exploited as a nanocarrier tool for gene delivery and imaging (Shiota et al. 2007; Ohyabu et al. 2009; Yoshioka et al. 2011).

1.6 Summary

The concomitant, yet independent, discoveries of mortalin have reflected its multiple functionality. Mortalin is indeed a two-billion year old resident of the mitochondria, the living descendant of the first DnaK in endosymbiotic alpha-proteobacter that become a power-generating organelle and acquired multiple roles reflecting an example of molecular-evolutionary protein economics. Furthermore, this promiscuous chaperone is partnered with a larger array of binding partners, from transcription factors, cell receptors, cytoskeleton elements and many others. Like a "molecular megalomaniac", mortalin has expanded its biological roles. The qualitative and quantitative nature of mortalin responses, as seen from the upregulation in various tumors to the presence of sick (oxidized) forms in neurodegenerative diseases, render future experimental and clinical studies to unravel its mechanistic aspects. From the understanding that the typical tools used for the basic study of mortalin functions in the laboratory, such as recombinant mortalin, mini-mortalins, ribozymes and siRNA, internalizing-mortalin antibodies, etc., are also expected to evolve into the next generation chaperonotherapeutics—from mortalin peptide-based cytotoxics, vaccines, tumor senescence-modulators and youth rejuvenators. As we come to appreciate the rapidly growing chaperonology of mortalin, this stress protein has undoubtedly emerged as an extremely versatile molecule. The importance of mortalin in cell biology is underscored by its high degree of structural and phylogenetic conservation, and the fact that no cell survives in its absence. As newer genomic technologies, like chaperonomics and systems biology, offer fresher perspectives, it is anticipated that our present knowledge on the physiological roles of this chaperone may still be limited. Nonetheless, the present book portraits the wealth of information on this versatile stress molecule and picture the mechanisms on how it plays essential roles in stress and survival.

References

Ananthan J, Goldberg AL, Voellmy R (1986) Abnormal proteins serve as eukaryotic stress signals and trigger the activation of heat shock genes. Science 232:522–524

Anderson S, Bankier AT, Barrell BG, de Bruijn MH, Coulson AR, Drouin J, Eperon IC, Nierlich DP, Roe BA, Sanger F et al (1981) Sequence and organization of the human mitochondrial genome. Nature 290:457–465

Berner YN, Stern F (2004) Energy restriction controls aging through neuroendocrine signal transduction. Ageing Res Rev 3:189–198

Bhattacharyya T, Karnezis AN, Murphy SP, Hoang T, Freeman BC, Phillips B, Morimoto RI (1995) Cloning and subcellular localization of human mitochondrial hsp70. J Biol Chem 270:1705–1710

Boxenbaum H (1991) Gompertz mortality analysis: aging, longevity hormesis and toxicity. Arch Gerontol Geriatr 13:125–137

Brunner M, Schneider HC, Lill R, Neupert W (1995) Dissection of protein translocation across the mitochondrial outer and inner membranes. Cold Spring Harb Symp Quant Biol 60:619–627

Burkholder WF, Zhao X, Zhu X, Hendrickson WA, Gragerov A, Gottesman ME (1996) Mutations in the C-terminal fragment of DnaK affecting peptide binding. Proc Natl Acad Sci U S A 93:10632–10637

Calabrese V, Scapagnini G, Giuffrida Stella AM, Bates TE, Clark JB (2001) Mitochondrial involvement in brain function and dysfunction: relevance to aging, neurodegenerative disorders and longevity. Neurochem Res 26:739–764

Carette J, Lehnert S, Chow TY (2002) Implication of PBP74/mortalin/GRP75 in the radio-adaptive response. Int J Radiat Biol 78:183–190

Cechetto JD, Gupta RS (2000) Immunoelectron microscopy provides evidence that tumor necrosis factor receptor-associated protein 1 (TRAP-1) is a mitochondrial protein which also localizes at specific extramitochondrial sites. Exp Cell Res 260:30–39

Chardonnet S, Decottignies P, Amar L, Le Caer JP, Davis S, Laroche S, Le Marechal P (2007) New mortalin and histidyl tRNA synthetase isoforms point out a pitfall in proteomic analysis of Egr1 genetically modified mice. Proteomics 7:289–298

Choi J, Forster MJ, McDonald SR, Weintraub ST, Carroll CA, Gracy RW (2004) Proteomic identification of specific oxidized proteins in ApoE-knockout mice: relevance to Alzheimer's disease. Free Radic Biol Med 36:1155–1162

Craven SE, French D, Ye W, de Sauvage F, Rosenthal A (2005) Loss of Hspa9b in zebrafish recapitulates the ineffective hematopoiesis of the myelodysplastic syndrome. Blood 105:3528–3534

Dahlseid JN, Lill R, Green JM, Xu X, Qiu Y, Pierce SK (1994) PBP74, a new member of the mammalian 70-kDa heat shock protein family, is a mitochondrial protein. Mol Biol Cell 5:1265–1275

Deocaris CC, Kaul SC, Wadhwa R (2006) On the brotherhood of the mitochondrial chaperones mortalin and heat shock protein 60. Cell Stress Chaperones 11:116–128

Deocaris CC, Kaul SC, Wadhwa R (2007a) Heat shock chaperone mortalin and carcinogenesis. In: Calderwood SK, Sherman MY, Ciocca DR (eds) Heat shock proteins in cancer. Springer, Netherlands, pp 141–158

Deocaris CC, Kaul SC, Wadhwa R (2007b) Mortalin—a driver on the crossroads of stress, aging and carcinogenesis. In: Sreedhar AS, Srinivas UK (eds) Stress response: a molecular biology approach. Research Signpost, pp 135–158

Deocaris CC, Widodo N, Shrestha BG, Kaur K, Ohtaka M, Yamasaki K, Kaul SC, Wadhwa R (2007c) Mortalin sensitizes human cancer cells to MKT-077-induced senescence. Cancer Lett 252:259–269

Deocaris CC, Takano S, Priyandoko D, Kaul Z, Yaguchi T, Kraft DC, Yamasaki K, Kaul SC, Wadhwa R (2008) Glycerol stimulates innate chaperoning, proteasomal and stress-resistance functions: implications for geronto-manipulation. Biogerontology 9:269–282

Domanico SZ, DeNagel DC, Dahlseid JN, Green JM, Pierce SK (1993) Cloning of the gene encoding peptide-binding protein 74 shows that it is a new member of the heat shock protein 70 family. Mol Cell Biol 13:3598–3610

Dundas SR, Lawrie LC, Rooney PH, Murray GI (2005) Mortalin is over-expressed by colorectal adenocarcinomas and correlates with poor survival. J Pathol 205:74–81

Ecroyd H, Carver JA (2008) Unraveling the mysteries of protein folding and misfolding. IUBMB Life 60:769–774

Ellis J (1987) Proteins as molecular chaperones. Nature 328:378–379

Elstner M, Andreoli C, Ahting U, Tetko I, Klopstock T, Meitinger T, Prokisch H (2008) MitoP2: an integrative tool for the analysis of the mitochondrial proteome. Mol Biotechnol 40:306–315

Gabaldon T, Huynen MA (2003) Reconstruction of the proto-mitochondrial metabolism. Science 301:609

Gifondorwa DJ, Robinson MB, Hayes CD, Taylor AR, Prevette DM, Oppenheim RW, Caress J, Milligan CE (2007) Exogenous delivery of heat shock protein 70 increases lifespan in a mouse model of amyotrophic lateral sclerosis. J Neurosci 27:13173–13180

Gupta S, Knowlton AA (2005) HSP60, Bax, apoptosis and the heart. J Cell Mol Med 9:51–58

Gupta RS, Ramachandra NB, Bowes T, Singh B (2008) Unusual cellular disposition of the mitochondrial molecular chaperones Hsp60, Hsp70 and Hsp10. Novartis Found Symp 291:59–68; discussion 69–73, 137–140

Hargitai J, Lewis H, Boros I, Racz T, Fiser A, Kurucz I, Benjamin I, Vigh L, Penzes Z, Csermely P et al (2003) Bimoclomol, a heat shock protein co-inducer, acts by the prolonged activation of heat shock factor-1. Biochem Biophys Res Commun 307:689–695

Hartl FU (1991) Heat shock proteins in protein folding and membrane translocation. Semin Immunol 3:5–16

Helmbrecht K, Zeise E, Rensing L (2000) Chaperones in cell cycle regulation and mitogenic signal transduction: a review. Cell Prolif 33:341–365

Hori O, Ichinoda F, Tamatani T, Yamaguchi A, Sato N, Ozawa K, Kitao Y, Miyazaki M, Harding HP, Ron D et al (2002) Transmission of cell stress from endoplasmic reticulum to mitochondria: enhanced expression of Lon protease. J Cell Biol 157:1151–1160

Houenou LJ, Li L, Lei M, Kent CR, Tytell M (1996) Exogenous heat shock cognate protein Hsc 70 prevents axotomy-induced death of spinal sensory neurons. Cell Stress Chaperones 1:161–166
Huynen MA, de Hollander M, Szklarczyk R (2009) Mitochondrial proteome evolution and genetic disease. Biochim Biophys Acta 1792:1122–1129
Jin J, Hulette C, Wang Y, Zhang T, Pan C, Wadhwa R, Zhang J (2006) Proteomic identification of a stress protein, mortalin/mthsp70/GRP75: relevance to Parkinson disease. Mol Cell Proteomics 5:1193–1204
Jin J, Li GJ, Davis J, Zhu D, Wang Y, Pan C, Zhang J (2007) Identification of novel proteins interacting with both alpha-synuclein and DJ-1. Mol Cell Proteomics 6:845–859
Kaul SC, Wadhwa R, Matsuda Y, Hensler PJ, Pereira-Smith OM, Komatsu Y, Mitsui Y (1995) Mouse and human chromosomal assignments of mortalin, a novel member of the murine hsp70 family of proteins. FEBS Lett 361:269–272
Kaul SC, Duncan EL, Englezou A, Takano S, Reddel RR, Mitsui Y, Wadhwa R (1998) Malignant transformation of NIH3T3 cells by overexpression of mot-2 protein. Oncogene 17:907–911
Kaul SC, Reddel RR, Mitsui Y, Wadhwa R (2001) An N-terminal region of mot-2 binds to p53 in vitro. Neoplasia 3:110–114
Kaul SC, Aida S, Yaguchi T, Kaur K, Taira K, Wadhwa R (2005) Activation of wild type p53 function by its mortalin-binding cytoplasmically localizing carboxy-terminus peptides. J Biol Chem 280:39373–39379
Kaul SC, Deocaris CC, Wadhwa R (2007) Three faces of mortalin: a housekeeper, guardian and killer. Exp Gerontol 42:263–274
Kimura K, Tanaka N, Nakamura N, Takano S, Ohkuma S (2007) Knockdown of mitochondrial heat shock protein 70 promotes progeria-like phenotypes in caenorhabditis elegans. J Biol Chem 282:5910–5918
Kirkwood TB, Shanley DP (2000) Caloric restriction, hormesis and life history plasticity. Hum Exp Toxicol 19:338–339
Knowlton AA, Gupta S (2003) HSP60, Bax, and cardiac apoptosis. Cardiovasc Toxicol 3:263–268
Laskey RA, Honda BM, Mills AD, Finch JT (1978) Nucleosomes are assembled by an acidic protein which binds histones and transfers them to DNA. Nature 275:416–420
Li HM, Niki T, Taira T, Iguchi-Ariga SM, Ariga H (2005) Association of DJ-1 with chaperones and enhanced association and colocalization with mitochondrial Hsp70 by oxidative stress. Free Radic Res 39:1091–1099
Lindquist S, Craig EA (1988) The heat-shock proteins. Annu Rev Genet 22:631–677
Liu Y, Liu W, Song XD, Zuo J (2005) Effect of GRP75/mthsp70/PBP74/mortalin overexpression on intracellular ATP level, mitochondrial membrane potential and ROS accumulation following glucose deprivation in PC12 cells. Mol Cell Biochem 268:45–51
Lu WJ, Lee NP, Kaul SC, Lan F, Poon RT, Wadhwa R, Luk JM (2011a) Induction of mutant p53-dependent apoptosis in human hepatocellular carcinoma by targeting stress protein mortalin. Int J Cancer 129(8):1806–1814
Lu WJ, Lee NP, Kaul SC, Lan F, Poon RT, Wadhwa R, Luk JM (2011b) Mortalin-p53 interaction in cancer cells is stress dependent and constitutes a selective target for cancer therapy. Cell Death Differ 18:1046–1056
Ma Z, Izumi H, Kanai M, Kabuyama Y, Ahn NG, Fukasawa K (2006) Mortalin controls centrosome duplication via modulating centrosomal localization of p53. Oncogene 25:5377–5390
Macario AJ, Conway de Macario E (2002) Sick chaperones and ageing: a perspective. Ageing Res Rev 1:295–311
Macario AJ, Conway de Macario E (2007) Chaperonopathies and chaperonotherapy. FEBS Lett 581:3681–3688
Margulis L (1975) Symbiotic theory of the origin of eukaryotic organelles; criteria for proof. Symp Soc Exp Biol 1975(29):21–38
Massa SM, Longo FM, Zuo J, Wang S, Chen J, Sharp FR (1995) Cloning of rat grp75, an hsp70-family member, and its expression in normal and ischemic brain. J Neurosci Res 40:807–819

Matsuoka T, Wada J, Hashimoto I, Zhang Y, Eguchi J, Ogawa N, Shikata K, Kanwar YS, Makino H (2005) Gene delivery of tim44 reduces mitochondrial superoxide production and ameliorates neointimal proliferation of injured carotid artery in diabetic rats. Diabetes 54:2882–2890

Merrick BA, Walker VR, He C, Patterson RM, Selkirk JK (1997) Induction of novel Grp75 isoforms by 2-deoxyglucose in human and murine fibroblasts. Cancer Lett 119:185–190

Michikawa Y, Baba T, Arai Y, Sakakura T, Kusakabe M (1993) Structure and organization of the gene encoding a mouse mitochondrial stress-70 protein. FEBS Lett 336:27–33

Michishita E, Nakabayashi K, Suzuki T, Kaul SC, Ogino H, Fujii M, Mitsui Y, Ayusawa D (1999) 5-Bromodeoxyuridine induces senescence-like phenomena in mammalian cells regardless of cell type or species. J Biochem 126:1052–1059

Montesano GN, Chirico G, Pirozzi G, Costantino E, Landriscina M, Esposito F (2007) Tumor necrosis factor-associated protein 1 (TRAP-1) protects cells from oxidative stress and apoptosis. Stress 10:342–350

Naylor DJ, Hoogenraad NJ, Hoj PB (1996) Isolation and characterisation of a cDNA encoding rat mitochondrial GrpE, a stress-inducible nucleotide-exchange factor of ubiquitous appearance in mammalian organs. FEBS Lett 396:181–188

Ohyabu Y, Kaul Z, Yoshioka T, Inoue K, Sakai S, Mishima H, Uemura T, Kaul SC, Wadhwa R (2009) Stable and nondisruptive in vitro/in vivo labeling of mesenchymal stem cells by internalizing quantum dots. Hum Gene Ther 20:217–224

Ornatsky OI, Connor MK, Hood DA (1995) Expression of stress proteins and mitochondrial chaperonins in chronically stimulated skeletal muscle. Biochem J 311:119–123

Orsini F, Migliaccio E, Moroni M, Contursi C, Raker VA, Piccini D, Martin-Padura I, Pelliccia G, Trinei M, Bono M et al (2004) The life span determinant p66Shc localizes to mitochondria where it associates with mitochondrial heat shock protein 70 and regulates trans-membrane potential. J Biol Chem 279:25689–25695

Osorio C, Sullivan PM, He DN, Mace BE, Ervin JF, Strittmatter WJ, Alzate O (2007) Mortalin is regulated by APOE in hippocampus of AD patients and by human APOE in TR mice. Neurobiol Aging 28:1853–1862

Passarella S, Marra E, Atlante A, Barile M, Doonan S, Quagliariello E (1990) Uptake of aspartate aminotransferase into mitochondria in vitro causes efflux of malate dehydrogenase and vice versa. Biochim Biophys Acta 1022:273–282

Pilzer D, Fishelson Z (2005) Mortalin/GRP75 promotes release of membrane vesicles from immune attacked cells and protection from complement-mediated lysis. Int Immunol 17:1239–1248

Pilzer D, Gasser O, Moskovich O, Schifferli JA, Fishelson Z (2005) Emission of membrane vesicles: roles in complement resistance, immunity and cancer. Springer Semin Immunopathol 27:375–387

Pilzer D, Saar M, Koya K, Fishelson Z (2009) Mortalin inhibitors sensitize K562 leukemia cells to complement-dependent cytotoxicity. Int J Cancer 126:1428–1435

Pizzatti L, Sa LA, de Souza JM, Bisch PM, Abdelhay E (2006) Altered protein profile in chronic myeloid leukemia chronic phase identified by a comparative proteomic study. Biochim Biophys Acta 1764:929–942

Polyak K, Li Y, Zhu H, Lengauer C, Willson JK, Markowitz SD, Trush MA, Kinzler KW, Vogelstein B (1998) Somatic mutations of the mitochondrial genome in human colorectal tumours. Nat Genet 20:291–293

Powers MV, Workman P (2007) Inhibitors of the heat shock response: biology and pharmacology. FEBS Lett 581:3758–3769

Ran Q, Wadhwa R, Kawai R, Kaul SC, Sifers RN, Bick RJ, Smith JR, Pereira-Smith OM (2000) Extramitochondrial localization of mortalin/mthsp70/PBP74/GRP75. Biochem Biophys Res Commun 275:174–179

Rehling P, Brandner K, Pfanner N (2004) Mitochondrial import and the twin-pore translocase. Nat Rev Mol Cell Biol 5:519–530

Ritossa F (1962) A new puffing pattern induced by a temperature shock and DNP in Drosophila. Experientia 18:571–573

Sadekova S, Lehnert S, Chow TY (1997) Induction of PBP74/mortalin/Grp75, a member of the hsp70 family, by low doses of ionizing radiation: a possible role in induced radioresistance. Int J Radiat Biol 72:653–660

Saretzki G, Armstrong L, Leake A, Lako M, von Zglinicki T (2004) Stress defense in murine embryonic stem cells is superior to that of various differentiated murine cells. Stem Cells 22:962–971

Schneider HC, Berthold J, Bauer MF, Dietmeier K, Guiard B, Brunner M, Neupert W (1994) Mitochondrial Hsp70/MIM44 complex facilitates protein import. Nature 371:768–774

Seyb KI, Ansar S, Bean J, Michaelis ML (2006) beta-Amyloid and endoplasmic reticulum stress responses in primary neurons: effects of drugs that interact with the cytoskeleton. J Mol Neurosci 28:111–123

Sherman M, Multhoff G (2007) Heat shock proteins in cancer. Ann N Y Acad Sci 1113:192–201

Shi M, Jin J, Wang Y, Beyer RP, Kitsou E, Albin RL, Gearing M, Pan C, Zhang J (2008) Mortalin: a protein associated with progression of Parkinson disease? J Neuropathol Exp Neurol 67:117–124

Shiota M, Ikeda Y, Kaul Z, Itadani J, Kaul SC, Wadhwa R (2007) Internalizing antibody-based targeted gene delivery for human cancer cells. Hum Gene Ther 18:1153–1160

Singh B, Soltys BJ, Wu ZC, Patel HV, Freeman KB, Gupta RS (1997) Cloning and some novel characteristics of mitochondrial Hsp70 from Chinese hamster cells. Exp Cell Res 234:205–216

Sirk D, Zhu Z, Wadia JS, Shulyakova N, Phan N, Fong J, Mills LR (2007) Chronic exposure to sub-lethal beta-amyloid (Abeta) inhibits the import of nuclear-encoded proteins to mitochondria in differentiated PC12 cells. J Neurochem 103:1989–2003

Smith DL, Evans CA, Pierce A, Gaskell SJ, Whetton AD (2002) Changes in the proteome associated with the action of Bcr-Abl tyrosine kinase are not related to transcriptional regulation. Mol Cell Proteomics 1:876–884

Soltys BJ, Gupta RS (1999) Mitochondrial-matrix proteins at unexpected locations: are they exported? Trends Biochem Sci 24:174–177

Soltys BJ, Gupta RS (2000) Mitochondrial proteins at unexpected cellular locations: export of proteins from mitochondria from an evolutionary perspective. Int Rev Cytol 194:133–196

Takano S, Wadhwa R, Yoshii Y, Nose T, Kaul SC, Mitsui Y (1997) Elevated levels of mortalin expression in human brain tumors. Exp Cell Res 237:38–45

Takano S, Wadhwa R, Mitsui Y, Kaul SC (2001) Identification and characterization of molecular interactions between glucose-regulated proteins (GRPs) mortalin/GRP75/peptide-binding protein 74 (PBP74) and GRP94. Biochem J 357:393–398

Tatsuta T (2009) Protein quality control in mitochondria. J Biochem 146:455–461

Tissieres A, Mitchell HK, Tracy UM (1974) Protein synthesis in salivary glands of Drosophila melanogaster: relation to chromosome puffs. J Mol Biol 84:389–398

Um JH, Kang CD, Hwang BW, Ha MY, Hur JG, Kim DW, Chung BS, Kim SH (2003a) Involvement of DNA-dependent protein kinase in regulation of the mitochondrial heat shock proteins. Leuk Res 27:509–516

Um JH, Kim SJ, Kim DW, Ha MY, Jang JH, Chung BS, Kang CD, Kim SH (2003b) Tissue-specific changes of DNA repair protein Ku and mtHSP70 in aging rats and their retardation by caloric restriction. Mech Ageing Dev 124:967–975

Van Laar VS, Dukes AA, Cascio M, Hastings TG (2008) Proteomic analysis of rat brain mitochondria following exposure to dopamine quinone: implications for Parkinson disease. Neurobiol Dis 29:477–489

Voloboueva LA, Duan M, Ouyang Y, Emery JF, Stoy C, Giffard RG (2007) Overexpression of mitochondrial Hsp70/Hsp75 protects astrocytes against ischemic injury in vitro. J Cereb Blood Flow Metab 28:1009–1016

Wadhwa R, Kaul SC, Ikawa Y, Sugimoto Y (1993a) Identification of a novel member of mouse hsp70 family. Its association with cellular mortal phenotype. J Biol Chem 268:6615–6621

Wadhwa R, Kaul SC, Sugimoto Y, Mitsui Y (1993b) Induction of cellular senescence by transfection of cytosolic mortalin cDNA in NIH 3T3 cells. J Biol Chem 268:22239–22242

Wadhwa R, Akiyama S, Sugihara T, Reddel RR, Mitsui Y, Kaul SC (1996) Genetic differences between the pancytosolic and perinuclear forms of murine mortalin. Exp Cell Res 226:381–386

Wadhwa R, Takano S, Robert M, Yoshida A, Nomura H, Reddel RR, Mitsui Y, Kaul SC (1998) Inactivation of tumor suppressor p53 by mot-2, a hsp70 family member. J Biol Chem 273:29586–29591

Wadhwa R, Sugihara T, Yoshida A, Nomura H, Reddel RR, Simpson R, Maruta H, Kaul SC (2000) Selective toxicity of MKT-077 to cancer cells is mediated by its binding to the hsp70 family protein mot-2 and reactivation of p53 function. Cancer Res 60:6818–6821

Wadhwa R, Ando H, Kawasaki H, Taira K, Kaul SC (2003a) Targeting mortalin using conventional and RNA-helicase-coupled hammerhead ribozymes. EMBO Rep 4:595–601

Wadhwa R, Yaguchi T, Hasan MK, Taira K, Kaul SC (2003b) Mortalin-MPD (mevalonate pyrophosphate decarboxylase) interactions and their role in control of cellular proliferation. Biochem Biophys Res Commun 302:735–742

Wadhwa R, Takano S, Kaur K, Aida S, Yaguchi T, Kaul Z, Hirano T, Taira K, Kaul SC (2005) Identification and characterization of molecular interactions between mortalin/mtHsp70 and HSP60. Biochem J 391:185–190

Wadhwa R, Takano S, Kaur K, Deocaris CC, Pereira-Smith OM, Reddel RR, Kaul SC (2006) Upregulation of mortalin/mthsp70/Grp75 contributes to human carcinogenesis. Int J Cancer 118:2973–2980

Wadhwa R, Ryu J, Gao R, Choi IK, Morrow G, Kaur K, Kim I, Kaul SC, Yun CO, Tanguay RM (2010) Proproliferative functions of Drosophila small mitochondrial heat shock protein 22 in human cells. J Biol Chem 285:3833–3839

Walker C, Bottger S, Low B (2006) Mortalin-based cytoplasmic sequestration of p53 in a nonmammalian cancer model. Am J Pathol 168:1526–1530

Widodo N, Deocaris CC, Kaur K, Hasan K, Yaguchi T, Yamasaki K, Sugihara T, Ishii T, Wadhwa R, Kaul SC (2007) Stress chaperones, mortalin, and pex19p mediate 5-aza-2′ deoxycytidine-induced senescence of cancer cells by DNA methylation-independent pathway. J Gerontol A Biol Sci Med Sci 62:246–255

Yang H, Zhou X, Liu X, Yang L, Chen Q, Zhao D, Zuo J, Liu W (2011a) Mitochondrial dysfunction induced by knockdown of mortalin is rescued by Parkin. Biochem Biophys Res Commun 410:114–120

Yang L, Guo W, Zhang Q, Li H, Liu X, Yang Y, Zuo J, Liu W (2011b) Crosstalk between Raf/MEK/ERK and PI3K/AKT in suppression of Bax conformational change by Grp75 under glucose deprivation conditions. J Mol Biol 414(5):654–666

Yi X, Luk JM, Lee NP, Peng J, Leng X, Guan XY, Lau GK, Beretta L, Fan ST (2008) Association of mortalin (HSPA9) with liver cancer metastasis and prediction for early tumor recurrence. Mol Cell Proteomics 7:315–325

Yokoyama K, Fukumoto K, Murakami T, Harada S, Hosono R, Wadhwa R, Mitsui Y, Ohkuma S (2002) Extended longevity of Caenorhabditis elegans by knocking in extra copies of hsp70F, a homolog of mot-2 (mortalin)/mthsp70/Grp75. FEBS Lett 516:53–57

Yoneda T, Benedetti C, Urano F, Clark SG, Harding HP, Ron D (2004) Compartment-specific perturbation of protein handling activates genes encoding mitochondrial chaperones. J Cell Sci 117:4055–4066

Yoo JY, Ryu J, Gao R, Yaguchi T, Kaul SC, Wadhwa R, Yun CO (2010) Tumor suppression by apoptotic and anti-angiogenic effects of mortalin-targeting adeno-oncolytic virus. J Gene Med 12:586–595

Yoshioka T, Mishima H, Kaul Z, Ohyabu Y, Sakai S, Ochiai N, Kaul SC, Wadhwa R, Uemura T (2011) Fate of bone marrow mesenchymal stem cells following the allogeneic transplantation of cartilaginous aggregates into osteochondral defects of rabbits. J Tissue Eng Regen Med 5:437–443

Zhang Y, Wada J, Hashimoto I, Eguchi J, Yasuhara A, Kanwar YS, Shikata K, Makino H (2006) Therapeutic approach for diabetic nephropathy using gene delivery of translocase of inner mitochondrial membrane 44 by reducing mitochondrial superoxide production. J Am Soc Nephrol 17:1090–1101

Zhao Q, Wang J, Levichkin IV, Stasinopoulos S, Ryan MT, Hoogenraad NJ (2002) A mitochondrial specific stress response in mammalian cells. Embo J 21:4411–4419

Chapter 2
Mortalin's Machinery

Custer C. Deocaris, Sunil C. Kaul and Renu Wadhwa

Abstract Mortalin/mtHsp70 performs a wide array of cellular functions and has been implicated in aging, cancer and neurodegenerative diseases. Similar to other Hsp70s, its ability to chaperone misfolded proteins and bind to a myriad of clients is derived from its N-terminal nucleotide-binding domain (NBD) regulating substrate affinity of the C-terminal substrate-binding domain (SBD) in a nucleotide- and co-chaperone-dependent mechanism. To understand the structural dynamics of its allostery making this relevant to mortalin's cellular function, this chapter describes key structural features of these two domains as well as provide an appreciation as to possibly how a single amino acid change, Gly to Arg in the SBD that can be viewed so minor, is able to metamorphose from a life-extending species of mortalin (mot-2) into one that induces senescence and even inhibits tumor growth (mot-1).

Keywords Mortalin · Chaperone · Structure · Binding domains · Functions

2.1 Introduction

Having around 40–60% identity between its eukaryotic and prokaryotic members, the Hsp70 superfamily comprises one of the most highly conserved proteins in biology (Hunt and Morimoto 1985; Karlin and Brocchieri 1998). These chaperones are also strictly required for cellular viability in every organism. Despite their diversity, it has been observed that each type does not have overlapping or redundant functions. In yeast, among its highly identical Hsp70s (amino acid identity: 60–80%), none of the members could be functionally interchanged for another.

Within the mitochondrion, a key component of the organellar import machinery and protein quality control is mortalin (mtHsp70). With a molecular weight of 73,913 Da, this mitochondrial chaperone is composed of 679 amino acids and shares

C. C. Deocaris (✉)
Department of Food Science and Technology, College of Home Economics,
University of the Philippines, Diliman, Quezon City, Philippines
e-mail: cdeocaris@gmail.com

S. C. Kaul · R. Wadhwa
National Institute of Advanced Industrial Science and Technology (AIST),
Central 4, 1-1-1 Higashi, Tsukuba, Ibaraki 305-8562, Japan

S. C. Kaul, R. Wadhwa (eds.), *Mortalin Biology: Life, Stress and Death*,
DOI 10.1007/978-94-007-3027-4_2, © Springer Science+Business Media B.V. 2012

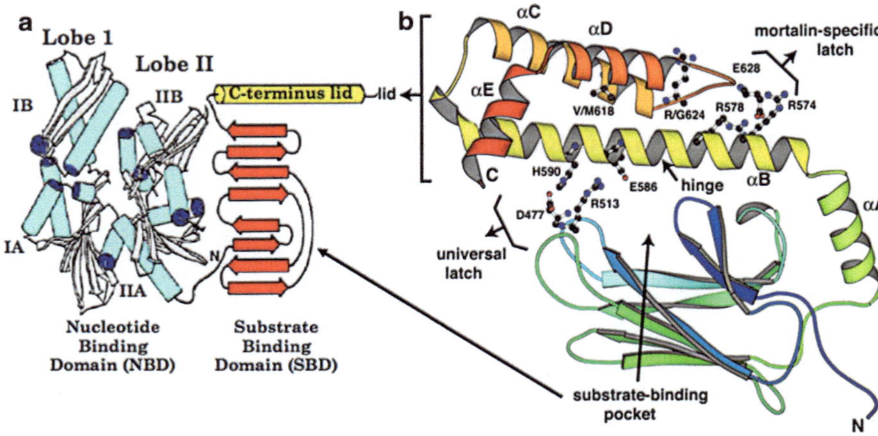

Fig. 2.1 Ribbon diagrams of mortalin domains. **a** N-terminal Nucleotide Binding Domain (NBD) shows the 2 sub-domains: IA and IB (Lobe 1), and IIA, and IIB (Lobe 2). The folding of these lobes, makes up for a deep pocket that binds to ATP (Deocaris et al. 2007b). Substrate-binding domain (SBD) is shown here as two sets of four-stranded anti-parallel β-sheets (*red*) resembling a twisted sandwich that forms the substrate-binding site. **b** The "substrate lid" is composed of alpha-helices A–C. Electrostatic interactions among Arg574, Arg578, and Asp628 constitute the "mortalin-specific latch", and among Asp477, Arg513, Glu586, and His590 that forms the "common latch" which has been found to be evolutionarily-conserved among all Hsp70s. (Modified from Deocaris et al. 2006a)

high degree of identity with other Hsp70s, including *Escherichia coli* DnaK (51%), *Saccharomyces cerevisiae* SSC1p (65%), the constitutive cytosolic Hsp70 from rat, Hsc70 (46%), and the rat endoplasmic reticulum isoform, BiP (49%) (Webster et al. 1994). The molecular structure of mortalin/mtHsp70 has not yet been elucidated, as previous attempts to crystallize it had been unsuccessful. Given the strict evolutionary conservation among Hsp70s, the "mortalin machinery" can readily be modeled after the well-known allosteric dynamics of two canonical inter-connected modules: the N-terminal Nucleotide Binding Domain (also called the ATPase domain) and C-terminal substrate-binding domain (SBD) (Fig. 2.1a, b). The interaction of a substrate with the SBD, the affinity and kinetics of substrate binding are being controlled by the activity of NBD. Exchange of ADP for ATP in the NBD causes significant local conformational changes that ultimately lead to the opening of the lid sub-domain. This action is concomitant to the decrease in binding affinity for substrates. Upon ATP hydrolysis, the protein is returned to the ADP-bound configuration ready for another round of substrate binding and release. With the interplay between these domains which constitute a basis for the chaperoning activity of mortalin, we offer a hypothesis on how its natural allelic forms in mouse may possess contradictory biological functions: one which is associated with aging (mortalin 1, or mot-1) and the other which confers lifespan-extension (mortalin 2, or mot-2) (Deocaris et al. 2006b).

2.2 N-terminal Nucleotide Binding Domain (NBD)

The NBD is the chaperone's control unit. To appreciate the function of this domain, we made a model through a homology-based method using the crystal structure of ATP-bound human Hsp70 (1S3X) (Deocaris et al. 2007b) as this region contains a high degree of sequential (>80%) and structural conservation (Mayer and Bukau 1998). Mortalin's NBD is similar to the ATPases of several other functionally diverse proteins such actin, hexokinase, etc. (Bork et al. 1992).

As shown in Fig. 2.1a, the 44-kDa structure folds into a pair of lobes each with 2 sub-domains: subdomains IA and IB (in lobe I), and, IIA and IIB (in lobe II). The folding of these lobes, stabilized by metal-binding, creates a deep pocket for the catalysis of ATP (Sriram et al. 1997). As ATP lodges into this central pocket, particularly at the interface between sub-domains IIA and IIB, the geometric and energetic effects of its binding and hydrolysis is rapidly communicated throughout the entire NBD (Smock et al. 2010). NMR readout of *E. coli* DnaK showed that in the ADP-bound state, NDB and SBD are both dissociated and largely independent (Bertelsen et al. 2009). In contrast, upon ATP binding, conformational rearrangement in the NBD weakens binding of substrates within the C-terminal SBD and promotes release of the substrate. Of note, the inter-domain linker connecting the NBD and SBD is highly conserved underscoring the importance of this short segment in transmitting the allosteric changes (Bertelsen et al. 2009; Bhattacharya et al. 2009; Mayer 2011; Mayer and Bukau 2005).

The turnover rate of Hsp70 ATPase is rather slow (0.02–0.2 min^{-1}) and is insufficient to assist chaperoning during cellular stress. Therefore, it is essential that co-factors modulate mortalin's ATPase to fine-tune the regulation between its ATP- and ADP-bound states (Ha and McKay 1994, 1995). It has been established that the precise functioning of the NBD critically involves its interaction with two co-chaperone families: the J-domain proteins that catalyze ATP hydrolysis (Craig et al. 2006) and the nucleotide exchange factors (NEFs) that help in the replacement of ADP with ATP (Kabani 2009; Xu et al. 2008). Studies on *E. coli* Hsp70, DnaK, first suggested that the ATPase activity of the protein could be cyclically stimulated by co-chaperones DnaJ and GrpE. DnaJ permits the hydrolysis of Hsp70-bound ATP allowing the ADP-bound Hsp70 to interact more strongly with unfolded proteins. The nucleotide exchange factor GrpE enables the recycling of Hsp70 back intoan ATP-bound state permitting the efficient release of its substrate. Multiple GrpE-like proteins and a unique human GrpE homologue, HMGE, have been reported to be restricted to the mitochondria, and form chaperone-pairing with mortalin/mtHsp70 (Deocaris et al. 2006a). Catalysis of ADP-ATP exchange, and hydrolysis by J-domain proteins and NEFs, i.e. the GrpEs, has to be balanced to facilitate equilibrium between substrate binding and release.

a) *J-domain proteins.* DnaJ proteins are 80-amino acid small proteins containing the N-terminal J domain essential for stimulating ATPase activity. Transport of preproteins into the mitochondrial matrix is mediated by the presequence translocase–associated motor (PAM) that has four essential subunits: (1) mortalin,

(2) the peripheral membrane protein, Tim44, (3) the nucleotide exchange factor, Mge1 and (4) J-domain protein PAM18 that stimulates the ATPase activity of mortalin. PAM18 is required for the interaction of mortalin with Tim44 and protein translocation into the matrix, see review by Deocaris et al. (2006a). The function of other co-chaperones, such as mtDnaJ (Zhao et al. 2002), dj2 and dj3 in regulating mortalin is still obscure although it is interesting to note that these cytosolic DnaJs are also translocated in the mitochondria (Goswami et al. 2010; Terada et al. 1997; Terada and Mori 2000). Another co-chaperone that stimulates ATPase activity of mortalin is the Hsp70 escort protein (Hep). Previously, it has been shown that Hep is critical to the solubility of mortalin and inhibits self-association of the chaperone (Zhai et al. 2008). The binding sites of Mge1 and Tim44 to mortalin have been assigned at the variable region along C-terminal alpha-helical lid (Strub et al. 2003).

b) *NEFs.* Nucleotide exchange factor Mge1 and multiple GrpE-like proteins are restricted to the mitochondria and form chaperone pairing with mortalin (Choglay et al. 2001; Naylor et al. 1998). Association of GrpE with ADP-bound Hsp70 hastens nucleotide exchange by as much as 5000-fold and reduces affinity for ADP by 200-fold (Packschies et al. 1997). Essential during the translocation of precursor proteins into the mitochondrion, Mge1 has been demonstrated to enhance the otherwise low intrinsic ATPase activity of mortalin *via* the release of ADP and Pi (Dekker and Pfanner 1997). This results in the dissociation of Tim44-mortalin complex and the enhanced binding of mortalin to the transiting pre-protein (Schneider et al. 1996). Subdomain IA, which contains both N- and C-terminus and links to the remaining part of mortalin, is a target of the J-domain proteins (Jiang et al. 2007; Xu et al. 2008). This subdomain is highly conserved but more rigid than the other subdomains (Flaherty et al. 1990). NEFs target, in most cases, subdomain IIB. Subdomain IIB is characterized by its high mobility, especially at the β-sheet E and the exposed loop connecting the two strands of this sheet. Upon binding by GrpE, there is suppression in its mobility.

MKT-077, mitochondriotopic drug that binds to and abrogates mortalin function, has been deduced to target the alpha-helix within sub-domain IIB of the catalytic cleft based on immunoaffinity studies with truncated versions of mortalin (Wadhwa et al. 2000). Using various spectrophotometric tools to analyze protein structure, i.e. circular dichroism polarimetry, ANS and tryptophan fluorescence, we have deduced that binding of MKT-077 induces not just (local) conformational changes, but also global tertiary conformational shifts in mortalin. As reported, the presence of MKT-077 ablated mortalin chaperone function based on *in vitro* (insulin aggregation assay) and *in vivo* (luciferase test) assays (Deocaris et al. 2007b).

2.3 C-terminal Substrate Binding Domain (SBD)

There is a wider sequence variation in the SBD, compared to the ATPase domain signifying the diversification of Hsp70 substrates and clients (Rudiger et al. 1997, 2000). Comprising of five distinct helical domains, mortalin's SBD is 18-kDa in weight.

The SBD is composed of two sets of four-stranded anti-parallel β-sheets forming a twisted sandwich. The most predominant feature of SBD is its "substrate lid" that "swallows and seals" substrates during an ADP-bound state (Zhu et al. 1996). The 10-kDa substrate "lid" is composed of a helix kinked in the middle and bent upwards. This is a common structural feature among Hsp70s allowing it to flip-flop with the substrates ingress and egress (Fig. 2.1b). A consensus motif recognized by DnaK has been identified by screening a phage-display library. Consistent with the observed features of signal sequences of proteins destined to the mitochondrial compartment, Gragerov and Gottesman (1994) found that peptides containing internal alipathic residues, such as NRLLTG, and those that contain basic (not acidic) residues favor binding to this prokaryotic Hsp70. Additionally, these sequences are found buried deep in the α-strands of folded peptides and are generally exposed only when misfolded. Based on various protein folding algorithms employed, the distal end of the predominantly alpha-helical lid that is followed by an aperiodic glycine/proline rich region may adopt disorder in conformation. While the function of this variable C-terminal domain remains unclear, a disordered secondary structural state may permit the binding to co-factors specific to a type of Hsp70 to expand its bioactivity (Horton et al. 2001; Stahl et al. 1999) and also elicit immunomodulatory functions (Deocaris et al. 2005; Pockley et al. 2008).

2.4 Mortalin-1 *Versus* Mortalin-2: A Case of a Substrate Lid "Overbite"

Discussion on the structure-function relationship of mortalin is not complete without mentioning the differences between the 2 species of mortalin in mice: the senescence-inducing mouse mot-1, an allelic variant that differs by only two amino acids in the SBD from mot-2, the functional equivalent of human mortalin (Kaul et al. 1998, 2003, 2005, 2007). Structurally, the two proteins and their single amino acid mutants displayed different mobilities on SDS-polyacrylamide gel presenting an evidence for their different secondary structures (Fig. 2.2a–d). Based on the interaction, immunohistochemical and reporter assays, it was deduced that only mot-2 interacts with and inactivates p53 function (Kaul et al. 2000). However, primary structural difference between mot-1 and mot-2 at amino acid residues 618 or 624 did not concern the interaction of mortalin with p53. Later, it was determined that the interaction of mortalin and p53 involved N-terminal region of the mortalin protein (Kaul et al. 2001). However, interestingly, when assessed for chaperoning function of different mortalin constructs, we found that mot-2 served as a better chaperone compared to mot-1 (Deocaris et al. 2008; unpublished). In addition to the roles of ADP-ATP exchange in propagating allosteric changes in the SBD, another major feature is the presence of "locking" mechanisms that help facilitate the entry/exit of protein substrates. These so-called electrostatic "latches" between the lid and substrate-binding region are important in holding protein clients clasped into the SBD. Previous studies identified evolutionary conserved "latches" (common to all Hsp70s) consisting of

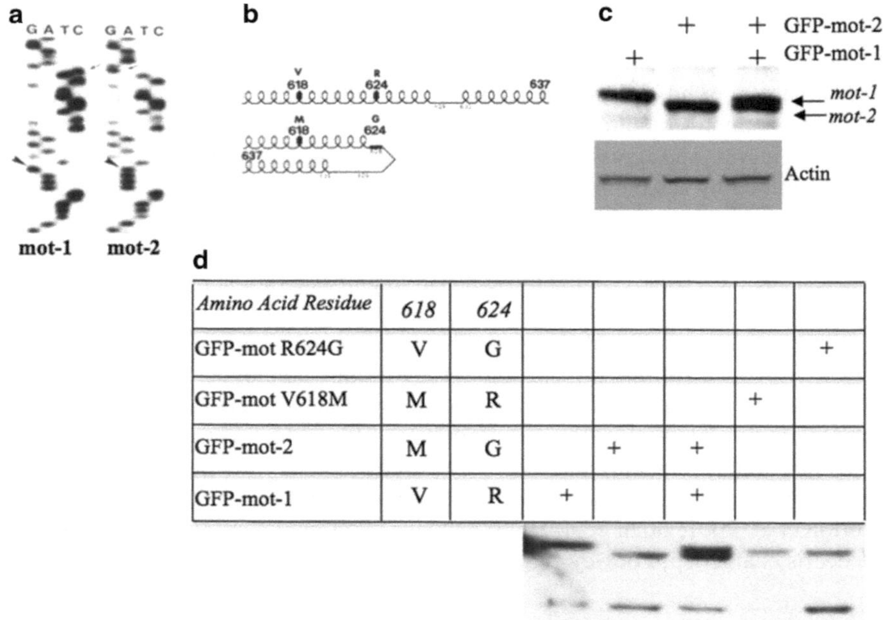

Fig. 2.2 Structural difference between mouse mot-1 and mot-2. Nucleotide changes G to A and C to G **a** resulting in the amino acid changes V618 M and R624G **b** are shown. Mobility of GFP tagged mot-1 and mot-2 proteins **c** and of the single amino acid mutants **d** are shown Order of mobility (slow to fast) was: mot-1 (V618, R624), mutant V618 M (M618, R624), mutant R624G (V618, G624) and mot-2 (M618, G624). (Modified from Wadhwa et al. 1993b; Kaul et al. 2000)

Asp477, Arg513, Glu586 and His590 (Mayer et al. 2000). In our homology work, we discovered an additional "latch" on the opposite end, consisting of Arg574, Arg578 and Glu628 that are mortalin-specific (see Fig. 2.1b). A replacement of Gly624 (in mot-2), located at the C-terminus of α-helix C, by Arg (in mot-1) is likely to extend the α-helix C. In contrast, Gly, a strong helix breaker, shortens the L3 (C-D) loop (Fig. 2.1b). The latter should perturb the structure of the "mortalin-specific latch", presumably pulling apart the electrostatic attraction and in effect weakening the attractive forces as residues fall beyond the scope of the electrostatic field, a case of a chaperone "overbite". From these, we view that the single amino acid change could therefore influence acquisition of substrates consistent with the ageing-inducing property of murine mot-1 (Nakabayashi et al. 1997; Wadhwa et al. 1993b) vis-à-vis the known life-span extending and immortalization features of mouse and human mot-2 chaperones (Kaul et al. 1998, 2003; Yokoyama et al. 2002). Using mot-1 and mot-2 recombinant proteins and the antibodies that react specifically to either of them, we isolated mot-1 and mot-2 binding proteins from human cell lysates. Analysis of these complexes by 2-D gel electrophoresis and mass spectrometry has revealed some proteins that bind preferentially either to mot-1 or to mot-2 (Wadhwa et al.; unpublished). The unique biology of the different mortalin species demonstrates the flexibility of cellular function vis-à-vis the minor structural changes in the protein. In the various systems of natural aging and escaped aging (with immortal and tumor cells), this phenomenon can also be demonstrated. As an example, when mortalin becomes oxidized, it may be transformed into a "sick chaperone" that can even induce the building up a catastrophe of misfolded proteins. Our data indicated that H_2O_2-oxidized mortalin, more than just remaining benign and non-functional, promotes aggregation of misfolded proteins at concentrations in excess relative to mortalin. Consistent to this, the oxidized form of mortalin has been reported abundant in tissues from Alzhemier's and Parkinson's Diseases (Choi et al. 2004).

The basic paradigm of molecular chaperones is that they recognize and selectively bind non-native, but not native proteins (or even themselves) to form relatively stable complexes. It is interesting to note that the trajectory of our understanding on the cellular roles of mortalin is quite unlike the way other Hsp70s have gone (Kaul et al. 2007). Other chaperones was seen as only helpers in maintaining proteome integrity, however, mortalin was initially discovered as a "mortality" associated gene (Wadhwa et al. 1993a), a stress-associated protein targeted by toxicants (Bruschi et al. 1993), then as an antigen-processing protein (Domanico et al. 1993) and finally as the main motor component for mitochondrial biogenesis (Webster et al. 1994). Unexpectedly, mortalin has been shown participate in regulating the dynamic organization of the cytoskeleton (Cicchillitti et al. 2009) and in cellular uptake mechanisms via the lipid rafts (Wittrup et al. 2010). It, thus, appears that chaperonization may be just one mechanism on how this molecule executes its diverse functions within the mitochondria and beyond (Gupta et al. 2008). In addition to several specialized review articles on the structure-function relationship of chaperones (Baker et al. 2007; Genevaux et al. 2007; Hohfeld et al. 2001; Kim et al. 2006; Liberek et al. 2008; Mogk and Bukau 2004; Mayer and Bukau 2005; Saibil 2008; Slepenkov and Witt 2002; Tomkiewicz et al. 2007; Young et al. 2003), there have been few recent reviews published specifically on mortalin (Deocaris et al. 2006a, 2007a, 2009; Kaul et al. 2007). The present review provides just a basic appreciation of the pro-aging or anti-aging properties vis-à-vis to mortalin's amazing molecular architecture.

References

Baker MJ, Frazier AE, Gulbis JM, Ryan MT (2007) Mitochondrial protein-import machinery: correlating structure with function. Trends Cell Biol 17:456–464

Bertelsen EB, Chang L, Gestwicki JE, Zuiderweg ER (2009) Solution conformation of wild-type E. coli Hsp70 (DnaK) chaperone complexed with ADP and substrate. Proc Natl Acad Sci U S A 106:8471–8476

Bhattacharya A, Kurochkin AV, Yip GN, Zhang Y, Bertelsen EB, Zuiderweg ER (2009) Allostery in Hsp70 chaperones is transduced by subdomain rotations. J Mol Biol 388:475–490

Bork P, Sander C, Valencia A (1992) An ATPase domain common to prokaryotic cell cycle proteins, sugar kinases, actin, and hsp70 heat shock proteins. Proc Natl Acad Sci U S A 89:7290–7294

Bruschi SA, West KA, Crabb JW, Gupta RS, Stevens JL (1993) Mitochondrial HSP60 (P1 protein) and a HSP70-like protein (mortalin) are major targets for modification during S-(1,1,2,2-tetrafluoroethyl)-L-cysteine-induced nephrotoxicity. J Biol Chem 268:23157–23161

Choi J, Forster MJ, McDonald SR, Weintraub ST, Carroll CA, Gracy RW (2004) Proteomic identification of specific oxidized proteins in ApoE-knockout mice: relevance to Alzheimer's disease. Free Radic Biol Med 36:1155–1162

Choglay AA, Chapple JP, Blatch GL, Cheetham ME (2001) Identification and characterization of a human mitochondrial homologue of the bacterial co-chaperone GrpE. Gene 267:125–134

Cicchillitti L, Di Michele M, Urbani A, Ferlini C, Donat MB, Scambia G, Rotilio D (2009) Comparative proteomic analysis of paclitaxel sensitive A2780 epithelial ovarian cancer cell line and its resistant counterpart A2780TC1 by 2D-DIGE: the role of ERp57. J Proteome Res 8:1902–1912

Craig EA, Huang P, Aron R, Andrew A (2006) The diverse roles of J-proteins, the obligate Hsp70 co-chaperone. Rev Physiol Biochem Pharmacol 156:1–21

Deocaris CC, Taira K, Kaul SC, Wadhwa R (2005) Mimotope-hormesis and mortalin/grp75/mthsp70: a new hypothesis on how infectious disease-associated epitope mimicry may explain low cancer burden in developing nations. FEBS Lett 579:586–590

Deocaris CC, Kaul SC, Wadhwa R (2006a) On the brotherhood of the mitochondrial chaperones mortalin and heat shock protein 60. Cell Stress Chaperones 11:116–128

Deocaris CC, Yamasaki K, Kaul SC, Wadhwa R (2006b) Structural and functional differences between mouse mot-1 and mot-2 proteins that differ in two amino acids. Ann NY Acad Sci 1067:220–223

Deocaris CC, Widodo N, Ishii T, Kaul SC, Wadhwa R (2007a) Functional significance of minor structural and expression changes in stress chaperone mortalin. Ann NY Acad Sci 1119:165–175

Deocaris CC, Widodo N, Shrestha BG, Kaur K, Ohtaka M, Yamasaki K, Kaul SC, Wadhwa R (2007b) Mortalin sensitizes human cancer cells to MKT-077-induced senescence. Cancer Lett 252:259–269

Deocaris CC, Takano S, Priyandoko D, Kaul Z, Yaguchi T, Kraft DC, Yamasaki K, Kaul SC, Wadhwa R (2008) Glycerol stimulates innate chaperoning, proteasomal and stress-resistance functions—implications for geronto-manipulation. Biogerontology 9:269–282

Deocaris CC, Kaul SC, Wadhwa R (2009) The versatile stress protein mortalin as a chaperone therapeutic agent. Protein Pept Lett 16:517–529

Domanico SZ, Denagel DC, Dahlseid JN, Green JM, Pierce SK (1993) Cloning of the gene encoding peptide-binding protein 74 shows that it is a new member of the heat shock protein 70 family. Mol Cell Biol 13:3598–3610

Dekker PJ, Pfanner N (1997) Role of mitochondrial GrpE and phosphate in the ATPase cycle of matrix Hsp70. J Mol Biol 270:321–327

Flaherty KM, Deluca-Flaherty C, Mckay DB (1990) Three-dimensional structure of the ATPase fragment of a 70 K heat-shock cognate protein. Nature 346:623–628

Genevaux P, Georgopoulos C, Kelley WL (2007) The Hsp70 chaperone machines of Escherichia coli: a paradigm for the repartition of chaperone functions. Mol Microbiol 66:840–857

Goswami AV, Chittoor B, D'silva P (2010) Understanding the functional interplay between mammalian mitochondrial Hsp70 chaperone machine components. J Biol Chem 285:19472–19482

Gragerov A, Gottesman ME (1994) Different peptide binding specificities of hsp70 family members. J Mol Biol 241:133–135

Gupta RS, Ramachandra NB, Bowes T, Singh B (2008) Unusual cellular disposition of the mitochondrial molecular chaperones Hsp60, Hsp70 and Hsp10. Novartis Found Symp 291:59–68

Ha JH, Mckay DB (1994) ATPase kinetics of recombinant bovine 70 kDa heat shock cognate protein and its amino-terminal ATPase domain. Biochemistry 33:14625–14635

Ha JH, Mckay DB (1995) Kinetics of nucleotide-induced changes in the tryptophan fluorescence of the molecular chaperone Hsc70 and its subfragments suggest the ATP-induced conformational change follows initial ATP binding. Biochemistry 34:11635–11644

Hohfeld J, Cyr DM, Patterson C (2001) From the cradle to the grave: molecular chaperones that may choose between folding and degradation. EMBO Rep 2:885–890

Horton LE, James P, Craig EA, Hensold JO (2001) The yeast hsp70 homologue Ssa is required for translation and interacts with Sis1 and Pab1 on translating ribosomes. J Biol Chem 276:14426–14433

Hunt C, Morimoto RI (1985) Conserved features of eukaryotic hsp70 genes revealed by comparison with the nucleotide sequence of human hsp70. Proc Natl Acad Sci U S A 82:6455–6459

Jiang J, Maes EG, Taylor AB, Wang L, Hinck AP, Lafer EM, Sousa R (2007) Structural basis of J cochaperone binding and regulation of Hsp70. Mol Cell 28:422–433

Kabani M (2009) Structural and functional diversity among eukaryotic Hsp70 nucleotide exchange factors. Protein Pept Lett 16:623–660

Karlin S, Brocchieri L (1998) Heat shock protein 70 family: multiple sequence comparisons, function, and evolution. J Mol Evol 47:565–577

Kaul SC, Duncan EL, Englezou A, Takano S, Reddel RR, Mitsui Y, Wadhwa R (1998) Malignant transformation of NIH3T3 cells by overexpression of mot-2 protein. Oncogene 17:907–911

Kaul SC, Takano S, Reddel RR, Mitsui Y, Wadhwa R (2000) Transcriptional inactivation of p53 by deletions and single amino acid changes in mouse mot-1 protein. Biochem Biophys Res Commun 279:602–606

Kaul SC, Reddel RR, Mitsui Y, Wadhwa R (2001) An N-terminal region of mot-2 binds to p53 in vitro. Neoplasia 3:110–114

Kaul SC, Yaguchi T, Taira K, Reddel RR, Wadhwa R (2003) Overexpressed mortalin (mot-2)/mthsp70/GRP75 and hTERT cooperate to extend the in vitro lifespan of human fibroblasts. Exp Cell Res 286:96–101

Kaul SC, Aida S, Yaguchi T, Kaur K, Wadhwa R (2005) Activation of wild type p53 function by its mortalin-binding, cytoplasmically localizing carboxyl terminus peptides. J Biol Chem 280:39373–39379

Kaul SC, Deocaris CC, Wadhwa R (2007) Three faces of mortalin: a housekeeper, guardian and killer. Exp Gerontol 42:263–274

Kim HP, Morse D, Choi AM (2006) Heat-shock proteins: new keys to the development of cytoprotective therapies. Expert Opin Ther Targets 10:759–769

Liberek K, Lewandowska A, Zietkiewicz S (2008) Chaperones in control of protein disaggregation. EMBO J 27:328–335

Mayer MP (2011) Gymnastics of molecular chaperones. Mol Cell 39:321–331

Mayer MP, Bukau B (1998) Hsp70 chaperone systems: diversity of cellular functions and mechanism of action. Biol Chem 379:261–268

Mayer MP, Bukau B (2005) Hsp70 chaperones: cellular functions and molecular mechanism. Cell Mol Life Sci 62:670–684

Mayer MP, Rudiger S, Bukau B (2000) Molecular basis for interactions of the DnaK chaperone with substrates. Biol Chem 381:877–885

Mogk A, Bukau B (2004) Molecular chaperones: structure of a protein disaggregate. Curr Biol 14:R78–R80

Nakabayashi K, Ogata T, Fujii M, Tahara H, Ide T, Wadhwa R, Kaul SC, Mitsui Y, Ayusawa D (1997) Decrease in amplified telomeric sequences and induction of senescence markers by introduction of human chromosome 7 or its segments in SUSM-1. Exp Cell Res 235:345–353

Naylor DJ, Stines AP, Hoogenraad NJ, Hoj PB (1998) Evidence for the existence of distinct mammalian cytosolic, microsomal, and two mitochondrial GrpE-like proteins, the co-chaperones of specific Hsp70 members. J Biol Chem 273:21169–21177

Packschies L, Theyssen H, Buchberger A, Bukau B, Goody RS, Reinstein J (1997) GrpE accelerates nucleotide exchange of the molecular chaperone DnaK with an associative displacement mechanism. Biochemistry 36:3417–3422

Pockley AG, Muthana M, Calderwood SK (2008) The dual immunoregulatory roles of stress proteins. Trends Biochem Sci 33:71–79

Rudiger S, Germeroth L, Schneider-Mergener J, Bukau B (1997) Substrate specificity of the DnaK chaperone determined by screening cellulose-bound peptide libraries. EMBO J 16:1501–1507
Rudiger S, Mayer MP, Schneider-Mergener J, Bukau B (2000) Modulation of substrate specificity of the DnaK chaperone by alteration of a hydrophobic arch. J Mol Biol 304:245–251
Saibil HR (2008) Chaperone machines in action. Curr Opin Struct Biol 18:35–42
Schneider HC, Westermann B, Neupert W, Brunner M (1996) The nucleotide exchange factor MGE exerts a key function in the ATP-dependent cycle of mt-Hsp70-Tim44 interaction driving mitochondrial protein import. EMBO J 15:5796–5803
Slepenkov SV, Witt SN (2002) The unfolding story of the Escherichia coli Hsp70 DnaK: is DnaK a holdase or an unfoldase? Mol Microbiol 45:1197–1206
Smock RG, Rivoire O, Russ WP, Swain JF, Leibler S, Ranganathan R, Gierasch LM (2010) An interdomain sector mediating allostery in Hsp70 molecular chaperones. Mol Syst Biol 6:414
Sriram M, Osipiuk J, Freeman B, Morimoto R, Joachimiak A (1997) Human Hsp70 molecular chaperone binds two calcium ions within the ATPase domain. Structure 5:403–414
Stahl B, Tobaben S, Sudhof TC (1999) Two distinct domains in hsc70 are essential for the interaction with the synaptic vesicle cysteine string protein. Eur J Cell Biol 78:375–381
Strub A, Zufall N, Voos W (2003) The putative helical lid of the Hsp70 peptide-binding domain is required for efficient preprotein translocation into mitochondria. J Mol Biol 334:1087–1099
Terada K, Kanazawa M, Bukau B, Mori M (1997) The human DnaJ homologue dj2 facilitates mitochondrial protein import and luciferase refolding. J Cell Biol 139:1089–1095
Terada K, Mori M (2000) Human DnaJ homologs dj2 and dj3, and bag-1 are positive cochaperones of hsc70. J Biol Chem 275:24728–24734
Tomkiewicz D, Nouwen N, Driessen AJ (2007) Pushing, pulling and trapping—modes of motor protein supported protein translocation. FEBS Lett 581:2820–2828
Wadhwa R, Kaul SC, Ikawa Y, Sugimoto Y (1993a) Identification of a novel member of mouse hsp 70 family: its association with cellular mortal phenotype. J Biol Chem 268:6615–6621
Wadhwa R, Kaul SC, Sugimoto Y, Mitsui Y (1993b) Induction of cellular senescence by transfection of cytosolic mortalin cDNA in NIH 3T3 cells. J Biol Chem 268:22239–22242
Wadhwa R, Sugihara T, Yoshida A, Nomura H, Reddel RR, Simpson R, Maruta H, Kaul SC (2000) Selective toxicity of MKT-077 to cancer cells is mediated by its binding to the hsp70 family protein mot-2 and reactivation of p53 function. Cancer Res 60:6818–6821
Webster TJ, Naylor DJ, Hartman DJ, Hoj PB, Hoogenraad NJ (1994) cDNA cloning and efficient mitochondrial import of pre-mtHSP70 from rat liver. DNA Cell Biol 13:1213–1220
Wittrup A, Zhang SH, Svensson KJ, Kucharzewska P, Johansson MC, Mörgelin M, Belting M (2010) Magnetic nanoparticle-based isolation of endocytic vesicles reveals a role of the heat shock protein GRP75 in macromolecular delivery. Proc Natl Acad Sci U S A 10:13342–13347
Xu Z, Page RC, Gomes MM, Kohli E, Nix JC, Herr AB, Patterson C, Misra S (2008) Structural basis of nucleotide exchange and client binding by the Hsp70 cochaperone Bag2. Nat Struct Mol Biol 15:1309–1317
Yokoyama K, Fukumoto K, Murakami T, Harada S, Hosono R, Wadhwa R, Mitsui Y, Ohkuma S (2002) Extended longevity of Caenorhabditis elegans by knocking in extra copies of hsp70 F, a homolog of mot-2 (mortalin)/mthsp70/Grp75. FEBS Lett 516:53–57
Young JC, Barral JM, Hartl FU (2003) More than folding: localized functions of cytosolic chaperones. Trends Biochem Sci 28:541–547
Zhai P, Stanworth C, Liu S, Silberg JJ (2008) The human escort protein Hep binds to the ATPase domain of mitochondrial hsp70 and regulates ATP hydrolysis. J Biol Chem 283:26098–26106
Zhao Q, Wang J, Levichkin IV, Stasinopoulos S, Ryan MT, Hoogenraad NJ (2002) A mitochondrial specific stress response in mammalian cells. EMBO J 21:4411–4419
Zhu X, Zhao X, Burkholder WF, Gragerov A, Ogata CM, Gottesman ME, Hendrickson WA (1996) Structural analysis of substrate binding by the molecular chaperone DnaK. Science 272:1606–1614

Chapter 3
The Role of Mortalin in Iron Homeostasis

Wen-I Luo and James A. Cowan

Abstract The role of chaperones in prokaryotic and eukaryotic iron-sulfur cluster assembly pathways is discussed in the context of the complex sequence of events that are involved in [2Fe–2S] cluster biogenesis. Building from a background description of the functions of the component proteins and the various stages of cluster assembly and transfer to target proteins, similarities and differences in chaperone functions are presented with a particular focus on the unique characteristics of mortalin, relative to other chaperones.

Keywords Iron-sulfur cluster · Iron homeostasis · Chaperone · Hsp70 · Mortalin · ISU · IscU

3.1 Iron-Sulfur (Fe–S) Cluster Biogenesis

3.1.1 Fe–S Cluster Biosynthesis and Iron Homeostasis

Iron is one of the most abundant transition metal ions in most if not all living organisms, and maintenance of iron homeostasis—uptake, transport, distribution and export—is vital for life processes (Lill and Muhlenhoff 2005). The cellular chemistry of iron is highly regulated since excess iron is toxic; either through adventitious binding and inhibition of cellular biomolecules, or through generation of unwanted reactive oxygen species via Fenton chemistry (Papanikolaou and Pantopoulos 2005). Cellular iron can be divided into two classes: heme-iron and non-heme iron. The former is a prosthetic group composed of a tetrapyrrolic ring with chelated iron bound to the inner core. A fair amount of cellular iron exists in the form of heme-iron and heme iron proteins participate in crucial biological functions such as oxygen binding and electron transfer. Non-heme iron can be classified into several classes including

J. A. Cowan (✉)
Evans Laboratory of Chemistry, The Ohio State University, 100 West 18th Avenue, Columbus, Ohio 43210, USA
e-mail: cowan@chemistry.ohio-state.edu

W.-I. Luo
Ohio State Biochemistry Program, The Ohio State University, Columbus, OH 43210, USA

S. C. Kaul, R. Wadhwa (eds.), *Mortalin Biology: Life, Stress and Death*,
DOI 10.1007/978-94-007-3027-4_3, © Springer Science+Business Media B.V. 2012

Fe–S clusters, Fe-O-Fe clusters (oxo-bridge), and mononuclear nitrogen/oxygen ligated iron. Among these the Fe–S cluster family is the most prominent, both in size and diversity of functional uses (Frausto da Silva and Williams 2001).

Fe–S clusters are one of the most primitive biological metal cofactors and are crucial for enzymes that participate in key biological events, such as mitochondrial respiration, photosynthesis, gene regulation and nitrogen fixation. The iron-sulfur world theory proposes that Fe–S clusters are among the most ancient catalysts and evolved from the reducing volcanic environment (Wachtershauser 2007; Martin and Russell 2003). Structurally versatile, protein-bound Fe–S clusters span a range of multinuclear [2Fe–2S], [3Fe–4S], [4Fe–4S] and [8Fe–7S] centers. In biological systems, Fe–S clusters are involved in various functions that include electron transfer, catalysis, oxygen (and oxidants) sensing, and sulfur exchange (Barras et al. 2005; Beinert 2000; Frausto da Silva and Williams 2001; Johnson et al. 2005; Lill and Muhlenhoff 2005; Lill 2009; Lill and Muhlenhoff 2006; Mansy and Cowan 2004; Marquet et al. 2007; Py and Barras 2010; Rawat and Stemmler 2011; Xu and Moller 2011). Defects in Fe–S cluster biogenesis *in vivo* can result in calamitous effects (Sheftel and Lill 2009; Sheftel et al. 2010; Ye and Rouault 2010).

3.1.2 Overview of Fe–S Cluster Biosynthesis Machinery

Since the very first Fe–S protein, ferredoxin (*Clostridium pasteurianum* and spinach) was characterized in the 1960s, studies regarding the biosynthesis of Fe–S clusters have flourished (San Pietro and Lang 1958; Mortenson et al. 1962); however, the mechanistic details of Fe–S cluster biogenesis are still being elaborated.

Free iron ion and inorganic sulfide are toxic to biological systems; therefore a complex molecular machinery composed of multiple proteins is required to synthesize Fe–S clusters *in vivo*. So far, three major operons (*nif, suf, isc)* have been found to encode various components of the Fe–S cluster biogenesis apparatus and are generally designated as NIF (nitrogen fixation), SUF (mobilization of sulfur) and ISC (iron-sulfur cluster) machineries (Godman and Balk 2008; Takahashi and Nakamura 2002; Tokumoto and Takahashi 2001; Tokumoto et al. 2002). Although there is some variation among the protein components of these systems, they all require a cysteine desulfurase as a sulfur donor in addition to scaffold proteins (Fontecave et al. 2008). Certain bacterial species possess more than one of these systems, while in non-photosynthetic eukaryotic systems the mitochondrial ISC apparatus represents the general housekeeping Fe–S cluster provider. Some protein components are interchangeable among these systems, but this is not always the case (Dos Santos et al. 2007; Tokumoto et al. 2004).

The NIF machinery is responsible for the biosynthesis of Fe–S clusters in nitrogenases. Extensive studies of the NIF machinery have focused on the nitrogenase from *Azotobacter vinelandii* (Jacobson et al. 1989). The [2Fe–2S] cluster assembly scaffold protein, NifU, and the cysteine desulfurase, NifS, were the first two proteins to be characterized as required components for formation of Fe–S clusters (Johnson et al. 2005). The deletion of either *nifU* or *nifS* gene resulted in a decline of the activity of both the Fe and MoFe proteins, but did not fully abolish their activity. This finding led to another discovery that there exists a "housekeeping" machinery that can compensate for the loss of either NifU or NifS (Zheng et al. 1998). The discovery also suggested that, in some organisms, there may exist multiple Fe–S cluster assembling systems that co-exist and are regulated depending on the biological circumstances.

Another Fe–S cluster system, the SUF apparatus, was later identified in *Escherichia coli* (Takahashi and Tokumoto 2002). SUF is found broadly in plastids, bacteria and pathogens such as *Mycobacterium tuberculosis* (Huet et al. 2005). While details of the SUF machinery remain unclear, it has been suggested that this fulfills a housekeeping function that operates under conditions of oxidative stress and of iron starvation.

The SUF system is composed of proteins that are encoded by the *suf* operon and gene products include: *suf*ABCDSE (Fontecave et al. 2005). The regulation of the *suf* operon and the housekeeping *isc* operon is executed by transcription repressors, such as IscR, which is sensitive to the status of physiological iron homeostasis (Rincon-Enriquez et al. 2008; Schwartz et al. 2001).

In the SUF machinery, sulfur is provided by the cysteine desulfurase, SufS, with its partner, SufE. Due to the requirement for an additional protein SufE, the mechanism of sulfur transfer in the SUF machinery is distinct from other cysteine desulfurases, such as NifS in NIF and IscS/Nfs1 in ISC systems (Loiseau et al. 2003). In terms of Fe–S cluster assembly, it is suggested that Fe–S clusters are assembled by a U-type scaffold protein, SufU, and an A-type scaffold protein, SufA (Albrecht et al. 2010; Sendra et al. 2007). SufU is found to be the indispensible scaffold protein, relative to SufA (Albrecht et al. 2010). SufC is an ABC-type ATPase, and its ATPase activity is shown to be accelerated by SufB (Eccleston et al. 2006). Structurally, SufB, SufC and SufD form a stable complex, $SufBC_2D$, and flavin is a required redox cofactor in this complex (Wollers et al. 2010). *In vitro* studies have shown that this $SufBC_2D$ complex is able to interact with SufA and is involved in Fe–S cluster assembly (Chahal et al. 2009; Gupta et al. 2009). Studies also suggest the SufES complex to transfer sulfur to $SufBC_2D$ through an interaction between SufE and SufB (Layer et al. 2007).

The ISC apparatus is believed to constitute the general housekeeping machinery in Fe–S cluster biosynthesis under normal growth conditions. As a result of evolutionary endosymbiosis, in the case of eukaryotic systems the formation of Fe–S clusters occurs in mitochondria. Some amitochondrial organisms also possess an ISC-like machinery through particular organelles, such as hydrogenosomes in *Trichomonas vaginalis* and mitosomes in *Encephalitozoon cuniculi* (Goldberg et al. 2008; Dolezal et al. 2007; Sutak et al. 2004; Tachezy et al. 2001). A phylogenetic comparison of mitochondria Hsp70 and genes encoding chaperones in microspordia *Nosema locustae* shows that they might share the same evolutionary root, but the loss of mitochondria occurred later (Germot et al. 1997).

Fe–S cluster biosynthesis in the eukaryotic mitochondrial ISC system is very similar to that promoted by the bacterial ISC machinery; however, there are still some differences across the systems. The ISC machinery is composed of proteins that are encoded by the *isc* operon. Gene products of the *isc* operon include:

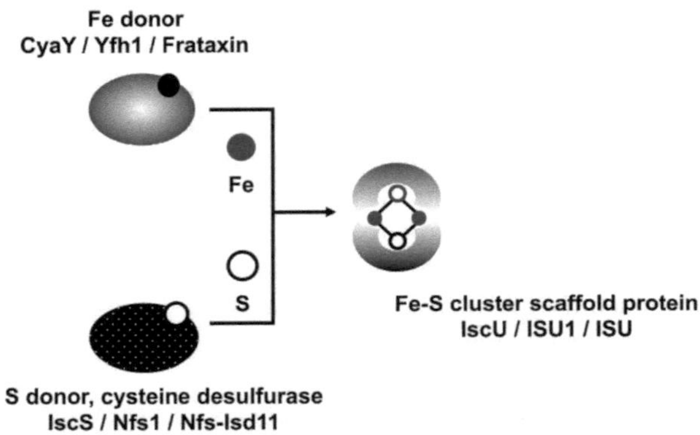

Fig. 3.1 Model for Fe–S cluster assembly in the ISC machinery. (1) Iron is provided by the iron donor; (2) sulfane sulfur is transferred by cysteine desulfurase; and (3) U-type scaffold protein assembles [2Fe–2S] clusters through 3 conserved cysteine residues at the active site. Chaperones have been implicated both in the assembly reaction and cluster stabilization, as well as subsequent transfer to an apo target protein

*isc*RSUA-*hsc*BA-*fdx*, where *isc*R encodes the transcription repressor that regulates the expression of *isc*RSUA, *isc*S encodes the cysteine desulfurase, *isc*U and *isc*A encode scaffold proteins, *hsc*A and *hsc*B are identified as genes that encode the chaperone and the co-chaperone, respectively; and *fdx* encodes ferredoxin that participates in electron transfer (Zheng et al. 1998).

3.1.3 Iron-Sulfur Cluster (ISC) Machinery

Several components are involved in Fe–S cluster biosynthesis promoted by the ISC machinery. The key proteins include: (1) the iron donor: including the bacterial frataxin ortholog CyaY, yeast frataxin Yfh1, and human frataxin; (2) the sulfur donor, cysteine desulfurase: including the bacterial IscS, yeast Nfs1p, human Nfs1 (human Nfs1 is in complex with Isd11); and (3) the assembly scaffold proteins: in the case of U-type scaffold proteins (described later), these include bacterial IscU, yeast ISU1, human ISU (Fig. 3.1). Other protein components are also involved in the assembly process in order to assist in the context of electron transfer, delivery of iron and sulfide, or cluster transfer to target proteins (Mühlenhoff et al. 2003; Lill 2009).

3.1.3.1 Iron Donors and Cysteine Desulfurases

In eukaryotes, Yfh1/frataxin is suggested to be the iron donor protein for mitochondrial Fe–S cluster biogenesis (Stemmler et al. 2010; Layer et al. 2006). *In vitro* studies show that Yfh1/frataxin binds iron through acidic amino acid residues on the $\alpha 1$ and $\beta 1$ domain (Yoon and Cowan 2003; Huang et al. 2008; Correia et al. 2010; Leidgens et al. 2010; Foury et al. 2007; He et al. 2004). In yeast, Yfh1 defects cause iron accumulation and the inactivation of multiple Fe–S proteins (Stemmler et al. 2010; Layer et al. 2006), whereas in humans it is known that a genetic defect of frataxin (specifically, with the expansion of a GAA triplet repeats in the first intron of the frataxin gene) leads to Friedrich's ataxia. Iron accumulation is observed in the brains and hearts of Friedrich ataxia patients (Wilson 2006). A number of biochemical and genetics studies have shown that Yfh1/frataxin interacts with the scaffold protein, IscU/ISU (Cook et al. 2010; Wang and Craig 2008; Ramazzotti et al. 2004); however, in addition to serving as a likely iron donor in Fe–S cluster biogenesis, several other functional roles for frataxin have been proposed, including (1) iron chaperone; (2) iron storage protein; (3) iron donor to the iron carrier proteins, Mrs3p and Mrs4p; (4) iron donor to ferrochelatase in heme synthesis; and (5) as an iron regulator (Aloria et al. 2004; Bencze et al. 2006, 2007; Chen et al. 2002; Gakh et al. 2002, 2008; Schoenfeld et al. 2005; Subramanian et al. 2011; Zhang et al. 2005, 2006). Overall, service as an iron donor in Fe–S cluster biogenesis is a generally accepted role for frataxin.

Studies of 56 genomes have revealed that the frataxin/CyaY gene shares an identical phylogenetic distribution as the chaperone and co-chaperone genes *hsc*A and *hsc*B (Huynen et al. 2001) [note: eukaryotic orthologs of *hsc*A do not reside in the mitochondria]. This co-evolution provides strong support for the idea that frataxin is specifically and directly involved in Fe–S cluster biogenesis in the ISC machinery, in addition to a general role in iron homeostasis. Also, in yeast *Saccharomyces cerevisiae*, it has been shown that the chaperone, Ssq1, is crucial for the maturation of Yfh1 (Voisine et al. 2000). Direct interaction between Yfh1 and the scaffold protein, Isu1, has been shown by biochemical and genetic studies. Through the design of various mutants it has been shown that the iron-dependent interaction between Yfh1 and Isu1 is attained through the β-sheet surface of Yfh1 by means of several key residues: N122, K123, Q124 and W131 (Wang and Craig 2008; Leidgens et al. 2010; Cook et al. 2010). Human frataxin has also been found to significantly stimulate the activity of the human Nfs1-Isd11 complex delivering a sulfide to ISU during cluster assembly (Shan et al. 2007).

An additional common characteristic of the aforementioned Fe–S cluster biosynthesis machinery is the existence of cysteine desulfurase. In the bacterial ISC system, it is found that inactivation of *isc*S gene seriously impairs production of downstream Fe–S proteins such as ferredoxin. Cysteine desulfurases are pyridoxal phosphate (PLP)-dependent enzymes that extract sulfur from a cysteine substrate through the formation of a persulfide bond and with the release of alanine. All cysteine desulfurases possess a conservative cysteine residue at the active site (Zheng et al. 1994; Smith et al. 2005). There are two classes of cysteine desulfurases: group I cysteine desulfurases have a consensus motif, SSGSAC(T/S)S, as in the case of NifS, IscS and Nfs; while group II enzymes have a consensus motif, RXGHHCA, as in the case of SufS (Mihara and Esaki 2002). Cysteine desulfurases not only serve as sulfur donors in Fe–S cluster biogenesis, but also function as universal sulfur providers for thiamines, tRNAs and molybdopterins (Kessler 2006; Mihara and Esaki 2002; Mühlenhoff et al. 2004; Nakai et al. 2004; Paris et al. 2010). RNA interference experiments reveal that the depletion of mitochondrial Nfs causes severe loss of activity for Fe–S enzymes, changes in mitochondrial morphology and cell growth retardation; all of which support an indispensible role for Nfs in cellular chemistry (Fosset et al. 2006; Biederbick et al. 2006).

The interaction of cysteine desulfurases with either the scaffold protein IscU/ISU or SufE has been demonstrated *in vitro* through various biophysical experiments. It is believed that cysteine desulfurase participates in Fe–S cluster biogenesis through this interaction (Heller 2010; Nuth and Cowan 2009; Outten et al. 2003; Shi et al. 2010). *In vivo* studies in yeast show that gene depletion of Nfs1p results in a significant loss of iron incorporation into ISU1 (Muhlenhoff et al. 2003). This shows that cysteine desulfurase is required to stabilize iron binding to the scaffold protein presumably through [2Fe–2S] cluster formation.

In the eukaryotic mitochondrial ISC machinery, the cysteine desulfurase, Nfs1, functions in partnership with another mitochondrial inner-membrane associated protein, Isd11, which is required for Fe–S cluster biogenesis (Shi et al. 2009). Bioinformatic analysis shows that Isd11 most likely represents an evolutionary innovation for eukaryotes during endosymbiosis (Richards and van der Giezen 2006), although the exact functional role for Isd11 remains unclear. *In vivo*, a depletion of Isd11 results in a severe loss of activity of downstream Fe–S proteins. Results from pull-down and co-immunoprecipitation assays suggest that Nfs1 and Isd11 form a complex. Although Nfs1 can maintain its activity in the absence of Isd11, Nfs1 tends to aggregate (Adam et al. 2006; Wiedemann et al. 2006).

Recent *in vitro* and *in vivo* studies suggest that frataxin interacts with an ISU/Nfs1/Isd11 complex, instead of individual proteins to assemble Fe–S cluster (Shan et al. 2007; Tsai and Barondeau 2010; Schmucker et al. 2011). The presence of frataxin accelerates Fe–S cluster formation and this suggests that frataxin might also serve to activate *de novo* Fe–S cluster biogenesis (Tsai and Barondeau 2010).

3.1.3.2 Scaffold Proteins

There are three classes of Fe–S cluster scaffold proteins: U-type, A-type and NFU-type scaffold proteins (Schilke et al. 1999). The role of the latter is complicated by the fact that discrete forms of NFU protein have been identified with potentially distinct roles. The thioredoxin motif has been implicated both with persulfide reductase activity, releasing inorganic sulfide, and also [2Fe–2S] cluster formation (Liu and Cowan 2007; Liu et al. 2009). Among all scaffold proteins, the U-type family of scaffold proteins is the best characterized. Examples include NifU in NIF, IscU in bacterial ISC, and ISU in mitochondrial ISC (in yeast *S. cerevisiae*, there are two cognate ISU proteins: ISU1 and ISU2, where ISU1 is the major scaffold protein). A common hallmark in the sequence of all U-type scaffold proteins is the presence of three conserved cysteine residues that participate in [2Fe–2S] cluster formation in the presence of cysteine desulfurase and an iron donor.

Mutation of either ISU1 or ISU2 in *S. cerevisiae* results in the reduction of the downstream [4Fe–4S] cluster containing aconitase activity (Garland et al. 1999). Results from RNA interference suppression of human ISU shows the inactivation of downstream aconitase, but the activation of iron regulatory proteins (IRPs) (Tong and Rouault 2006). These results illustrate the role of ISU in promotion of intracellular iron homeostasis. Furthermore, the deletion of both genes is lethal. In addition, *in vitro* studies show that IscU/ISU is able to assemble [2Fe–2S] clusters and then transfer these to apo-proteins such as apo-ferredoxin and apo-IscA *in vitro* (Mansy et al. 2002; Wu et al. 2002c; Agar et al. 2000; Bonomi et al. 2005). It has been demonstrated that ISU1/ISU directly interacts with Yfh1/frataxin and cysteine desulfurases (Nuth and Cowan 2009; Leidgens et al. 2010; He et al. 2004; Foury et al. 2007). Taken together these facts show the essential role for IscU/ISU in the ISC machinery.

To date, the role of A-type scaffold proteins is uncertain. It has previously been proposed that A-type proteins can serve as an iron donor (Ding and Clark 2004; Ding et al. 2004). However, this hypothesis has garnered less support as a result of extensive evidence to support the role of A-type proteins as alternative Fe–S scaffold proteins. Yeast ISA and bacterial IscA in the ISC machinery, as well as SufA in the SUF machinery, can all bind [2Fe–2S] clusters and transfer these to apo-proteins (Bonomi et al. 2005; Wu and Cowan 2003; Wu et al. 2002a, b; Ollagnier-de-Choudens et al. 2001). Also, it has been shown in bacterial systems that holo-IscU is able to transfer an Fe–S cluster to apo-IscA, but not vice versa (Ollagnier-de-Choudens et al. 2004). Moreover, in yeast *S. cerevisiae*, mutation studies of two A-type scaffold proteins, ISA1 and ISA2, yield no evidence that they are essential, although the deletion causes iron accumulation, the decline of downstream aconitase activity and defects in cell growth (Jensen and Culotta 2000; Pelzer et al. 2000). Both biochemical and genetic studies support the U-type scaffold protein as the main cluster assembly protein. A recent study in *Bacillus subtilis*, which employs a SUF system shows that SufU is essential for cell growth while SufA is unnecessary under minimal growth conditions. A recently identified A-type scaffold protein, ErpA, has been shown to transfer Fe–S clusters to a protein participating in isoprenoid biosynthesis, apo-IspG (Loiseau et al. 2007).

Aside from assembling Fe–S clusters, scaffold proteins have recently found to be a target of cobalt and copper stress. In *E. coli*, excess cobalt can react with holo IscU and holo SufA. It is suggested that cobalt can outcompete iron during cluster formation, thereby forming a connection between cobalt and iron metabolism (Ranquet et al. 2007). Genetic analysis shows that cobalt stress causes the up-regulation of multiple *isc* genes and the induction of the *suf* operon (Fantino et al. 2010), thereby establishing a connection between cobalt and iron homeostasis. It has also been shown in *B. subtilis* that excess copper can directly target SufU and destabilize Fe–S cluster formation as a reflection of copper toxicity. Data suggests that this feedback regulation is not the consequence of reactive oxygen species generated by copper, but results from the direct interaction between copper and SufU (Chillappagari et al. 2010), providing another link connecting copper and iron homeostasis. Both cases demonstrate the inter-relationship between metals in biological systems and further emphasize the crucial role of Fe–S cluster biogenesis in iron homeostasis.

3.1.3.3 Other Components

Ferredoxins and ferredoxin reductases are also required for efficient Fe–S cluster biogenesis (Mühlenhoff et al. 2002; Li et al. 2001; Lange et al. 2000). In yeast *S. cerevisiae* the ferredoxin and the ferredoxin reductase are termed Yah1 and Arh1, respectively. In humans there exists two forms of mitochondrial ferredoxin, namely ferredoxin 1 (Fdx1; adrenodoxin) and ferredoxin 2 (Fdx2). RNA interference experiments show that human Fdx1 might not be involved in Fe–S cluster biogenesis, but is able to reduce cytochrome P450 and mediate other biological functions such as steroid biosynthesis. On the contrary, Fdx2 is essential for heme and Fe–S cluster biosynthesis (Sheftel et al. 2010). Both ferredoxin and ferredoxin reductase are found to be essential for the maturation of downstream Fe–S cluster containing proteins. The attenuation of either Yah1 or Arh1 leads to cellular iron accumulation. However, the specific roles for these proteins, and in particular the direction of electron flow to or from the nascent scaffold-bound cluster, remains to be established.

Genetics studies in yeast *S. cerevisiae* suggest a role for a mitochondrial monothiol glutaredoxin, Grx5, in Fe–S cluster biosynthesis (Rodríguez-Manzaneque et al. 2002). The deletion of the *grx5* gene results in higher sensitivity toward oxidative stress, iron accumulation and inactivation of mitochondrial Fe–S proteins (Linares et al. 2009). This phenotype can be rescued by overexpression of Ssq1 (a chaperone in the ISC system, described later) and ISA2. In a zebra fish model, Grx5 deficiency results in anemia; whereas in Grx5-deficient patients, severe anemia and iron overload are observed (Wingert et al. 2000; Ye et al. 2010). A simulation study shows that, Grx5 might act on Nfs1 or the Nfs1-scaffold protein complex; nevertheless, a recent study of a Δ *grx5* yeast mutant shows that *isa2*, but not *isu1* or *ssc1*, can suppress the growth defect (Alves et al. 2004). Another experiment also shows that Grx5 is involved in Fe-S cluster biogenesis through interaction with ISA (Kim et al. 2010). To date, it is unclear whether the monothiol glutaredoxin participates in Fe–S cluster delivery or assembly.

A recent *in vitro* study shows that nucleotides adenosine triphosphate (ATP) and guanosine triphosphate (GTP) are needed to enhance the Fe–S cluster biosynthesis in isolated mitochondria. Both *in vivo* and *in vitro* studies further suggest that, hydrolysis of GTP is necessary for cluster insertion to aconitase, Aco1, and ferredoxin, Yah1 (Amutha et al. 2008). However, mechanistic details on the role of the nucleotide have yet to be clarified.

The export of *de novo* Fe–S clusters from mitochondria is executed by an ABC (ATP binding cassette) transporter, termed Atm1 in yeast and ABCB7 in humans (Lill and Kispal 2001). In yeast, Atm1-deficient mutants show retarded growth phenotype and hampered maturation of downstream Fe–S proteins; however, there is no damage to Fe–S cluster biosynthesis caused by Atm1-depletion (Miao et al. 2009). In humans, patients who suffer from X-linked sideroblastic anemia or cerebellar ataxia (XLSA/A) inherit mutations in the ABCB7 transporter and experience iron accumulation (Lill and Kispal 2001). Characterization of yeast Atm1 suggests that the ATPase activity of Atm1 is inhibited by vanadate, as found for other ABC transporters (Kuhnke et al. 2006; Chen and Cowan 2006). The ATPase activity of Atm1 is stimulated by sulfhydryl-containing compounds (Kuhnke et al. 2006).

3.2 Hsp70 Chaperones in the ISC Machinery

In the ISC machinery it has been acknowledged that an Hsp70 family chaperone and a co-chaperone are involved (Vickery and Cupp-Vickery 2007). Similar to all Hsp70 chaperones, the ISC chaperone contains two domains; namely a nucleotide binding domain (NBD) and a substrate binding domain (SBD). Interdomain communication is regulated through nucleotide ATP hydrolysis (Bukau and Horwich 1998; Morano 2007; Buchberger et al. 1995; Revington et al. 2005; Swain et al. 2007; Vogel et al. 2006a, b; Rist et al. 2006; Chang et al. 2007). Structurally, the binding of adenosine diphosphate (ADP), the product of ATP hydrolysis in the NBD, enhances substrate binding in the SBD; whereas the binding of ATP causes the release of substrate from the SBD (Liu et al. 2003; Jiang et al. 2006; Bertelsen et al. 2009; Chang et al. 2008; Farr et al. 1998). All Hsp70 chaperones tend to bind to short hydrophobic peptides, and substrate binding (either with short peptides, polypeptides or native proteins) induces weak ATPase activity (Craig and Marszalek 2002; Ha and McKay 1994; Montgomery et al. 1999). J-type co-chaperones that have an N-terminal J-domain containing a conserved HPD motif are found to interact with Hsp70 chaperones and are involved in the substrate binding/release cycle (Acebrón et al. 2008; Laufen et al. 1999; Silberg and Vickery 2000, Silberg et al. 2004; Schmid et al. 1994; Russell et al. 1998).

3.2.1 Hsp70 Chaperones in the Bacterial and Yeast ISC Machinery

The most extensively studied chaperone pairs in the ISC apparatus are *E. coli* HscA and its co-chaperone, HscB (Silberg et al. 1998; Vickery et al. 1997). Although *E. coli* HscA belongs to the Hsp70 family, unlike the housekeeping *E. coli* DnaK, HscA is not essential for the refolding of misfolded proteins under stress, and cannot compensate defects from DnaK mutants (Hesterkamp and Bukau 1998). Moreover, kinetics studies show that in the substrate binding-release cycle, the rate-limiting step for HscA is ATP hydrolysis instead of ADP release in the case of DnaK (Silberg and Vickery 2000). Both HscA and DnaK share 41% sequence identity. In yeast *S. cerevisiae*, Ssq1, the chaperone is found to be crucial for Fe-S cluster biogenesis; whereas another Hsp70 chaperone, Ssc1, is found to be involved in nascent protein folding through three mitochondrial import machinery. Phylogenetic studies show that Ssq1 is more related to DnaK than to HscA (Schilke et al. 2006). Moreover, structurally, the HscB co-chaperone shows significant differences relative to DnaJ.

Both HscB and DnaJ share a similar J-domain; nevertheless, the C-terminal domains differ from each other. These cases demonstrate the intricacy of chaperone systems in Fe–S cluster biogenesis across diverse systems.

Although the existence of an ISC-type Hsp70 chaperone is common to all systems, the evolutionary development of these chaperones is very different from one system to another. Genetic analysis shows that in bacterial ISC systems the inactivation of the *hsc*A gene causes a severe reduction in the availability of Fe–S cluster for the downstream ferredoxin, both in the presence or the absence of excess iron; whereas inactivation of the *hsc*B gene only causes a slight effect on the formation of holo-ferredoxin (Tokumoto and Takahashi 2001; Takahashi and Nakamura 1999). More interestingly, unlike HscA and Ssq1, many eukaryotes do not possess such a specialized Hsp70 chaperone for Fe–S cluster biogenesis, but utilize a multifunctional mtHsp70, Ssc1, as the working chaperone for Fe–S cluster biogenesis. Indeed, in yeast *S. cerevisiae*, an organism that holds three Hsp70 chaperones: Ssq1, Ssc1 and Ecm 10, Ssc1 is found to be able to partially replace Ssq1 in Fe–S cluster biogenesis when Ssq1 is impaired (Craig and Marszalek 2002). One example for the use of a multifunctional chaperone in the eukaryotic ISC machinery is the human system, which utilizes mortalin, an Ssc1 homologue. It is proposed that both Ssq1 and Ssc1 function by the same mechanism.

In yeast *S. cerevisiae* there exists three mitochondrial Hsp70 chaperones (Ssq1, Ssc1 and Ecm 10). *In vivo* studies show that, Ssc1 interacts with the integral membrane protein, Tim44, in the mitochondrial import machinery by both the NBD and SBD (D'Silva et al. 2004; Liu et al. 2003). As with most Hsp70 chaperones that are involved in proteostasis, Ssc1 is able to assist protein folding and to respond to stress conditions by aiding protein refolding and degradation (Liu et al. 2001). Ecm1 and Ssc1 share 82% identity; however, the deletion of Ecm1 yields no evident change in cellular phenotype. Ssq1 shares 52% sequence identity with Ssc1; however, instead of participating in proteostasis, the main function of Ssq1 is involvement in Fe–S cluster biogenesis (Dutkiewicz et al. 2006). In yeast, mutations of Ssq1 and its co-chaperone, Jac1, are found to cause iron accumulation and impaired activity of downstream Fe–S cluster containing proteins, such as aconitase and succinate dehydrogenase (Kim et al. 2001).

Studies of *E. coli* HscA show that IscU is able to raise the ATPase activity of HscA (Silberg et al. 2001) while screening of a phage display library demonstrates that the binding motif LPPVK of IscU can be recognized by HscA (Hoff et al. 2002, 2003; Tapley et al. 2006; Tapley and Vickery 2004). Studies of yeast *S. cerevisiae* also support Ssq1 binding to ISU1, with the assistance of the co-chaperone, Jac1. Similar to HscA, Ssq1 recognizes the LPPVK motif at ISU1. The observation that the overexpression of Jac1 is able to rescue the effect caused by the ISU1 mutant suggests that Ssq1 might target the Jac1-ISU1 complex instead of ISU1 alone (Schilke et al. 2006; Dutkiewicz et al. 2004). Also, studies from the Ssq1 mutant, F462S, show that F462 is likely to be involved in Ssq1 and ISU1 binding (Knieszner et al. 2005). In the case of DnaK, the chaperone functions together with its co-chaperone, DnaJ and the nucleotide exchange factor, GrpE. HscA functions with its co-chaperone, HscB. However, there is no apparent interaction between HscA and GrpE. Meanwhile,

Ssq1 requires a GrpE ortholog, Mge1, to stimulate ATPase activity (Lutz et al. 2001; Dutkiewicz et al. 2003; Silberg et al. 1998). It is shown that involvement of either HscB or Jac1 co-chaperone can stimulate the interaction between Hsp70 chaperones and U-type scaffold proteins in the presence of ATP (Bonomi et al. 2008; Hoff et al. 2000; Kim et al. 2001).

3.2.2 Role of Hsp70 Chaperones in the ISC Machinery

To date, the role of Hsp70 chaperones in ISC machinery is still a topic of intense studies. Based on the interaction between HscA and IscU, as well as the interaction between Ssq1 and ISU1, it has been widely accepted that Hsp70 chaperones are involved in Fe–S cluster biosynthesis through interaction with U-type scaffold proteins. However, many questions still remain open. For example, what mechanism controls the interaction among Hsp70 chaperones, co-chaperones and scaffold proteins? Does the chaperone target the scaffold protein, or the co-chaperone-scaffold protein complex? How is that achieved? It is demonstrated that following the interaction of co-chaperone and scaffold protein, the ATPase activity of the chaperone is enhanced. Recent NMR studies of HscB-IscU interaction show that the surface region of the C-terminal domain of HscB stabilizes the ordered state of apo-IscU. It is also shown that multiple residues on HscB are important for complex formation (Füzéry et al. 2008, 2011; Kim et al. 2009).

Another key question stems from the role of chaperones in cluster biogenesis; namely, Fe–S cluster assembly, relative to Fe–S cluster transfer? *In vitro* experiments show that Ssq1 can induce Fe–S cluster formation on ISU1, but the absence of any stimulatory effect from either ATP or Jac1 leads to uncertainty as to whether the induction of cluster formation is directly facilitated by the chaperone. *In vitro* studies of *Thermotoga maritima* DnaK (*Tm* DnaK) suggest a role for chaperone in the promotion of Fe–S cluster stability (Wu et al. 2005). Another *in vitro* study of the *A. vinelandii* HscA/HscB pair suggests promotion of cluster transfer from holo-IscU to apo-ferredoxin in the presence of excess ATP (Chandramouli and Johnson 2006). Possibly, different classes of chaperone function in distinct mechanistic roles.

An interesting observation stems from studies of yeast *S. cerevisiae*, where both Ssc1 and Ssq1 were found to be involved in Yfh1 import and maturation (Voisine et al. 2000). By use of radiolabeled Yfh1, the level of intermediate and mature forms of Yfh1 was compared. The processing of frataxin is less effective in *ssq1* gene-depleted mitochondria than the processing in wild-type mitochondria. This defect can be rescued by enhanced expression levels of Ssc1 (Voisine et al. 2000). In addition, *in vivo* studies show that human mortalin interacts with frataxin (Shan et al. 2007), although the functional role is unclear.

In vivo studies in yeast indicate that, the deletion of *ssq1* leads to enhanced levels of mRNA and protein for ISU1, and increases the amount of Fe–S cluster accumulation on ISU1 (Andrew et al. 2008; Mühlenhoff et al. 2003). This result suggests that a high level of ISU1 is required for cell growth in the absence of Ssq1, consistent with a role of chaperone in stabilizing ISU-bound cluster, and also shows that there exists a stringent regulatory mechanism to control Fe–S cluster biosynthesis. More exploration of the regulatory roles of chaperones is necessary.

3.3 Mortalin in the ISC Machinery

3.3.1 Mortalin in the Human ISC Machinery

Mortalin (also termed PBP74, mtHsp75 and GRP75) is related to multiple human diseases, such as cancer, Parkinson's and Alzheimer's (Deocaris 2008; Hsu et al. 2008; Iosefson and Azem 2010; Jin et al. 2006; Marchenko et al. 2000; de Mena et al. 2009; Shi et al. 2008; Takano et al. 1997; Takashima et al. 2003; Wadhwa et al. 1998, 2006). The direct link between mortalin and these diseases is still under investigation; however, it is proposed that mortalin could be a therapeutic target (Deocaris 2009). Mortalin resides in multiple cellular locations, but is found mainly in mitochondria (Bhattacharyya et al. 1995; Ran et al. 2000; Wadhwa et al. 2002). Moreover, mortalin is known to have multiple functions that include intracellular trafficking, antigen processing, regulation of cell proliferation, aging, and tumorigenesis. Mortalin interacts with multiple substrates, including p53, fibroblast growth factor-1 (FGF-1), and Tim44 in the mitochondrial import machinery (Kaul et al. 2002, 2007). Mortalin also interacts transiently with nascent mitochondrial proteins (Deocaris 2006; Mizzen et al. 1991).

It has been suggested that mortalin, an Ssc1 homologue, is involved in Fe–S cluster biogenesis in the human mitochondrial ISC machinery where the Ssq1 homologue is lacking, and so mortalin is very likely to be the Hsp70 chaperone that participates in Fe–S cluster biogenesis. The direct interaction between frataxin and mortalin was demonstrated through screening for proteins that interact directly with frataxin. Two approaches were utilized for the identification of mitochondrial binding partners for frataxin. First, the analysis of binding partners by use of immunoprecipitation experiments with anti-frataxin antibody against mitochondrial protein extracts. Proteins associated with frataxin were identified by mass spectrometry and included mortalin and Isd11. To further confirm the interactions, cells were transfected with constructs of flag-tagged frataxin and myc-tagged mortalin. Bi-directional co-immunoprecipitation results supported a direct interaction between frataxin and mortalin (Shan et al. 2007). These results strongly support the participation of mortalin in human mitochondrial Fe–S cluster biogenesis.

The co-chaperone that interacts with human ISU and mortalin has recently been identified. Similar to mortalin, this type III co-chaperone, Hsc20 is found to reside mainly in mitochondria. RNA interference experiments show that depletion of Hsc20 decreases the activity of downstream Fe–S cluster containing proteins such as aconitase and succinate dehydrogenase; providing further support for Hsc20 as the co-chaperone for mortalin (Uhrigshardt et al. 2010).

Unlike bacterial HscA and yeast Ssq1, progress in biochemical studies of human mortalin has been relatively slow, in spite of the significance of this chaperone. The

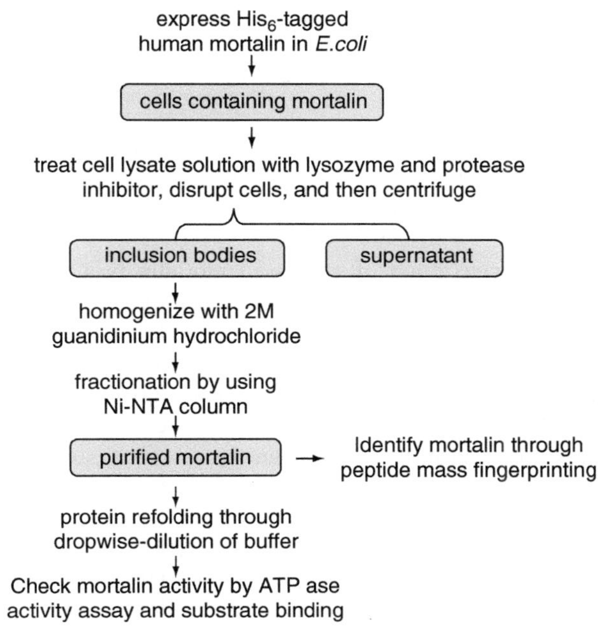

Fig. 3.2 Purification scheme for His_6-tagged mortalin. Mortalin resides in inclusion bodies, and can be solubilized in the presence of chaotropic agents. Following an initial purification, the purified mortalin is refolded by dropwise-dilution of buffer with rapid stirring

main issue stems from difficulties in obtaining sufficient amounts of human mortalin, which tends to aggregate during purification and lacks good solubility. It is found that, a small protein, Hep1, is able to increase the solubility of mortalin through the interaction with both NBD and the interdomain linker (Zhai et al. 2008; Blamowska et al. 2010). Deletion of Hep1 causes both Ssq1 and Ssc1 to form insoluble aggregates (Sichting et al. 2005). Nevertheless, details regarding the enhancement are unclear and the exact function of Hep1 has yet to be identified.

Recently, a protocol for the overexpression and purification of mortalin was reported and the characterization of this crucial chaperone was able to proceed (Fig. 3.2) (Luo et al. 2010). Expression of human mortalin in *E. coli* from a pET28b(+) expression vector encoding an N-terminal His_6-tagged protein resulted in the production of relatively large amounts of mortalin (tens of mg) in the form of inclusion bodies that facilitated its purification and subsequent biochemical and biophysical studies. The solubilization of mortalin from inclusion bodies was achieved by using chaotropic reagents under mild conditions. Protein identification was performed through peptide mass fingerprinting of an excised sample band following gel eletrophoresis. After a slow refolding step, the protein sample was obtained. Maintenance of the secondary structure for purified mortalin was demonstrated through circular dichroism spectroscopy. Functional reconstitution was demonstrated both by ATPase activity measurements and the binding of the substrate, ISU. A fluorescence probe tagged peptide that contains the plausible binding motif, LPPVK was used for the substrate binding assay, and binding was monitored through fluorescence anisotropy.

Table 3.1 Purified and refolded mortalin exhibits a similar secondary structure pattern to that exhibited by *Tm* DnaK. Secondary structure elements are determined by both CD spectroscopy and primary sequence analysis (nnpredict program)

	Human mortalin		*Tm* DnaK	
	α-helix	β-strand	α-helix	β-strand
CD spectroscopy	24.0	19.0	27.0	15.0
Primary sequence analysis	30.6	13.6	33.6	14.3

Characterization of expressed mortalin was conducted through spectroscopic methods, including: (1) spectroscopic signatures of Hsp70 chaperones; (2) secondary structure determination; and (3) kinetic properties. One universal hallmark for all Hsp70 family chaperones is that, following ATP addition, the solvent exposed tryptophan in the NBD exhibits a characteristic shift in the emission maximum following a change in environment. After ATP addition to mortalin, a 3 nm blue-shift was observed in the tryptophan fluorescence, suggesting the residue to be buried in a more hydrophobic environment. In terms of the secondary structure, the purified and refolded mortalin shows a similar structural pattern as that for *Tm* DnaK (Table 3.1). Single turnover kinetics of ATPase activity shows that mortalin exhibits slow turnover with a rate constant for ATP hydrolysis, k_{hyd}, of 6.0×10^4 ($\pm 0.7 \times 10^4$) s^{-1}. This data is also consistent with k_{hyd} values obtained for both *E. coli* and *Tm* DnaK, 3.7×10^4 ($\pm 0.3 \times 10^4$) s^{-1} and 6.0×10^4, respectively (Luo et al. 2010; Montgomery et al. 1999). These results suggest that these Hsp70 chaperones share the same characteristic slow basal turnover. Detailed kinetic rate constants for discrete steps in the molecular mechanism of the mortalin ATPase cycle have yet to be determined.

3.3.2 *Structural and Functional Relationships Between Mortalin and Scaffold Proteins*

In the case of *E. coli* DnaK and yeast Ssq1, the nucleotide exchange factors GrpE and Mge1, respectively, are necessary for substrate binding (Dutkiewicz et al. 2003). However, in the case of *E. coli* HscA the existence of a nucleotide exchange factor was not found to be essential. The role of a nucleotide exchange factor in facilitating mortalin interaction with ISU has not yet been determined; however, the cited results suggest that chaperones involved in Fe–S cluster biogenesis might function by distinct pathways in different species, and that mechanistic questions concerning the mortalin-ISU interaction may not be resolved by using model chaperones from other species. Additional biochemical studies are needed to resolve this question.

Similar to other Hsp70 chaperones, it is suggested that mortalin binds with ISU through its SBD; however, the binding interface needs to be identified to provide insights into the interaction. Determination of the structure of full-length mortalin and/or a mortalin-ISU complex will be helpful in assessing the role of mortalin in regulating Fe–S cluster biogenesis.

3.4 Concluding Remarks

Human mortalin, a multi-functional Hsp70 chaperone, is involved in Fe–S cluster biogenesis through interaction with the Fe–S cluster scaffold protein, ISU. Although participation in cluster biogenesis is widely accepted, many of the key functional details remain unclear. Extensive studies in the case of *E. coli* HscA and yeast Ssq1 have shed some light on plausible scenarios, but the relevance of using chaperones from other cellular systems to address key structural and mechanistic questions for the human homolog is debatable. Several important questions, therefore, remain to be addressed. Answers to these will require structural and mechanistic studies that address the specific roles of various chaperones in promoting the individual steps involved in cluster assembly pathways (including sulfur and iron delivery), as well as the stabilization of both key metal-cluster intermediates and the final cluster product, and elucidation of the role of chaperones in promoting cluster delivery to various target apo-proteins. Also required is the dissection of how these various roles and interactions depend on the identity of the operon gene products with which the chaperone is engaged (*isc*, *suf* and *nif*). The role of mortalin, relative to other chaperones, in the maintenance of iron homeostasis and regulatory roles for iron-dependent gene expression are all topics worthy of further exploration; particularly in light of the connections of Fe–S cluster biogenesis to the maintenance of a balanced concentration of redox oxygen species. Furthermore, the link between mortalin and several major human diseases might ultimately reflect core origins in cellular iron chemistry, and the consequences of impaired iron homeostasis and implications for the generation of reactive oxygen species.

Acknowledgments The authors thank the National Institutes of Health for generous support of this research through grant # AI072443.

References

Acebrón SP, Fernández-Sáiz V, Taneva SG, Moro F, Muga A (2008) DnaJ recruits DnaK to protein aggregates. J Biol Chem 283:1381–1390

Adam AC, Bornhövd C, Prokisch H, Neupert W, Hell K (2006) The Nfs1 interacting protein Isd11 has an essential role in Fe/S cluster biogenesis in mitochondria. EMBO J 25:174–183

Agar JN, Zheng L, Cash, VL, Dean DR, Johnson MK (2000) Role of the IscU protein in iron-sulfur cluster biosynthesis: IscS-mediated assembly of a [Fe2S2] cluster in IscU. J Am Chem Soc 122:2136–2137

Albrecht AG, Netz DJ, Miethke M, Pierik AJ, Burghaus O, Peuckert F, Lill R, Marahiel MA (2010) SufU is an essential iron-sulfur cluster scaffold protein in Bacillus subtilis. J Bacteriol 192:1643–1651

Aloria K, Schilke B, Andrew A, Craig EA (2004) Iron-induced oligomerization of yeast frataxin homologue Yfh1 is dispensable in vivo. EMBO Rep 5:1096–1101

Alves R, Herrero E, Sorribas A (2004) Predictive reconstruction of the mitochondrial iron-sulfur cluster assembly metabolism. II. Role of glutaredoxin Grx5. Proteins 57:481–492

Amutha B, Gordon DM, Gu Y, Lyver ER, Dancis A, Pain D (2008) GTP is required for iron-sulfur cluster biogenesis in mitochondria. J Biol Chem 283:1362–1371

Andrew AJ, Song JY, Schilke B, Craig EA (2008) Posttranslational regulation of the scaffold for Fe-S cluster biogenesis, Isu. Mol Biol Cell 19:5259–5266
Barras F, Loiseau L, Py B (2005) How Escherichia coli and Saccharomyces cerevisiae build Fe/S proteins. Adv Microb Physiol 50:41–101
Beinert H (2000) Iron-sulfur proteins: ancient structures, still full of surprises. J Biol Inorg Chem 5:2–15
Bencze KZ, Kondapalli KC, Cook JD, McMahon S, Millàn-Pacheco C, Pastor N, Stemmler TL (2006) The structure and function of frataxin. Crit Rev Biochem Mol Biol 41:269–291
Bencze KZ, Yoon T, Millàn-Pacheco C, Bradley PB, Pastor N, Cowan JA, Stemmler TL (2007) Human frataxin: iron and ferrochelatase binding surface. Chem Commun 18:1798–1800
Bertelsen EB, Chang L, Gestwicki JE, Zuiderweg ER (2009) Solution conformation of wild-type E. coli Hsp70 (DnaK) chaperone complexed with ADP and substrate. Proc Natl Acad Sci U S A 106:8471–8476
Bhattacharyya T, Karnezis AN, Murphy SP, Hoang T, Freeman BC, Phillips B, Morimoto RI (1995) Cloning and subcellular localization of human mitochondrial hsp70. J Biol Chem 270:1705–1710
Biederbick A, Stehling O, Rösser R, Niggemeyer B, Nakai Y, Elsässer HP, Lill R (2006) Role of human mitochondrial Nfs1 in cytosolic iron-sulfur protein biogenesis and iron regulation. Mol Cell Biol 26:5675–5687
Blamowska M, Sichting M, Mapa K, Mokranjac D, Neupert W, Hell K (2010) ATPase domain and interdomain linker play a key role in aggregation of mitochondrial Hsp70 chaperone Ssc1. J Biol Chem 285:4423–4431
Bonomi F, Iametti S, Ta D, Vickery LE (2005) Multiple turnover transfer of [2Fe2S] clusters by the iron-sulfur cluster assembly scaffold proteins IscU and IscA. J Biol Chem 280:29513–29518
Bonomi F, Iametti S, Morleo A, Ta D, Vickery LE (2008) Studies on the mechanism of catalysis of iron-sulfur cluster transfer from IscU[2Fe2S] by HscA/HscB chaperones. Biochemistry 47:12795–12801
Buchberger A, Theyssen H, Schröder H, McCarty JS, Virgallita G, Milkereit P, Reinstein J, Bukau B (1995) Nucleotide-induced conformational changes in the ATPase and substrate binding domains of the DnaK chaperone provide evidence for interdomain communication. J Biol Chem 270:16903–16910
Bukau B, Horwich AL (1998) The Hsp70 and Hsp60 chaperone machines. Cell 92:351–366
Chahal HK, Dai Y, Saini A, Ayala-Castro C, Outten FW (2009) The SufBCD Fe-S scaffold complex interacts with SufA for Fe-S cluster transfer. Biochemistry 48:10644–10653
Chandramouli K, Johnson MK (2006) HscA and HscB stimulate [2Fe-2S] cluster transfer from IscU to apoferredoxin in an ATP-dependent reaction. Biochemistry 45:11087–11095
Chang HC, Tang YC, Hayer-Hartl M, Hartl FU (2007) SnapShot: molecular chaperones, part I. Cell 128:212
Chang YW, Sun YJ, Wang C, Hsiao CD (2008) Crystal structures of the 70-kDa heat shock proteins in domain disjoining conformation. J Biol Chem 283:15502–15511
Chen CA, Cowan JA (2006) Characterization of Saccharomyces cerevisiae Atm1p: functional studies of an ABC7 type transporter. Biochim Biophys Acta 1760:1857–1865
Chen OS, Hemenway S, Kaplan J (2002) Inhibition of Fe-S cluster biosynthesis decreases mitochondrial iron export: evidence that Yfh1p affects Fe-S cluster synthesis. Proc Natl Acad Sci U S A 99:12321–12326
Chillappagari S, Seubert A, Trip H, Kuipers OP, Marahiel MA, Miethke M (2010) Copper stress affects iron homeostasis by destabilizing iron-sulfur cluster formation in Bacillus subtilis. J Bacteriol 192:2512–2524
Cook JD, Kondapalli KC, Rawat S, Childs WC, Murugesan Y, Dancis A, Stemmler TL (2010) Molecular details of the yeast frataxin-Isu1 interaction during mitochondrial Fe-S cluster assembly. Biochemistry 49:8756–8765
Correia AR, Wang T, Craig EA, Gomes CM (2010) Iron-binding activity in yeast frataxin entails a trade off with stability in the alpha1/beta1 acidic ridge region. Biochem J 426:197–203

Craig EA, Marszalek J (2002) A specialized mitochondrial molecular chaperone system: a role in formation of Fe/S centers. Cell Mol Life Sci 59:1658–1665
de Mena L, Coto E, Sánchez-Ferrero E, Ribacoba R, Guisasola LM, Salvador C, Blázquez M, Alvarez V (2009) Mutational screening of the mortalin gene (HSPA9) in Parkinson's disease. J Neural Transm 116:1289–1293
Deocaris CC, Kaul SC, Wadhwa R (2006) On the brotherhood of the mitochondrial chaperones mortalin and heat shock protein 60. Cell Stress Chaperones 11:116–128
Deocaris CC, Kaul SC, Wadhwa R (2008) From proliferative to neurological role of an hsp70 stress chaperone, mortalin. Biogerontology 9, 391–403
Deocaris CC, Kaul SC, Wadhwa R (2009) The versatile stress protein mortalin as a chaperone therapeutic agent. Protein Pept Lett 16, 517–529
Ding H, Clark RJ (2004) Characterization of iron binding in IscA, an ancient iron-sulphur cluster assembly protein. Biochem J 379:433–440
Ding H, Clark RJ, Ding B (2004) IscA mediates iron delivery for assembly of iron-sulfur clusters in IscU under the limited accessible free iron conditions. J Biol Chem 279:37499–37504
Dolezal P, Dancis A, Lesuisse E, Sutak R, Hrdý I, Embley TM, Tachezy J (2007) Frataxin, a conserved mitochondrial protein, in the hydrogenosome of Trichomonas vaginalis. Eukaryot Cell 6:1431–1438
Dos Santos PC, Johnson DC, Ragle BE, Unciuleac MC, Dean DR (2007) Controlled expression of nif and isc iron-sulfur protein maturation components reveals target specificity and limited functional replacement between the two systems. J Bacteriol 189:2854–2862
D'Silva P, Liu Q, Walter W, Craig EA (2004) Regulated interactions of mtHsp70 with Tim44 at the translocon in the mitochondrial inner membrane. Nat Struct Mol Biol 11:1084–1091
Dutkiewicz R, Schilke B, Knieszner H, Walter W, Craig EA, Marszalek J (2003) Ssq1, a mitochondrial Hsp70 involved in iron-sulfur (Fe/S) center biogenesis. Similarities to and differences from its bacterial counterpart. J Biol Chem 278:29719–29727
Dutkiewicz R, Schilke B, Cheng S, Knieszner H, Craig EA, Marszalek J (2004) Sequence-specific interaction between mitochondrial Fe-S scaffold protein Isu and Hsp70 Ssq1 is essential for their in vivo function. J Biol Chem 279:29167–29174
Dutkiewicz R, Marszalek J, Schilke B, Craig EA, Lill R, Mühlenhoff U (2006) The Hsp70 chaperone Ssq1p is dispensable for iron-sulfur cluster formation on the scaffold protein Isu1p. J Biol Chem 281:7801–7808
Eccleston JF, Petrovic A, Davis CT, Rangachari K, Wilson RJ (2006) The kinetic mechanism of the SufC ATPase: the cleavage step is accelerated by SufB. J Biol Chem 281:8371–8378
Fantino JR, Py B, Fontecave M, Barras F (2010) A genetic analysis of the response of Escherichia coli to cobalt stress. Environ Microbiol 12:2846–2857
Farr CD, Slepenkov SV, Witt SN (1998) Visualization of a slow, ATP-induced structural transition in the bacterial molecular chaperone DnaK. J Biol Chem 273:9744–9748
Fontecave M, Choudens SO, Py B, Barras F (2005) Mechanisms of iron-sulfur cluster assembly: the SUF machinery. J Biol Inorg Chem 10:713–721
Fontecave M, Ollagnier-de-Choudens, S (2008) Iron-sulfur cluster biosynthesis in bacteria: mechanisms of cluster assembly and transfer. Arch Biochem Biophys 474:226–237
Fosset C, Chauveau MJ, Guillon B, Canal F, Drapier JC, Bouton C (2006) RNA silencing of mitochondrial m-Nfs1 reduces Fe-S enzyme activity both in mitochondria and cytosol of mammalian cells. J Biol Chem 281:25398–25406
Foury F, Pastore A, Trincal M (2007) Acidic residues of yeast frataxin have an essential role in Fe-S cluster assembly. EMBO Rep 8:194–199
Frausto da Silva JJR, Williams RJP (2001) The biological chemistry of the elements. Oxford University Press, New York.
Füzéry AK, Tonelli M, Ta DT, Cornilescu G, Vickery LE, Markley JL (2008) Solution structure of the iron-sulfur cluster cochaperone HscB and its binding surface for the iron-sulfur assembly scaffold protein IscU. Biochemistry 47:9394–9404

Füzéry AK, Oh JJ, Ta DT, Vickery LE, Markley JL (2011) Three hydrophobic amino acids in Escherichia coli HscB make the greatest contribution to the stability of the HscB-IscU complex. BMC Biochem 12:3

Gakh O, Adamec J, Gacy AM, Twesten RD, Owen WG, Isaya G (2002) Physical evidence that yeast frataxin is an iron storage protein. Biochemistry 41:6798–6804

Gakh O, Smith DY 4th, Isaya G (2008) Assembly of the iron-binding protein frataxin in Saccharomyces cerevisiae responds to dynamic changes in mitochondrial iron influx and stress level. J Biol Chem 283:31500–31510

Garland SA, Hoff K, Vickery LE, Culotta VC (1999) Saccharomyces cerevisiae ISU1 and ISU2: members of a well-conserved gene family for iron-sulfur cluster assembly. J Mol Biol 294:897–907

Germot A, Philippe H, Le Guyader H (1997) Evidence for loss of mitochondria in Microsporidia from a mitochondrial-type HSP70 in Nosema locustae. Mol Biochem Parasitol 87:159–168

Godman J, Balk J (2008) Genome analysis of Chlamydomonas reinhardtii reveals the existence of multiple, compartmentalized iron-sulfur protein assembly machineries of different evolutionary origins. Genetics 179:59–68

Goldberg AV, Molik S, Tsaousis AD, Neumann K, Kuhnke G, Delbac F, Vivares CP, Hirt RP, Lill R, Embley TM (2008) Localization and functionality of microsporidian iron-sulphur cluster assembly proteins. Nature 452:624–628

Gupta V, Sendra M, Naik SG, Chahal HK, Huynh BH, Outten FW, Fontecave M, Ollagnier de Choudens S (2009) Native Escherichia coli SufA, coexpressed with SufBCDSE, purifies as a [2Fe-2S] protein and acts as an Fe-S transporter to Fe-S target enzymes. J Am Chem Soc 131:6149–6153

Ha JH, McKay DB (1994) ATPase kinetics of recombinant bovine 70 kDa heat shock cognate protein and its amino-terminal ATPase domain. Biochemistry 33:14625–14635

He Y, Alam SL, Proteasa SV, Zhang Y, Lesuisse E, Dancis A, Stemmler TL (2004) Yeast frataxin solution structure, iron binding, and ferrochelatase interaction. Biochemistry 43:16254–16262

Heller K (2010) Synopsis. Multiple sulfur acceptors dock at IscS. PLoS Biol 8:e1000353

Hesterkamp T, Bukau B (1998) Role of the DnaK and HscA homologs of Hsp70 chaperones in protein folding in E. coli. EMBO J 17:4818–4828

Hoff KG, Silberg JJ, Vickery LE (2000) Interaction of the iron-sulfur cluster assembly protein IscU with the Hsc66/Hsc20 molecular chaperone system of Escherichia coli. Proc Natl Acad Sci U S A 97:7790–7795

Hoff KG, Ta DT, Tapley TL, Silberg JJ, Vickery LE (2002) Hsc66 substrate specificity is directed toward a discrete region of the iron-sulfur cluster template protein IscU. J Biol Chem 277, 27353–27359

Hoff KG, Cupp-Vickery JR, Vickery LE (2003) Contributions of the LPPVK motif of the iron-sulfur template protein IscU to interactions with the Hsc66-Hsc20 chaperone system. J Biol Chem 278:37582–37589

Hsu WM, Lee H, Juan HF, Shih YY, Wang BJ, Pan CY, Jeng YM, Chang HH, Lu MY, Lin KH, Lai HS, Chen WJ, Tsay YG, Liao YF, Hsieh FJ (2008) Identification of GRP75 as an independent favorable prognostic marker of neuroblastoma by a proteomics analysis. Clin Cancer Res 14:6237–6245

Huang J, Dizin E, Cowan JA (2008) Mapping iron binding sites on human frataxin: implications for cluster assembly on the ISU Fe-S cluster scaffold protein. J Biol Inorg Chem 13:825–836

Huet G, Daffé M, Saves I (2005) Identification of the Mycobacterium tuberculosis SUF machinery as the exclusive mycobacterial system of [Fe-S] cluster assembly: evidence for its implication in the pathogen's survival. J Bacteriol 187:6137–6146

Huynen MA, Snel B, Bork P, Gibson TJ (2001) The phylogenetic distribution of frataxin indicates a role in iron-sulfur cluster protein assembly. Hum Mol Genet 10:2463–2468

Iosefson O, Azem A (2010) Reconstitution of the mitochondrial Hsp70 (mortalin)-p53 interaction using purified proteins—identification of additional interacting regions. FEBS Lett 584:1080–1084

Jacobson MR, Brigle KE, Bennett LT, Setterquist RA, Wilson MS, Cash VL, Beynon J, Newton WE, Dean DR (1989) Physical and genetic map of the major nif gene cluster from Azotobacter vinelandii. J Bacteriol 171:1017–1027

Jensen LT, Culotta VC (2000) Role of Saccharomyces cerevisiae ISA1 and ISA2 in iron homeostasis. Mol Cell Biol 20:3918–3927

Jiang J, Lafer EM, Sousa R (2006) Crystallization of a functionally intact Hsc70 chaperone. Acta Crystallogr Sect F Struct Biol Cryst Commun 62:39–43

Jin J, Hulette C, Wang Y, Zhang T, Pan C, Wadhwa R, Zhang J (2006) Proteomic identification of a stress protein, mortalin/mthsp70/GRP75: relevance to Parkinson disease. Mol Cell Proteomics 5:1193–1204

Johnson DC, Dean DR, Smith AD, Johnson MK (2005) Structure, function, and formation of biological iron-sulfur clusters. Annu Rev Biochem 74:247–281

Kaul SC, Taira K, Pereira-Smith OM, Wadhwa R (2002) Mortalin: present and prospective. Exp Gerontol 37:1157–1164

Kaul SC, Deocaris CC, Wadhwa R (2007) Three faces of mortalin: a housekeeper, guardian and killer. Exp Gerontol 42:263–274

Kessler D (2006) Enzymatic activation of sulfur for incorporation into biomolecules in prokaryotes. FEMS Microbiol Rev 30:825–840

Kim R, Saxena S, Gordon DM, Pain D, Dancis A (2001) J-domain protein, Jac1p, of yeast mitochondria required for iron homeostasis and activity of Fe-S cluster proteins. J Biol Chem 276, 17524–17532

Kim JH, Füzéry AK, Tonelli M, Ta DT, Westler WM, Vickery LE, Markley JL (2009) Structure and dynamics of the iron-sulfur cluster assembly scaffold protein IscU and its interaction with the cochaperone HscB. Biochemistry 48:6062–6071

Kim KD, Chung WH, Kim HJ, Lee KC, Roe JH (2010) Monothiol glutaredoxin Grx5 interacts with Fe-S scaffold proteins Isa1 and Isa2 and supports Fe-S assembly and DNA integrity in mitochondria of fission yeast. Biochem Biophys Res Commun 392:467–472

Knieszner H, Schilke B, Dutkiewicz R, D'Silva P, Cheng S, Ohlson M, Craig EA, Marszalek J (2005) Compensation for a defective interaction of the hsp70 ssq1 with the mitochondrial Fe-S cluster scaffold isu. J Biol Chem 280:28966–28972

Kuhnke G, Neumann K, Mühlenhoff U, Lill R (2006) Stimulation of the ATPase activity of the yeast mitochondrial ABC transporter Atm1p by thiol compounds. Mol Membr Biol 23:173–184

Lange H, Kaut A, Kispal G, Lill R (2000) A mitochondrial ferredoxin is essential for biogenesis of cellular iron-sulfur proteins. Proc Natl Acad Sci U S A 97:1050–1055

Laufen T, Mayer MP, Beisel C, Klostermeier D, Mogk A, Reinstein J, Bukau B (1999) Mechanism of regulation of hsp70 chaperones by DnaJ cochaperones. Proc Natl Acad Sci U S A 96:5452–5457

Layer G, Ollagnier-de Choudens S, Sanakis Y, Fontecave M (2006) Iron-sulfur cluster biosynthesis: characterization of Escherichia coli CYaY as an iron donor for the assembly of [2Fe-2S] clusters in the scaffold IscU. J Biol Chem 281:16256–16263

Layer G, Gaddam SA, Ayala-Castro CN, Ollagnier-de Choudens S, Lascoux D, Fontecave M, Outten FW (2007) SufE transfers sulfur from SufS to SufB for iron-sulfur cluster assembly. J Biol Chem 282:13342–13350

Leidgens S, De Smet S, Foury F (2010) Frataxin interacts with Isu1 through a conserved tryptophan in its beta-sheet. Hum Mol Genet 19:276–286

Li J, Saxena S, Pain D, Dancis A (2001) Adrenodoxin reductase homolog (Arh1p) of yeast mitochondria required for iron homeostasis. J Biol Chem 276:1503–1509

Lill R (2009) Function and biogenesis of iron-sulphur proteins. Nature 460:831–838

Lill R, Kispal G (2001) Mitochondrial ABC transporters. Res Microbiol 152:331–340

Lill R, Muhlenhoff U (2005) Iron-sulfur-protein biogenesis in eukaryotes. Trends Biochem Sci 30:133–141

Lill R, Muhlenhoff U (2006) Iron-sulfur protein biogenesis in eukaryotes: components and mechanisms. Annu Rev Cell Dev Biol 22:457–486

Lill R, Dutkiewicz R, Elsässer HP, Hausmann A, Netz DJ, Pierik AJ, Stehling O, Urzica E, Mühlenhoff U (2006) Mechanisms of iron-sulfur protein maturation in mitochondria, cytosol and nucleus of eukaryotes. Biochim Biophys Acta 1763:652–667

Linares GR, Xing W, Govoni KE, Chen ST, Mohan S (2009) Glutaredoxin 5 regulates osteoblast apoptosis by protecting against oxidative stress. Bone 44:795–804

Liu Q, Krzewska J, Liberek K, Craig EA (2001) Mitochondrial Hsp70 Ssc1: role in protein folding. J Biol Chem 276:6112–6118

Liu Q, D'Silva P, Walter W, Marszalek J, Craig EA (2003) Regulated cycling of mitochondrial Hsp70 at the protein import channel. Science 300:139–141

Liu Y, Cowan JA (2007) Iron sulfur cluster biosynthesis: human NFU mediates sulfide delivery to ISU in the final step of [2Fe-2S] cluster assembly. Chem Commun 30:3192–31940

Liu Y, Qi W, Cowan JA (2009) Iron-sulfur cluster biosynthesis: functional characterization of the N- and C-terminal domains of human NFU. Biochemistry 48:973–980

Loiseau L, Ollagnier-de-Choudens S, Nachin L, Fontecave M, Barras F (2003) Biogenesis of Fe-S cluster by the bacterial Suf system: SufS and SufE form a new type of cysteine desulfurase. J Biol Chem 278:38352–38359

Loiseau L, Gerez C, Bekker M, Ollagnier-de Choudens S, Py B, Sanakis Y, Teixeira de Mattos J, Fontecave M, Barras F (2007) ErpA, an iron sulfur (Fe-S) protein of the A-type essential for respiratory metabolism in Escherichia coli. Proc Natl Acad Sci U S A 104:13626–13631

Luo WI, Dizin E, Yoon T, Cowan JA (2010) Kinetic and structural characterization of human mortalin. Protein Expr Purif 72:75–81

Lutz T, Westermann B, Neupert W, Herrmann JM (2001) The mitochondrial proteins Ssq1 and Jac1 are required for the assembly of iron sulfur clusters in mitochondria. J Mol Biol 307:815–825

Mansy SS, Cowan JA (2004) Iron-sulfur cluster biosynthesis: toward an understanding of cellular machinery and molecular mechanism. Acc Chem Res 37:719–725

Mansy SS, Wu G, Surerus KK, Cowan JA (2002) Iron-sulfur cluster biosynthesis. Thermatoga maritima IscU is a structured iron-sulfur cluster assembly protein. J Biol Chem 277:21397–21404

Marchenko ND, Zaika A, Moll UM (2000) Death signal-induced localization of p53 protein to mitochondria. A potential role in apoptotic signaling. J Biol Chem 275:16202–16212

Marquet A, Bui BT, Smith AG, Warren MJ (2007) Iron-sulfur proteins as initiators of radical chemistry. Nat Prod Rep 24:1027–1040

Martin W, Russell MJ (2003) On the origins of cells: a hypothesis for the evolutionary transitions from abiotic geochemistry to chemoautotrophic prokaryotes, and from prokaryotes to nucleated cells. Philos Trans R Soc B-Biol Sci 358:59–83

Miao R, Kim H, Koppolu UM, Ellis EA, Scott RA, Lindahl PA (2009) Biophysical characterization of the iron in mitochondria from Atm1p-depleted Saccharomyces cerevisiae. Biochemistry 48:9556–9568

Mihara H, Esaki N (2002) Bacterial cysteine desulfurases: their function and mechanisms. Appl Microbiol Biotechnol 60:12–23

Mizzen LA, Kabiling AN, Welch WJ (1991) The two mammalian mitochondrial stress proteins, grp 75 and hsp 58, transiently interact with newly synthesized mitochondrial proteins. Cell Regul 2:165–179

Montgomery DL, Morimoto RI, Gierasch LM (1999) Mutations in the substrate binding domain of the Escherichia coli 70 kDa molecular chaperone, DnaK, which alter substrate affinity or interdomain coupling. J Mol Biol 286:915–932

Morano KA (2007) New tricks for an old dog: the evolving world of Hsp70. Ann N Y Acad Sci 1113:1–14

Mortenson LE, Valentine RC, Carnahan JE (1962) An electron transport factor from Clostridium pasteurianum. Biochem Biophys Res Commun 7:448–452

Mühlenhoff U, Richhardt N, Gerber J, Lill R (2002) Characterization of iron-sulfur protein assembly in isolated mitochondria. A requirement for ATP, NADH, and reduced iron. J Biol Chem 277:29810–29816

Mühlenhoff U, Gerber J, Richhardt N, Lill R (2003) Components involved in assembly and dislocation of iron-sulfur clusters on the scaffold protein Isu1p. EMBO J 22:4815–4825
Mühlenhoff U, Balk J, Richhardt N, Kaiser JT, Sipos K, Kispal G, Lill R (2004) Functional characterization of the eukaryotic cysteine desulfurase Nfs1p from Saccharomyces cerevisiae. J Biol Chem 279:36906–36915
Nakai Y, Umeda N, Suzuki T, Nakai M, Hayashi H, Watanabe K, Kagamiyama H (2004) Yeast Nfs1p is involved in thio-modification of both mitochondrial and cytoplasmic tRNAs. J Biol Chem 279:12363–12368
Nuth M, Cowan JA (2009) Iron-sulfur cluster biosynthesis: characterization of IscU-IscS complex formation and a structural model for sulfide delivery to the [2Fe-2S] assembly site. J Biol Inorg Chem 14:829–839
Ollagnier-de-Choudens S, Mattioli T, Takahashi Y, Fontecave M (2001) Iron-sulfur cluster assembly: characterization of IscA and evidence for a specific and functional complex with ferredoxin. J Biol Chem 276:22604–22607
Ollagnier-de-Choudens S, Sanakis Y, Fontecave M (2004) SufA/IscA: reactivity studies of a class of scaffold proteins involved in [Fe-S] cluster assembly. J Biol Inorg Chem 9:828–838
Outten FW, Wood MJ, Munoz FM, Storz G (2003) The SufE protein and the SufBCD complex enhance SufS cysteine desulfurase activity as part of a sulfur transfer pathway for Fe-S cluster assembly in Escherichia coli. J Biol Chem 278:45713
Papanikolaou G, Pantopoulos K (2005) Iron metabolism and toxicity. Toxicol Appl Pharmacol 202:199–211
Paris Z, Changmai P, Rubio MA, Zíková A, Stuart KD, Alfonzo JD, Lukes J (2010) The Fe/S cluster assembly protein Isd11 is essential for tRNA thiolation in Trypanosoma brucei. J Biol Chem 285:22394–22402
Pelzer W, Mühlenhoff U, Diekert K, Siegmund K, Kispal G, Lill R (2000) Mitochondrial Isa2p plays a crucial role in the maturation of cellular iron-sulfur proteins. FEBS Lett 476:134–139
Py B, Barras F (2010) Building Fe-S proteins: bacterial strategies. Nat Rev Microbiol 8:436–446
Ramazzotti A, Vanmansart V, Foury F (2004) Mitochondrial functional interactions between frataxin and Isu1p, the iron-sulfur cluster scaffold protein, in Saccharomyces cerevisiae. FEBS Lett 557:215–220
Ran Q, Wadhwa R, Kawai R, Kaul SC, Sifers RN, Bick RJ, Smith JR, Pereira-Smith OM (2000) Extra-mitochondrial localization of mortalin/mthsp70/ PBP74/GRP75. Biochem Biophys Res Commun 275:174–179
Ranquet C, Ollagnier-de-Choudens S, Loiseau L, Barras F, Fontecave M (2007) Cobalt stress in Escherichia coli. The effect on the iron-sulfur proteins. J Biol Chem 282:30442–30451
Rawat S, Stemmler TL (2011) Key players and their role during mitochondrial iron-sulfur cluster biosynthesis. Chemistry 17:746–753
Revington M, Zhang Y, Yip GN, Kurochkin AV, Zuiderweg ER (2005) NMR investigations of allosteric processes in a two-domain Thermus thermophilus Hsp70 molecular chaperone. J Mol Biol 349:163–183
Richards TA, Van Der Giezen M (2006) Evolution of the Isd11-IscS complex reveals a single alpha-proteobacterial endosymbiosis for all eukaryotes. Mol Biol Evol 23:1341–1344
Rincon-Enriquez G, Crété P, Barras F, Py B (2008) Biogenesis of Fe/S proteins and pathogenicity: IscR plays a key role in allowing Erwinia chrysanthemi to adapt to hostile conditions. Mol Microbiol 67:1257–1273
Rist W, Graf C, Bukau B, Mayer MP (2006) Amide hydrogen exchange reveals conformational changes in hsp70 chaperones important for allosteric regulation. J Biol Chem 281:16493–16501
Rodríguez-Manzaneque MT, Tamarit J, Bellí G, Ros J, Herrero E (2002) Grx5 is a mitochondrial glutaredoxin required for the activity of iron/sulfur enzymes. Mol Biol Cell 13:1109–1121
Russell R, Jordan R, McMacken R (1998) Kinetic characterization of the ATPase cycle of the DnaK molecular chaperone. Biochemistry 37:596–607
San Pietro A, Lang HM (1958) Photosynthetic pyridine nucleotide reductase. I. Partial purification and properties of the enzyme from spinach. J Biol Chem 231:211–229

Schilke B, Voisine C, Beinert H, Craig E (1999) Evidence for a conserved system for iron metabolism in the mitochondria of Saccharomyces cerevisiae. Proc Natl Acad Sci U S A 96:10206–10211

Schilke B, Williams B, Knieszner H, Pukszta S, D'Silva P, Craig EA, Marszalek J (2006) Evolution of mitochondrial chaperones utilized in Fe-S cluster biogenesis. Curr Biol 16:1660–1665

Schmid D, Baici A, Gehring H, Christen P (1994) Kinetics of molecular chaperone action. Science 263:971–973

Schmucker S, Martelli A, Colin F, Page A, Wattenhofer-Donzé M, Reutenauer L, Puccio H (2011) Mammalian frataxin: an essential function for cellular viability through an interaction with a preformed ISCU/NFS1/ISD11 iron-sulfur assembly complex. PLoS One 6: e16199

Schoenfeld RA, Napoli E, Wong A, Zhan S, Reutenauer L, Morin D, Buckpitt AR, Taroni F, Lonnerdal B, Ristow M, Puccio H, Cortopassi GA (2005) Frataxin deficiency alters heme pathway transcripts and decreases mitochondrial heme metabolites in mammalian cells. Hum Mol Genet 14:3787–3799

Schwartz CJ, Giel JL, Patschkowski T, Luther C, Ruzicka FJ, Beinert H, Kiley PJ (2001) IscR, an Fe-S cluster-containing transcription factor, represses expression of Escherichia coli genes encoding Fe-S cluster assembly proteins. Proc Natl Acad Sci U S A 98:14895–14900

Sendra M, Ollagnier de Choudens S, Lascoux D, Sanakis Y, Fontecave M (2007) The SUF iron-sulfur cluster biosynthetic machinery: sulfur transfer from the SUFS-SUFE complex to SUFA. FEBS Lett 581:1362–1368

Shan Y, Napoli E, Cortopassi G (2007) Mitochondrial frataxin interacts with ISD11 of the NFS1/ISCU complex and multiple mitochondrial chaperones. Hum Mol Genet 16:929–941

Sheftel AD, Lill R (2009) The power plant of the cell is also a smithy: the emerging role of mitochondria in cellular iron homeostasis. Ann Med 41:82–99

Sheftel A, Stehling O, Lill R (2010) Iron-sulfur proteins in health and disease. Trends Endocrinol Metab 21:302–314

Sheftel AD, Stehling O, Pierik AJ, Elsässer HP, Mühlenhoff U, Webert H, Hobler A, Hannemann F, Bernhardt R, Lill R (2010) Humans possess two mitochondrial ferredoxins, Fdx1 and Fdx2, with distinct roles in steroidogenesis, heme, and Fe/S cluster biosynthesis. Proc Natl Acad Sci U S A 107:11775–11780

Shi M, Jin J, Wang Y, Beyer RP, Kitsou E, Albin RL, Gearing M, Pan C, Zhang J (2008) Mortalin: a protein associated with progression of Parkinson disease? J Neuropathol Exp Neurol 67:117–124

Shi Y, Ghosh MC, Tong WH, Rouault TA (2009) Human ISD11 is essential for both iron-sulfur cluster assembly and maintenance of normal cellular iron homeostasis. Hum Mol Genet 18:3014–3025

Shi R, Proteau A, Villarroya M, Moukadiri I, Zhang L, Trempe JF, Matte A, Armengod ME, Cygler M (2010) Structural basis for Fe-S cluster assembly and tRNA thiolation mediated by IscS protein-protein interactions. PLoS Biol 8:e1000354

Sichting M, Mokranjac D et al (2005) Maintenance of structure and function of mitochondrial Hsp70 chaperones requires the chaperone Hep1. EMBO J 24:1046–1056

Silberg JJ, Vickery LE (2000) Kinetic characterization of the ATPase cycle of the molecular chaperone Hsc66 from Escherichia coli. J Biol Chem 275:7779–7786

Silberg JJ, Hoff KG et al (1998) The Hsc66-Hsc20 chaperone system in Escherichia coli: chaperone activity and interactions with the DnaK-DnaJ-grpE system. J Bacteriol 180:6617–6624

Silberg JJ, Hoff KG et al (2001) The Fe/S assembly protein IscU behaves as a substrate for the molecular chaperone Hsc66 from Escherichia coli. J Biol Chem 276:1696–1700

Silberg JJ, Tapley TL et al (2004) Regulation of the HscA ATPase reaction cycle by the co-chaperone HscB and the iron-sulfur cluster assembly protein IscU. J Biol Chem 279:53924–53931

Smith AD, Frazzon J et al (2005) Role of conserved cysteines in mediating sulfur transfer from IscS to IscU. FEBS Lett 579:5236–5240

Stemmler TL, Lesuisse E et al (2010) Frataxin and mitochondrial FeS cluster biogenesis. J Biol Chem 285, 26737–26743

Subramanian P, Rodrigues AV et al (2011) Iron chaperones for mitochondrial Fe-S cluster biosynthesis and ferritin iron storage. Curr Opin Chem Biol 15:312-318 (Jan 31 issue)

Sutak R, Dolezal P et al (2004) Mitochondrial-type assembly of FeS centers in the hydrogenosomes of the amitochondriate eukaryote Trichomonas vaginalis. Proc Natl Acad Sci U S A 101:10368–10373

Swain JF, Dinler G et al (2007) Hsp70 chaperone ligands control domain association via an allosteric mechanism mediated by the interdomain linker. Mol Cell 26:27–39

Tachezy JL, Sanchez B et al (2001) Mitochondrial type iron-sulfur cluster assembly in the amitochondriate eukaryotes Trichomonas vaginalis and Giardia intestinalis, as indicated by the phylogeny of IscS. Mol Biol Evol 18:1919–1928

Takahashi Y, Nakamura M (1999) Functional assignment of the ORF2-iscS-iscU-iscA-hscB-hscA-fdx-ORF3 gene cluster involved in the assembly of Fe-S clusters in Escherichia coli. J Biochem 126:917–926

Takahashi Y, Tokumoto U (2002) A third bacterial system for the assembly of iron-sulfur clusters with homologs in archaea and plastids. J Biol Chem 277:28380–28383

Takano S, Wadhwa R et al (1997) Elevated levels of mortalin expression in human brain tumors. Exp Cell Res 237:38–45

Takashima M, Kuramitsu Y et al (2003) Proteomic profiling of heat shock protein 70 family members as biomarkers for hepatitis C virus-related hepatocellular carcinoma. Proteomics 3:2487–2493

Tapley TL, Vickery LE (2004) Preferential substrate binding orientation by the molecular chaperone HscA. J Biol Chem 279:28435–28442

Tapley TL, Cupp-Vickery JR et al (2006) Structural determinants of HscA peptide-binding specificity. Biochemistry 45:8058–8066

Tokumoto U, Takahashi Y (2001) Genetic analysis of the isc operon in Escherichia coli involved in the biogenesis of cellular iron-sulfur proteins. J Biochem 130:63–71

Tokumoto U, Nomura S et al (2002) Network of protein-protein interactions among iron-sulfur cluster assembly proteins in Escherichia coli. J Biochem 131:713–719

Tokumoto U, Kitamura S et al (2004) Interchangeability and distinct properties of bacterial Fe-S cluster assembly systems: functional replacement of the isc and suf operons in Escherichia coli with the nifSU-like operon from Helicobacter pylori. J Biochem 136:199–209

Tong WH, Rouault TA (2006) Functions of mitochondrial ISCU and cytosolic ISCU in mammalian iron-sulfur cluster biogenesis and iron homeostasis. Cell Metab 3:199–210

Tsai CL, Barondeau DP (2010) Human frataxin is an allosteric switch that activates the Fe-S cluster biosynthetic complex. Biochemistry 49:9132–9139

Uhrigshardt H, Singh A, Kovtunovych G, Ghosh M, Rouault TA (2010) Characterization of the human HSC20, an unusual DnaJ type III protein, involved in iron-sulfur cluster biogenesis. Hum Mol Genet 19:3816–3834

Vickery LE, Cupp-Vickery JR (2007) Molecular chaperones HscA/Ssq1 and HscB/Jac1 and their roles in iron-sulfur protein maturation. Crit Rev Biochem Mol Biol 42:95–111

Vickery LE, Silberg JJ, Ta DT (1997) Hsc66 and Hsc20, a new heat shock cognate molecular chaperone system from Escherichia coli. Protein Sci 6:1047–1056

Vogel M, Bukau B, Mayer MP (2006a) Allosteric regulation of Hsp70 chaperones by a proline switch. Mol Cell 21:359–367

Vogel M, Mayer MP, Bukau B (2006b) Allosteric regulation of Hsp70 chaperones involves a conserved interdomain linker. J Biol Chem 281:38705–38711

Voisine C, Schilke B, Ohlson M, Beinert H, Marszalek J, Craig EA (2000) Role of the mitochondrial Hsp70 s, Ssc1 and Ssq1, in the maturation of Yfh1. Mol Cell Biol 20:3677–3684

Wachtershauser G (2007) On the chemistry and evolution of the pioneer organism. Chem Biodivers 4:584–602

Wadhwa R, Takano S, Robert M, Yoshida A, Nomura H, Reddel RR, Mitsui Y, Kaul SC (1998) Inactivation of tumor suppressor p53 by mot-2, a hsp70 family member. J Biol Chem 273:29586–29591

Wadhwa R, Taira K, Kaul SC (2002) An Hsp70 family chaperone, mortalin/mthsp70/PBP74/Grp75: what, when, and where? Cell Stress Chaperones 7:309–316

Wadhwa R, Takano S, Kaur K, Deocaris CC, Pereira-Smith OM, Reddel RR, Kaul SC (2006) Upregulation of mortalin/mthsp70/Grp75 contributes to human carcinogenesis. Int J Cancer 118:2973–2980

Wang T, Craig EA (2008) Binding of yeast frataxin to the scaffold for Fe-S cluster biogenesis, Isu. J Biol Chem 283:12674–12679

Wiedemann N, Urzica E, Guiard B, Müller H, Lohaus C, Meyer HE, Ryan MT, Meisinger C, Mühlenhoff U, Lill R, Pfanner N (2006) Essential role of Isd11 in mitochondrial iron-sulfur cluster synthesis on Isu scaffold proteins. EMBO J 25:184–195

Wilson RB (2006) Iron dysregulation in Friedreich ataxia. Semin Pediatr Neurol 13:166–175

Wingert RA, Galloway JL, Barut B, Foott H, Fraenkel P, Axe JL, Weber GJ, Dooley K, Davidson AJ, Schmid B, Paw BH, Shaw GC, Kingsley P, Palis J, Schubert H, Chen O, Kaplan J, Zon LI, Tübingen (2000) Screen Consortium (2005) Deficiency of glutaredoxin 5 reveals Fe-S clusters are required for vertebrate haem synthesis. Nature 436:1035–1039

Wollers S, Layer G, Garcia-Serres R, Signor L, Clemancey M, Latour JM, Fontecave M, Ollagnier de Choudens S (2010) Iron-sulfur (Fe-S) cluster assembly: the SufBCD complex is a new type of Fe-S scaffold with a flavin redox cofactor. J Biol Chem 285:23331–23341

Wu SP, Cowan JA (2003) Iron-sulfur cluster biosynthesis. A comparative kinetic analysis of native and Cys-substituted ISA-mediated [2Fe-2S]2+ cluster transfer to an apoferredoxin target. Biochemistry 42:5784–5791

Wu G, Mansy SS, Hemann C, Hille R, Surerus KK, Cowan JA (2002a) Iron-sulfur cluster biosynthesis: characterization of Schizosaccharomyces pombe Isa1. J Biol Inorg Chem 7:526–532

Wu G, Mansy SS, Wu Sp SP, Surerus KK, Foster MW, Cowan JA (2002b) Characterization of an iron-sulfur cluster assembly protein (ISU1) from Schizosaccharomyces pombe. Biochemistry 41:5024–5032

Wu SP, Wu G, Surerus KK, Cowan JA (2002c) Iron-sulfur cluster biosynthesis. Kinetic analysis of [2Fe-2S] cluster transfer from holo ISU to apo Fd: role of redox chemistry and a conserved aspartate. Biochemistry 41:8876–8885

Wu SP, Mansy SS, Cowan JA (2005) Iron-sulfur cluster biosynthesis. Molecular chaperone DnaK promotes IscU-bound [2Fe-2S] cluster stability and inhibits cluster transfer activity. Biochemistry 44:4284–4293

Xu XM, Moller SG (2011) Iron-sulfur clusters: biogenesis, molecular mechanisms, and their functional significance. Antioxid Redox Signal 15:271–307

Ye H, Rouault TA (2010) Human iron-sulfur cluster assembly, cellular iron homeostasis, and disease. Biochemistry 49:4945–4956

Ye H, Jeong SY, Ghosh MC, Kovtunovych G, Silvestri L, Ortillo D, Uchida N, Tisdale J, Camaschella C, Rouault TA (2010) Glutaredoxin 5 deficiency causes sideroblastic anemia by specifically impairing heme biosynthesis and depleting cytosolic iron in human erythroblasts. J Clin Invest 120:1749–1761

Yoon T, Cowan JA (2003) Iron-sulfur cluster biosynthesis. Characterization of frataxin as an iron donor for assembly of [2Fe-2S] clusters in ISU-type proteins. J Am Chem Soc 125:6078–6084

Zhai P, Stanworth C, Liu S, Silberg JJ (2008) The human escort protein Hep binds to the ATPase domain of mitochondrial hsp70 and regulates ATP hydrolysis. J Biol Chem 283:26098–26106

Zhang Y, Lyver ER, Knight SA, Lesuisse E, Dancis A (2005) Frataxin and mitochondrial carrier proteins, Mrs3p and Mrs4p, cooperate in providing iron for heme synthesis. J Biol Chem 280:19794–19807

Zhang Y, Lyver ER, Knight SA, Pain D, Lesuisse E, Dancis A (2006) Mrs3p, Mrs4p, and frataxin provide iron for Fe-S cluster synthesis in mitochondria. J Biol Chem 281:22493–22502

Zheng L, White RH, Cash VL, Dean DR (1994) Mechanism for the desulfurization of L-cysteine catalyzed by the nifS gene product. Biochemistry 33:4714–4720

Zheng L, Cash VL, Flint DH, Dean DR (1998) Assembly of iron-sulfur clusters. Identification of an iscSUA-hscBA-fdx gene cluster from Azotobacter vinelandii. J Biol Chem 273:13264–13272

Chapter 4
Functional Characteristics of Mortalin

Walter A. Baseler, Tara L. Croston and John M. Hollander

Abstract Mortalin, also termed mitochondria heat shock protein 70 (mthsp70), peptide binding protein 74 (PBP74), or glucose regulated protein 75 (Grp75), is a member of the Hsp70 chaperone family and has been shown to possess unique functional characteristics in a number of different subcellular loci. Functional roles can be classified into two major classes based upon subcelluar location. The first class of functions involve those that occur in the mitochondrion which include participation in the import of nuclear-encoded cytoplasmically-delivered proteins, nascent protein folding and protein degradation within the mitochondrion, and interaction with submitochondrial constituents. The second class involves extramitochondrial functions that include interaction with protein 53 (p53), centrosomes, growth factors, immune system constituents, proteins of the endoplasmic reticulum, and metabolic constituents. The purpose of this review is to describe some of the major functional roles of mortalin action in various subcellular compartments. Further, evaluation of pathology on mortalin proteomic status will be examined in an effort to highlight the potential importance of mortalin in disease initiation and progression.

Keywords Mortalin · Mitochondria · Proteomics

4.1 Introduction

Since its identification in mouse in 1993, the protein mortalin has been the intense focus of numerous researchers interested in elucidating its various functional roles and cellular sites (Wadhwa et al. 1993a, b). Further, owing to the influence of pathology on the dynamic nature of the protein, mortalin has and continues to gain interest to researchers focused on identifying the protein's impact on disease states and the potential usefulness of the protein as a therapeutic tool and/or a pathological biomarker. The goal of this chapter is to examine some of the functional roles of mortalin and the influence of disease states on the dynamic regulation of this protein.

J. M. Hollander (✉) · W. A. Baseler · T. L. Croston
West Virginia University School of Medicine, Division of Exercise Physiology,
Center for Cardiovascular and Respiratory Sciences, 1 Medical Center Drive,
Morgantown, WV 26506, USA
e-mail: jhollander@hsc.wvu.edu

S. C. Kaul, R. Wadhwa (eds.), *Mortalin Biology: Life, Stress and Death*,
DOI 10.1007/978-94-007-3027-4_4, © Springer Science+Business Media B.V. 2012

4.2 Mortalin Overview

Since its discovery, the protein mortalin has gained considerable attention for its dynamic and differential regulation in pathological models (Domanico et al. 1993; Wadhwa et al. 1993a, b). Owing to its varied subcellular distribution, cDNAs encoding for the cytoplasmic form of the protein (mot-1) and the perinuclear form of the protein (mot-2) were identified in normal and immortalized mouse fibroblasts (Wadhwa et al. 1993b). The different proteins have been described and attributed to two distinct isoforms with opposing functions. Where the mouse cytoplasmic form of mortalin, mot-1 is associated with enhanced propensity for cellular senescence, the mouse perinuclear form of mortalin, mot-2, is associated with enhanced malignancy transformation (Kaul et al. 1998; Wadhwa et al. 1993b). Cloning and identification of human mortalin cDNA suggested that the phenomenon of multiple mortalin gene products, as identified in mouse does not necessarily exist in human. Mortalin cloned and identified from human appeared to be a single gene product that possesses similarities to the mouse mot-2 gene product, and has been subsequently termed hmot-2 (Kaul et al. 1998). A number of names, in addition to mortalin, have been utilized to define the protein, which include peptide binding protein 74 (PBP74), glucose regulated protein 75 (Grp75), and mitochondrial heat shock protein 70 (mthsp70). The gene product in human has been assigned to chromosome 5q31.1 and gene name HSPA9B (Kaul et al. 1995, 2007). Analyses of the human mortalin gene indicate an approximately 2,800 base pair genomic fragment that is transcribed from an approximately 18,000 base pair region, consisting of 17 exons (Xie et al. 2000). Mortalin has a protein sequence length of 679 amino acids with a molecular weight of 73,913 daltons and possesses an N-terminal 46 amino acid long mitochondrial targeting signal (Kaul et al. 2007).

Of particular interest is the high homology of mortalin with various heat shock protein 70 family members, including the constitutive (hsc70) and inducible (hsp70i) forms of hsp70. As a result, the protein has been placed into the Hsp70 family of proteins and termed mitochondrial heat shock protein 70 (mthsp70). Interestingly, it appears as though the protein is not inducible under conditions of heat stress (Kaul et al. 2002). Studies have been undertaken to determine the subcellular localization of mortalin using a number of different techniques including microscopy, protein tagging with specific antibodies, cell fractionation, and organelle-specific markers (Ran et al. 2000; Wadhwa et al. 1995). These techniques have revealed a number of potential localization sites for mortalin including endoplasmic reticulum, cytoplasmic vesicles, cytosol, and the mitochondrion (Dahlseid et al. 1994; Singh et al. 1997; Soltys and Gupta, 1999). Thus far, it appears as though the mitochondrion is the principle locale for mortalin. Because of its multiple subcellular locales, functional characteristics of the protein are numerous in scope. The following sections will discuss these various functions by addressing some of mortalin's actions that are specific for the mitochondrion, as well as those external to the mitochondrion. Due to mortalin's dynamic nature, examination of its proteomic regulation under pathological conditions will also be addressed.

4.3 Functional Roles of Mortalin

Currently, there are a multitude of studies examining the functional roles of mortalin, which can be further characterized into mitochondrial and extra-mitochondrial categories. Mortalin has been shown to be essential for the viability of mitochondria by regulating nuclear-encoded mitochondrial protein import, protein folding, and protein degradation. Further, mortalin has extramitochondrial functions in multiple cellular compartments including the cytoplasm, centrosomes, cellular membranes, and the endoplasmic reticulum. Mortalin certainly is a dynamic protein with literature showing it is capable of regulating important cellular processes such as cellular proliferation, apoptosis, aging, immune regulation and calcium homeostasis. The goal of this chapter is to describe some of the reported functions of mortalin and to examine proteomic alterations resulting from disease states.

4.3.1 Mortalin-Mitochondrial Functions

Studied extensively in *Saccharomyces cerevisiae* mitochondria, mortalin was characterized as an essential component for the import of nuclear-encoded proteins into the various mitochondria sub-compartments. Alterations to this critical protein has lead to catastrophic damage to the organelle and the cell itself including altered protein import, decreased antioxidant defenses, enhanced misfolding/degradation of proteins, and enhanced cellular apoptosis in various pathologies. To obtain a full understanding of mortalin, this section will discuss the process of mitochondrial import, essential mortalin constituents, and the functions of mortalin within the mitochondria.

4.3.1.1 Mitochondrial Protein Import

Currently, there are an estimated 1500 proteins in the human mitochondrion, with only 13 transcribed and translated in the organelle itself (Anderson et al. 1981; Perocchi et al. 2006; Calvo et al. 2006). The vast majority of proteins (>99%) are nuclear-encoded and imported into the mitochondrion through a complex mechanism of translocation (Chacinska et al. 2009). Protein import into the matrix is a process involving direct interaction between the imported protein, the outer mitochondrial membrane (OMM), inner mitochondrial membrane (IMM), and the presequence translocase-associated motor (PAM) creating a "supercomplex", which generates a singular avenue for translocation once the protein is delivered to the mitochondrion (Chacinska et al. 2003). Proteins destined for the mitochondria are known as pre-proteins (Folsch et al. 1996) and have specific targeting signals with cleavable N-terminal pre-sequences being the most prevalent (Folsch et al. 1996; Otera et al. 2007). Figure 4.1 is a diagram depicting the mechanism of nuclear-encoded protein import into the mitochondrion. Nuclear-encoded preproteins are transported

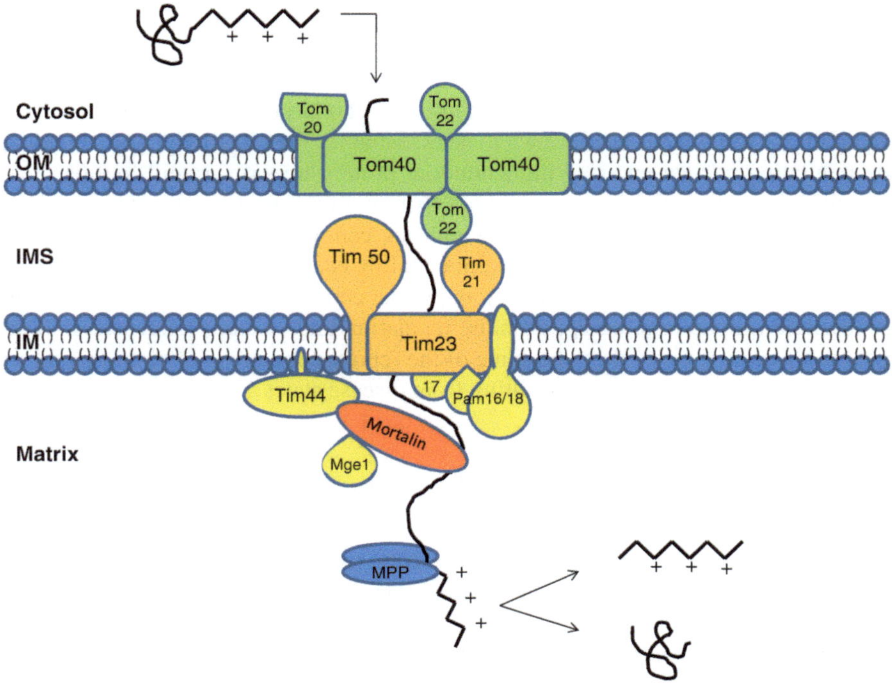

Fig. 4.1 Mitochondrial import of matrix-targeted precursor proteins. N-terminal preproteins are recognized by OMM cytosolic receptors Tom20 and Tom22. Preproteins then pass through the pore forming protein Tom40 for entrance into the inner membrane space of the mitochondria. Matrix-destined proteins then bind to the inner membrane space side of Tom22 as well as interact with Tim21 and Tim50 receptor domains allowing for direct insertion into the Tim23 complex. Membrane potential ($\Delta\Psi m$) initially drives the preprotein into the Tim23 pore. The Tim23 core complex then recruits PAM to pull the preprotein through the IMM and into the matrix. The PAM central component, mortalin, binds to Tim44 on the matrix side of the IMM, attaches to the translocating polypeptide, and disassociates from the complex through ATP hydrolysis, pulling the protein through the Tim23 channel. N-terminal targeting signals are then proteolytically removed by matrix-processing peptidase (MPP). Figure adapted from van der Laan et al. (2010) Biochimica et Biophysica Acta 1803:732–739

via cytosolic chaperones hsc70, hsp90, and mitochondrial import stimulation factor (MSF-L) to the translocase of the outer membrane (TOM) complex, which has multiple receptors that help recognize and guide pre-proteins to and through the OMM to enter into the mitochondrion (Emr et al. 1986; Hill et al. 1998; Yano et al. 2004; Young et al. 2003). Tom20 is the main receptor for N-terminal presequence recognition and subsequent binding to the cytosolic side of the OMM. This protein has a binding groove for the hydrophobic face of the polypeptide pre-sequence allowing for proper attachment to the mitochondrion. Tom22 is negatively charged and cooperates with Tom20 to help keep the incoming protein unfolded by attaching itself to the positively charged pre-sequence. Finally, β-barrel protein Tom40, along with its three subunits Tom5, Tom6, and Tom7 (small Toms) form the import pore of the TOM complex through which translocating polypeptides are guided into the inner membrane space. Once through the OMM, pre-proteins bind simultaneously to the inner membrane space domain of Tom22 and Tim50, as well as IMM protein Tim21, allowing for pre-proteins to be guided to the translocase of the inner membrane 23 (TIM23) complex. The TIM23 complex is the main mitochondrial import pore that allows entrance into and through the IMM (Chacinska et al. 2005; Neupert and Herrmann 2007). Tim23 has a direct role in the translocation of preproteins into different areas of the mitochondrion including the inner membrane space, IMM, and matrix (Chacinska et al. 2005; Distler et al. 2008). The majority of pre-proteins that travel through the TIM23 complex are dependent upon mitochondrial membrane potential ($\Delta\Psi m$) and the PAM complex. Mortalin is an essential import protein and the central subunit of the PAM complex. In a well studied mechanism, mortalin will anchor to Tim44 on the matrix side of the IMM binding itself to a translocating pre-protein, "trapping" and "pulling" it through the IMM in an ATP-dependent manner (Voos et al. 1999). Once into the mitochondrial matrix, the N-terminal pre-sequence is subsequently cleaved by matrix processing peptidase (MPP) and the unfolded translocating protein is refolded into its native confirmation by co-chaperones hsp60 and hsp10 (Gupta 1995; Singh et al. 1990).

Pre-sequence Translocase-associated Motor (PAM) Complex

Mortalin contains three key regions that include an ATPase domain, peptide binding domain, and a carboxy-terminal segment, all of which are essential for proper import functionality (Krimmer et al. 2000). Mutations to mortalin in multiple species have proven to have detrimental effects upon matrix-associated import. As an example, a mutation to the Tim44 binding domain of mortalin reduced mitochondrial translocation activity significantly (Lim et al. 2001). Additionally, inactivation of ssc1 (mortalin yeast equivalent) by temperature-sensitive mutation inhibited the unfolding, translocation, and subsequent folding of imported preproteins *in vitro* (Gambill et al. 1993; Kang et al. 1990). Interestingly, yeast cells containing low intrinsic ATP production from their mitochondria are still viable, however a deletion of Ssc1 causes complete death of the cell denoting the vast importance of this critical protein for not only the mitochondrial import process, but for overall cellular viability (Craig et al. 1987).

Mortalin driven protein import is regulated by the four other essential members of the PAM complex, which include Tim44, Pam16 (Tim14), Pam18 (Tim16), GrpE (Mge1) and the non-essential protein Tim17. Tim44 is an adaptor protein that anchors itself to the IMM side of the mitochondrial matrix, serving as a pedestal for mortalin (Ungermann et al. 1996). Recent studies have highlighted Tim44 as a translocon with multiple functions including recruitment and coordination of PAM complex constituents, allowing for efficient protein translocation (Schiller et al. 2008). A mutation of the J-related segment of Tim44 in *Saccharomyces cerevisiae* significantly decreased mitochondrial protein import viability, presumably through a dysfunctional Tim44/mortalin interaction (Merlin et al. 1999).

Matrix protein GrpE is the nucleotide exchange factor that promotes the release of ADP from mortalin. Liberation of ADP disassociates the preprotein from mortalin and allows for rapid binding of new ATP to further drive mitochondrial protein translocation (Dekker and Pfanner 1997). Examination of a conditional mutant allele of Mge1 (eukaryotic GrpE homolog) revealed decreased mortalin binding efficiency to translocating pre-proteins and enhanced binding of fully imported pre-proteins highlighting the importance of GrpE regulatory effect upon mortalin (Laloraya et al. 1995).

Mortalin is also regulated by j-like protein Pam16 and j-protein Pam18, which function collectively to regulate protein import activity. Pam18 uses its j-domain to stimulate the ATPase activity of mortalin and is critical for proper PAM complex function. Additionally, Pam16 acts as a negative regulator of Pam18 by forming a heterodimer which controls its ability to interact with the pre-sequence translocase and is known to effect the formation of the mortalin-Tim44 complex (D'Silva et al. 2005; Frazier et al. 2004). Further, the destabilization of the Pam16-Pam18 complex not only exhibited a decrease in mitochondrial protein import, but also had lethal effects upon yeast cell viability (D'Silva et al. 2005; Mokranjac et al. 2006). Finally, although not essential for cell viability, a sixth member of the complex was identified as Pam17. Genetic and biochemical analyses indicated that Pam17 interacts with Tim44 in a complementary manner to assist in protein import (Hutu et al. 2008; Schiller 2009). Also, Tim17 has been shown to interact with several other PAM complex proteins, including Tim23, prior to pre-protein translocation and its interaction is required for proper Pam16-Pam18 complex formation (van der Laan et al. 2005). Interestingly, mitochondria lacking Pam17 has shown impaired mitochondrial protein import and PAM complex formation, although import itself was not completely abolished (van der Laan et al. 2005). With multiple proteins comprising the PAM complex, some with multiple regulatory functions, mortalin driven mitochondrial protein import is one of the more complicated processes that occur inside the mitochondrion.

Brownian Ratchet Versus Power Stroke Models

Functionally, mortalin is characterized as the primary motor subunit which aids in the unfolding, transport, and refolding of mitochondrially-targeted proteins containing N-terminal prsequences through the mitochondrial IMM to target destinations. The vast majority of mitochondrial matrix proteins are imported through a mortalin-dependent mechanism. However, the motor subunit has also been shown to aid in the import of mitochondrial IMM and IMS proteins as well. Oxidase assembly 1 (oxa1), a mitochondrial IMM translocase, was proven to be imported into the matrix via mortalin and subsequently reinserted back into the mitochondrial IMM (Bohnert et al. 2010). Likewise, IMS space proteins containing bipartite signaling sequences (matrix and IMS), such as cytochrome b_2, are partially pulled into the matrix while being actively unfolded by mortalin. The bipartite signaling sequences are subsequently cleaved allowing the functional portion of the protein to stay in the IMS where is actively folded into its native confirmation (Gartner et al. 1995).

Although mortalin is widely accepted as an essential constituent for mitochondrial protein import, how it exerts its effect upon translocating proteins is still unclear. Currently, there are two different models proposed for mortalin's mechanism of action; the Brownian ratchet model and the power stroke model (Fig. 4.2, Neupert and Brunner 2002). Specifically, the Brownian ratchet mechanism proposes the N-terminal pre-protein of the polypeptide is translocated into the matrix initially by the pulling force of mitochondrial membrane potential ($\Delta\Psi$m). Once through the IMM, the pre-protein binds to mortalin, ATP hydrolysis occurs, and mortalin is released from the import docking subunit Tim44. Mortalin acts to "trap" the translocating protein by preventing its retrograde movement back into the IMM (Liu et al. 2003; Okamoto et al. 2002). Additional mortalin/Tim44 complexes bind to deeper segments of the incoming polypeptide further assisting in the movement of the protein into the matrix. Nucleotide exchange factor GrpE, catalyzes the release of ADP from mortalin, facilitating the disassociation of the motor subunit from the protein (Liu et al. 2003; Okamoto et al. 2002). This mechanism is thought to alleviate the dependence upon $\Delta\Psi$m for mitochondrial protein import (Strub et al. 2000).

The power stroke model begins in a similar fashion with mortalin attached to the docking subunit Tim44. However, mortalin does not dissociate from Tim44, rather it undergoes a conformational change fueled by binding to the incoming polypeptide followed by ATP hydrolysis, which actively propels the protein through the IMM and into the matrix. GrpE again functions as a nucleotide exchange factor, facilitating the release of mortalin from the translocating protein. This method indicates that multiple molecules of mortalin would be needed to thrust the protein into the mitochondrial matrix (Chauwin et al. 1998). Indeed, blue native complex analyses of arrested TIM-TOM-PAM supercomplexes estimate 6–7 mortalin molecules are available per imported protein at any given time (Dekker et al. 1997).

Recent literature supports the notion that there is not a singular mechanism of action pertaining to the motor function of mortalin, but likely both models are employed in conjunction (Krayl et al. 2007). As an example, pre-proteins must be in an unfolded confirmation to fit through the pores of the OMM and IMM. Mortalin will exert the necessary amount of force needed to push the N-terminal portion of the imported protein into the matrix. This motion actively unfolds C-terminal hydrophobic domains at the OMM pore during the import process allowing for linear movement into the target destination (Voisine et al. 1999). Mortalin is the likely driving force for the linearization of proteins due to the absence of unfolding enzymes present at the OMM (Voos and Rottgers 2002). Additionally, the unfolding of proteins during mitochondrial import is accelerated when the pre-sequence is long enough to span the IMM and OMM, thereby contacting mortalin for expedited entry into the matrix (Matouschek et al. 1997). In this scenario, the power stroke model is the most logical mechanism of import for tightly folded proteins due to a large amount of force needed to linearize the protein prior to translocation. However, proteins lacking tightly folded domains do not need to be forcefully straightened. Therefore, the Brownian ratchet model would seem to be the most efficient mechanism of action

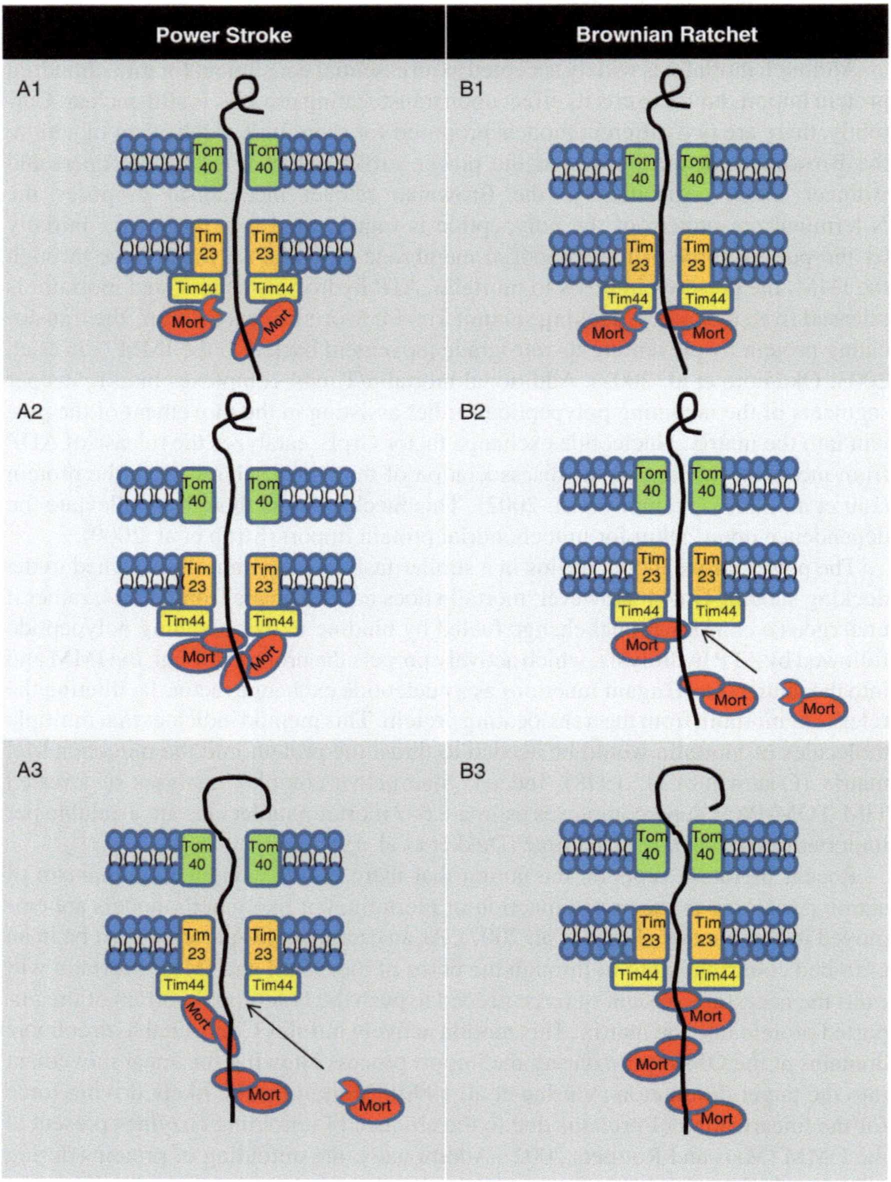

Fig. 4.2 Mitochondrial protein import models. In the power stroke model, the linearized and translocating protein is recognized by mortalin once it enters the matrix (A1). Binding of the polypeptide in conjunction with ATP hydrolysis causes a conformational change in mortalin, pulling the protein through the IMM (A2). Further, an additional mortalin/Tim44 complex will bind and thrust the protein further through the IMM (A3). Conversely, in the Brownian ratchet model, the linearized and translocating protein will bind in a similar fashion as the power stroke model, after which the translocating preprotein moves through the IMM by utilizing the eletrophoretic force of membrane potential ($\Delta\Psi$m) (B1). Unlike the power stroke model, mortalin binding does not propel the protein further into the matrix, rather mortalin acts to inhibit backsliding of the polypeptide into the IMM and inner membrane space (B2). Multiple mortalin subunits will then bind to the translocating peptide until it fully enters into the matrix for subsequent protein refolding (B3). Figure adapted from Neupert and Brunner (2002) Nat Rev Mol Cell Biol 3:555–565

(Wilcox et al. 2005). Thus, the mechanism by which mortalin exerts itself upon mitochondrial protein import likely depends upon the composition of the translocating protein itself.

4.3.1.2 Mitochondrial Protein Folding

Upon pre-protein translocation into the mitochondrial matrix, mortalin also plays a role by assisting in the refolding of imported proteins. Seminal experimentation utilizing the unique folding properties of dihydrofolate reductase and a novel temperature-sensitive Ssc1 mutant (yeast mortalin equivalent) which causes irreversible binding of mortalin and the translocating peptide, Kang et al. determined improper protein folding and enhanced propensity for proteinase K mediated degradation (Kang et al. 1990). Further, Manning-Krieg et al. concluded that mortalin acts in a sequential manner with fellow mitochondrial matrix chaperones hsp60 and hsp10 to properly fold newly imported proteins. Additionally, functional analyses revealed mortalin assisted folding is complemented by GrpE and co-chaperone Mdj1 (Manning-Krieg et al. 1991). Knockout studies of Mdj1 in *Saccharomyces cerevisiae* resulted in the loss of mitochondrial DNA and decreased cellular viability. Interestingly, Mdj1 dysregulation did not affect mitochondrial protein import, but the folding of proteins was adversely affected, indicating the importance of the Dnaj family homolog in the folding process (Rowley et al. 1994). It was surmised that mortalin, GrpE, and Mdj1 form a complex which keep the fully translocated pre-protein in a loosely folded conformation until GrpE-mediated ATP hydrolysis occurs, releasing the pre-protein and transferring it to hsp60/hsp10 complex where the protein could be subsequently folded into its native conformation (Manning-Krieg et al. 1991).

4.3.1.3 Mitochondrial Protein Degradation

In addition to its role in mitochondrial protein import, mortalin and co-chaperone Mdj1 also assist in mitochondrial protein degradation (Savel'ev et al. 1998). Specifically, mortalin stabilizes misfolded or damaged polypeptides in an unfolded conformation making them highly susceptible to proteolytic degradation from mitochondrial proteases such as m-AAA and PIM1 (Saveliev et al. 1999; Wagner et al. 1997). To confirm this phenomenon, mammalian apocytochrome P450scc (apo-p450scc) was mislocated and imported into yeast mitochondrial matrix undergoing rapid degradation through an ATP-dependent PIM1 proteolytic mechanism. PIM1 mediated proteolysis of apo-p450scc was abolished when mortalin underwent a heat-sensitive mutation rendering it inactive and unable to assist in the protein degradation process (Savel'ev et al. 1998). This function is absolutely essential as degradation is necessary for mitochondrial genomic integrity and for the proper intron excision of mitochondrially-encoded electron transport chain proteins (van Dyck et al. 1994, 1998).

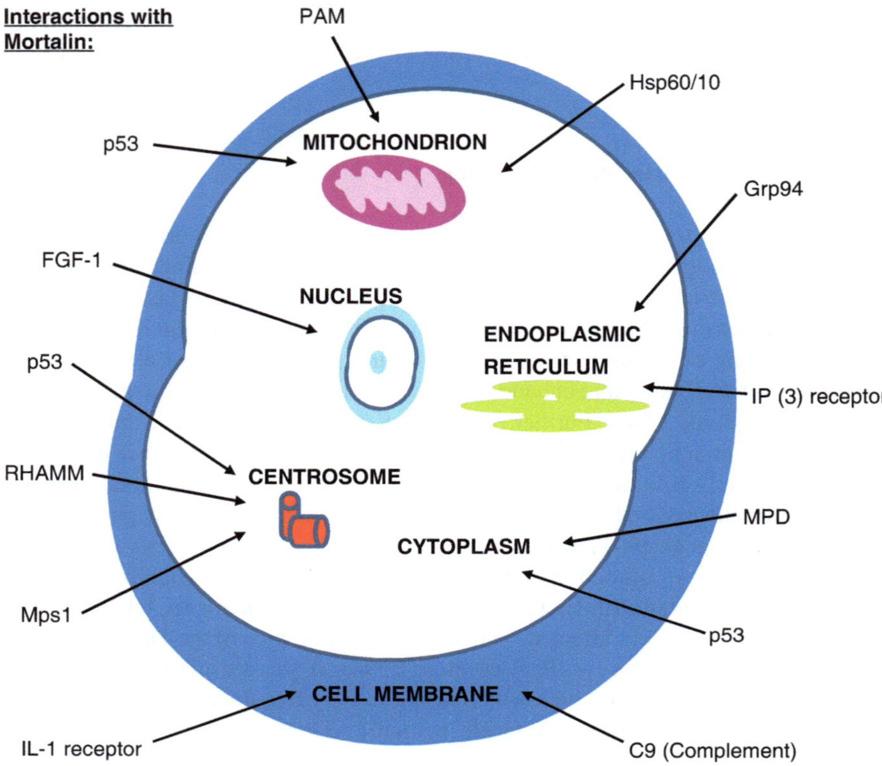

Fig. 4.3 Mortalin binding partners within the cell. Graphic depiction of the numerous proteins and protein complexes that interacts with mortalin. The various subcellular locations, in which mortalin action occurs, are indicated. The major sites of mortalin interaction and function are the mitochondrion, nucleus, endoplasmic reticulum, centrosomes, cellular membrane, and the cytoplasm

4.3.2 Mortalin-Extramitochondrial Functions

Though mortalin has been traditionally studied for its impactful roles inside of the mitochondrion, literature has highlighted extra-mitochondrial locations and functions for this protein including, but not limited to, the cytosol, centrosomes, cellular membranes, and the endoplasmic reticulum (Fig. 4.3). The binding of mortalin with numerous proteins highlight diverse functions for this multifaceted protein including the regulation of cellular proliferation, participation in the apoptotic cascade, and as a constituent of the immune system. The following section will focus on a few of the most recognized mortalin/protein interactions as well as discuss their functional implications.

4.3.2.1 Mortalin and Protein 53 (p53)

Pertaining to cellular proliferation, mortalin has been shown to possess the unique ability to bind to the tumor suppressor protein p53. Known as the "guardian of the genome", p53 is multifunctional, playing a direct role in DNA repair, cell cycle arrest and cellular apoptotic initiation (Lane 1992). It is speculated that greater than 50% of human cancers correlate with a mutation to p53 (Hainaut and Hollstein 2000). Interestingly, mortalin/p53 interaction has been shown in multiple cellular locations including the cytoplasm, centrosomes, and the mitochondrion (Ma et al. 2006; Marchenko et al. 2000; Moll et al. 1992, 1995; Wadhwa et al. 2006). Biochemical analyses by Kaul et al. determined that the p53 binding site of mortalin was mapped to the C-terminal amino acid region 312–352, which was termed the cytoplasmic sequestration domain. Additionally, the binding domain of p53 was determined to be the N-terminal amino acid region 253–282 through his-tagged mutational deletion studies (Kaul et al. 2001).

Mortalin/p53 interaction was first identified in the cytoplasm of tumor cells, highlighting the potential importance of this interaction with cancer progression (Moll et al. 1992, 1995; Moll and Schramm 1998; Wadhwa et al. 2006). Mechanistically, p53 is sequestered from the nucleus to the cytoplasm by mortalin, which is significant because the nucleus is the site of action for p53's tumor suppressive functions within the cell (Moll et al. 1992). When MKT-077, a cationic rhodacyanine dye and selective inhibitor of mortalin was administered to cancerous cells, mortalin/p53 binding was abolished and p53 mediated transcriptional activation function was restored further establishing the direct role mortalin has upon p53 regulation (Wadhwa et al. 2000). Additionally, a mortalin-targeting Adon virus (mot-Adon) was constructed and employed in human cancer cells, which silenced mortalin and enhanced p53 mediated apoptosis (Yoo et al. 2010).

p53 augmentation has also been associated with aging related to senescence in multiple organisms mostly due to its role in terminal cell cycle arrest (Donehower 2002). Experimentation was performed in which glycerol fed worms exhibited increases in mortalin, decreases in p53 activity and lower levels of age-associated fluorescence, denoting the importance of the mortalin/p53 interaction in the aging process (Deocaris et al. 2006, 2008).

Further, experimentation from the Moll laboratory elucidated a novel finding which suggests p53 will localize to the mitochondrion during times of stress and directly influence mitochondrial derived cellular apoptosis (Marchenko et al. 2000; Mihara et al. 2003). When p53 is targeted to the mitochondrion, it will interact with the Bcl2 protein Bax to induce permeabilization of the OMM resulting in cytochrome *c* release, caspase 3/9 activation, and subsequent apoptosis of the cell (Chipuk et al. 2004; Mihara et al. 2003). Mortalin, being highly expressed in the mitochondrion, has been shown to form mortalin/p53 heterodimers similar to those seen in the cytoplasm (Marchenko et al. 2000). An interesting hypothesis is that mortalin can sequester mitochondrial p53 into the mitochondrial matrix initiating protein degradation and thereby functioning in an anti-apoptotic fashion. However, further experimentation is certainly warranted to validate such a proposition.

4.3.2.2 Mortalin and Centrosomes

Mortalin has been shown to localize with duplicated centrosomes (Ma et al. 2006). Centrosomes are microtubule organizing centers of mammalian cells and function to create a microtubule structure, which interacts with chromosomes to build the mitotic spindle during mitosis (Stearns 2001). Mortalin was found to co-localize with centrosomes in late G1, S, and G2 phases with disassociation occurring during mitosis. Mortalin overexpression diminished p53 suppression of centrosome duplication, further cementing mortalin's critical role in regulating cell cycle division (Ma et al. 2006). Additionally, mortalin is associated with Mps1, a downstream kinase essential for initiation of centrosome duplication (Kanai et al. 2007). It is postulated that phosphorylation of mortalin amino acids' Thr62 and Ser65 via Mps1 superactivates Mps1 in a feedback manner accelerating centrosome duplication. Further, immunoprecipitation assays determined mortalin will interact with the receptor for hyaluronan mediated motility (RHAMM), a known microtubule and centrosomal protein (Kuwabara et al. 2006). RHAMM is an integral protein, which has the ability to influence cell migration, ras transformation, tumor progression, and metastasis (Hall et al. 1995). Functionally, mortalin was determined to bind with RHAMM during interphase and may play a role in the stabilization of microtubules (Kuwabara et al. 2006).

4.3.2.3 Mortalin and Fibroblast Growth Factor-1 (FGF-1)

Fibroblast growth factors (FGF) are a group of proteins shown to be involved in cellular proliferation and differentiation, angiogenesis, wound healing, as well as embryonic development. Interestingly, mortalin has been shown to interact with fibroblast growth factor-1 (FGF-1) through *in vitro* studies utilizing rat L6 cells, rat-1 cell fibroblasts, and BALB/c3T3 (Mizukoshi et al. 1999, 2001). Tyrosine phosphorylation of mortalin, especially in the early G1 phase enhanced binding of mortalin-FGF-1 complexes by 4 fold (Mizukoshi et al. 2001). Confocal microscopy determined mortalin-FGF-1 binding occurred in juxtanuclear regions 12 h after G1 onset, which is similar to the distribution of fibroblast growth factors receptors FGFR-1 and FGFR-4. Therefore, colocalization of FGF-1 and mortalin may regulate the ability of FGF-1 to bind FGFR's, regulating the growth factor's function (Mizukoshi et al. 2001).

4.3.2.4 Mortalin and the Immune System

Playing an intricate role in innate and acquired immunity, the complement system aids in the extermination of pathogens, as well as the removal of immune complexes and apoptotic cells (Walport 2001a, b). Mechanistically, the complement forms the membrane attack complex (MAC), which consists of subunits C5b, C6, C7, C8, and C9. The components of the MAC assemble on the cell surface creating a pore, inflicting permanent damage and leading to lysis of a pathogen or necrosis of targeted cells (Pilzer et al. 2010). Cancer cells have shown increased resistance to complement-dependent cytotoxicity, in part due to the inability of the MAC to form. Pilzer et al. determined that mortalin was capable of binding to complement subunits C8 and C9 leading to mortalin and the complement membrane attack complexes to be released in membrane vesicles from human erythroleukemia K562 cells. The cellular "shedding" of mortalin and complement subunits decreased MAC formation and increased cellular survival (Pilzer and Fishelson 2005). In a recent study, K562 and colorectal carcinoma HCT116 cells were pretreated with the specific mortalin inhibitor MKT-077. The results of this study revealed decreases in mortalin-C9, binding as well as stark increases in MAC-induced cell death in MKT-077 cells highlighting mortalin's regulatory role upon complement induced lysis in tumor cells (Pilzer et al. 2010). Mortalin has also been shown to have additional roles in immune-regulation including binding with the Interleukin-1 (IL-1) receptor in an ATP dependent manner (Sacht et al. 1999). IL-1 is a proinflammatory cytokine, which possesses multiple functions regarding regulation of local and systemic immune responses (Bankers-Fulbright et al. 1996; Dinarello 1997). Presently, there are two known IL-1 receptors and both were colocalized with mortalin irrespective of IL-1 binding. Functionally, it is believed that mortalin-IL-1 receptor binding may play a role in the internalization of the receptor, thereby regulating IL-1 function upon targeted cells (Sacht et al. 1999).

4.3.2.5 Mortalin and the Endoplasmic Reticulum

The endoplasmic reticulum (ER) forms a network of tubules, vesicles, and cisternae, which perform in concert to execute critical cellular functions including protein synthesis, lipid synthesis, carbohydrate metabolism, and regulation of calcium content. Given their proximity in eukaryotic cells, the ER and mitochondrion form a dynamic endomembrane network which work together to control certain aspects of metabolic flow, protein transport, intracellular Ca^{2+} signaling, and cell death (Brough et al. 2005; Ferri and Kroemer 2001; Szabadkai et al. 2006; Szabadkai and Rizzuto 2004). Pertaining to Ca^{2+} regulation, mortalin was shown to be a physical link between mitochondrial OMM protein voltage dependent anion channel (VDAC1) and the ER Ca^{2+}-release channel inositol 1, 4, 5-triphosphate receptor (IP(3)R) (Szabadkai et al. 2006). Interestingly, a knockdown of mortalin abolished Ca^{2+} accumulation inside of the mitochondrion highlighting mortalin's regulatory role in this cellular process. Additionally, Takano et al. identified glucose regulated protein 94 (Grp94) as another binding partner with mortalin (Takano et al. 2001). Grp94 is an ER protein that functions in the processing and transport of secreted protein, as well as endoplasmic reticulum-associated degradation (Christianson et al. 2008). However, the biological function of such an association at this present time is unknown and warrants further investigation.

4.3.2.6 Mortalin and Mevalonate Pyrophosphate Decarboxylase (MPD)

Mevalonate pyrophosphate decarboxylase (MPD) is an essential enzyme to the metabolic mevalonate pathway, which is critical for a number of cellular processes including protein prenylation and cellular growth control (Massy et al. 1996). Specifically MPD directed prenylation is known to activate the essential cellular growth and survival protein Ras. Mortalin/MPD interactions were initially proven through multiple experiments including two-hybrid and co-immunoprecipitation assays in mammalian cells (Wadhwa et al. 2003). Interestingly, overexpression of mortalin decreased Ras levels and phosphorylated ERK2, which was subsequently restored by co-expression of MPD indicating an association between mortalin, MPD, and Ras. Functionally, mortalin is hypothesized to act as a proliferation controller, regulating MPD stimulated prenylation of Ras leading to the inhibition of cellular growth (Wadhwa et al. 2003).

4.4 Mortalin Proteomics in Pathology

Because of the diverse functional roles implicated for mortalin and its suggested dynamic nature, an interest in understanding how the protein is influenced during pathological states has emerged. Of particular interest is the observation that in terms of proteomic regulation, mortalin appears to be dynamically influenced by both physiological and pathological input. As a result, mortalin status and its associated functions are predicted to be critical determinants of disease initiation and progression. This section will focus on some of the proteomic changes in mortalin content that have been observed and result from pathological stimuli. Emphasis will be placed on those pathologies in which mortalin dynamics have been observed to be most pronounced. These observed dynamics might be of particular relevance for discovery based studies where proteomic changes in mortalin may provide the potential basis for biomarker identification or therapeutic interventions in clinical settings. Further, the dynamic nature of mortalin in disease states may suggest an intersection or commonality between various pathological etiologies, in which mortalin exists at the crossroads.

4.4.1 Mortalin and Cancer

The impact of cancer on mortalin content has been actively pursued in a number of different cancer cell models and types. This section will briefly review some of the proteomic differences observed in a number of cancer models as means for describing commonality in terms of dynamic regulation of mortalin proteomic response. It is important to point out that the specific studies discussed represent only a fraction of the experimentation evaluating the influence of cancer on mortalin content. Examining a human neuroblastoma cell line (SH-SY5Y) and utilizing two-dimensional differential gel electrophoresis, Hsu and colleagues observed an up-regulation of mortalin following differentiation. Because differentiation of neuroblastoma tumors are associated with benign ganglioneuroma or high probability of regression leading to favorable prognoses, theses authors concluded that mortalin expression is an independent favorable prognostic indicator (Hsu et al. 2008). These proteomic findings suggest that mortalin may be of use as a clinical biomarker for neuroblastoma malignancy or progression. In contrast, a number of studies have suggested that mortalin expression may be correlated with increased cancer incidence, progression, and poor prognoses in other cancer models. Pizzatti et al. examined bone marrow cell protein profiles from chronic myeloid leukemia patients and compared them to healthy bone marrow donors using a comparative proteomic approach based on two-dimensional polyacrylamide gel electrophoresis. These authors observed a consistent increase in mortalin levels of chronic myeloid leukemia patients as compared to control. The authors go on to suggest a mechanistic role for mortalin in mediating antiapoptotic effects (Pizzatti et al. 2006). Similar outcomes have been noted in colorectal adenocarcinoma patients, in which a colorectal cancer tissue microarray linked to a patient database was examined. Mortalin expression was correlated with colorectal neoplasia and poor patient survival rates (Dundas et al. 2005). Examining hepatocellular carcinoma-associated hepatitis C virus using two-dimensional gel electrophoresis and matrix-assisted laser desorption/ionization-time of flight mass spectrometry, Takashima et al. observed increased levels of nine proteins, which included mortalin. The authors speculated that these proteins might play important roles in the pathogenesis of hepatocellular carcinoma (Takashima et al. 2003). Niforou and colleagues examined the proteomic make-up of a human osteosarcoma U2OS cell line using two-dimensional gel electrophoresis and matrix-assisted laser desorption ionization-mass spectrometry, coupled with peptide mass fingerprint and post source decay. The authors identified 11 protooncogenes related to the cancerous state, which included mortalin (Niforou et al. 2008). Taken together, these studies indicate a common feature in which mortalin expression profiles are influenced in cancer cell models, indicating a central role for this protein in the pathogenesis of various cancers. Further, these studies suggest that determination of mortalin expression levels may be of utility in the clinical evaluation of cancer initiation, progression, and mortality prognoses.

4.4.2 Mortalin and Neurological Disease

A number of studies have elucidated the impact of various neurological disorders and the associated impact on mortalin protein content. These studies have included proteomic evaluation of seizure disorder models, analyses of brain proteomes in genetically altered mouse models, and proteomic evaluation of multiple sclerosis patients. To date, the greatest number of studies examining mortalin changes resulting from neurological disorders have focused on Parkinson's disease and Alzheimer's disease. Parkinson's disease results from progressive neurodegeneration in the substantia nigra pars compacta and the formation of Lewy bodies. Though it is not entirely clear, mitochondrial and proteasomal dysfunction has been suggested to be contributors in the pathogenesis. Using a shotgun multidimensional protein identification technology approach to quantitatively profile mitochondrial proteins, Jin and colleagues examined the proteomic differences in the substantia nigra pars compacta of Parkinson's patients as compared to age-matched controls (Jin et al. 2006). Among the relevant findings reported was a significant decrease in mortalin content (approximately two fold) in Parkinson's patients as compared to controls. Performing complementary analyses in dopaminergic neurons these authors determined that manipulation of mortalin content influences Parkinson's disease-type pathogenesis via effects on mitochondrial, proteasomal, and oxidative stress factors. Because dopaminergic neurons are implicated as contributing to Parkinson's disease pathogenesis, studies have been undertaken to elucidate the role played by dopamine and dopamine metabolites. Utilizing a two-dimensional difference in gel electrophoresis technique in combination with cysteine- and lysine-reactive fluorescent dyes, Van Laar et al. evaluated protein alterations in rat brain mitochondria following *in vitro* exposure to the reactive dopamine quinine (Van Laar et al. 2008). Their analyses revealed a decrease in several important mitochondrial proteins including mortalin. These authors suggest that specific proteins, such as mortalin, may be particularly susceptible to dopamine oxidation, which may be associated with Parkinson's disease. Further, loss of mortalin could have a major impact on nuclear-encoded mitochondrial protein importation at times of stress, which may occur as a result of the Parkinson's pathology.

Mortalin has also been implicated in the molecular and cellular events underlying Alzheimer's disease. Currently, four genes have been linked to Alzheimer's disease, one of which is apolipoprotein E (apoE). ApoE has been identified as a polymorphic risk factor associated with the onset and progression of Alzheimer's disease (Osorio et al. 2007). Polymorphisms in the *APOE* gene have been linked to as much as 65–90% of Alzheimer's patient cases, depending upon age of prognosis. The apoE protein possesses three common isoforms, apoE2, apoE3, and apoE4, which differ in the number and location of cysteine and arginine residues. It is believed that the apoE4 isoform, which possesses two arginines at positions 112 and 158 respectively, is associated with an increased risk of neurodegeneration in Alzheimer's disease, though it is unclear as to the molecular mechanisms involved. Using a two-dimensional difference in gel electrophoresis approach coupled with protein identification via matrix-assisted laser desorption/ionization-time of flight mass spectrometry/mass spectrometry, Osorio and colleagues examined the proteomic make-up of human hippocampi as well as *APOE4* targeted replacement mice hippocampi (Osorio et al. 2007). These authors identified multiple mortalin isoforms in both model systems that were differentially expressed as a result of the pathology and may be the result of post-translational modifications such as phosphorylation and oxidation. These authors speculate that post-translational modification of mortalin may affect its binding and translocation properties, which would have implications for cellular function and ultimately disease progression. Choi and colleagues performed similar studies using an

ApoE-knockout mouse model, which display equivalent neuronal alterations to those of the *APOE4* targeted replacement mouse model, by assessing oxidatively-modified proteins (Buttini et al. 1999). These experiments were carried out by determining the content of protein carbonylation in young and old ApoE-knockout mice using a two-dimensional electrophoresis approach coupled with immunostaining utilizing an anti-dinitrophenyl antibody to visualize carbonyl content (oxidized proteins) (Choi et al. 2004). These authors identified six protein targets of oxidation that resulted from the absence of the ApoE protein, one of which was mortalin. Because only six of the more than 300 proteins identified exhibited increased oxidation in the ApoE-knockout mice, these authors speculate that the identified proteins possess critical roles for the disease progression. As a whole, these studies suggest that mortalin content or post-translational modification status is associated with the progression of certain neurological diseases. Further, mortalin status may be predictive of neurodegenerative disease progression, and its molecular manipulation may offer an avenue for clinical intervention.

4.4.3 Mortalin and Cardiovascular Disease

Mortalin status has been examined in a number of different cardiovascular-related pathologies and cell types. Further, proteomic evaluations have indicated a pathological influence on the cardiovascular system that may be associated with dynamic changes in mortalin protein content. Pshezhetsky examined the treatment of ouabain on the remodeling of vascular smooth muscle cells (Pshezhetsky 2007). Ouabain is a sodium pump inhibitor that has been used for the treatment of heart failure through its inhibitory role on apoptosis. Using two-dimensional gel electrophoresis followed by tandem mass spectroscopy this author determined that ouabain treatment induced the expression of soluble and membrane-bound proteins with one specific protein being identified as mortalin. These authors suggest that mortalin may play a role in apoptosis through an inactivation of p53. Further investigation indicated that mortalin mRNA and protein levels were increased in vascular smooth muscle cells after ouabain treatment (Taurin et al. 2002). Together, these findings illustrate the important function of mortalin during the inhibition of apoptosis in vascular smooth muscle cells.

Phenotypic changes associated with the transition of cardiac hypertrophy into heart failure have been linked to a number of stimulating factors, which include endothelin-1. Using an *in vitro* model of primary cell culture, Agnetti et al. performed proteomic analyses on neonatal rat cardiomyocytes to determine specific changes during the initial stages of cardiac hypertrophy (Agnetti et al. 2008). Neonatal rat cardiomyocytes were treated with endothelin-1 for 48 h to induce cardiac hypertrophy and to identify the initial molecular events triggering cardiac hypertrophy. Using two-dimensional polyacrylamide gel electrophoresis and mass spectroscopy analysis, the authors observed a 9-fold decrease in mortalin content. These authors suggest that the proteomic signatures resulting from endothelin-1 treatment in neonatal rat cardiomyocytes, which include impact on molecular chaperones such as mortalin, indicate a role for mitochondrial impairment at the early stages of cardiac hypertrophy. These findings also implicate mortalin loss as a potential contributor to the initial pathogenesis of cardiac hypertrophy resulting from enhanced endothelin-1 presence.

Myocardial infarction has been shown to induce the 70-kDa family of heat shock proteins (HSP70). Kilgore et al. hypothesized that the surviving myocardium would synthesize the HSP70 family of protein in a regional and temporal fashion (Kilgore et al. 1996). A rat model of myocardial infarction was used to determine the synthesis of three HSP70 family members, including mortalin, in various myocardial regions (infarcted zone, border zone and septum). Assessment of damaged tissue was examined at various time periods to determine spatial and temporal alterations. After 24 h of recovery, Western blot analyses indicated decreased mortalin content in the tissue bordering the infarction. After 5–7 days of recovery, mortalin protein concentration was two-fold higher than other HSP70 family members in both the septal tissue and infarcted area. After 14–21 days, mortalin concentration returned to normal levels, similar to those seen in control animals. These authors concluded that the myocardium might be limited in its ability to increase HSP70 family members rapidly following infarction. Nevertheless an increase in HSP70 family members will occur 5–7 days following infarction. Further, the decrease in mortalin content in the infarcted zone, 24 h following coronary ligation, indicates a direct impact on the mitochondrion. These results suggest that enhancing mortalin content may provide benefit from the initial negative effects presented to the ischemic zone. Such a hypothesis was addressed by Williamson et al., who examined the cardioprotective benefit of mortalin overexpression on neonatal rat cardiomyocytes following hypoxia/reoxygenation insult (Williamson et al. 2008). Using an adenoviral vector encoding for mortalin these authors observed benefits to cell viability and mitochondrial function following hypoxia/reoxygenation insult. Taken together, these results indicate that mortalin proteomic content is influenced by cardiac ischemia and that interventions aimed at enhancing mortalin content, particularly to those cells directly impacted by ischemic conditions, may provide cardioprotective benefit in clinical settings.

Age-related changes to cardiac proteomic profiles have been examined in an effort to determine reasons for the hearts decreased ability to respond to stress. Dai and colleagues investigated the proteomic make-up of aged left ventricular tissue in order to identify specific alterations in protein content (Dai et al. 2008). The left ventricles of young, middle-aged and old male mouse hearts were examined following two-dimensional polyacrylamide gel electrophoresis and mass spectroscopy analysis. The analyses revealed a decrease in mortalin with increasing age suggesting that mortalin loss is correlated with the deleterious effects associated with advancing age. These authors conclude that changes in specific proteins during aging, such as mortalin, may predispose the aged heart to stress and limit its ability to withstand pathological challenge.

4.4.4 Mortalin and Diabetes Mellitus

Interest has arisen in elucidating the impact of diabetes mellitus on broad-scale proteomic make-up in the heart. Several studies have been undertaken to determine whether proteomic profiles are influenced in both type 1 and type 2 diabetic settings. In a study utilizing streptozotocin as a means for pancreatic beta cell destruction, Hamblin et al. examined proteomic profiles of rat left ventricular tissue eight weeks following exposure (Hamblin et al. 2007). Two-dimensional difference in gel electrophoresis coupled with lysine-reactive fluorescent dyes were employed for proteomic assessment. Among the 1000 protein spots identified, was mortalin, which was increased in diabetic myocardium by approximately 40% as compared to control. In contrast Turko and colleagues, examined isolated rat heart mitochondria following streptozotocin exposure at multiple time points (1 week and 4 week post streptozotocin exposure) utilizing two-dimensional polyacrylamide gel electrophoresis followed by matrix-assisted laser desorption ionization/time-of-flight mass spectrometry (Turko and Murad 2003). Among the proteins identified was mortalin, which was observed to be significantly decreased. These authors go on to suggest that if diabetes mellitus adversely affects mitochondrial chaperones such as mortalin, it could have ramifications for mitochondrial refolding and contribute to mitochondrial dysfunction. To complement these findings these authors examined protein nitration in alloxan-susceptible and alloxan-resistant mouse lines (Turko et al. 2003). These two strains of mice are characterized by their responses to alloxan in which alloxan-susceptible mice become hyperglycemic following alloxan treatment, yielding an enhanced oxidative stress profile, while alloxan-resistant mice remain normoglycemic. Using matrix-assisted laser desorption ionization/time-of-flight mass spectrometric analysis of in-gel tryptic digest of nitrotyrosine immunopositive spots, the authors identified seven proteins that were nitrotyrosine positive following four weeks of alloxan treatment, one of which was mortalin. These results suggest that these proteins may be susceptible to protein nitration, which could lead to dysfunctional mitochondria during diabetes mellitus.

Mitochondria exist in distinct subcellular locations, and as a result, may be differentially influenced by pathologies, such as diabetes mellitus. Cardiac mitochondria are characterized by specific subcellular spatial locations within the cell, including mitochondria located at the sarcolemma, subsarcolemmal mitochondria (SSM), and mitochondria situated between the myofibrils, interfibrillar mitochondria (IFM). Using a streptozotocin-induced type 1 diabetes mellitus model, Dabkowski et al. determined that IFM display greater dysfunctional profiles with type 1 diabetic insult, including enhanced oxidative stress, decreased electron transport chain function, and decreased cardiolipin content (Dabkowski et al. 2009). Using multiple proteomic platforms, including isobaric tag for relative and absolute quantitation coupled with matrix-assisted laser desorption ionization/time-of-flight mass spectrometry and two-dimensional difference in-gel electrophoresis, Baseler et al. determined that IFM displayed enhanced proteomic dysregulation compared to SSM in a type 1 diabetic setting, with the majority of proteins lost coming from nuclear-encoded sources that rely upon import into the mitochondrion. One of the identified proteins that displaying reduced content as a result of type 1 diabetes mellitus was mortalin. Because over 99% of mitochondrial proteins are nuclear-encoded and must be imported, alterations to this process could have implications for mitochondrial function. Indeed, these authors observed decreases in time-dependent *in vitro* mitochondrial protein import into the isolated diabetic IFM, indicating a decrease in the ability of IFM to import proteins into the matrix subcompartment (Baseler et al. 2011). Because mortalin is essential for nuclear-encoded mitochondrial protein import into the matrix and to a lesser extent, IMM, decreases in this protein may precipitate proteomic dysregulation within the type 1 diabetic IFM.

Similar proteomic analyses were performed in *db/db* mice, a common model used for the analyses of type 2 diabetes mellitus (Dabkowski et al. 2010). As in the type 1 diabetes mellitus studies, proteomic evaluations of isolated mitochondrial subpopulations from *db/db* and littermate control mice were evaluated. In contrast to the type 1 diabetes mellitus findings, SSM appeared to be most impacted morphologically and functionally by type 2 diabetes mellitus. Using an isobaric tag for relative and absolute quantitation approach coupled with matrix-assisted laser desorption ionization/time-of-flight mass spectrometry, subpopulation-specific proteomic changes were assessed. Again, in contrast with the findings from the type 1 diabetes mellitus studies, SSM displayed the most dynamic proteomic changes. Of greatest interest was the observation that mortalin content was decreased in SSM, the mitochondrial subpopulation most impacted by type 2 diabetes mellitus, yet remained unchanged in IFM. Taken together, these findings indicate that mitochondrial proteomes are differentially regulated during diabetes mellitus. Further, those mitochondrial proteomes most impacted by the pathology are directly correlated with the degree of mitochondrial dysfunction, and perhaps most interestingly, mortalin loss. These studies suggest that loss of mortalin is associated with mitochondrial dysfunction in the diabetic heart, regardless of diabetes type (type 1 vs. type 2), and the findings may offer justification for the molecular manipulation of mortalin during diabetes mellitus.

Acknowledgments This work was supported by the National Institutes of Health from the National Institutes of Diabetes and Digestive and Kidney Diseases [DP2DK083095]. Walter Baseler is a recipient of an NIH Predoctoral Fellowship [T32HL090610] and American Heart Association Predoctoral Fellowship [10PRE3420006]. Tara Croston is a recipient of an NIH Pre-doctoral Fellowship [T32HL090610].

References

Agnetti G, Bezstarosti K, Dekkers DH, Verhoeven AJ, Giordano E, Guarnieri C, Caldarera CM, Van Eyk JE, Lamers JM (2008) Proteomic profiling of endothelin-1-stimulated hypertrophic cardiomyocytes reveals the increase of four different desmin species and alpha-B-crystallin. Biochim Biophys Acta 1784:1068–1076

Anderson S, Bankier AT, Barrell BG, de Bruijn MH, Coulson AR, Drouin J, Eperon IC, Nierlich DP, Roe BA, Sanger F, Schreier PH, Smith AJ, Staden R, Young IG (1981) Sequence and organization of the human mitochondrial genome. Nature 290:457–465

Bankers-Fulbright JL, Kalli KR, McKean DJ (1996) Interleukin-1 signal transduction. Life Sci 59:61–83
Baseler WA, Dabkowski ER, Williamson CL, Croston TL, Thapa D, Powell MJ, Razunguzwa TT, Hollander JM (2011) Proteomic alterations of distinct mitochondrial subpopulations in the type 1 diabetic heart: contribution of protein import dysfunction. Am J Physiol Regul Integr Comp Physiol 300:R186–200
Bohnert M, Rehling P, Guiard B, Herrmann JM, Pfanner N, van der Laan M (2010) Cooperation of stop-transfer and conservative sorting mechanisms in mitochondrial protein transport. Curr Biol 20:1227–1232
Brough D, Schell MJ, Irvine RF (2005) Agonist-induced regulation of mitochondrial and endoplasmic reticulum motility. Biochem J 392:291–297
Buttini M, Orth M, Bellosta S, Akeefe H, Pitas RE, Wyss-Coray T, Mucke L, Mahley RW (1999) Expression of human apolipoprotein E3 or E4 in the brains of Apoe-/- mice: isoform-specific effects on neurodegeneration. J Neurosci 19:4867–4880
Calvo S, Jain M, Xie X, Sheth SA, Chang B, Goldberger OA, Spinazzola A, Zeviani M, Carr SA, Mootha VK (2006) Systematic identification of human mitochondrial disease genes through integrative genomics. Nat Genet 38:576–582
Chacinska A, Rehling P, Guiard B, Frazier AE, Schulze-Specking A, Pfanner N, Voos W, Meisinger C (2003) Mitochondrial translocation contact sites: separation of dynamic and stabilizing elements in formation of a TOM-TIM-preprotein supercomplex. EMBO J 22:5370–5381
Chacinska A, Lind M, Frazier AE, Dudek J, Meisinger C, Geissler A, Sickmann A, Meyer HE, Truscott KN, Guiard B, Pfanner N, Rehling P (2005) Mitochondrial presequence translocase: switching between TOM tethering and motor recruitment involves Tim21 and Tim17. Cell 120:817–829
Chacinska A, Koehler CM, Milenkovic D, Lithgow T, Pfanner N (2009) Importing mitochondrial proteins: machineries and mechanisms. Cell 138:628–644
Chauwin JF, Oster G, Glick BS (1998) Strong precursor-pore interactions constrain models for mitochondrial protein import. Biophys J 74:1732–1743
Chipuk JE, Kuwana T, Bouchier-Hayes L, Droin NM, Newmeyer DD, Schuler M, Green DR (2004) Direct activation of Bax by p53 mediates mitochondrial membrane permeabilization and apoptosis. Science 303:1010–1014
Choi J, Forster MJ, McDonald SR, Weintraub ST, Carroll CA, Gracy RW (2004) Proteomic identification of specific oxidized proteins in ApoE-knockout mice: relevance to Alzheimer's disease. Free Radic Biol Med 36:1155–1162
Christianson JC, Shaler TA, Tyler RE, Kopito RR (2008) OS-9 and GRP94 deliver mutant alpha1-antitrypsin to the Hrd1-SEL1L ubiquitin ligase complex for ERAD. Nat Cell Biol 10:272–282
Craig EA, Kramer J, Kosic-Smithers J (1987) SSC1, a member of the 70-kDa heat shock protein multigene family of Saccharomyces cerevisiae, is essential for growth. Proc Natl Acad Sci U S A 84:4156–4160
D'Silva PR, Schilke B, Walter W, Craig EA (2005) Role of Pam16's degenerate J domain in protein import across the mitochondrial inner membrane. Proc Natl Acad Sci U S A 102:12419–12424
Dabkowski ER, Williamson CL, Bukowski VC, Chapman RS, Leonard SS, Peer CJ, Callery PS, Hollander JM (2009) Diabetic cardiomyopathy-associated dysfunction in spatially distinct mitochondrial subpopulations. Am J Physiol Heart Circ Physiol 296:H359–369
Dabkowski ER, Baseler WA, Williamson CL, Powell M, Razunguzwa TT, Frisbee JC, Hollander JM (2010) Mitochondrial dysfunction in the type 2 diabetic heart is associated with alterations in spatially distinct mitochondrial proteomes. Am J Physiol Heart Circ Physiol 299:H529–540
Dahlseid JN, Lill R, Green JM, Xu X, Qiu Y, Pierce SK (1994) PBP74, a new member of the mammalian 70-kDa heat shock protein family, is a mitochondrial protein. Mol Biol Cell 5:1265–1275
Dai Q, Escobar GP, Hakala KW, Lambert JM, Weintraub ST, Lindsey ML (2008) The left ventricle proteome differentiates middle-aged and old left ventricles in mice. J Proteome Res 7:756–765

Dekker PJ, Pfanner N (1997) Role of mitochondrial GrpE and phosphate in the ATPase cycle of matrix Hsp70. J Mol Biol 270:321–327

Dekker PJ, Martin F, Maarse AC, Bomer U, Muller H, Guiard B, Meijer M, Rassow J, Pfanner N (1997) The Tim core complex defines the number of mitochondrial translocation contact sites and can hold arrested preproteins in the absence of matrix Hsp70-Tim44. EMBO J 16:5408–5419

Deocaris CC, Yamasaki K, Kaul SC, Wadhwa R (2006) Structural and functional differences between mouse mot-1 and mot-2 proteins that differ in two amino acids. Ann N Y Acad Sci 1067:220–223

Deocaris CC, Kaul SC, Wadhwa R (2008) From proliferative to neurological role of an hsp70 stress chaperone, mortalin. Biogerontology 9:391–403

Dinarello CA (1997) Induction of interleukin-1 and interleukin-1 receptor antagonist. Semin Oncol 24:S9-81–93

Distler AM, Kerner J, Hoppel CL (2008) Proteomics of mitochondrial inner and outer membranes. Proteomics 8:4066–4082

Domanico SZ, DeNagel DC, Dahlseid JN, Green JM, Pierce SK (1993) Cloning of the gene encoding peptide-binding protein 74 shows that it is a new member of the heat shock protein 70 family. Mol Cell Biol 13:3598–3610

Donehower LA (2002) Does p53 affect organismal aging? J Cell Physiol 192:23–33

Dundas SR, Lawrie LC, Rooney PH, Murray GI (2005) Mortalin is over-expressed by colorectal adenocarcinomas and correlates with poor survival. J Pathol 205:74–81

Emr SD, Vassarotti A, Garrett J, Geller BL, Takeda M, Douglas MG (1986) The amino terminus of the yeast F1-ATPase beta-subunit precursor functions as a mitochondrial import signal. J Cell Biol 102:523–533

Ferri KF, Kroemer G (2001) Organelle-specific initiation of cell death pathways. Nat Cell Biol 3:E255–263

Folsch H, Guiard B, Neupert W, Stuart RA (1996) Internal targeting signal of the BCS1 protein: a novel mechanism of import into mitochondria. EMBO J 15:479–487

Frazier AE, Dudek J, Guiard B, Voos W, Li Y, Lind M, Meisinger C, Geissler A, Sickmann A, Meyer HE, Bilanchone V, Cumsky MG, Truscott KN, Pfanner N, Rehling P (2004) Pam16 has an essential role in the mitochondrial protein import motor. Nat Struct Mol Biol 11:226–233

Gambill BD, Voos W, Kang PJ, Miao B, Langer T, Craig EA, Pfanner N (1993) A dual role for mitochondrial heat shock protein 70 in membrane translocation of preproteins. J Cell Biol 123:109–117

Gartner F, Bomer U, Guiard B, Pfanner N (1995) The sorting signal of cytochrome b2 promotes early divergence from the general mitochondrial import pathway and restricts the unfoldase activity of matrix Hsp70. EMBO J 14:6043–6057

Gupta RS (1995) Evolution of the chaperonin families (Hsp60, Hsp10 and Tcp-1) of proteins and the origin of eukaryotic cells. Mol Microbiol 15:1–11

Hainaut P, Hollstein M (2000) p53 and human cancer: the first ten thousand mutations. Adv Cancer Res 77:81–137

Hall CL, Yang B, Yang X, Zhang S, Turley M, Samuel S, Lange LA, Wang C, Curpen GD, Savani RC, Greenberg AH, Turley EA (1995) Overexpression of the hyaluronan receptor RHAMM is transforming and is also required for H-ras transformation. Cell 82:19–26

Hamblin M, Friedman DB, Hill S, Caprioli RM, Smith HM, Hill MF (2007) Alterations in the diabetic myocardial proteome coupled with increased myocardial oxidative stress underlies diabetic cardiomyopathy. J Mol Cell Cardiol 42:884–895

Hill K, Model K, Ryan MT, Dietmeier K, Martin F, Wagner R, Pfanner N (1998) Tom40 forms the hydrophilic channel of the mitochondrial import pore for preproteins. Nature 395:516–521

Hsu WM, Lee H, Juan HF, Shih YY, Wang BJ, Pan CY, Jeng YM, Chang HH, Lu MY, Lin KH., Lai HS, Chen WJ, Tsay YG, Liao YF, Hsieh FJ (2008) Identification of GRP75 as an independent favorable prognostic marker of neuroblastoma by a proteomics analysis. Clin Cancer Res 14:6237–6245

Hutu DP, Guiard B, Chacinska A, Becker D, Pfanner N, Rehling P, van der Laan M (2008) Mitochondrial protein import motor: differential role of Tim44 in the recruitment of Pam17 and J-complex to the presequence translocase. Mol Biol Cell 19:2642–2649
Jin J, Hulette C, Wang Y, Zhang T, Pan C, Wadhwa R, Zhang J (2006) Proteomic identification of a stress protein, mortalin/mthsp70/GRP75: relevance to Parkinson disease. Mol Cell Proteomics 5:1193–1204
Kanai M, Ma Z, Izumi H, Kim SH, Mattison CP, Winey M, Fukasawa K (2007) Physical and functional interaction between mortalin and Mps1 kinase. Genes Cells 12:797–810
Kang PJ, Ostermann J, Shilling J, Neupert W, Craig EA, Pfanner N (1990) Requirement for hsp70 in the mitochondrial matrix for translocation and folding of precursor proteins. Nature 348:137–143
Kaul SC, Wadhwa R, Matsuda Y, Hensler PJ, Pereira-Smith OM, Komatsu Y, Mitsui Y (1995) Mouse and human chromosomal assignments of mortalin, a novel member of the murine hsp70 family of proteins. FEBS Lett 361:269–272
Kaul SC, Dunca EL, Englezou A, Takano S, Reddel RR, Mitsui Y, Wadhwa R (1998) Malignant transformation of NIH3T3 cells by overexpression of mot-2 protein. Oncogene 17:907–911
Kaul SC, Reddel RR, Mitsui Y, Wadhwa R (2001) An N-terminal region of mot-2 binds to p53 in vitro. Neoplasia 3:110–114
Kaul SC, Taira K, Pereira-Smith OM, Wadhwa (2002). Mortalin: present and prospective. Exp Gerontol 37:1157–1164
Kaul SC, Deocaris CC, Wadhwa (2007). Three faces of mortalin: a housekeeper, guardian and killer. Exp Gerontol 42:263–274
Kilgore JL, Musch TI, Ross CR (1996) Regional distribution of HSP70 proteins after myocardial infarction. Basic Res Cardiol 91:283–288
Krayl M, Lim JH, Martin F, Guiard B, Voos W (2007) A cooperative action of the ATP-dependent import motor complex and the inner membrane potential drives mitochondrial preprotein import. Mol Cell Biol 27:411–425
Krimmer T, Rassow J, Kunau WH, Voos W, Pfanner N (2000) Mitochondrial protein import motor: the ATPase domain of matrix Hsp70 is crucial for binding to Tim44, while the peptide binding domain and the carboxy-terminal segment play a stimulatory role. Mol Cell Biol 20:5879–5887
Kuwabara H, Yoneda M, Hayasaki H, Nakamura T, Mori H (2006) Glucose regulated proteins 78 and 75 bind to the receptor for hyaluronan mediated motility in interphase microtubules. Biochem Biophys Res Commun 339:971–976
Laloraya S, Dekker PJ, Voos W, Craig EA, Pfanner N (1995) Mitochondrial GrpE modulates the function of matrix Hsp70 in translocation and maturation of preproteins. Mol Cell Biol 15:7098–7105
Lane DP (1992) Cancer. p53, guardian of the genome. Nature 358:15–16
Lim JH, Martin F, Guiard B, Pfanner N, Voos W (2001) The mitochondrial Hsp70-dependent import system actively unfolds preproteins and shortens the lag phase of translocation. EMBO J 20:941–950
Liu Q, D'Silva P, Walter W, Marszalek J, Craig EA (2003) Regulated cycling of mitochondrial Hsp70 at the protein import channel. Science 300:139–141
Ma Z, Izumi H, Kanai M, Kabuyama Y, Ahn NG, Fukasawa K (2006) Mortalin controls centrosome duplication via modulating centrosomal localization of p53. Oncogene 25:5377–5390
Manning-Krieg UC, Scherer PE, Schatz G (1991) Sequential action of mitochondrial chaperones in protein import into the matrix. EMBO J 10:3273–3280
Marchenko ND, Zaika A, Moll UM (2000) Death signal-induced localization of p53 protein to mitochondria. A potential role in apoptotic signaling. J Biol Chem 275:16202–16212
Massy ZA, Keane WF, Kasiske BL (1996) Inhibition of the mevalonate pathway: benefits beyond cholesterol reduction? Lancet 347:102–103
Matouschek A, Azem A, Ratliff K, Glick BS, Schmid K, Schatz G (1997) Active unfolding of precursor proteins during mitochondrial protein import. EMBO J 16:6727–6736

Merlin A, Voos W, Maarse AC, Meijer M, Pfanner N, Rassow J (1999) The J-related segment of tim44 is essential for cell viability: a mutant Tim44 remains in the mitochondrial import site, but inefficiently recruits mtHsp70 and impairs protein translocation. J Cell Biol 145:961–972
Mihara M, Erster S, Zaika A, Petrenko O, Chittenden T, Pancoska P, Moll UM (2003) p53 has a direct apoptogenic role at the mitochondria. Mol Cell 11:577–590
Mizukoshi E, Suzuki M, Loupatov A, Uruno T, Hayashi H, Misono T, Kaul SC, Wadhwa R, Imamura T (1999) Fibroblast growth factor-1 interacts with the glucose-regulated protein GRP75/mortalin. Biochem J 343:461–466
Mizukoshi E, Suzuki M, Misono T, Loupatov A, Munekata E, Kaul SC, Wadhwa R, Imamura T (2001) Cell-cycle dependent tyrosine phosphorylation on mortalin regulates its interaction with fibroblast growth factor-1. Biochem Biophys Res Commun 280:1203–1209
Mokranjac D, Bourenkov G, Hell K, Neupert W, Groll M (2006) Structure and function of Tim14 and Tim16, the J and J-like components of the mitochondrial protein import motor. EMBO J 25:4675–4685
Moll UM, Schramm L (1998). p53-an acrobat in tumorigenesis. Crit Rev Oral Biol Med 9:23–37
Moll UM, Riou G, Levine AJ (1992) Two distinct mechanisms alter p53 in breast cancer: mutation and nuclear exclusion. Proc Natl Acad Sci U S A 89:7262–7266
Moll UM, LaQuaglia M, Benard J, Riou G (1995) Wild-type p53 protein undergoes cytoplasmic sequestration in undifferentiated neuroblastomas but not in differentiated tumors. Proc Natl Acad Sci U S A 92:4407–4411
Neupert W, Brunner M (2002) The protein import motor of mitochondria. Nat Rev Mol Cell Biol 3:555–565
Neupert W, Herrmann JM (2007) Translocation of proteins into mitochondria. Annu Rev Biochem 76:723–749
Niforou KM, Anagnostopoulos AK, Vougas K, Kittas C, Gorgoulis VG, Tsangaris GT (2008) The proteome profile of the human osteosarcoma U2OS cell line. Cancer Genomics Proteomics 5:63–78
Okamoto K, Brinker A, Paschen SA, Moarefi I, Hayer-Hartl M, Neupert W, Brunner M (2002) The protein import motor of mitochondria: a targeted molecular ratchet driving unfolding and translocation. EMBO J 21:3659–3671
Osorio C, Sullivan PM, He DN, Mace BE, Ervin JF, Strittmatter WJ, Alzate O (2007) Mortalin is regulated by APOE in hippocampus of AD patients and by human APOE in TR mice. Neurobiol Aging 28:1853–1862
Otera H, Taira Y, Horie C, Suzuki Y, Suzuki H, Setoguchi K, Kato H, Oka T, Mihara K (2007) A novel insertion pathway of mitochondrial outer membrane proteins with multiple transmembrane segments. J Cell Biol 179:1355–1363
Perocchi F, Jensen LJ, Gagneur J, Ahting U, von Mering C, Bork P, Prokisch H, Steinmetz LM (2006) Assessing systems properties of yeast mitochondria through an interaction map of the organelle. PLoS Genet 2:e170
Pilzer D, Fishelson Z (2005) Mortalin/GRP75 promotes release of membrane vesicles from immune attacked cells and protection from complement-mediated lysis. Int Immunol 17:1239–1248
Pilzer D, Saar M, Koya K, Fishelson Z (2010) Mortalin inhibitors sensitize K562 leukemia cells to complement-dependent cytotoxicity. Int J Cancer 126:1428–1435
Pizzatti L, Sa LA, de Souza JM, Bisch PM, Abdelhay E (2006) Altered protein profile in chronic myeloid leukemia chronic phase identified by a comparative proteomic study. Biochim Biophys Acta 1764:929–942
Pshezhetsky AV (2007) Proteomic analysis of vascular smooth muscle cells treated with ouabain. Methods Mol Biol 357:253–269
Ran Q, Wadhwa R, Kawai R, Kaul SC, Sifers RN, Bick RJ, Smith JR, Pereira-Smith OM (2000) Extramitochondrial localization of mortalin/mthsp70/PBP74/GRP75. Biochem Biophys Res Commun 275:174–179

Rowley N, Prip-Buus C, Westermann B, Brown C, Schwarz E, Barrell B, Neupert W (1994) Mdj1p, a novel chaperone of the DnaJ family, is involved in mitochondrial biogenesis and protein folding. Cell 77:249–259

Sacht G, Brigelius-Flohe R, Kiess M, Sztajer H, Flohe L (1999) ATP-sensitive association of mortalin with the IL-1 receptor type I. Biofactors 9:49–60

Savel'ev AS, Novikova LA, Kovaleva IE, Luzikov VN, Neupert W, Langer T (1998) ATP-dependent proteolysis in mitochondria. m-AAA protease and PIM1 protease exert overlapping substrate specificities and cooperate with the mtHsp70 system. J Biol Chem 273:20596–20602

Saveliev AS, Kovaleva IE, Novikova LA, Isaeva LV, Luzikov V (1999). Can foreign proteins imported into yeast mitochondria interfere with PIM1p protease and/or chaperone function? Arch Biochem Biophys 363:373–376

Schiller D (2009) Pam17 and Tim44 act sequentially in protein import into the mitochondrial matrix. Int J Biochem Cell Biol 41:2343–2349

Schiller D, Cheng YC, Liu Q, Walter W, Craig EA (2008) Residues of Tim44 involved in both association with the translocon of the inner mitochondrial membrane and regulation of mitochondrial Hsp70 tethering. Mol Cell Biol 28:4424–4433

Singh B, Soltys BJ, Wu ZC, Patel HV, Freeman KB, Gupta RS (1997) Cloning and some novel characteristics of mitochondrial Hsp70 from Chinese hamster cells. Exp Cell Res 234:205–216

Singh B, Patel HV, Ridley RG, Freema KB, Gupta RS (1990) Mitochondrial import of the human chaperonin (HSP60) protein. Biochem Biophys Res Commun 169:391–396

Soltys BJ, Gupta RS (1999) Mitochondrial-matrix proteins at unexpected locations: are they exported? Trends Biochem Sci 24:174–177

Stearns T (2001) Centrosome duplication. a centriolar pas de deux. Cell 105:417–420

Strub A, Lim JH, Pfanner N, Voos W (2000) The mitochondrial protein import motor. Biol Chem 381:943–949

Szabadkai G, Rizzuto R (2004) Participation of endoplasmic reticulum and mitochondrial calcium handling in apoptosis: more than just neighborhood? FEBS Lett 567:111–115

Szabadkai G, Bianchi K, Varnai P, De Stefani D, Wieckowski MR, Cavagna D, Nagy AI, Balla T, Rizzuto R (2006) Chaperone-mediated coupling of endoplasmic reticulum and mitochondrial Ca2+channels. J Cell Biol 175:901–911

Takano S, Wadhwa R, Mitsui Y, Kaul SC (2001) Identification and characterization of molecular interactions between glucose-regulated proteins (GRPs) mortalin/GRP75/peptide-binding protein 74 (PBP74) and GRP94. Biochem J 357:393–398

Takashima M, Kuramitsu Y, Yokoyama Y, Iizuka N, Toda T, Sakaida I, Okita K, Oka M, Nakamura K (2003) Proteomic profiling of heat shock protein 70 family members as biomarkers for hepatitis C virus-related hepatocellular carcinoma. Proteomics 3:2487–2493

Taurin S, Seyrantepe V, Orlov SN, Tremblay TL, Thibault P, Bennett MR, Hamet P, Pshezhetsky AV (2002) Proteome analysis and functional expression identify mortalin as an antiapoptotic gene induced by elevation of [Na+]i/[K+]i ratio in cultured vascular smooth muscle cells. Circ Res 91:915–922

Turko IV, Murad F (2003) Quantitative protein profiling in heart mitochondria from diabetic rats. J Biol Chem 278:35844–35849

Turko IV, Li L, Aulak KS, Stuehr DJ, Chang JY, Murad F (2003) Protein tyrosine nitration in the mitochondria from diabetic mouse heart. Implications to dysfunctional mitochondria in diabetes. J Biol Chem 278:33972–33977

Ungermann C, Guiard B, Neupert W, Cyr DM (1996) The delta psi- and Hsp70/MIM44-dependent reaction cycle driving early steps of protein import into mitochondria. EMBO J 15:735–744

van der Laan M, Chacinska A, Lind M, Perschil I, Sickmann A, Meyer HE, Guiard B, Meisinger C, Pfanner N, Rehling P (2005) Pam17 is required for architecture and translocation activity of the mitochondrial protein import motor. Mol Cell Biol 25:7449–7458

van der Laan M, Huto, DP, Rehling P (2010) On the mechanism of preprotein import by the mitochondrial presequence translocase. Biochim Biophys Acta 1803:732–739

Van Dyck L, Pearce DA, Sherman F (1994) PIM1 encodes a mitochondrial ATP-dependent protease that is required for mitochondrial function in the yeast Saccharomyces cerevisiae. J Biol Chem 269:238–242

Van Dyck L, Neupert W, Langer T (1998) The ATP-dependent PIM1 protease is required for the expression of intron-containing genes in mitochondria. Genes Dev 12:1515–1524

Van Laar VS, Dukes AA, Cascio M, Hastings TG (2008) Proteomic analysis of rat brain mitochondria following exposure to dopamine quinone: implications for Parkinson disease. Neurobiol Dis 29:477–489

Voisine C, Craig EA, Zufall N, von Ahsen O, Pfanner N, Voos W (1999) The protein import motor of mitochondria: unfolding and trapping of preproteins are distinct and separable functions of matrix Hsp70. Cell 97:565–574

Voos W, Rottgers K (2002) Molecular chaperones as essential mediators of mitochondrial biogenesis. Biochim Biophys Acta 1592:51–62

Voos W, Martin H, Krimmer T, Pfanner N (1999) Mechanisms of protein translocation into mitochondria. Biochim Biophys Acta 1422:235–254

Wadhwa R, Kaul SC, Ikawa Y, Sugimoto Y (1993a) Identification of a novel member of mouse hsp70 family. Its association with cellular mortal phenotype. J Biol Chem 268:6615–6621

Wadhwa R, Kaul SC, Sugimoto Y, Mitsui Y (1993b) Induction of cellular senescence by transfection of cytosolic mortalin cDNA in NIH 3T3 cells. J Biol Chem 268:22239–22242

Wadhwa R, Pereira-Smith OM, Reddel RR, Sugimoto Y, Mitsui Y, Kaul SC (1995) Correlation between complementation group for immortality and the cellular distribution of mortalin. Exp Cell Res 216:101–106

Wadhwa R, Sugihara T, Yoshida A, Nomura H, Reddel RR, Simpson R, Maruta H, Kaul SC (2000) Selective toxicity of MKT-077 to cancer cells is mediated by its binding to the hsp70 family protein mot-2 and reactivation of p53 function. Cancer Res 60:6818–6621

Wadhwa R, Yaguchi T, Hasan MK, Taira K, Kaul SC (2003) Mortalin-MPD (mevalonate pyrophosphate decarboxylase) interactions and their role in control of cellular proliferation. Biochem Biophys Res Commun 302:735–742

Wadhwa R, Takano S, Kaur K, Deocaris CC, Pereira-Smith OM, Reddel RR, Kaul SC (2006) Upregulation of mortalin/mthsp70/Grp75 contributes to human carcinogenesis. Int J Cancer 118:2973–2980

Wagner I, van Dyck L, Savel'ev AS, Neupert W, Langer T (1997) Autocatalytic processing of the ATP-dependent PIM1 protease: crucial function of a pro-region for sorting to mitochondria. EMBO J 16:7317–7325

Walport MJ (2001a) Complement. First of two parts. N Engl J Med 344:1058–1066

Walport MJ (2001b) Complement. Second of two parts. N Engl J Med 344:1140–1144

Wilcox AJ, Choy J, Bustamante C, Matouschek A (2005) Effect of protein structure on mitochondrial import. Proc Natl Acad Sci U S A 102:15435–15440

Williamson CL, Dabkowski ER, Dillmann WH, Hollander JM (2008) Mitochondria protection from hypoxia/reoxygenation injury with mitochondria heat shock protein 70 overexpression. Am J Physiol Heart Circ Physiol 294:H249–256

Xie H, Hu Z, Chyna B, Horrigan SK, Westbrook CA (2000) Human mortalin (HSPA9): a candidate for the myeloid leukemia tumor suppressor gene on 5q31. Leukemia 14:2128–2134

Yano M, Terada K, Mori M (2004) Mitochondrial import receptors Tom20 and Tom22 have chaperone-like activity. J Biol Chem 279:10808–10813

Yoo JY, Ryu J, Gao R, Yaguchi T, Kaul SC, Wadhwa R, Yun CO (2010) Tumor suppression by apoptotic and anti-angiogenic effects of mortalin-targeting adeno-oncolytic virus. J Gene Med 12:586–595

Young JC, Hoogenraad NJ, Hartl FU (2003) Molecular chaperones Hsp90 and Hsp70 deliver preproteins to the mitochondrial import receptor Tom70. Cell 112:41–50

Part II
Mortalin in Evolution

Chapter 5
Mortalin and *Drosophila* DmHsp22: Two Mitochondrial Chaperones Regulating Aging and Carcinogenesis

Marie Le Pécheur, Geneviève Morrow and Robert M. Tanguay

Abstract Heat shock proteins (Hsps) are generally involved in stress response and protection from molecular damages. While they can promote cell survival by their involvement in protein folding, turnover and trafficking, their mis-expression can also promote tumorigenesis. Specific members of Hsps have distinct intracellular localization. Mitochondria contain several members of the major HSP sub-families that have important functions in maintaining mitochondrial functions and homeostasis. This chapter emphasizes the contribution of two different mitochondrial chaperones, *Drosophila melanogaster*'s small heat shock protein Hsp22 (DmHsp22) and the mitochondrial Hsp70/mortalin (mtHsp70) in the regulation of aging and carcinogenesis.

Keywords Aging · Cancer · *Drosophila* Hsp22 · Mitochondria · Mortalin

5.1 Introduction

Heat shock proteins (Hsps) were discovered over 50 years ago in *Drosophila* and are present in all organisms. They were initially identified by their induction under heat stress but they are also induced upon other stress conditions such as inflammation, exposure to toxins, hypoxia, and starvation. As molecular chaperones, Hsps are key factors involved in repair of cellular damages and cell survival. Hsps are divided in different sub-families based on their induction by heat stress and their molecular weights: HSP100/ClpB, HSP90, HSP70, HSP60, HSP40 and the small HSP (sHSP) (Kampinga et al. 2009; Morrow et al. 2006; Orejuela et al. 2007). Each sub-family has its own structural characteristics and co-factors or co-chaperones and different members can therefore be involved in different steps of the protein folding and turnover mechanisms. For example, small heat shock proteins (sHsps) are ATP-independent chaperones that create reservoirs of unfolded proteins while the interaction of both the oligomeric Hsp60 and the monomeric Hsp70 with unfolded

R. M. Tanguay (✉) · M. Le Pécheur · G. Morrow
Laboratory of Cell and Developmental Genetics, Department of Molecular Biology, Medical Biochemistry and Pathology, Institut de Biologie Intégrative et des Systèmes and PROTEO, Université Laval, G1V 0A6 Québec, Canada
e-mail: robert.tanguay@ibis.ulaval.ca

S. C. Kaul, R. Wadhwa (eds.), *Mortalin Biology: Life, Stress and Death*,
DOI 10.1007/978-94-007-3027-4_5, © Springer Science+Business Media B.V. 2012

proteins necessitates the presence of ATP. Distinct chaperone networks exist in each sub-cellular compartment to appropriately maintain proper protein homeostasis and to fulfill organelles-specific needs.

Mitochondria are unique organelles in the cell because they possess their own DNA. Their main function is to generate ATP, the energy molecule, by a very efficient energy process called oxidative phosphorylation (OXPHOS). During the necessary oxygen-dependent ATP production by the respiratory chain, there is also generation of reactive oxygen species (ROS) as a natural by-product. ROS can directly alter soluble and membrane-bound proteins, and therefore can affect the respiratory complexes, the proteins involved in their assembly and the import of substrates and nuclear encoded mitochondrial proteins (Das et al. 2001; Sohal and Dubey 1994). To limit the cellular damages caused by ROS, mitochondria have a number of natural antioxidant mechanisms including MnSOD, glutathione peroxidase and glutathione reductase (Green and Kroemer 2004). In addition to these antioxidants that eliminate ROS, mitochondrial chaperones prevent protein aggregation and promote their correct folding and the clearance of modified proteins to prevent further destabilization of mitochondrial homeostasis (Hendrick and Hartl 1995; Parcellier et al. 2003). Members of the sHSP, HSP70, HSP60 and HSP100/CLPB sub-families have been identified in mitochondria (Leidhold et al. 2006).

The present review summarizes our current knowledge about two mitochondrial chaperones, mortalin and *Drosophila melanogaster* Hsp22 (DmHsp22), in the aging process as well as in cellular senescence, apoptosis and cancer.

5.2 Both Mortalin and DmHsp22 are Mitochondrial Chaperones

5.2.1 mtHsp70/Mortalin: A Major Mitochondrial Hsp

Mortalin (mtHsp70/Hsp75/Grp75/TRAP-1/HspA9) is a highly conserved member of the HSP70 family constitutively expressed and not inducible by heat stress. It is predominantly found in mitochondria but has also been associated with multiple subcellular compartments including cytosol, endoplasmic reticulum and cytoplasmic vesicles. Mortalin is involved in various functions such as intracellular trafficking, antigen processing and regulation of cell proliferation, aging, differentiation, apoptosis and tumorigenesis through its interactions with several partners (Carette et al. 2002; Craig 1989; Dundas et al. 2005; Voos and Rottgers 2002; Wadhwa et al. 2002a). Mortalin is a 679 amino acid long protein with a molecular weight of 74 kDa, which undergoes Ca^{2+} -independent autophosphorylation and has a high degree of identity with other members of the HSP70 family, including *Escherichia Coli* DnaK (51%) *Saccharomyces cerevisiae* SSCIP1 (65%), the constitutive cytosolic Hsp70 from rat (Hsc70) (46%) and the endoplasmic isoform Bip (49%) (Wadhwa et al. 2005). Based on its evolutionary conservation within the HSP70 family, mortalin consists of two main domains, the N-terminal ATPase domain and the C-terminal peptide binding domain (Deocaris et al. 2006b). Upon ATP binding, mortalin undergoes a specific conformational change that is needed for its chaperone function (von Ahsen et al. 1995). The crystal structure of intact mortalin has not been elucidated so far, but the recently resolved crystal structures of truncated bovine Hsc70 and *Geobacillus kaustophilus* DnaK as well as the NMR structure of the full-length *E. Coli* DnaK have provided precious informations on structural mechanisms (Chang et al. 2008; Jiang et al. 2005). Moreover, several important residues involved in allosteric regulation of mortalin have recently been identified (Rist et al. 2006; Vogel et al. 2006a, b). In human, mortalin has been implicated with several major human diseases including neurodegenerative diseases and cancer (Kaul et al. 2007, 2002).

5.2.2 DmHsp22: A Unique sHsp in *Drosophila melanogaster*

Small Hsps are molecular chaperones distributed in numerous species, from bacteria to humans, and are characterized by a relatively low monomeric molecular mass (15–43 kda) and a conserved C-terminal "α-crystallin" domain organized in a β-sheet sandwich. The sHsps form large oligomeric structures which constantly exchange units, presumably dimers, possess chaperone-like activity and are involved in various functions such as stress resistance, protein folding and apoptosis (Haslbeck et al. 2005). They recognize unfolded proteins, prevent their irreversible aggregation and facilitate refolding of bound substrates in cooperation with ATP-dependent molecular chaperones such as Hsp70 and Hsp40. While the deletion of sHsps genes does not cause thermo-sensitivity, their over-expression increases thermo-tolerance in numerous organisms suggesting the involvement of sHsps in the control of aggregation upon heat shock (Kitagawa et al. 2000; Nakamoto et al. 2000). Small Hsps such as human HspB1 have been implicated in apoptosis through their ability to prevent the formation of apoptosome and the subsequent activation of caspase 9 (Garrido et al. 2006; Rocchi et al. 2006). In humans, mutations in sHsps have also been associated with a number of neurodegenerative diseases (Dierick et al. 2005; Orejuela et al. 2007).

In *Drosophila melanogaster*, there are 12 sHsps, with the 4 main ones (DmHsp22, DmHsp23, DmHsp26 and DmHsp27) showing distinct intracellular localization and developmental expression pattern (reviewed in Michaud et al. 2002). These four main sHsps also share three conserved domains of homology including the "α-crystallin" domain (Ingolia and Craig 1982). DmHsp22 is a 174 amino acid long protein with a molecular weight of 19.7 kDa, which localizes in the mitochondrial matrix (Morrow et al. 2000), whereas DmHsp23 and DmHsp26 are in the cytosol, and DmHsp27 in the nucleus (Beaulieu et al. 1989; Marin et al. 1993, 1996; Michaud et al. 2008). During development, each of these sHsps shows a specific pattern of expression in various tissues and cells. However, they are coordinately expressed following stress although with small exceptions. For example, DmHsp22 is strongly induced by heat shock in testes while DmHsp23 and DmHsp27 although present are not induced (Michaud et al. 1997). Both DmHsp22 and DmHsp23 are involved in protection from the disturbance of normal redox state in *D. melanogaster*. Indeed, *Dm*Hsp23 plays important role in hypoxia tolerance (Azad et al. 2009), whereas DmHsp22 has a protective role in hyperoxia (Gruenewald et al. 2009). Recently, it was shown that DmHsp22 and DmHsp23 are both up-regulated during the recovery from prolonged chill coma suggesting that both DmHsp22 and DmHsp23 also contribute to adaptive responses to fluctuating thermal conditions (Colinet et al. 2010). All four main sHsps show chaperone-like activity in numerous *in vitro* protein protection assays (Heikkila et al. 2007; Horwitz 1992; Jakob et al. 1993; Morrow et al. 2006). The *in vitro* chaperone-like activity of DmHsp22 is relatively high in comparison to the others sHsps of Drosophila. The reason for these differences in cell-specific developmental expression and in chaperone-like activity between sHsps remains unknown, but it has been suggested that each sHsp have specific substrates to accomplish their function (Morrow et al. 2006).

Although mortalin and DmHsp22 belong to two different families of Hsps, they are localized in the same sub-cellular compartment suggesting they may interact with the same substrate(s) and contribute to common functions within the mitochondria.

5.3 Both DmHsp22 and Mortalin Interact with Hsp60

Hsp60 is a mitochondrial chaperonin and is the most important component of the protein-folding system within the matrix. Accordingly, yeast cells carrying a null mutation of *hsp60* are non-viable due to severe defects in folding of mitochondrial proteins. It was shown that newly imported proteins interact with Hsp60 upon their entry into the mitochondrial matrix. Although mitochondria have their own DNA and protein translation system, the majority of mitochondrial proteins are coded by nuclear DNA, synthesized in cytosol and then imported in the organelle. This process involves two translocases, one across the mitochondrial outer membrane (TOM) and the other in the mitochondrial inner membrane (TIM). The molecular architecture of these translocation channels requires that proteins have to cross the membrane in a completely extended conformation (Schwartz et al. 1999; Schwartz and Matouschek 1999). On the matrix side, mortalin interacts with Tim44 and Tim33 to facilitate the import and initiate the refolding of imported proteins (D'Silva et al. 2004; Schneider et al. 1994; Voos and Rottgers 2002). These proteins are then transferred by mortalin to Hsp60 to allow their refolding to their native state (Deocaris et al. 2006a; Hartl et al. 1992; Langer and Neupert 1991). It was shown that the N-terminal extension of mortalin is involved in its interaction with Hsp60 both *in vitro* and *in vivo* (Wadhwa et al. 2005) and that this interaction is critical to the maintenance of the mitochondrial proteome integrity (Cheng et al. 1989; Stetler et al. 2010). Although predominantly detected in mitochondria, Hsp60 and mortalin were also found in extra-mitochondrial sites in a variety of cells and tissues and therefore could also interact together outside the mitochondria (Cicconi et al. 2004; Ran et al. 2000).

While the interaction of mortalin with Hsp60 involves its role in mitochondrial protein import, the interaction of DmHsp22 with the chaperonin seems to be linked to the intrinsic chaperone function of sHsps. Indeed, sHsps form large oligomeric structures with their substrates to prevent misfolded proteins to form toxic aggregates and/or to drive unwanted interactions with non-native partners. However, sHsps cannot promote by themselves the refolding of their substrate and therefore need the help of ATP-dependant chaperones. According to this model, an interaction between DmHsp22 and Hsp60 has recently been observed by immunoprecipitation using mitochondrial protein extracts of both Drosophila and mammalian cells expressing DmHsp22.

5.4 Both DmHsp22 and Mortalin have Beneficial Effects in the Aging Process

The free radical theory of aging suggests that the accumulation of macromolecular damages induced by toxic ROS plays a central role in the aging process (Finkel and Holbrook 2000; Harman 1956; Stadtman 1992). Indeed, during aging, mitochondrial malfunctions and their consequences decrease the ATP level and increase the ROS level leading to the production of abnormal proteins and inhibiting their clearance. Therefore, aging appears to be an imbalance between damages and repair of macromolecules including DNA and proteins (Campisi and Vijg 2009). Proteins are particularly sensitive to aging-related damages such as cleavage, covalent modifications, oxidative lesions, glycations, crosslinkings and denaturation that have negative effects on their function (Semba et al. 2010; Stadtman 2006).

Numerous data have shown that the induction of various chaperones is impaired in aged organisms. This would not be due to a decreased level of expression of heat shock factor 1 (Hsf1), the transcriptional factor responsible for the induction of most Hsps, during aging, but rather to a decrease of its activation and binding to the heat shock elements found in *hsps* promoters (Locke and Tanguay 1996). The resulting age-induced alteration of the chaperone network favours the accumulation of proteins aggregates and switches the balance between damage and repair on the wrong side, therefore promoting the toxic effects of misfolded/damaged proteins. Interestingly, over-expression of chaperones often delays the onset or diminishes the symptoms of age-related diseases and can even lead to increased longevity (Csermely and Soti 2002; Morrow et al. 2010).

The level of mortalin naturally decreases in senescent human fibroblasts and in aged *Caenorhabditis elegans* (Kaul et al. 2004; Nakamura et al. 2007). However, its over-expression leads to an extended lifespan in nematodes and human cells (Kaul et al. 2003; Yokoyama et al. 2002). Studies on brain ischemia-like injuries suggest that the protective effect of mortalin over-expression would come from a suppression of ROS accumulation as well as a decrease in ROS production, leading to the preservation of mitochondrial functions (Giffard et al. 2008, 2009; Zuo et al. 2005). While the over-expression of mortalin is beneficial, its absence is detrimental. Indeed, in *Saccharomyces cerevisiae* the knock-out of Sscl (yeast's mortalin homolog) is lethal and its mutations cause aggregation of mitochondria (Craig et al. 1987; Endo et al. 2001). Moreover, *C. elegans* worms treated with a *hsp70F* (nematode's mortalin homolog) siRNA, display a reduction of Hsp60 level leading to abnormal mitochondrial morphology, diminution of ATP levels and accelerated aging phenotype (Nakamura et al. 2007). All together, these observations suggest that the level of mortalin is decisive for cell survival (Strub et al. 2003).

In Drosophila, DmHsp22 is normally expressed during the metamorphosis from larvae to pupae and during aging (King and Tower 1999). While DmHsp22 is detectable around 30 days of age, its mRNA is present as soon as day 6. Interestingly, flies selected for an increased longevity phenotypes have been shown to accumulate *dmhsp22* and *dmhsp23* mRNA sooner in life (Kurapati et al. 2000). Also arguing for a beneficial role of DmHsp22 in the aging process, its over-expression resulted in a 32% increase in mean lifespan (Morrow et al. 2004b) while a decrease of approximately 40% in lifespan was observed in the absence of DmHsp22 (Morrow et al. 2004a). In addition to having an increased longevity, flies over-expressing DmHsp22 were also more resistant to oxidative and thermal stress, and retained their locomotor activity longer (healthspan). Interestingly, the over-expression of DmHsp22 also resulted in an increased lifespan of human primary fibroblasts (Wadhwa et al. 2010). Indeed, while control cells entered senescence at 45 population doubling (PD), DmHsp22 expressing cells went up to 65 PD before senescence appeared as found by growth, cell morphology and beta-galactosidase assay.

While DmHsp22 and mortalin belong to different HSP sub-families, they have distinct functions and are differently expressed during the normal course of aging, they share the ability to increase lifespan, suggesting that the maintenance of mitochondrial homeostasis *per se* is more important than the way to achieve it.

5.5 Both DmHsp22 and Mortalin Interact with p53 and Induce Tumour Formation

The constant control of cell proliferation is important, and cells that escape such control are prone to form tumours. Genes that can affect tumour formation are generally involved in DNA repair systems as well as in cell growth and apoptosis. Due to their role in proteostasis and cell survival, a tight control of Hsps is important since they can become chaperones of cell death by helping in the maturation of proteins involved in the apoptotic pathway (Garrido et al. 2001; Samali et al. 1999) or, on the other hand, prevent apoptosis by activating anti-apoptotic mediators (Arnaudeau et al. 2009; Jaattela et al. 1998; Samali et al. 1999; Takayama et al. 1997; Zylicz et al. 2001).

Mortalin was identified as one of the important anti-apoptotic genes in a proteome analysis of ouabain-treated vascular smooth muscle cells (VSMCs) (Taurin et al. 2002). Moreover, its over-expression suppresses apoptosis from various stressors such as arsenate in rat, glucose starvation and ischemia reperfusion in Chinese hamster lung (CHL) (Deocaris et al. 2006a; Gao et al. 2003; Lau et al. 2004). Furthermore, mortalin has been shown to interact with the tumour suppressor p53, causing its cytoplasmic sequestration and protecting cells from ATP depletion and energy deprivation associated with cell death by preventing the rapid rise in mitochondrial

Table 5.1 Similar effects of DmHsp22 and mortalin in oncogenesis

Assay	Property cell	Control vector	Vecteur-DmHsp22	Vecteur-mtHsp70	Vecteur-Hsp60
Cell transformation	Colony formation	–	++	++	–
Chemotaxis	Invasion	–	++	++	ND
Wound scratch	Invasion	–	++	++	ND
Balb/c nude mice	Tumour formation	–	++	++	–
Interaction with p53		–	++	++	–

MCF7 and/or A549 cancer cells were injected by either the empty control vector, the DmHsp22, mtHsp70 or Hsp60 vectors and various malignant properties were measured (Wadhwa et al. 2010). The interaction with p53 was measured by co-immunoprecipitation and is described in Iosefson and Azem (2010) and Wadhwa et al. (2010)

ROS (Wadhwa et al. 1998, 2002b; Zuo et al. 2005; Taurin et al. 2002). However, the association of mortalin with p53 in mitochondria has been observed in p53-induced apoptosis (Marchenko et al. 2000) and would therefore not account for mortalin anti-apoptotic effect by itself. Very similar to mortalin, DmHsp22 was found to interact with p53, causing its sequestration and therefore inactivation of its transcriptional activity (Wadhwa et al. 2010).

Impairment of apoptosis is a common characteristic of cancer cell and the p53 inactivation represents one of the major mechanisms rendering cell independent to normal regulatory factors. Tumour cells often have elevated level of Hsps, which is consistent with a background of genetic instability, therefore facilitating malignant transformation and cancer progression. High levels of Hsp70 and the small Hsp27 (HspB1) have been associated with different tumours and often correlate with the severity of the tumour. It is also well documented that Hsps over-expression is associated with a poor prognosis in terms of survival or response to therapy in different types of cancers (Ciocca and Calderwood 2005; Ciocca et al. 2010). However, it is still unclear how the Hsps become over-expressed in cancers. It was hypothesised that the hostile tumour microenvironment (low glucose, pH, hypoxia, nutrient deprivation) is responsible for their induction.

Accordingly, the level of mortalin expression was found to be elevated in many human tumours and in all tumour-derived cells and *in vitro* immortalized cells studied. Over-expression of mortalin is sufficient to increase the malignancy of breast carcinoma and quantitative estimation of mortalin expression shows that the tumor cells with higher level of mortalin have more aggressive tumor phenotype (i.e. metastasis) (Lu et al. 2011; Wadhwa et al. 2006). Interestingly, DmHsp22 has similar effects to mortalin in human cancer cells. Indeed, expression of DmHsp22 in cancer cells increases malignant properties in assays of cell transformation and causes tumour formation very similar to the one induced by mortalin upon injection in nude mice. Consistent with these findings, cells expressing DmHsp22 were more resistant to various anticancer drugs (Wadhwa et al. 2010). Table 5.1 summarizes the similar effects of DmHsp22 and mortalin in oncogenesis.

5.6 Conclusions and Perspectives

Hsps consist of a super-family of proteins with chaperone properties involved in various biological processes such as stress resistance, cell proliferation, apoptosis, cancers and neurodegenerative diseases. Hsps are generally induced by various stresses, but they are also expressed in a cell- and developmental stage-specific fashion in the absence of stress, suggesting that they play different roles in cell physiology. So, the capacity of Hsps to have various functions in cells might have important repercussions in pathological situations such as neurodegenerative diseases and cancers. Oncogenesis and apoptosis are closely related processes and defaults in apoptosis are involved in the development and progression of many cancers. Hsps are often highly and constitutively expressed in cancer cells and tumors and they are associated with poor prognosis. So, it appears that chaperone control of cell homeostasis is not random in cells but is compartmentalized in sub-cellular organelles and is dependent on the microenvironment. Among the Hsp super-family, some have essential functions in maintaining mitochondrial homeostasis.

DmHsp22 and mortalin are chaperone proteins involved in protein quality control, which helps cells to adapt to changing environmental situations. Interestingly, studies using different members of the sHSP sub-family in *D. melanogaster* and in *C. elegans* have shown that over-expression of many sHsps can have positive effects on extension of lifespan and more-importantly on extension of health-span. Moreover it was also shown that a small mitochondrial chaperone from an insect could have cross-species effects in human cells by slowing the senescence process by its chaperoning power and specific inactivation of tumor suppressor protein p53 that is centrally involved in the regulation of cell proliferation in normal and cancer cells. The core protein of the import complex, mortalin, is responsible for the coupling of ATP hydrolysis with peptide movement and translocation. Mortalin stabilizes partially unfolded proteins resulting in more efficient refolding and preventing aggregation. The mechanism(s) involved in the protection against age-associated damages may include prevention of misfolded proteins aggregation and removal of damaged proteins, but it cannot be excluded at this time that these proteins also act at other levels. These two mitochondrial Hsps can have similar effect on the fate of an organelle although they belong to two different HSP sub-families and have different functions. This suggests that the maintenance of mitochondrial homeostasis is one of the ultimate goal of the cell, which is not surprising taking into account the role of mitochondria in energy production and apoptosis.

Acknowledgments RMT is supported by a grant from the Canadian Institutes of Health Research (CIHR). MLP received a post-doctoral fellowship from PROTEO.

References

Arnaudeau S, Arboit P, Bischof P, Shin-ya K, Tomida A, Tsuruo T, Irion O, Cohen M (2009) Glucose-regulated protein 78: a new partner of p53 in trophoblast. Proteomics 9:5316–5327

Azad P, Zhou D, Russo E, Haddad GG (2009) Distinct mechanisms underlying tolerance to intermittent and constant hypoxia in Drosophila melanogaster. PLoS One 4:e5371
Beaulieu JF, Arrigo AP, Tanguay RM (1989) Interaction of Drosophila 27,000 Mr heat-shock protein with the nucleus of heat-shocked and ecdysone-stimulated culture cells. J Cell Sci 92:29–36
Campisi J, Vijg J (2009) Does damage to DNA and other macromolecules play a role in aging? If so, how? J Gerontol A-Biol 64:175–178
Carette J, Lehnert S, Chow TY (2002) Implication of PBP74/mortalin/GRP75 in the radio-adaptive response. Int J Radiat Biol 78:183–190
Chang YW, Sun YJ, Wang C, Hsiao CD (2008) Crystal structures of the 70-kDa heat shock proteins in domain disjoining conformation. J Biol Chem 283:15502–15511
Cheng MY, Hartl FU, Martin J, Pollock RA, Kalousek F, Neupert W, Hallberg EM, Hallberg RL, Horwich AL (1989) Mitochondrial heat-shock protein hsp60 is essential for assembly of proteins imported into yeast mitochondria. Nature 337:620–625
Cicconi R, Delpino A, Piselli P, Castelli M, Vismara D (2004) Expression of 60 kDa heat shock protein (Hsp60) on plasma membrane of Daudi cells. Mol Cell Biochem 259:1–7
Ciocca DR, Calderwood SK (2005) Heat shock proteins in cancer: diagnostic, prognostic, predictive, and treatment implications. Cell Stress Chaperones 10:86–103
Ciocca DR, Fanelli MA, Cuello-Carrion FD, Castro GN (2010) Heat shock proteins in prostate cancer: from tumorigenesis to the clinic. Int J Hyperthermia 26:737–747
Colinet H, Lee SF, Hoffmann A (2010) Knocking down expression of Hsp22 and Hsp23 by RNA interference affects recovery from chill coma in Drosophila melanogaster. J Exp Biol 213:4146–4150
Craig EA (1989) Essential roles of 70 kDa heat inducible proteins. Bioessays 11:48–52
Craig EA, Kramer J, Kosicsmithers J (1987) Ssc1, a Member of the 70-Kda heat-shock protein multigene family of Saccharomyces cerevisiae, is essential for growth. Proc Natl Acad Sci U S A 84:4156–4160
Csermely P, Soti C (2002) Chaperones come of age. Cell Stress Chaperonones 7:186–190
D'Silva P, Liu Q, Walter W, Craig EA (2004) Regulated interactions of mtHsp70 with Tim44 at the translocon in the mitochondrial inner membrane. Nat Struct Mol Biol 11:1084–1091
Das N, Levine RL, Orr WC, Sohal RS (2001) Selectivity of protein oxidative damage during aging in Drosophila melanogaster. Biochem J 360:209–216
Deocaris CC, Kaul SC, Wadhwa R (2006a) On the brotherhood of the mitochondrial chaperones mortalin and heat shock protein 60. Cell Stress Chaperones 11:116–128
Deocaris CC, Yamasaki K, Kaul SC, Wadhwa R (2006b) Structural and functional differences between mouse mot-1 and mot-2 proteins that differ in two amino acids. Ann NY Acad Sci 1067:220–223
Dierick I, Irobi J, De Jonghe P, Timmerman V (2005) Small heat shock proteins in inherited peripheral neuropathies. Ann Med 37:413–422
Dundas SR, Lawrie LC, Rooney PH, Murray GI (2005) Mortalin is over-expressed by colorectal adenocarcinomas and correlates with poor survival. J Pathol 205:74–81
Endo T, Kawai A, Nishikawa S, Hirata A (2001) Loss of the mitochondrial Hsp70 functions causes aggregation of mitochondria in yeast cells. J Cell Sci 114:3565–3574
Finkel T, Holbrook NJ (2000) Oxidants, oxidative stress and the biology of ageing. Nature 408:239–247
Gao CX, Zhang SQ, Yin Z, Liu W (2003) [Molecular chaperone GRP75 reprove cells from injury caused by glucose deprivation]. Shi Yan Sheng Wu Xue Bao. 36:381–387
Garrido C, Gurbuxani S, Ravagnan L, Kroemer G (2001) Heat shock proteins: endogenous modulators of apoptotic cell death. Biochem Biophys Res Commun 286:433–442
Garrido C, Brunet M, Didelot C, Zermati Y, Schmitt E, Kroemer G (2006) Heat shock proteins 27 and 70: anti-apoptotic proteins with tumorigenic properties. Cell Cycle 5:2592–2601
Giffard RG, Voloboueva LA, Duan M, Ouyang Y, Emery JF, Stoy C (2008) Overexpression of mitochondrial Hsp70/Hsp75 protects astrocytes against ischemic injury in vitro. J Cerebr Blood F Met 28:1009–1016

Giffard RG, Xu LJ, Voloboueva LA, Ouyang YB, Emery JF (2009) Overexpression of mitochondrial Hsp70/Hsp75 in rat brain protects mitochondria, reduces oxidative stress, and protects from focal ischemia. J Cerebr Blood F Met 29:365–374

Green DR, Kroemer G (2004) The pathophysiology of mitochondrial cell death. Science 305:626–629

Gruenewald C, Botella JA, Bayersdorfer F, Navarro JA, Schneuwly S (2009) Hyperoxia-induced neurodegeneration as a tool to identify neuroprotective genes in Drosophila melanogaster. Free Radic Biol Med 46:1668–1676

Harman D (1956) Aging: a theory based on free radical and radiation chemistry. J Gerontol 11:298–300

Hartl FU, Martin J, Neupert W (1992) Protein folding in the cell: the role of molecular chaperones Hsp70 and Hsp60. Annu Rev Biophys Biomol Struct 21:293–322

Haslbeck M, Franzmann T, Weinfurtner D, Buchner J (2005) Some like it hot: the structure and function of small heat-shock proteins. Nat Struct Mol Biol 12:842–846

Heikkila JJ, Kaldis A, Morrow G, Tanguay RM (2007) The use of the Xenopus oocyte as a model system to analyze the expression and function of eukaryotic heat shock proteins. Biotechnol Adv 25:385–395

Hendrick JP, Hartl FU (1995) The role of molecular chaperones in protein folding. FASEB J 9:1559–1569

Horwitz J (1992) Alpha-crystallin can function as a molecular chaperone. Proc Natl Acad Sci U S A 89:10449–10453

Ingolia TD, Craig EA (1982) Four small Drosophila heat shock proteins are related to each other and to mammalian alpha-crystallin. Proc Natl Acad Sci U S A 79:2360–2364

Iosefson O, Azem A (2010) Reconstitution of the mitochondrial Hsp70 (mortalin)-p53 interaction using purified proteins – Identification of additional interacting regions. FEBS Lett 584:1080–1084

Jaattela M, Wissing D, Kokholm K, Kallunki T, Egeblad M (1998) Hsp70 exerts its anti-apoptotic function downstream of caspase-3-like proteases. EMBO J 17:6124–6134

Jakob U, Gaestel M, Engel K, Buchner J (1993) Small heat shock proteins are molecular chaperones. J Biol Chem 268:1517–1520

Jiang J, Prasad K, Lafer EM, Sousa R (2005) Structural basis of interdomain communication in the Hsc70 chaperone. Mol Cell 20:513–524

Kampinga HH, Hageman J, Vos MJ, Kubota H, Tanguay RM, Bruford EA, Cheetham ME, Chen B, Hightower LE (2009) Guidelines for the nomenclature of the human heat shock proteins. Cell Stress Chaperones 14:105–111

Kaul SC, Taira K, Pereira-Smith OM, Wadhwa R (2002) Mortalin: present and prospective. Exp Gerontol 37:1157–1164

Kaul SC, Yaguchi T, Taira K, Reddel RR, Wadhwa R (2003) Overexpressed mortalin (mot-2)/mthsp70/GRP75 and hTERT cooperate to extend the in vitro lifespan of human fibroblasts. Exp Cell Res 286:96–101

Kaul SC, Wadhwa R, Takano S, Taira K (2004) Reduction in mortalin level by its antisense expression causes senescence-like growth arrest in human immortalized cells. J Gene Medicine 6:439–444

Kaul SC, Deocaris CC, Wadhwa R (2007) Three faces of mortalin: a housekeeper, guardian and killer. Exp Gerontol 42:263–274

King V, Tower J (1999) Aging-specific expression of Drosophila hsp22. Dev Biol 207:107–118

Kitagawa M, Matsumura Y, Tsuchido T (2000) Small heat shock proteins, IbpA and IbpB, are involved in resistances to heat and superoxide stresses in Escherichia coli. FEMS Microbiol Lett 184:165–171

Kurapati R, Passananti HB, Rose MR, Tower J (2000) Increased hsp22 RNA levels in Drosophila lines genetically selected for increased longevity. J Gerontol A Biol Sci Med Sci 55:B552–559

Langer T, Neupert W (1991) Heat shock proteins hsp60 and hsp70: their roles in folding, assembly and membrane translocation of proteins. Curr Top Microbiol Immunol 167:3–30

Lau AT, He QY, Chiu JF (2004) A proteome analysis of the arsenite response in cultured lung cells: evidence for in vitro oxidative stress-induced apoptosis. Biochem J 382:641–650

Leidhold C, von Janowsky B, Becker D, Bender T, Voos W (2006) Structure and function of Hsp78, the mitochondrial ClpB homolog. J Struct Biol 156:149–164

Locke M, Tanguay RM (1996) Increased HSF activation in muscles with a high constitutive Hsp70 expression. Cell Stress Chaperones 1:189–196

Lu WJ, Lee NP, Kaul SC, Lan F, Poon RT, Wadhwa R, Luk JM (2011) Mortalin-p53 interaction in cancer cells is stress dependent and constitutes a selective target for cancer therapy. Cell Death Differ 18:1046–1056

Marchenko ND, Zaika A, Moll UM (2000) Death signal-induced localization of p53 protein to mitochondria. A potential role in apoptotic signaling. J Biol Chem 275:16202–16212

Marin R, Valet JP, Tanguay RM (1993) hsp23 and hsp26 exhibit distinct spatial and temporal patterns of constitutive expression in Drosophila adults. Dev Genet 14:69–77

Marin R, Demers M, Tanguay RM (1996) Cell-specific heat-shock induction of Hsp23 in the eye of Drosophila melanogaster. Cell Stress Chaperones 1:40–46

Michaud S, Marin R, Westwood JT, Tanguay RM (1997) Cell-specific expression and heat-shock induction of Hsps during spermatogenesis in Drosophila melanogaster. J Cell Sci 110:1989–1997

Michaud S, Morrow G, Marchand J, Tanguay RM (2002) Drosophila small heat shock proteins: cell and organelle-specific chaperones? Prog Mol Subcell Biol 28:79–101

Michaud S, Lavoie S, Guimond MO, Tanguay RM (2008) The nuclear localization of Drosophila Hsp27 is dependent on a monopartite arginine-rich NLS and is uncoupled from its association to nuclear speckles. Biochim Biophys Acta 1783:1200–1210

Morrow G, Inaguma Y, Kato K, Tanguay RM (2000) The small heat shock protein Hsp22 of Drosophila melanogaster is a mitochondrial protein displaying oligomeric organization. J Biol Chem 275:31204–31210

Morrow G, Battistini S, Zhang P, Tanguay RM (2004a) Decreased lifespan in the absence of expression of the mitochondrial small heat shock protein Hsp22 in Drosophila. J Biol Chem 279:43382–43385

Morrow G, Samson M, Michaud S, Tanguay RM (2004b) Overexpression of the small mitochondrial Hsp22 extends Drosophila life span and increases resistance to oxidative stress. FASEB J 18:598–599

Morrow G, Heikkila JJ, Tanguay RM (2006) Differences in the chaperone-like activities of the four main small heat shock proteins of Drosophila melanogaster. Cell Stress Chaperones 11:51–60

Morrow G, Kim HJ, Le Pecheur M, Kaul SC, Wadhwa R, Tanguay RM (2010) Protection from aging by small chaperones: A trade-off with cancer? Ann NY Acad Sci 1197:67–75

Nakamoto H, Suzuki N, Roy SK (2000) Constitutive expression of a small heat-shock protein confers cellular thermotolerance and thermal protection to the photosynthetic apparatus in cyanobacteria. FEBS Lett 483:169–174

Nakamura N, Kimura K, Tanaka N, Takano S, Ohkuma S (2007) Knockdown of mitochondrial heat shock protein 70 promotes progeria-like phenotypes in Calenorhabditis elegans. J Biol Chem 282:5910–5918

Orejuela D, Bergeron A, Morrow G, Tanguay RM (2007) Small heat shocks proteins in physiological and stress-related processes. In: Caderwood SK (ed) Cell stress proteins, vol 7. Springer, New York

Parcellier A, Gurbuxani S, Schmitt E, Solary E, Garrido C (2003) Heat shock proteins, cellular chaperones that modulate mitochondrial cell death pathways. Biochem Biophys Res Commun 304:505–512

Ran Q, Wadhwa R, Kawai R, Kaul SC, Sifers RN, Bick RJ, Smith JR, Pereira-Smith OM (2000) Extramitochondrial localization of mortalin/mthsp70/PBP74/GRP75. Biochem Biophys Res Commun 275:174–179

Rist W, Graf C, Bukau B, Mayer MP (2006) Amide hydrogen exchange reveals conformational changes in hsp70 chaperones important for allosteric regulation. J Biol Chem 281:16493–16501

Rocchi P, Jugpal P, So A, Sinneman S, Ettinger S, Fazli L, Nelson C, Gleave M (2006) Small interference RNA targeting heat-shock protein 27 inhibits the growth of prostatic cell lines and induces apoptosis via caspase-3 activation in vitro. BJU Int 98:1082–1089
Samali A, Cai J, Zhivotovsky B, Jones DP, Orrenius S (1999) Presence of a pre-apoptotic complex of pro-caspase-3, Hsp60 and Hsp10 in the mitochondrial fraction of jurkat cells. EMBO J 18:2040–2048
Schneider HC, Berthold J, Bauer MF, Dietmeier K, Guiard B, Brunner M, Neupert W (1994) Mitochondrial Hsp70/MIM44 complex facilitates protein import. Nature 371:768–774
Schwartz MP, Huang S, Matouschek A (1999) The structure of precursor proteins during import into mitochondria. J Biol Chem 274:12759–12764
Schwartz MP, Matouschek A (1999) The dimensions of the protein import channels in the outer and inner mitochondrial membranes. Proc Natl Acad Sci U S A 96:13086–13090
Semba RD, Nicklett EJ, Ferrucci L (2010) Does accumulation of advanced glycation end products contribute to the aging phenotype? J Gerontol A-Biol Sci 65:963–975
Sohal RS, Dubey A (1994) Mitochondrial oxidative damage, hydrogen peroxide release, and aging. Free Radic Biol Med 16:621–626
Stadtman ER (1992) Protein oxidation and aging. Science 257:1220–1224
Stadtman ER (2006) Protein oxidation and aging. Free Radical Res 40:1250–1258
Stetler RA, Gan Y, Zhang W, Liou AK, Gao Y, Cao G, Chen J (2010) Heat shock proteins: cellular and molecular mechanisms in the central nervous system. Prog Neurobiol 92:184–211
Strub A, Zufall N, Voos W (2003) The putative helical lid of the Hsp70 peptide-binding domain is required for efficient preprotein translocation into mitochondria. J Mol Biol 334:1087–1099
Takayama S, Bimston DN, Matsuzawa S, Freeman BC, Aime-Sempe C, Xie Z, Morimoto RI, Reed JC (1997) BAG-1 modulates the chaperone activity of Hsp70/Hsc70. EMBO J 16:4887–4896
Taurin S, Seyrantepe V, Orlov SN, Tremblay TL, Thibault P, Bennett MR, Hamet P, Pshezhetsky AV (2002) Proteome analysis and functional expression identify mortalin as an antiapoptotic gene induced by elevation of [Na +]i/[K +]i ratio in cultured vascular smooth muscle cells. Circ Res 91:915–922
Vogel M, Bukau B, Mayer MP (2006a) Allosteric regulation of Hsp70 chaperones by a proline switch. Mol Cell 21:359–367
Vogel M, Mayer MP, Bukau B (2006b) Allosteric regulation of Hsp70 chaperones involves a conserved interdomain linker. J Biol Chem 281:38705–38711
von Ahsen O, Voos W, Henninger H, Pfanner N (1995) The mitochondrial protein import machinery. Role of ATP in dissociation of the Hsp70.Mim44 complex. J Biol Chem 270:29848–29853
Voos W, Rottgers K (2002) Molecular chaperones as essential mediators of mitochondrial biogenesis. Biochim Biophys Acta 1592:51–62
Wadhwa R, Takano S, Robert, M, Yoshida, A, Nomura, H, Reddel, RR, Mitsui, Y, Kaul, SC (1998) Inactivation of tumor suppressor p53 by mot-2, a hsp70 family member. J Biol Chem 273:29586–29591
Wadhwa R, Taira K, Kaul SC (2002a) An Hsp70 family chaperone, mortalin/mthsp70/PBP74/Grp75: what, when, and where? Cell Stress Chaperones 7:309–316
Wadhwa R, Yaguchi T, Hasan MK, Mitsui Y, Reddel RR, Kaul SC (2002b) Hsp70 family member, mot-2/mthsp70/GRP75, binds to the cytoplasmic sequestration domain of the p53 protein. Exp Cell Res 274:246–253
Wadhwa R, Takano S, Kaur K, Aida S, Yaguchi T, Kaul Z, Hirano T, Taira K, Kaul SC (2005) Identification and characterization of molecular interactions between mortalin/mtHsp70 and HSP60. Biochem J 391:185–190
Wadhwa R, Takano S, Kaur K, Deocaris CC, Pereira-Smith OM, Reddel RR, Kaul SC (2006) Upregulation of mortalin/mthsp70/Grp75 contributes to human carcinogenesis. Int J Cancer 118:2973–2980
Wadhwa R, Ryu J, Gao R, Choi IK, Morrow G, Kaur K, Kim I, Kaul SC, Yun CO, Tanguay RM (2010) Proproliferative functions of Drosophila small mitochondrial heat shock protein 22 in human cells. J Biol Chem 285:3833–3839

Yokoyama K, Fukumoto K, Murakami T, Harada S, Hosono R, Wadhwa R, Mitsui Y, Ohkuma S (2002) Extended longevity of Caenorhabditis elegans by knocking in extra copies of hsp70 F, a homolog of mot-2 (mortalin)/mthsp70/Grp75. FEBS Lett 516:53–57

Zuo J, Liu Y, Liu W, Song XD (2005) Effect of GRP75/mthsp70/PBP74/mortalin overexpression on intracellular ATP level, mitochondrial membrane potential and ROS accumulation following glucose deprivation in PC12 cells. Mol Cell Biochem 268:45–51

Zylicz M, King FW, Wawrzynow A (2001) Hsp70 interactions with the p53 tumour suppressor protein. EMBO J 20:4634–4638

Chapter 6
Mortalin in Invertebrates and The Induction of Apoptosis by Wild-Type p53 Following Defeat of Mortalin-Based Cytoplasmic Sequestration in Cancerous Clam Hemocytes

Charles W. Walker, Ben Low and S. Anne Böttger

Abstract In this chapter we demonstrate the remarkable sequence homology between mortalin proteins from a broad array of invertebrate taxa, including the few species for which functional analyses have been conducted. We also discuss expression and functional data for full-length and truncated clam homologs for human mortalin and their function in cytoplasmic sequestration in cancerous clam hemocytes. Both clam proteins have N-terminal mitochondrial targeting and p53 binding domains, though the truncated variant is missing exon 3 containing the N-terminal ATP/ADP binding and ATPase domains. Both variants are over-expressed and complexed with p53 and both may be responsible for cytoplasmic sequestration of p53 in cancerous clam hemocytes. Clam hemocyte cancer is the only animal model thus far investigated where cytoplasmically sequestered wild-type p53 can be reactivated both *in vitro* and *in vivo* using both genotoxic and non-genotoxic therapies. Our results suggest that mortalin-based cytoplasmic sequestration of wild-type p53 in cancerous clam hemocytes can be reversed by treatment with antineoplastic drugs also employed against similar human diseases and will result either in transcription-based apoptosis when the nucleus is accessible or non-transcription-based apoptosis when nuclear access is blocked.

Keywords Cytoplasmic sequestration · Cancer · *Mya arenaria* · p53 · Mortalin Human

C. W. Walker (✉)
Department of Molecular, Cellular and Biomedical Sciences, Center For Marine Biology and Marine Biomedical Research Group, The University of New Hampshire, Durham, NH 03824, USA
e-mail: cwwalker@christa.unh.edu

B. Low
The Jackson Laboratory, 600 Main Street, Bar Harbor, Maine 04609, USA

S. A. Böttger
Department of Biology, West Chester University, West Chester, Pennsylvania 19383, USA

S. C. Kaul, R. Wadhwa (eds.), *Mortalin Biology: Life, Stress and Death*,
DOI 10.1007/978-94-007-3027-4_6, © Springer Science+Business Media B.V. 2012

6.1 Introduction

Mortalin is the mitochondrial heat shock 70 protein (Hsp70). Following synthesis in the cytoplasm a cleavable, N-terminal mitochondrial-targeting domain helps guide mortalin protein to mitochondria where it is imported to the mitochondrial matrix. Within mitochondria, mortalin participates with several additional proteins in translocating other mitochondrial proteins (e.g., Kreb's cycle enzymes, etc.) in an ATP dependent manner into the matrix through the TIM translocon (Hendrick and Hartl 1993; Kampinga and Craig 2010). A number of other functions have been attributed to mortalin including energy generation, chaperoning miss-folded proteins, stress response, carcinogenesis and involvement in diseases of old age (Akerfelt et al. 2010; Kaul et al. 2007). The extra-mitochondrial localization of mortalin in many cell types and its ability to bind cytoplasmic p53 in a number of normal and cancerous cells in the vertebrates and in cancerous soft shell clam hemocytes (CCH) suggest that it also has a chaperoning function in the cytoplasm (Deocaris et al. 2006; Gupta et al. 2008; Wadhwa et al. 1993a; Walker et al. 2006, 2009). In normal human cells, cytoplasmic mortalin binds to p53 by its protein-binding domain and may participate with a number of other proteins in a protein complex based on Hsp90 that delivers p53 to the nucleus along microtubules (Galigniana et al. 2004). As an early response to severe stress, human wild-type p53 protein is also translocated to mitochondria where it directly interacts with Bcl-2 family membrane proteins in the outer mitochondrial membrane to inactivate them and drive apoptosis (Yee and Vousden 2005). While mortalin is predicted to be involved in translocation of p53 to the nuclear membrane as a member of a multicomponent transport complex, it is unclear for any vertebrate or invertebrate cells whether mortalin is part of the chaperone complex involved in mitochondrial translocation.

When mortalin is overexpressed, it is responsible for tethering p53 protein in the cytoplasm (Wadhwa et al. 1993b, 2006). This phenomenon is called cytoplasmic sequestration in human and mouse cells and a similar phenotype exists in CCH and involves sequestration of wild-type p53 in the cytoplasm (Kelley et al. 2001; Walker et al. 2006, 2009). This phenomenon has also been observed in an unrelated group of human cancers, including undifferentiated neuroblastoma and colorectal adenocarcinoma in which mortalin is overexpressed and correlated with poor clinical outcome (Dundas et al. 2005; Moll et al. 1992, 1995, 1996). A study of primary and secondary glioblastomas suggested that mortalin and possibly other tethering molecules (e.g., cullin 7 or PARC) may also be responsible for cytoplasmic sequestration of p53 in these cancers (Nagpal et al. 2006; Nikolaev et al. 2003). As a result, p53 in these human and clam cancer cells is unable to enter the nucleus nor can it be translocated to mitochondria. These outcomes effectively block not only the transcriptional function of p53 within the nucleus, but also its non-transcriptional function at the mitochondrion and, as a result, prevents apoptosis.

In cancerous clam hemocytes (CCH), cytoplasmic sequestration of clam wild-type p53 protein can be reversed using the topoisomerase II poisons, mitoxantrone and etoposide (Walker et al. 2006). Cytoplasmic sequestration of p53 can also be reversed by treatment with MKT-077, a cationic inhibitor of mortalin (Walker et al. 2006). Under treatment with all of these compounds apoptosis of CCH occurs (Böttger et al. 2008). These results suggest that mortalin-based cytoplasmic sequestration of wild-type p53 in CCH can be reversed by treatment with antineoplastic drugs also employed against similar human diseases and will result either in transcription-based apoptosis (following nuclear translocation of p53) when the nucleus is accessible or non-transcription-based apoptosis (following mitochondrial translocation of p53) when nuclear access is blocked.

Here we present information on the structure of mortalin protein in a broad array of invertebrate taxa and on the few invertebrates for which functional analyses of mortalin have been conducted. We also discuss expression and functional data for full-length and truncated clam homologs for human mortalin in CCH and describe experiments that result in blocking cytoplasmic sequestration of clam p53 by mortalin during transcriptional and non-transcriptional induction of apoptosis in CCH.

6.2 Mortalin Homologs in Invertebrates

Reference sequences for mortalin include the human mortalin gene that is present as a single copy located on chromosome 5(5q31.1) and those of mice, Mot1 distributed throughout the cytoplasm and Mot2 localized near the nuclear membrane (Deocaris et al. 2006). While mortalin homologs are also found throughout other vertebrates (mammals, frogs, fish), distinct mortalin-like family members have also recently been identified for specific invertebrate species or can be mined from newly completed genomes for a number of invertebrate taxa (Figs. 6.1 and 6.2). Mortalin is present in the following invertebrate eukaryotes: invertebrate deuterostomes (Cephalochordata; Echinodermata and Hemichordata), lophotrochozoan and ecdysozoan protostomes (Mollusca; Insecta, Arachnida and Crustacea), Platyheminthia (free-living flatworms) and the Cnidarian (the sea anemone, *Nematostella*) (Figs. 6.1 and 6.2). In Fig. 6.2, the remarkable sequence conservation in these widely divergent invertebrate taxa is evident in predicted protein sequences for mortalin in the ATP/ADP binding, p53 binding and MKT-077 binding domains and the three characteristic Hsp70 and the leucine zipper domains. Functional data for mortalin exist for very few invertebrates: for nematode worms (He et al. 2010; Yokoyama et al. 2002), for echinoderms (Alvares et al. 2007; Mashanov et al. 2010), for molluscs (Kelley et al. 2001; Walker et al. 2006, 2009; Walker and Böttger 2008) and for a free-living flatworm (Conte et al. 2009). So far, these functional data fall into two broad categories: (a) activities related to p53 and apoptosis—discussed in relation to the free-living flatworm, *Dugesia japonica* in Chap. 8 of this volume (Conte et al. 2009) and the sea cucumber, *Holothuria glaberrima* (Mashanov et al. 2010) and (b) biomineralization—discussed in relation to the sea urchin, *Lytechinus variegatus* in Chap. 9 of this volume (Alvares et al. 2007). Functional data for mortalin in CCH in the bivalve mollusc, *Mya arenaria*, are presented in the following sections.

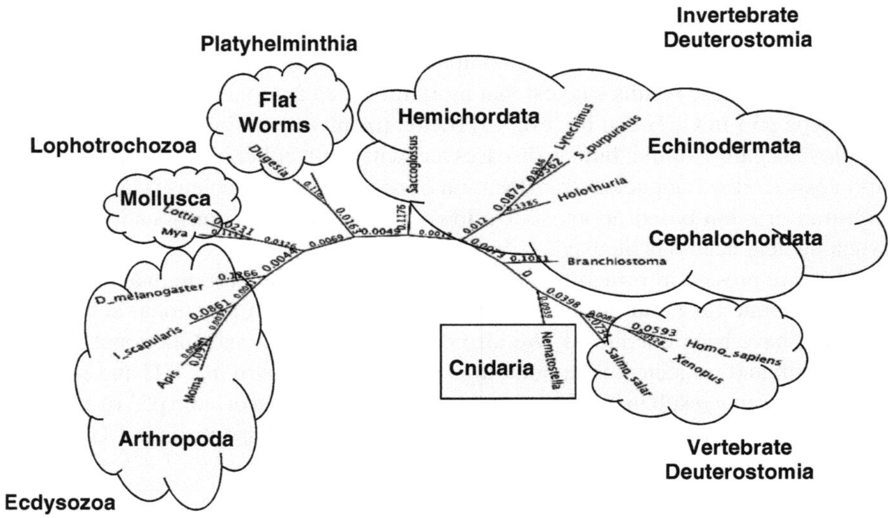

Fig. 6.1 Phylogenetic tree for invertebrate mortalins. Human (*Homo sapiens*) and clam (*Mya arenaria*) mortalin sequences were used to retrieve homologous gene models from the following genome projects located at the JGI Eukaryotic Genomics Web site (http://www.jgi.doe.gov/genome-projects/). GenBank assession numbers and other sources for the mortalin sequences included in this phylogenetic tree include: for **Vertebrate Deuterostomia**—*Homo sapiens* (AAH00478), *Xenopus laevis* (NP_001001229) and *Salmo salar* (ACN11027): for **Invertebrate Deuterostomia**—Cephalochordata—*Branchiostoma floridae* (XP_002592358); Echinodermata—*Strongylocentrotus purpuratus* (XP_802062), *Lytechinus variagatus* (ABK27326) and *Holothuria glaberrima* (ADR31818); Hemichordata—*Saccoglossus kowalevskii* (XP_002738188); for **Platyhelminthia**—*Dugesia japonica* (CAR66235); for **Lophotrochozoa Protostomia**—Mollusca—*Mya arenaria* (AY326398) and *Lottia gigantea* (JGI & Metazome) and for **Ecdysozoa Protostomia**—Arthropoda, Insecta—*Drosophila melanogaster* (NP_523741); *Apis mellifera* (NP_001153520), Arthropoda, Arachnida, *Ixodes scapularis* (XP_002402562) and Arthropoda; Crustacea *Moina mongolica* (ADA79523). This tree was produced using Geneious Pro 5.3.4 software, alignment of the sequences was performed using MUSCLE and the tree was generated using Jukes-Cantor neighbor joining with the **Cnidarian**, *Nematostella vectensis* (XP_001639786) as an out-group

6.3 Structure of Full-length and Truncated Variants of Clam Mortalin

A clam homolog for human mortalin was cloned from CCH using RACE techniques. The full-length clam mortalin clone (GenBank Accession # AY326398) is 2421 bp long, predicts a coding sequence of 692 aa and contains a short (72 bp) 5′ untranslated region (UTR) and a portion of the 3′ UTR (274 bp). Full-length clam mortalin is 91% conserved with human mortalin (HSPA9B, GenBank Accession # P38646). All significant functional domains are also conserved between clam and human versions of mortalin protein (Fig. 6.3). Within the N-terminal nucleotide-binding domain (Morano 2007), the first two Hsp70 motifs are 100% identical while the third is 93%

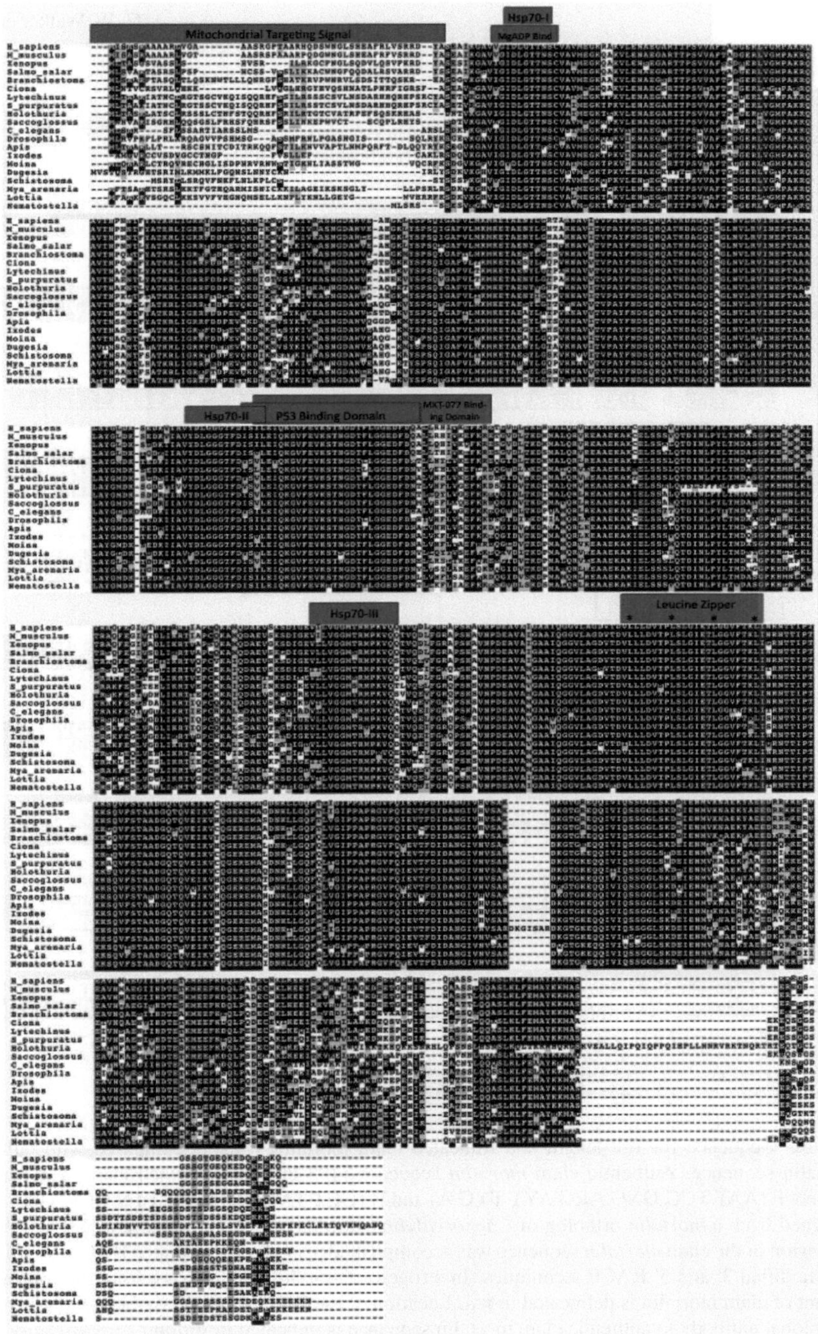

Fig. 6.2 Sequence comparison for representative vertebrate and available invertebrate sequences for predicted mortalin protein sequences. Sequences for all species were obtained as above. Notice the high sequence conservation in these widely divergent vertebrate and invertebrate taxa between predicted protein sequences for mortalin protein in the ATP/ADP binding, p53 binding and MKT-077 binding domains and the three characteristic Hsp70 and the leucine zipper domains

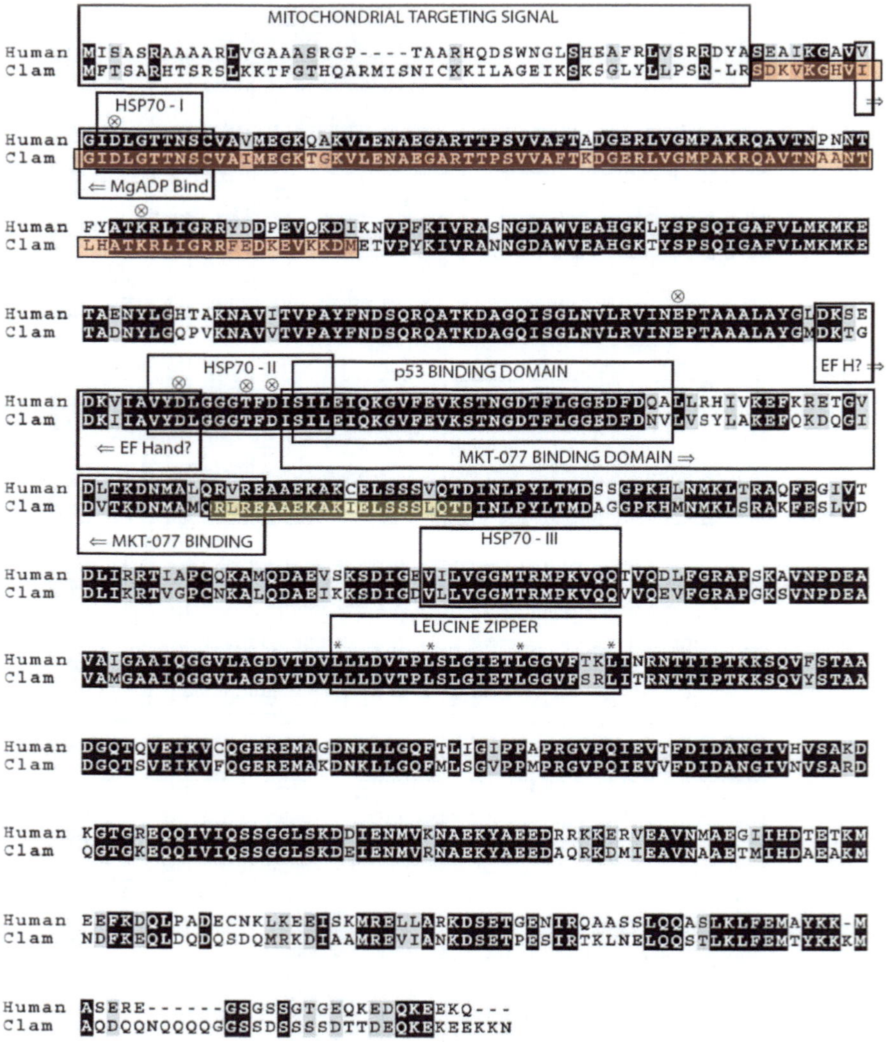

Fig. 6.3 Sequence for full-length and truncated clam mortalin variants compared with human mortalin sequence. Authentic clam *mortalin* sequence (193 bp) was obtained using degenerate primers F (AAT TCG GNG ARG AYT TYG A) and R (GCT TAC ATN CCR TTN GCR TCD AT) designed from a *mortalin* ortholog of *Caenorhabditis elegans* (GenBank accession, NP-504291). Extension of the clam *mortalin* sequence was accomplished using RLM-mediated RACE (Ambion) and modified 3′ and 5′ RACE techniques (Invitrogen). Exon 3 that is missing from the truncated variant of clam mortalin is delineated in *red*. Location of the 20mer peptide used in generating the polyclonal antibody to authentic clam mortalin sequence is indicated in *yellow*

```
MFTSARHTSRSLKKTFGTHQARMISNICKKILAGEIKSKSGLYLLPSRLR
SDKVKGHVIGIDLGTTNSCVAIMEGKTGKVLENAEGARTTPSVVAFTKDG
ERLVGMPAKRQAVTNAANTLHATKRLIGRRFEDKEVKKDMETVPYKIVRAN
NGDAWVEAHGKTYSPSQIGAFVLMKMKETADNYLGQPVKNAVVTVPAYFND
SQRQATKDAGQISGLNVLRVINEPTAAALAYGMDKTGDKIIAVYDLGGGT
FDISILEIQKGVFEVKSTNGDTFLGGEDFDNVLVSYLAKEFQKDQGIDVTK
DNMAMQRLREAAEKAKIELSSSLQTDINLPYLTMDAGGPKHMNMKLSRAKF
ESLVDDLIKRTVGPCNKALQDAEIKKSDIGDVLLVGGMTRMPKVQQVVQEV
FGRAPGKSVNPDEAVAMGAAIQGGVLAGDVTDVLLLDVTPLSLGIETLGGV
FSRLITRNTTIPTKKSQVYSTAADGQTSVEIKVFQGEREMAKDNKLLGQFM
LSGVPPMPRGVPQIEVVFDIDANGIVNVSARDQGTGKEQQIVIQSSGGLS
KDEIENMVRNAEKYAEEDAQRKDMIEAVNAAETMIHDAEAKMNDFKEQLDQ
DQSDQMRKDIAAMREVIANKDSETPESIRTKLNELQQSTLKLFEMTYKKK
MAQDQQNQQQQGGSSDSSSSDTTDEQKEKEEKKN
```

Fig. 6.4 Exon distribution in full-length clam mortalin. Exon junctions derived from an alignment of mortalin proteins from clam, human, *C. elegans* and *Drosophila*. One box indicates the location of an exon junction from human sequence only. Two boxes indicate human and at least one of the other sequences contained a junction at that location and three boxes indicate that all three sequences contained an exon-exon junction at that location. Exon 3 that is missing from the truncated variant of clam mortalin is delineated in *gray*. Location of the 20mer peptide used in generating the polyclonal antibody to authentic clam mortalin sequence is indicated in *yellow*

identical. The six amino acids identified as essential components of the ATPase active site are all retained in both the full-length clam mortalin variant and human mortalin (D^{62}, K^{124}, E^{225}, D^{247}, T^{252}, D^{254}) (McKay et al. 1994). In the C-terminal substrate-binding domain, the p53-binding domain is also 93% identical, and the MKT-077 binding domain is 75% identical, EF-hand motif (77%) and leucine zipper motif (92%) are all highly conserved with human sequence.

One splice variant was also identified (GenBank Accession # EF576660). This truncated variant is identical to the full-length clam mortalin variant except that 90 residues, representing exon 3, are missing in its N-terminal domain (residues 51–140) perhaps through alternative splicing. Genomic comparison of intron/exon structure of the truncated clam mortalin variant with distantly related homologs (human, fly, snail and worm) identified semi-conserved splice sites (Fig. 6.4). Two of these splice sites align with the points of divergence and convergence between the two clam variants. These data are consistent with the "exon cassette" theory of alternative splicing in which an exon is removed along with the neighboring introns during mRNA processing (Black 2003). Additionally, both full-length and truncated variants of clam mortalin contain a mitochondrial targeting sequence (MTS), which distinguishes them from other Hsp70 family members. Two of the six essential amino acids (D^{62}, K^{124}) are missing from the nucleotide-binding domain, as is the first Hsp70 motif. However, the p53 substrate-binding domain is completely intact.

6.4 Function of Mortalin in Soft Shell Clam Hemocyte Cancer

The soft shell clam, *Mya arenaria*, naturally develops a diffuse cancer of its hemocytes at high incidence (1–20% of populations) (Barber 1996; Walker et al. 2009). This fatal disease is found in clams at multiple sites along the coasts of all states between Maryland and Maine as well as in eastern Canada (Böttger et al. In prep.). Unlike normal clam hemolymph that contains phagocytic, terminally differentiated granular and non-granular hemocytes (Fig. 6.5a), fully cancerous clam hemolymph contains 100% CCH ($\sim 5 \times 10^8$ cells ml^{-1}) with an average diameter of 7–10 μm and a mean nuclear to cytoplasmic volume ratio of 1:1 (Fig. 6.5b). These nearly round, mitotic hemocytes have a monotonous appearance, lack large pseudopodia, attach only loosely with thin cytoplasmic projections to plastic or glass and are not motile nor phagocytic. CCH express a highly conserved homolog for human wild-type p53 protein (Kelley et al. 2001; Walker et al. 2009) that is rendered non-functional through sequestration in the cytoplasm by mortalin protein when the latter is over-expressed (Walker et al. 2006) (Fig. 6.5c, d). Expression of clam mortalin is 1,634 times higher in CCH than in normal clam hemocytes as determined by quantitative PCR (Böttger et al. 2008). Cytoplasmic sequestration does not occur in normal clam hemocytes (NCH; Fig. 6.5c) (Kelley et al. 2001; Walker et al. 2006).

6.4.1 *Mortalin and Cytoplasmic Sequestration of p53 in Clam Hemocyte Cancer*

In Western blot analyses of protein lysates prepared by NE/PER nuclear and cytoplasmic extraction (Pierce) and reacted with an antibody to authentic clam mortalin sequence (see 20 mer peptide used to generate this antibody in Figs. 6.3 and 6.4), the band that coincides with the truncated mortalin splice variant appears to be in the nuclear lysate (indicative of close proximity or even interaction with the nuclear membrane), although p53 is absent from the nucleus in these same cells (Fig. 6.6a, b). One explanation for this observation is that the nuclear localization signal of p53 binds to importin proteins that bring the complex to the nucleus and may even bind to the nuclear pore or begin to import the clam p53 protein, but that truncated clam mortalin (lacking the ATPase domain) cannot release p53 and as a result plugs the

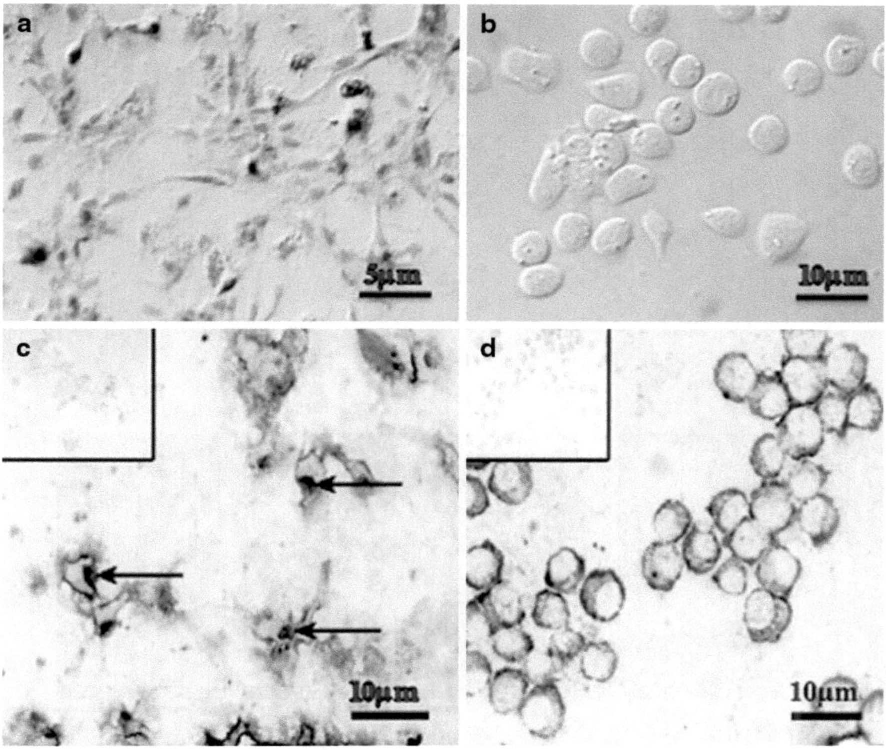

Fig. 6.5 Clam hemocyte cancer. **a** Interference image of freshly collected normal clam hemocytes (NCH) showing both granular and agranular, terminally differentiated hemocytes. **b** Interference image of freshly collected cancerous clam hemocytes (CCH). **c** NCH showing primarily nuclear localization of clam p53 (*arrows*) using an *M. arenaria* polyclonal antibody (Kelley et al. 2001). **d** CCH showing localization of p53 in the cytoplasm and absence from the nucleus using the same polyclonal antibody (see clear nuclei; insets in c and d are negative controls without primary antibody). Scale bars: a = 5 µm; b–d = 10 µm. Figure 6.5c, d from Kelley et al. 2001

nuclear pores. This may explain why the truncated variant of clam mortalin is found in a lysate that isolates only nuclear proteins. Phenotypically the result is a perinuclear distribution of mortalin and p53 family members in CCH that is identical to that seen in a subset of human and mouse cancers (Wadhwa et al. 1993b, 2006).

6.4.2 Mortalin in Transcriptional Induction of Apoptosis by p53 in Clam Hemocytes

Our observations have shown that genotoxic stress induced by treatment of CCH with etoposide promotes *de novo* transcription of wild-type p53 (602-fold increase over untreated CCH), translocation of p53 protein to the nucleus (Fig. 6.7a, b), followed by DNA damage (Fig. 6.7c, d) and apoptosis (Böttger et al. 2008).

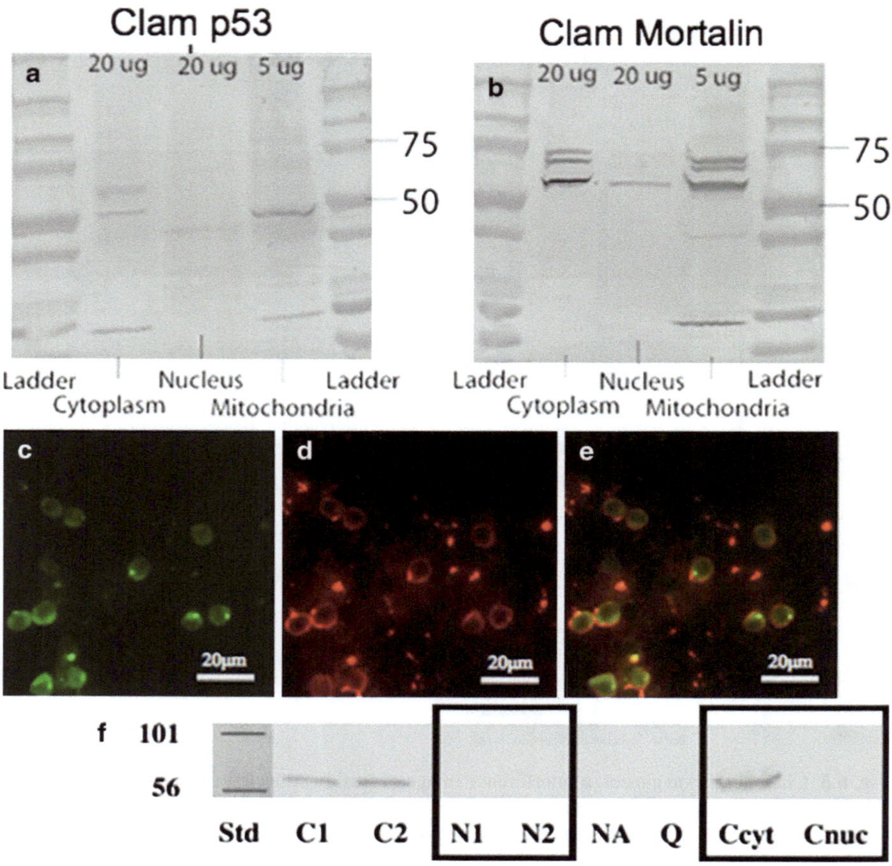

Fig. 6.6 Involvement of mortalin in cytoplasmic sequestration of p53 protein in cancerous clam hemocytes. **a** Western blots of protein lysates prepared by NE/PER nuclear and cytoplasmic extraction (Pierce) and reacted with clam p53 polyclonal antibody (Kelley et al. 2001). **b** Western blot analyses of protein lysates prepared by NE/PER nuclear and cytoplasmic extraction (Pierce) and reacted with the clam mortalin polyclonal antibody (see Fig. 6.3 for location of the peptide used to generate this antibody this antibody recognizes both full length and truncated variants of clam mortalin). **c** Quantum-dot immunocytochemical localization in cytoplasmically sequestered p53 in CCH. **d** Quantum-dot immunocytochemical localization in cytoplasmic mortalin in the same CCH. **e** Dual quantum-dot immunocytochemical localization of cytoplasmically sequestered p53 and mortalin in the same CCH—*yellow* indicates co-localization of the two proteins (Scale bar in 6c–e—20 µm). **f** Co-immunoprecipitation of clam p53 and mortalin in NCH and CCH. Std = protein standard, C1&2 = first and second eluate from CCH lysates, N1&2 = first and second eluate from NCH lysates, NA = first eluate from gel loaded with NCH and CCH lysate mix (negative control), Q = first elude from quenched gel loaded with NCH and CCH lysate mix (negative control), Ccyt = cytoplasmic protein extract from CCH, Cnuc = nuclear protein extract from CCH. Figure 6.6c–f from Walker et al. 2006

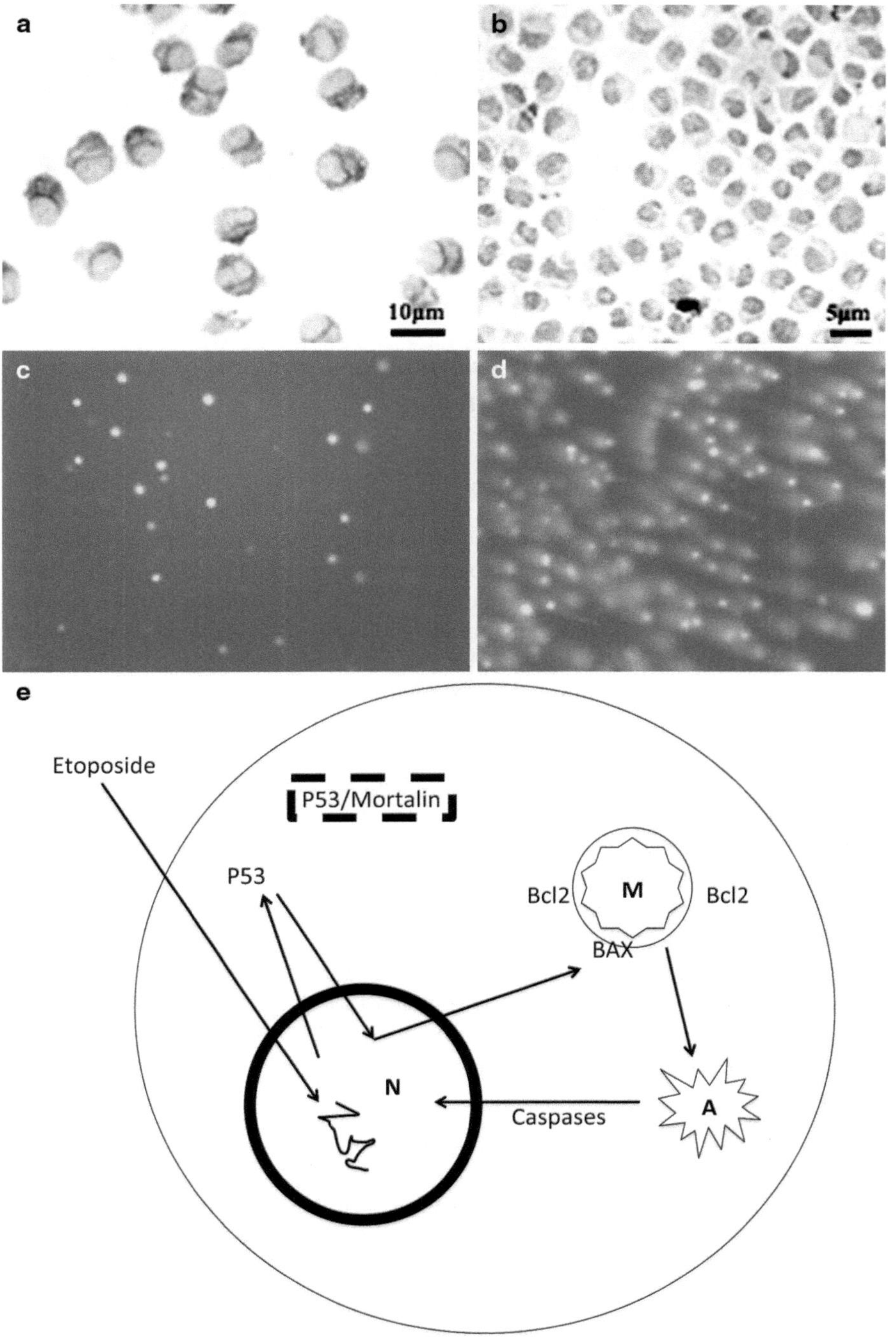

Fig. 6.7 Transcriptional initiation of apoptosis by soft shell clam p53. **a** Zero time control for CCH in which clam p53 protein is sequestered in the cytoplasm with a perinuclear distribution (notice the clear nuclei). **b** CCH from the same individual as 7a in which clam p53 protein has re-localized to the nucleus following treatment with 0.075 mM etoposide for 18 h. **c** Comet assay showing no

The most parsimonious interpretation of these data is that treatment with etoposide yields DNA damage that ultimately elevates p53 protein levels in the cytoplasm of CCH. *De novo* p53 protein overwhelms mortalin tethering, resulting in nuclear translocation of some p53 followed by DNA damage and apoptosis. Following treatment with etoposide, transcriptional targets of p53 would be activated to evaluate and repair DNA, or failing that p53 target proteins would induce the apoptosis of CCH that we have observed (Fig. 6.7e).

6.4.3 *Mortalin and Non-transcriptional Induction of Apoptosis by p53 in Clam Hemocytes*

In cells with severely damaged DNA, apoptosis usually results, while cells with repairable DNA can continue through the cell cycle. Recent studies in mouse and human cells have demonstrated that in addition to its role as a transcription factor, p53 has a direct, non-transcriptional apoptogenic role at the mitochondria (Becker et al. 2007, 2008; Marchenko et al. 2007; Talos et al. 2005). Wild type p53 exists as a dimer or oligomer at the mitochondrial membrane and binds via its DNA binding domain with Bcl2 forming an inhibitory complex. Such binding disrupts the anti-apoptotic Bcl2/Bax ratio and results in outer mitochondrial membrane permeability and the release of toxic peptides from the outer compartment of mitochondria resulting in apoptosis.

Treatment of CCH with MKT-077 only, a chemotherapeutic rhodocyanin, demonstrates that clam p53 (detected by a polyclonal antibody to clam authentic p53 sequence) (Kelley et al. 2001) that is normally found only in the cell cytoplasm is translocated to the nucleus and that DNA damage and apoptosis of CCH follow (Fig. 6.8a, b). When pre-treated with wheat germ agglutinin (WGA) to block nuclear pores, followed by treatment with MKT and reaction with clam p53 and mortalin polyclonal antibodies both proteins were found in the cytoplasm and at the mitochondria but not in the nucleus. Treatment of CCH with MKT-077 *in vitro* and *in vivo* resulted in translocation of clam p53 to mitochondria (Fig. 6.8a). Treatment with MKT-077 also induced apoptosis, as detected by the TUNEL assay (data not shown).

←

evidence of DNA damage in control NCH after 18 h following sham injection with carrier minus etoposide. **d** Comet assay showing extensive DNA damage in CCH from the same individual as in 8a following treatment with 0.075 mm etoposide for 18 h. **e** Model suggesting that treatment with etoposide yields DNA damage that ultimately elevates p53 levels in the cytoplasm of CCH. *De novo* p53 protein overwhelms mortalin tethering (indicated by box around p53/mortalin), resulting in nuclear translocation of some p53 followed by DNA damage and apoptosis. Following treatment with etoposide, transcriptional targets of p53 would be activated to evaluate and repair DNA, or failing that p53 target proteins would induce the apoptosis of CCH that we have observed. Pro-apoptotic Bax proteins from the general cytoplasm are then recruited to the mitochondria, releasing toxic compounds like caspases 9 and 3 that participate in apoptosis. N, nucleus; M, mitochondrion before apoptosis; A, apoptotic mitochondrion. (Scale bars for a, b = 10 μm)

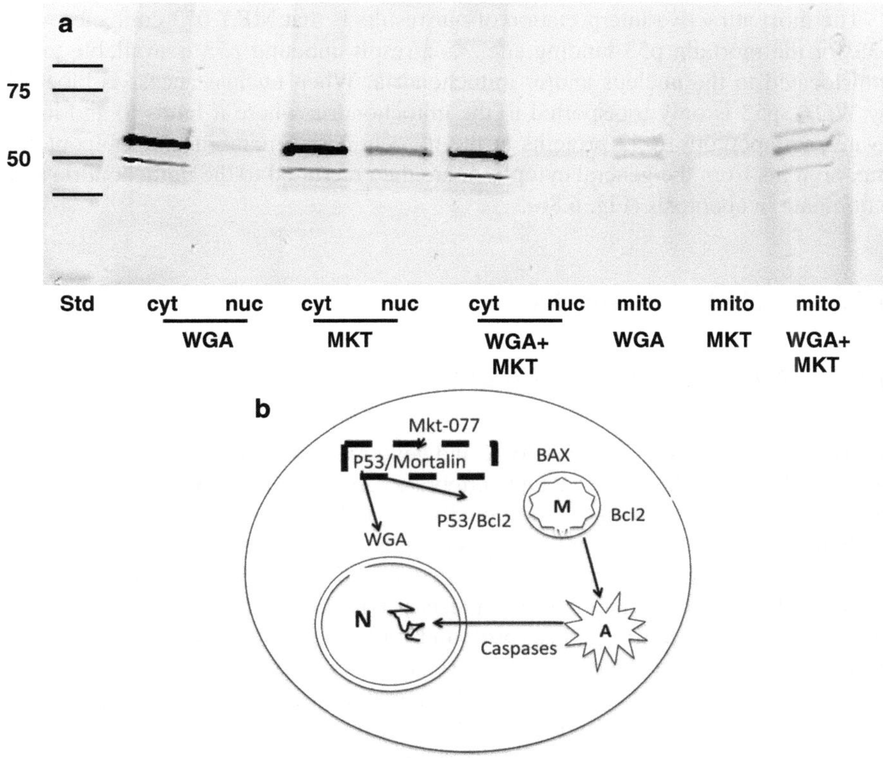

Fig. 6.8 Non-transcriptional initiation of apoptosis by soft shell clam p53. **a** Distribution of p53 in cytoplasmic, nuclear and mitochondrial protein extracts from CCH demonstrated using an *M. arenaria* polyclonal antibody (Kelley et al. 2001) following treatment with MKT-077 for 6 h. Treatment with MKT-077 occurred both with and without wheat germ agglutinin (WGA) pre-treatment. Pre-treatment of CCH with WGA to block nuclear import employed FITC-labeled WGA (Sigma) transfected into CCH using the Chariot protein delivery system (Active Motif). WGA was allowed to couple with the Chariot compound for 30 min at room temperature to form complexes that were incubated for 1 h with CCH suspended in culture medium at concentrations of 6×10^5 CCH/100 µl medium. Cytoplasmic, nuclear and mitochondrial protein extracts from CCH were treated as follows: for nuclear/cytoplasmic extracts—wheat germ agglutinin (WGA) only, 3.5 µM MKT-077 only, pre-treatment with WGA followed post-treatment with 3.5 µM MKT-077; and for mitochondrial extracts: WGA only, with 3.5 µM MKT-077 only and with 3.5 µM MKT-077 following pre-treatment with WGA. Treatment with wheat germ agglutinin (WGA) blocks nuclear access for p53 and results in its translocation to the mitochondrion where apoptosis of CCH follows. When normal clams are treated identically, p53 is transported to the mitochondria, but apoptosis does not result. Std = protein standard, nuc = nuclear protein, cyt = cytoplasmic protein. **b** A model suggesting that MKT-077 competes with p53 for the mortalin p53-binding site. As a result unbound p53 is available to be translocated to the nucleus and/or the mitochondria. Since nuclear access is blocked by WGA, p53 is only transported to the mitochondria where it binds to and inactivates antiapoptotic Bcl2 proteins in the mitochondrial membrane. Pro-apoptotic Bax proteins from the general cytoplasm are then recruited to the mitochondria, releasing toxic compounds like caspases 9 and 3 that participate in apoptosis. N, nucleus; M, mitochondrion before apoptosis; A, apoptotic mitochondrion. Figure 6.8a from Walker and Böttger 2008

The most attractive interpretation of our results is that MKT-077 competes with p53 for the mortalin p53-binding site. As a result unbound p53 is available to be translocated to the nucleus and/or mitochondria. When nuclear access is blocked by WGA, p53 is only transported to the mitochondria where it binds to and inactivates antiapoptotic Bcl2 proteins at the mitochondrial membrane. Pro-apoptotic Bax proteins from the general cytoplasm are then recruited to the mitochondria and participate in apoptosis (Fig. 6.8b).

6.5 Concluding Remarks

6.5.1 Relevance of Mortalin in CCH to Human Cancers

Analyzing the effects of non-genotoxic and genotoxic compounds on malignancy is currently limited to highly regulated vertebrate cancer models where tumors are induced (e.g., mouse; zebra fish). Additionally, while they may exist, no mouse models for human colon carcinoma or neuroblastoma have been identified that specifically address the role of mortalin in cytoplasmic sequestration of p53. Bruce Alberts has recently made a case for redefining cancer research. He makes the point that, "... often the discoveries that have the most profound impact on cancer treatments emanate from basic research on model organisms, rather than from studies of highly complex tumors". Clam hemocyte cancer is the only animal model thus far investigated where cytoplasmically sequestered wild-type p53 can be reactivated both *in vitro* and *in vivo* using both genotoxic and non-genotoxic therapies.

To survive in an otherwise healthy organism, cancer cells must develop blocks to both the transcriptional and non-transcriptional mitochondrial apoptotic-signaling pathways (Deng et al. 2007). Where cytoplasmic sequestration inhibits the action of wild-type p53, two molecular mechanisms have been proposed to account for such blocks in human neuroblastoma, colon carcinoma and clam hemocyte cancer cells. These blocks depend either upon: (a) overexpression of the mitochondrial Hsp70 protein mortalin, that complexes with and tethers wild-type p53 protein in the cytoplasm of human and CCH (Dundas et al. 2005; Wadhwa et al. 2006; Walker et al. 2006) or (b) hyperubiquination of wild-type p53 protein that prevents translocation of wild-type to the mitochondria (Becker et al. 2007, 2008; Marchenko et al. 2007). Evidence for hyperubiquination of wild-type p53 and its contribution to cytoplasmic sequestration of wild-type p53 comes from experiments with human neuroblastoma (e.g., IMR32, SKN-SH, CHP134, LAN5) and colorectal cancer cell lines (e.g., HCT116, RKO) (Moll et al. 1992, 1995, 1996). Normal clam hemocytes display both transcriptional and non-transcriptional apoptotic pathways and both of these pathways are functionally impaired in CCH (Walker et al. 2009; Walker and Böttger 2008).

No data are currently available for the ubiquination state of wild-type p53 in CCH. Among the questions that have not been addressed for human or clam cancer cells regarding the reactivation of cytoplasmically sequestered wild-type p53, include: (1)

are mortalin and hyperubiquination mechanisms identified in clam and human cancer cells respectively, common to both kinds of cancer cells and (2) are these two cancer promoting mechanisms interrelated and do they interact to promote cytoplasmic sequestration of wild-type p53?

6.5.2 *Mechanisms for Translocating p53 to Mitochondria: Is Mortalin Involved?*

Mechanisms governing the localization of p53 within mammalian cells are poorly understood. Both Hsp70 and Hsp90 participate with other proteins in multicomponent complexes that mediate translocation of cytoplasmic p53 to the nucleus using the molecular motor dynein (Galigniana et al. 2004). It is unknown whether the Hsp70 protein involved is mortalin. Even less understood is how severe stress can lead to mitochondrial translocation of p53 and non-transcriptional induction of apoptosis in CCH. Certainly the dynein/Hsp90/Hsp70 dependent complex just described might be involved and the mitochondrial chaperoning function of mortalin suggest close interaction between these two proteins. But, since many mitochondria are distributed peripherally in the cytoplasm and are nowhere near the nucleus, such dynein dependent protein complexes cannot be the only mechanism for translocating p53. While it is unknown even for human cancers what alternative multicomponent or other complexes involving mortalin and perhaps kinesin are involved in mitochondrial translocation, recent studies have implicated both the DNAJ-like protein Tid 1 (in mammals, a co-chaperone of Hsp70) and loss of hyperubiquination in this process for several human colon carcinoma and breast cancer cell lines (Ahn et al. 2010; Kampinga and Craig 2010; Marchenko et al. 2007). Further research is warranted on these lines to resolve the molecular role and kinetics of mortalin in mitochondrial functions for cell survival and death.

Acknowledgments This research was supported by National Cancer Institute grants (CA71008-01 and CA104112-01), UNH Sea Grant (R/FMD-166) and UNH Hatch Grant (353) to CWW.

References

Ahn BY, Trinh DL, Zajchowski LD, Lee B, Elwi AN, Kim SW (2010) Tid1 is a new regulator of p53 mitochondrial translocation and apoptosis in cancer. Oncogene 29:1155–1166

Akerfelt M, Morimoto RI, Sistonen L (2010) Heat shock factors: integrators of cell stress, development and lifespan. Nat Rev Mol Cell Biol 11(8):545–555

Alvares K, Dixit SN, Lux E, Barss J, Veis A (2007) The proteome of the developing tooth of the sea urchin, Lytechinus variegatus: mortalin is a constituent of the developing cell syncytium. J Exp Zool B Mol Dev Evol 308:357–370

Barber B (1996) Effects of gonadal neoplasms on oogenesis in softshell clams, *Mya arenaria*. J Invert Pathol 67:161–168

Becker K, Marchenko ND, Maurice M, Moll UM (2007) Hyperubiquitylation of wild-type p53 contributes to cytoplasmic sequestration in neuroblastoma. Cell Death Differ 14:1350–1360

Becker K, Marchenko ND, Palacios G, Moll UM (2008) A role of HAUSP in tumor suppression in a human colon carcinoma xenograft model. Cell Cycle 7:1205–1213

Black DL (2003) Mechanisms of alternative pre-messenger RNA splicing. Annu Rev Biochem 72:291–336

Böttger S, Jerszyk E, Low B, Walker C (2008) Genotoxic stress-induced expression of p53 and apoptosis in leukemic clam hemocytes with cytoplasmically sequestered p53. Cancer Res 68:777–782

Böttger SA, Taraska N, Lock N, Walker CW (In prep.). Frequencies of neoplasia in Mya arenaria along the East Coast of North America. Diseases of Aquatic Organisms.

Conte M, Deri P, Isolani ME, Mannini L, Batistoni R (2009) A mortalin-like gene is crucial for planarian stem cell viability. Dev Biol 334:109–118

Deng J, Carlson N, Takeyama K, Dal Cin P, Shipp M, Letai A (2007) BH3 profiling identifies three distinct classes of apoptotic blocks to predict response to ABT-737 and conventional chemotherapeutic agents. Cancer Cell 12:171–185

Dundas SR, Lawrie LC, Rooney PH, Murray GI (2005) Mortalin is over-expressed by colorectal adenocarcinomas and correlates with poor survival. J Pathol 205:74–81

Deocaris CC, Kaul SC, Wadhwa R (2006) On the brotherhood of the mitochondrial chaperones mortalin and heat shock protein 60. Cell Stress Chaperones 11:116–128

Galigniana MD, Harrell JM, O'Hagen HM, Ljungman M, Pratt WB (2004) Hsp90-binding immunophilins link p53 to dynein during p53 transport to the nucleus. J Biol Chem 279:22483–22489

Gupta RS, Ramachandra NB, Bowes T, Singh B (2008) Unusual cellular disposition of the mitochondrial molecular chaperones Hsp60, Hsp70 and Hsp10. Novartis Found Symp 291:59–68

He S, Yang L, Lv Z, Hu W, Cao J, Wei J, Sun X, Yang J, Zheng H, Wu Z (2010) Molecular and functional characterization of a mortalin-like protein from *Schistosoma japonicum* (SjMLP/hsp70) as a member of the HSP70 family. Parasitol Res 107:955–966

Hendrick JP, Hartl FU (1993) Molecular chaperone functions of heat-shock proteins. Annu Rev Biochem 62:349–384

Huang P, Oliff A (2001) Signaling pathways in apoptosis as potential targets for cancer therapy. Trends Cell Biol 11:343–348

Kampinga HH, Craig EA (2010) The HSP70 chaperone machinery: J proteins as drivers of functional specificity. Nat Rev Mol Cell Biol 11:579–592

Kaul SC, Deocaris CC, Wadhwa R (2007) Three faces of mortalin: a housekeeper, guardian and killer. Exp Gerontol 42:263–274

Kelley ML, Winge P, Heaney JD, Stephens RE, Farell JH, Van Beneden RJ, Reinisch CL, Lesser MP, Walker CW (2001) Expression of homologues for p53 and p73 in the softshell clam (Mya arenaria), a naturally-occurring model for human cancer. Oncogene 20:748–758

Marchenko ND, Wolff S, Erster S, Becker K, Moll UM (2007) Monoubiquitylation promotes mitochondrial p53 translocation. EMBO J 26:923–934

Mashanov VS, Zueva OR, Rojas-Catagena C, Garcia-Arraras JE (2010) Visceral regeneration in a sea cucumber involves extensive expression of survivin and mortalin homologs in the mesothelium. BMC Dev Biol 10:117

McKay DB, Wilbanks SM, Flaherty KM, Ha JH, O'Brien MC, Shrivanee LL (1994) Stress-70 proteins and their interaction with nucleotides. In: Morimoto RI (ed) The biology of heat shock proteins and molecular chaperones. Cold Spring, Harbor, pp 153–177

Moll UM, Riou G, Levine AJ (1992) Two distinct mechanisms alter p53 in breast cancer: mutation and nuclear exclusion. Proc Natl Acad Sci U S A 89:7262–7266

Moll UM, LaQuaglia M, Benard J, Riou G (1995) Wild-type p53 protein undergoes cytoplasmic sequestration in undifferentiated neuroblastomas but not in differentiated tumors. Proc Natl Acad Sci U S A 92:4407–4411

Moll UM, Ostermeyer AG, Haladay R, Winkfield B, Frazier M, Zambetti G (1996) Cytoplasmic sequestration of wild-type p53 protein impairs the G1 checkpoint after DNA damage. Mol Cell Biol 16:1126–1137

Morano KA (2007) New tricks for an old dog: the evolving world of Hsp70. Ann N Y Acad Sci 1113:1–14

Nagpal J, Jamoona A, Gulati ND, Mohan A, Braun A, Murali R, Jhanwar-Uniyal M (2006) Revisiting the role of p53 in primary and secondary glioblastomas. Anticancer Res 26:4633–4639

Nikolaev AY, Li M, Puskas N, Qin J, Gu W (2003) Parc: a cytoplasmic anchor for p53. Cell 112:29–40

Talos F, Petrenko O, Mena P, Moll UM (2005) Mitochondrially targeted p53 has tumor suppressor activities in vivo. Cancer Res 65:9971–9981

Wadhwa R, Kaul SC, Mitsui Y, Sugimoto Y (1993a) Differential subcellular distribution of mortalin in mortal and immortal mouse and human fibroblasts. Exp Cell Res 207:442–448

Wadhwa R, Kaul SC, Ikawa Y, Sugimoto Y (1993b) Identification of a novel member of mouse hsp70 family. Its association with cellular mortal phenotype. J Biol Chem 268:6615–6621

Wadhwa R, Takano S, Kaur K, Deocaris CC, Pereira-Smith OM, Reddel RR, Kaul SC (2006) Upregulation of mortalin/mthsp70/Grp75 contributes to human carcinogenesis. Int J Cancer 118:2973–2980

Walker CW, Böttger SA (2008) A naturally occurring cancer with molecular connectivity to human diseases. Cell Cycle 7:2286–2289

Walker C, Böttger S, Low B (2006) Mortalin-based cytoplasmic sequestration of p53 in a nonmammalian cancer model. Am J Pathol 168:1526–1530

Walker C, Böttger SA, Mulkern J, Jerszyk E, Litvaitis M, Lesser M (2009) Mass culture and characterization of tumor cells from a naturally occurring invertebrate cancer model: Applications for human and animal disease and environmental health. Biol Bull 216:23–39

Yee KS, Vousden KH (2005) Complicating the complexity of p53. Carcinogenesis 26:1317–1322

Yokoyama K, Fukumoto K, Murakami T, Harada S, Hosono R, Wadhwa R, Mitsui Y, Ohkuma S (2002) Extended longevity of Caenorhabditis elegans by knocking in extra copies of hsp70F, a homolog of mot-2 (mortalin)/mthsp70/Grp75. FEBS Lett 516:53–57

Chapter 7
Mortalin and Stem Cells: A Study from Planarians

Renata Batistoni

Abstract Adult stem cells are fundamental for physiological tissue renewal and regeneration after injury. The failure of stem cells to substitute dead or damaged cells is seen as one mechanism that limits the longevity of organisms. Planarian flatworms provide a unique model system to investigate in vivo the molecular machinery that controls adult stem cell functions in tissue renewal and repair. The extraordinary ability of planarians to regenerate whole animals from very small body fragments is in fact strictly associated with the unlimited growth potential of pluripotent stem cells, referred to as neoblasts. Neoblasts are distributed throughout the body and constantly undergo division to support continuous turnover of all specialized cell types. Moreover, upon amputation, these cells activate extensive proliferation to form the regenerative blastema, from which missing parts are regenerated. Damage or reduction in number of neoblasts deeply affects regeneration and planarian survival. Even though some progress has been made in identifying the genes that regulate self-renewal capacity and commitment of these cells, little is known about how these cells support continuous proliferative activity and long-term stability, maintaining control of cell division. I discuss evidence suggesting that *Djmot*, a planarian mortalin-like gene, significantly contributes to the mechanisms that regulate unlimited growth potential of stem cells in vivo, and I advance the hypothesis that its product plays a major role in keeping the activity of p53 under control in these cells.

Keywords *Djmot* · Molecular pathways · Planarians · Regeneration · Stem cells

7.1 Introduction

> If there were no regeneration there could be no life, if everything regenerated there would be no death. All organisms exist between these two extremes (Goss 1969).

Adult stem cells are defined as unspecialized cells that are capable of long term self-renewal and differentiation into specialized cell types. Their activity maintains tissue homeostasis and repair throughout the lifetime of an organism. In some species stem

---

R. Batistoni (✉)
Dipartimento di Biologia, Unità di Biologia Cellulare e dello Sviluppo, Università di Pisa, S.S.12 Abetone e Brennero 4, 56127 Pisa, Italy
e-mail: rbatistoni@biologia.unipi.it

S. C. Kaul, R. Wadhwa (eds.), *Mortalin Biology: Life, Stress and Death*,
DOI 10.1007/978-94-007-3027-4_7, © Springer Science+Business Media B.V. 2012

cells show remarkable regenerative potential regenerating limbs, tails or producing entire new individuals from small body fragments. Mammals, including humans, do not possess the same ability, even if some regenerative response to replace developing limb buds or digit tips has been observed in embryos and neonatal mice (Muneoka and Sassoon 1992; Muller et al. 1999). Gradual decline in the regenerative potential throughout lifetime has also been observed in aged mammals that do not replace tissues damaged by stress or injury as efficiently as in the young. Such changes have been attributed to an age-related decline in the ability of resident stem cells to engage in tissue homeostasis/repair. The molecular mechanisms associated with deficits in organ stem cell function are for the most part unknown. Knowledge of how adult stem cells change with advancing age, i.e. identifying, and ultimately manipulating, the genes that regulate stem cell number, replication rate, and self-renewal capacity, may provide novel therapeutic perspectives for age-related diseases and for improvement of tissue regeneration (Sharpless and Schatten 2009). Recent research supports the idea that tumor suppressor mechanisms may be important drivers of the functional decline of tissue stem cells, causing tissue degeneration and aging (Pearson and Sánchez Alvarado 2009). Conversely, failures in apoptosis or senescence can result in hyperproliferative diseases, such as cancer. How is biological age regulated in those cells capable of producing new progeny in animal models of regeneration? Understanding the mechanisms that underlie the unlimited growth potential of stem cells of these creatures has profound implications for the field of regenerative medicine, which seeks to identify and utilize regenerative mechanisms for repairing or replacing damaged human tissues. Here I present a suitable model organism in which this subject can be studied: the planarian flatworms.

7.2 The Extraordinary Stem Cell System of Planaria

In order to make the reader familiar with the planarian model system, it may be helpful to provide a brief description here. Planarians are free-living flatworms belonging to the phylum Plathyelminthes. Although these worms look nothing like a human, the systematic position as lower Bilateria places them in key position to vertically integrate molecular and genetic studies performed in planaria to mammals. Moreover, planarian genome shares a large number of genes with the human genome, including human sequences involved in various aspects of human biology and disease, that were not found in other classically used genetic systems, as *Caenorhabditis elegans* or *Drosophila melanogaster* (Spradling et al. 2006). The most fascinating aspect of planarian biology is represented by the amazing stem cell-based regenerative capabilities. Several entirely new individuals can be regenerated from the fragments of one (Fig. 7.1a, b). About a century ago, Morgan showed in one experiment that 1/279 of an intact planarian could regenerate a complete organism in a short time (Morgan 1901). During regeneration an outgrowth of tissue accumulates to the amputation surface under the wound epithelium as a blastema, a self-organizing structure that produces the missing body parts based on

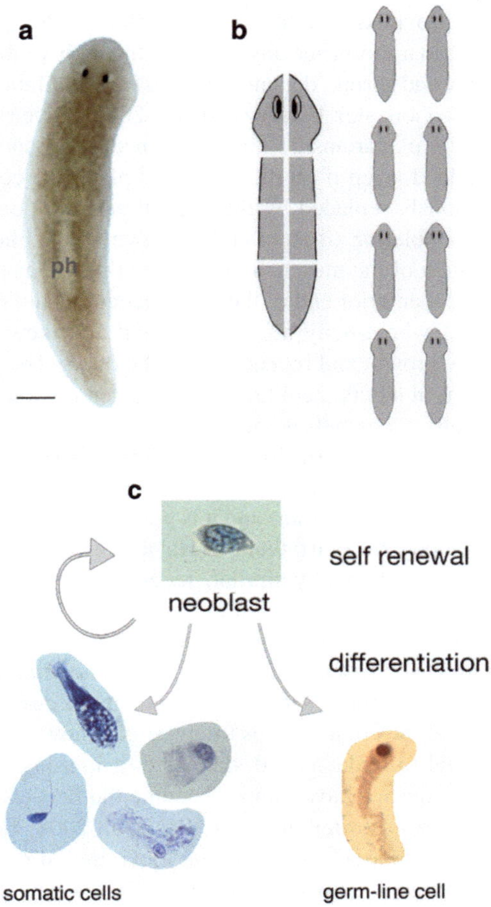

Fig. 7.1 The planarian model system. **a** Dorsal view of the morphological characteristics of *Dugesia japonica*, a planarian species used for molecular studies. Two eyespots can be seen in the head region, dorsal to the brain ganglia. The dorsoventrally flattened planarian body allows visualization of the pharynx, a protrusible muscular structure present on the ventral surface that is used as a feeding organ and for expelling waste substances (ph: pharynx). **b** Regeneration in planaria. Schematic diagram of the amputation procedure: small fragments cut at any level along all body axes can regenerate a new organism. Perfectly proportionate small planarians are formed when the fragments are regenerated. The "new" worms are clones of each other. Anterior is on the top. **c** A planarian stem cell (neoblast). Single cell types can be obtained by cell dissociation of entire worms (cell macerates: Baguñà and Romero 1981). Neoblasts are small, round or pear-shaped cells, morphologically characterized by a large nucleus-to-cytoplasm ratio relative to differentiated cells. Planarian stem cells can self-renew or produce all somatic cells and germ-line cells. Based on cell morphology, it has been estimated that planarians contain approximately 40 cell types (some representative cell types are shown). Scale bars: 1 mm in a, 10 μm in c

positional information provided from the stump (Adell et al. 2010). In addition to the capability to regenerate in a short time any lost part, the body of these animals is sufficiently plastic to grow and shrink, depending upon food availability (Pellettieri and Sánchez Alvarado 2007; González-Estévez and Saló 2010). Given the ability to renew their tissues indefinitely, planarians do not experience aging and have an apparent indefinite lifespan. Indeed, even if all differentiated planarian cells undergo aging, aged cells are continuously replaced by the exceptional stem cell system of these animals. A dynamic population of stem cells, known as neoblasts, is in fact distributed throughout much of the mesenchymal tissue (known as parenchyma), with the exception of the most anterior end of the cephalic region and the pharynx. These cells are responsible of whole-body homeostasis and are likewise responsible for forming new tissues by growth and regeneration (Handberg-Thorsager et al. 2008). Like stem cells resident in adults, neoblasts are considered somatic stem cells, but, like embryonic stem cells, these cells possess the ability to differentiate into any kind of cell of the body, including the germ-line, and are therefore regarded as pluripotent stem cells (Fig. 7.1c). Neoblasts (20–30% of the total cells) are the only cells possessing proliferative ability in planarians and it is possible to selectively disrupt them in vivo through whole-body X-ray irradiation (Hayashi et al. 2006, 2010). Although all neoblasts appear morphologically similar, recent ultrastructural and molecular findings provide evidence that, like in mammals, the planarian stem cell system, including progenitors and lineage-restricted stem cells, is complex and hierarchically ordered (Higuchi et al. 2007; Eisenhoffer et al. 2008). The remarkable regenerative properties of planarians, combined with the experimental accessibility of their stem cells in vivo, represent attractive attributes to investigate how adult stem cells work collectively in an animal to replace aged, damaged or missing organs and tissues. Among planarians, two species have rapidly become the best characterized: *Dugesia japonica* and *Schmidtea mediterranea* (Oviedo et al. 2008; Gentile et al. 2011). A variety of molecular techniques and genomic tools, including a fully sequenced planarian genome (http://genome.wustl.edu/genomes), are now available to dissect the mechanisms that regulate cellular and molecular dynamics of neoblasts (Rossi et al. 2008; Shibata et al. 2010). These cells can be visualised and quantified *in vivo,* and several genes that regulate stem cell maintenance, and differentiation have been functionally characterized by RNA interference (RNAi). The selective destruction of neoblasts by irradiation is a powerful experimental tool in these animals. Microarray approaches allowed comparison of the transcriptional profile of irradiated—free of stem cells—and unirradiated planarians, and identified several stem cell genes, belonging to different functional categories (Rossi et al. 2007). These studies also indicated that self-renewing neoblasts show a transcriptional profile that is different from the transcriptional profiles categorizing neoblast progeny fated to differentiate (Eisenhoffer et al. 2008). Molecular dissection of planarian stem cells also provided clear evidence that the term neoblasts includes distinct subpopulations of stem cells, each of which expresses particular markers and, consequently, could have different biological properties (Rossi et al. 2008).

7.3 Mortalin Gene Expression in Planarian Stem Cells

To obtain a complete picture of how stem cells function, it is important to understand how stem cells react to the impact of time and stress in their environment. Based on the observation that embryonic stem cells in mice (mES) are remarkably resistant to stress and also show high levels of heat shock proteins (HSP) expression, a role for these chaperones in the modulation of stem cell self-renewal and differentiation has been proposed (Prinsloo et al. 2009). We speculated that, similarly to pluripotent mES, also planarian stem cells could possess efficient stress defenses to maintain the stability of their genomes over many cell divisions. Interestingly, our recent work suggests that some HSP, including the heat shock protein-70 (HSP70) family stress chaperone, Mortalin (also known as mtHSP70/PBP74/GRP75), significantly contribute to the mechanisms that regulate stem cell activity in planarians (Conte et al. 2010).

Mortalin is a conserved mitochondrial protein, also localized in multiple extramitochondrial sites, including cytosol, cytoplasmic vescicles, endoplasmic reticulum and plasma membrane (Ran et al. 2000; Kaul et al. 2002). First identified in the cytoplasmic fraction of mouse mortal cells, this protein was then found to have differential distribution in cells with normal and immortal phenotype (Wadhwa et al. 1993, 2002; Deocaris et al. 2007, 2009). Mammalian Mortalin has been implicated in a variety of functions, including stress response, intracellular trafficking, control of cell proliferation and cell fate determination. In addition, its involvement in old age neurological pathologies has been recently demonstrated (Kaul et al. 2007; Deocaris et al. 2008; De Mena et al. 2009). Many cancer and immortal mammalian cells show high levels of mortalin expression. Overexpression of Mortalin significantly contributes to immortalization of the cells (Kaul et al. 1998; Wadhwa et al. 2006), while its knockdown causes growth arrest (Kaul et al. 2006). Furthermore, it has been demonstrated in *C. elegans* that knockdown of a *mortalin* homolog promotes progeria-like phenotypes, while its over-expression causes life span extension (Yokoyama et al. 2002; Kimura et al. 2007). These results indicate that a conserved mechanism, including Mortalin, may contribute to control the longevity both in mammalian and worm cells.

A gene (*Djmot*) encoding a heat-uninducible Mortalin-like protein of 680 amino acids was recently characterized in the planarian *D. japonica* (Conte et al. 2009) and an almost identical gene was also identified in the genome of *S. mediterranea*. In spite of the evolutionary distance, the planarian Mortalin appears very similar to the human Mortalin (Fig. 7.2). Distribution of *Djmot* transcripts was determined in *D. japonica* by in situ hybridization analysis (Fig. 7.3a, b). Expression of this gene appeared specifically localized in neoblasts and early progenitors, while no detectable hybridization signal could be found in differentiated cells. The neoblast-specific transcripts became dramatically downregulated in worms exposed to a dose of X-rays that specifically destroys stem cells, leaving postmitotic cells unaffected. Interestingly, this treatment produced transient induction of *Djmot* expression in intestine and nerve cells. This result demonstrates that, in these experimental conditions, transcription of this gene may be activated in postmitotic cells, probably

```
DjMot      -----------MVSTQSTRGHTSRIMLKMMKLFGQNSLNNYCKNIRLYSDKVKGHVIGID 49
HSPA9      MISASRAAAARLVGAAASRGPTAARHQDSWNGLSHEAFR-LVSRRDYASEAIKGAVVGID 59
                      :*.: ::** *:    .  : :.::::.   ..    *: :** *:***

DjMot      LGTTNSCVSVMDGKTPKVLENAEGKNYT-SVVAFTGDGERLVGTPAKRQAVINSANTLHA 108
HSPA9      LGTTNSCVAVMEGKQAKVLENAEGARTTPSVVAFTADGERLVGMPAKRQAVTNPNNTFYA 119
           ********:**:** .******** . * ******.******* ******* *. **::*

DjMot      TKRLIGRRFDEHEVQKDMKNVSYKIVRASNGDAWVEAQRKMYSPSQIGAFVLMKMKETAE 168
HSPA9      TKRLIGRRYDDPEVQKDIKNVPFKIVRASNGDAWVEAHGKLYSPSQIGAFVLMKMKETAE 179
           ********:*: *****:***.:**************: *:*******************

DjMot      NYLGTAVKNAVITVPAYFNDSQRQATKDAGQIAGLNVLRVINEPTAAALAYGLDKTDDKI 228
HSPA9      NYLGHTAKNAVITVPAYFNDSQRQATKDAGQISGLNVLRVINEPTAAALAYGLDKSEDKV 239
           **** :.*************************:**********************::**:

DjMot      IAVFDLGGGTFDVSVLEIQKGVFEVKSTSGDTFLGGEDFDNVLLQYIVSEFKKEQAVDIT 288
HSPA9      IAVYDLGGGTFDISILEIQKGVFEVKSTNGDTFLGGEDFDQALLRHIVKEFKRETGVDLT 299
           ***:********:*:*************.***********:.**::**.***:* .**:*

DjMot      KDPMAMQRVREAAEKAKIELSSAIQTDINLPYLTMDNTGPKHLSLKLTRSKLEQLVEHLI 348
HSPA9      KDNMALQRVREAAEKAKCELSSSVQTDINLPYLTMDSSGPKHLNMKLTRAQFEGIVTDLI 359
           ** **:*********** ****::************.:*****.:****:::* :* .**

DjMot      KRTIEPCEKVIKDAEVKTSEITEVLLVGGMTRMPRVQETVQKLFGRVPSKSVNPDEAVAM 408
HSPA9      RRTIAPCQKAMQDAEVSKSDIGEVILVGGMTRMPKVQQTVQDLFGRAPSKAVNPDEAVAI 419
           :*** **:*.::****..*:* **:*********:**:***.****.***:********:

DjMot      GAAIQGGVLAGDVTDVLLLDVTPLSLGIETLGGVFTKLINRNTTIPTKKSQVFSTAADGQ 468
HSPA9      GAAIQGGVLAGDVTDVLLLDVTPLSLGIETLGGVFTKLINRNTTIPTKKSQVFSTAADGQ 479
           ************************************************************

DjMot      TQVEIKVHQGEREMAADNKILGQFSLIGIPPAPRGVPQIEVTFDIDANGIVNVSARDKGI 528
HSPA9      TQVEIKVCQGEREMAGDNKLLGQFTLIGIPPAPRGVPQIEVTFDIDANGIVHVSAK---- 535
           ******* *******.***:****:**************************:***:

DjMot      SARDKGTGKEQQIVIQSSGGLSKDEIENMIKRAEENAENDKRKREMVEVVNQAEGIVYDT 588
HSPA9      ---DKGTGREQQIVIQSSGGLSKDDIENMVKNAEKYAEEDRRKKERVEAVNMAEGIIHDT 592
              *****:***************:****:*.**: **:*:**:* **.** ****::**

DjMot      ESKIDEFKSQLPAEECEKLKNLIQQVRDKLANKDTETAENIKKATNELQQSSLKLFEMAY 648
HSPA9      ETKMEEFKDQLPADECNKLKEEISKMRELLARKDSETGENIRQAASSLQQASLKLFEMAY 652
           *:*::***.****:**:***: *.::*: **.**:**.***::*:..***:*********

DjMot      KKMASERESKSSSGSSNDTPHQSESEKKEEQH 680
HSPA9      KKMASEREGSGSSG----TGEQKEDQKEEKQ- 679
           ********...***    * .*.*.:*:*:*
```

Fig. 7.2 Sequence comparison of DjMot with human Mortalin (HSPA9). The full-length *Djmot* sequence obtained in our laboratory encodes a putative Mortalin protein of 680 amino acids (DjMot). This protein shares 77.4% sequence identity with the human Mortalin (HSPA9)

to prevent planarian tissues from damage. Evidence for implication of Mortalin in the radio-adaptive response has also been found in human tumour cell lines, but little is currently known about the precise nature of the underlying mechanism (Carette et al. 2002; Wadhwa et al. 2002). A strong increase of *Djmot* expression was also observed in animals surviving long periods of food depletion. However, in starved planarians, the pattern did not change and remained similar to that observed in intact animals, i.e. confined to neoblasts and early progenitors. Thanks to the activation of autophagic processes, planarians are amazingly resistant to starvation, becoming progressively smaller than the regularly fed animals (González-Estévez and Saló 2010; Conte et al.

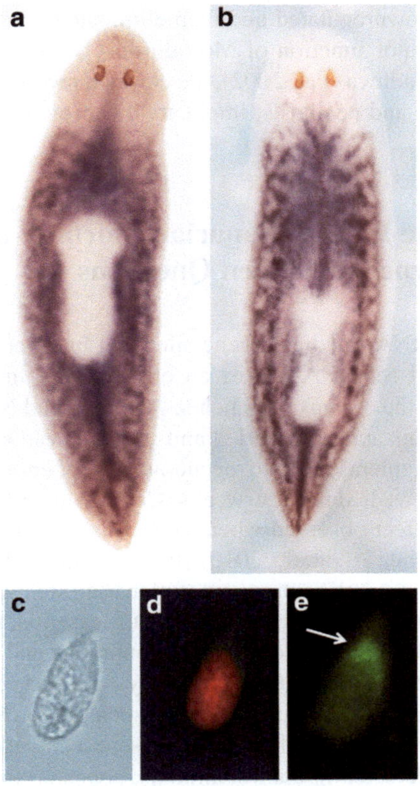

Fig. 7.3 Expression of DjMot in *D. japonica*. **a** Whole-mount in situ hybridization (WISH) using *Djmot* as a probe in an intact planarian. **b** WISH with *Djmcm2*, a gene specifically expressed in proliferating neoblasts (Salvetti et al. 2000) is shown for comparison. Dorsal view, anterior is on the top. WISH experiments were performed as described by Conte et al. (2009). Color development of the alkaline phosphate-coniugated anti-DIG-antibody was carried out with a mixture of BCIP/NBT (Sigma). **c** Phage contrast image of a planarian stem cell. **d** Nucleus staining with ethidium homodimer of the cell depicted in c. **e** Mamot antibody immunostaining of the cell depicted in c: the white arrow shows specific immunofluorescence signal detected in the scant cytoplasm. Scale bar: 1 mm in a and b, 10 μm in c–e

2011). Perfectly orchestrated proliferation, differentiation, autophagy, and apoptosis events remodel or restore the planarian body pattern, allowing these organisms to remain perfectly proportioned and in scale under stressful conditions, such as injury or starvation. As the number of proliferating neoblasts does not significantly change in starved animals (Baguñà 1976; Orii et al. 2005), overexpression of *mortalin* during starvation implies that this chaperone plays a precise, although still unknown, role to adapt planarian stem cells to adverse conditions.

Taken together, this study for the first time illustrates that planarian stem cells, similarly to mammalian ES, constitutively express high levels of *mortalin* transcripts.

These transcripts are downregulated and then eliminated when stem cells undergo differentiation. As a major function of Mortalin has been linked to the control of cellular immortality (Wadhwa et al. 2002), the elevated levels of expression detected both in mammalian ES and neoblasts, implicate this protein in the maintenance of both cell types.

7.4 Dissecting the Role of Planarian Mortalin in Stem Cells: Lessons Learned and Open Questions

The establishment of RNAi-mediated gene silencing by injection of adult planarians with double strand (ds)RNA causes an efficient and incremental effect that spreads to all tissues of the organism (Sánchez-Alvarado and Newmark 1999). After *Djmot*(RNAi), planarians appeared curled and showed head regression. Moreover, when amputated, the fragments failed to regenerate. The phenotypes were irreversible and all injected specimens died in a few weeks (Conte et al. 2009). As ventral curling and lysis is a typical effect observed when stem cells are eliminated from planarians (Pearson and Sánchez Alvarado 2010), these results indicate that *Djmot* is a key player of the self-preservation program of planarian stem cells. Further analysis with different molecular markers highlighted that the essential function(s) played by *Djmot* activity affected different neoblast subpopulations and also early differentiating progeny (Conte et al. 2009). Concomitantly with the permanent cell growth arrest observed in *Djmot*(RNAi) animals, cell dissociation provided evidence of the presence of a variable number of cells showing a senescent-like morphology. These cells, that were never detected in control animals, appeared flattened and enlarged, with condensed chromatin. Cellular senescence is known as an effective mechanism of tumor suppression, and a major role of HSPs in regulation of senescence programs has been recently recognized in cancer cells that constitutively express these chaperones at high levels (O'Callaghan-Sunol et al. 2007; Sherman et al. 2007; Gabai et al. 2009). Of specific interest in the suppression of cell senescence programs is the Mortalin-mediated functional inhibition and sequestration of p53, that has been implicated in the formation of cancer cells (Wadhwa et al. 2002). The p53-mortalin interactions have been the subject of detailed structure-function studies: presence of mortalin in the cytosol promotes sequestration of p53, thereby inhibiting its nuclear activity, essential for the prevention of cancer development. This mechanism efficiently prevents senescence and apoptosis, promoting immortalization of cells (Kaul et al. 2001, 2005, 2007; Walker et al. 2006; Iosefson and Azem 2010).

The p53-binding region is well conserved in DjMot, and a heterologous antibody specifically designed in this region (Mamot antibody: Walker et al. 2006), cross-reacts in neoblast-like cells (Conte et al. 2009, Fig. 7.3c–e). These data suggest that one possible functioning of DjMot in stem cells might be associated to the control of the planarian p53 pathway. Recently a p53 homolog (*Smed-p53*) has been characterized in the planarian *S. mediterranea*. This gene, expressed in stem cells and in newly made stem cell progeny, appears to play a dual role in adult stem cell lineages, as tumor suppressor and also as regulator of self-renewal. This dual role may represent an ancestral trait in a prototypical p53-like molecule, such as SMED-p53, that then split up the different functions among three paralogs (p53, p63 and p73) in the vertebrate lineage (Pearson and Sánchez Alvarado 2010) and is consistent with the role of DjMot as a binding partner. Based on our findings (Conte et al. 2009) I advance the hypothesis that the dramatic reduction of neoblasts and concomitant detection of senescent cells after *Djmot* knockdown could be the consequence of the massive release of p53 into the nucleus. What could be the specific physiological significance of the Mortalin-mediated sequestering of p53—a protein that regulates the cell cycle and hence functions as a tumor suppressor—in planarian stem cell lineages? It is well known that, following DNA damage, p53 transcriptionally activates p21, inducing G_1-phase arrest in mammalian cells. Our knowledge of the evolution of the p53 regulatory pathway is currently very restricted, but a link between cell cycle control and regeneration has been recently demonstrated in MRL mice, where lack of p21 expression converts a nonhealer into a regeneration-competent healer mouse (Bedelbaeva et al. 2010). Interestingly, no p21-like genes have been found in the planarian genome (Pearson and Sánchez Alvarado 2010). These findings suggest that mechanistic variations in the p53 pathway play a role for the regeneration competence in animals. Although most of the control of p53 appears to occur at the transcriptional level in planarians (Pearson and Sánchez Alvarado 2010), it is suggestive to speculate that the function of DjMot as p53-binding protein represents an ancestral post-translational regulatory mechanism, activating or suppressing p53 activity, that these organisms evolved to support the regenerative potential of their stem cells, escaping tumorigenesis.

7.5 Conclusions

Planarian stem cells proliferate indefinitely and their continuous proliferative activity probably involves protective strategies, including activation of members of the HSP family, specifically expressed in planarian stem cells. According to this working hypothesis we demonstrated that a planarian *hsp70* gene, *Djmot*—whose expression is elevated in stem cells and early progenitors -is crucial to support planarian stem cell proliferation, and inhibition of its activity causes irreversible growth arrest (Conte et al. 2009). I believe that, by antagonizing the p53 growth arrest pathway in planarian stem cells, DjMot may be a key factor required to achieve constitutive proliferation, reducing the incidence of tumor formation. Dissecting the possible link between growth arrest of stem cells in *Djmot*(RNAi) planarians and activation of p53 pathway remains the future main objective in this research.

Acknowledgments I greatly appreciate and thank colleagues and students who contributed to the main results mentioned in this review and also gratefully acknowledge the input provided through discussions and constructive criticisms.

References

Adell T, Cebrià F, Saló E (2010) Gradients in planarian regeneration and homeostasis. Cold Spring Harb Perspect Biol 2(1):a000505

Baguñà J (1976) Mitosis in the intact and regenerating planarian Dugesia mediterranea n. sp. I. Mitotic studies during growth, feeding and starvation. J Exp Zool 195:53–64

Baguñà J, Romero R (1981) Quantitative analysis of cell types during growth, degrowth and regeneration in the planarians Dugesia mediterranea and Dugesia tigrina. Hydrobiologia 84:181–194

Bedelbaeva K, Snyder A, Gourevitch D, Clark L, Zhang XM, Leferovich J, Cheverud JM, Lieberman P, Heber-Katz E (2010) Lack of p21 links cell cycle control and appendage regeneration in mice. Proc Natl Acad Sci U S A 107:5845–5850

Carette J, Lehnert S, Chow TY (2002) Implication of PBP74/mortalin/GRP75 in the radio-adaptive response. Int J Radiat Biol 78:183–190

Conte M, Deri P, Isolani ME, Mannini L, Batistoni R (2009) A mortalin-like gene is crucial for planarian stem cell viability. Dev Biol 334:109–118

Conte M, Deri P, Isolani ME, Mannini L, Batistoni R (2010) Characterization of hsp genes in planarian stem cells. Belg J Zool 140:137–143

Conte M, Isolani ME, Deri P, Mannini L, Batistoni R (2011) Expression of hsp90 mediates cytoprotective effects in the gastrodermis of planarians. Cell Stress Chaperones 16:33–39

De Mena L, Coto E, Sánchez-Ferrero E, Ribacoba R, Guisasola LM, Salvador C, Blázquez M, Alvarez V (2009) Mutational screening of the mortalin gene (HSPA9) in Parkinson's disease. J Neural Transm 116:1289–1293

Deocaris CC, Widodo N, Ishii T, Kaul SC, Wadhwa R (2007) Functional significance of minor structural and expression changes in stress chaperone mortalin. Ann NY Acad Sci 1119:165–175

Deocaris CC, Kaul SC, Wadhwa R (2008) From proliferative to neurological role of an hsp70 stress chaperone, mortalin. Biogerontology 9:391–403

Deocaris CC, Kaul SC, Wadhwa R (2009) The versatile stress protein mortalin as a chaperone therapeutic agent. Protein Pept Lett 16:517–529

Eisenhoffer GT, Kang H, Sánchez Alvarado A (2008) Molecular analysis of stem cells and their descendants during cell turnover and regeneration in the planarian Schmidtea mediterranea. Cell Stem Cell 3:327–339

Gabai VL, Yaglom JA, Waldman T, Sherman MY (2009) Heat shock protein hsp72 controls oncogene-induced senescence pathways in cancer cells. Mol Cell Biol 29:559–569

Gentile L, Cebrià F, Bartscherer K (2011) The planarian flatworm: an in vivo model for stem cell biology and nervous system regeneration. Dis Model Mech 4:12–19

González-Estévez C, Saló E (2010) Autophagy and apoptosis in planarians. Apoptosis 15:279–292

Goss RJ (1969) Principles of Regeneration. Academic, New York

Handberg-Thorsager M, Fernandez E, Saló E (2008) Stem cells and regeneration in planarians. Front Biosci 13:6374–6394

Hayashi T, Asami M, Higuchi S, Shibata N, Agata K (2006) Isolation of planarian X-ray-sensitive stem cells by fluorescente-activated cell sorting. Dev Growth Differ 48:371–380

Hayashi T, Shibata N, Okumura R, Kudome T, Nishimura O, Tarui H, Agata K (2010) Single-cell profiling of planarian stem cells using fluorescent activated cell sorting and its "index sorting" function for stem cell research. Dev Growth Differ 52:131–144

Higuchi S, Hayashi T, Hori I, Shibata N, Sakamoto H, Agata K (2007) Characterization and categorization of fluorescence activated cell sorted planarian stem cells by ultrastructural analysis. Dev Growth Differ 49:571–581

Kaul SC, Duncan EL, Englezou A, Takano S, Reddel RR, Mitsui Y, Wadhwa R (1998) Malignant transformation of NIH3T3 cells by overexpression of mot-2 protein. Oncogene 17:907–911

Kaul SC, Reddel RR, Mitsui Y, Wadhwa R (2001) An N-terminal region of mot-2 binds to p53 in vitro. Neoplasia 3:110–114

Kaul SC, Taira K, Pereira-Smith OM, Wadhwa R (2002) Mortalin: present and prospective. Exp Gerontol 37:1157–1164

Kaul SC, Aida S, Yaguchi T, Kaur K, Wadhwa R (2005) Activation of wild type p53 function by its mortalin-binding, cytoplasmically localizing carboxyl terminus peptides. J Biol Chem 280:39373–39379

Kaul Z, Yaguchi T, Kaul SC, Wadhwa R (2006) Quantum dot-based protein imaging and functional significance of two mitochondrial chaperones in cellular senescente and carcinogenesis. Ann N Y Acad Sci 1067:469–473

Kaul SC, Deocaris CC, Wadhwa R (2007) Three faces of mortalin: a housekeeper, guardian and killer. Exp Gerontol 4:263–274

Kimura K, Tanaka N, Nakamura N, Takano S, Ohkuma S (2007) Knockdown of mitochondrial heat shock protein 70 promotes progeria-like phenotypes in Caenorhabditis elegans. J Biol Chem 282:5910–5918

Iosefson O, Azem A (2010) Reconstitution of the mitochondrial Hsp70 (mortalin)-p53 interaction using purified proteins-identification of additional interacting regions. FEBS Lett 584:1080–1084

Morgan TH (1901) Regeneration. Macmillan, New York

Muller TL, Ngo-Muller V, Reginelli A, Taylor G, Anderson R, Muneoka K (1999) Regeneration in higher vertebrates: limb buds and digit tips. Semin Cell Dev Biol 10:405–413

Muneoka K, Sassoon D (1992) Molecular aspects of regeneration in developing vertebrate limbs. Dev Biol 152:37–49

O'Callaghan-Sunol C, Gabai VL, Sherman MY (2007) Hsp27 modulates p53 signaling and suppresses cellular senescence. Cancer Res 67:11779–11788

Orii H, Sakurai T, Watanabe K (2005) Distribution of the stem cells (neoblasts) in the planarian Dugesia japonica. Dev Genes Evol 215:143–157

Oviedo NJ, Nicolas CL, Adams DS, Levin M (2008) Planarians: a versatile and powerful model system for molecular studies of regeneration, adult stem cell regulation, aging, and behavior. Cold Spring Harb Protoc 11:10.1101/pdb.emo101

Pearson BJ, Sánchez Alvarado A (2009) Regeneration, stem cells, and the evolution of tumor suppression. Cold Spring Harb Symp Quant Biol 73:565–572

Pearson BJ, Sánchez Alvarado A (2010) A planarian p53 homolog regulates proliferation and self-renewal in adult stem cell lineages. Development 137:213–221

Pellettieri J, Sánchez Alvarado A (2007) Cell turnover and adult tissue homeostasis: from humans to planarians. Ann Rev Genet 41:83–105

Prinsloo E, Setati MM, Longshaw VM, Blatch GL (2009) Chaperoning stem cells: a role for heat shock proteins in the modulation of stem cell self-renewal and differentiation? Bioessays 31:370–377

Ran Q, Wadhwa R, Kawai R, Kaul SC, Sifers RN, Bick RJ, Smith JR, Pereira-Smith OM (2000) Extramitochondrial localization of mortalin/mthsp70/PBP74/GRP75. Biochem Biophys Res Commun 275:174–179

Rossi L, Salvetti A, Marincola FM, Lena A, Deri P, Mannini L, Batistoni R, Wang E, Gremigni V (2007) Deciphering the molecular machinery of stem cells: a look at the neoblast gene expression profile. Genome Biol 8:R62

Rossi L, Salvetti A, Batistoni R, Deri P, Gremigni V (2008) Planarians, a tale of stem cells. Cell Mol Life Sci 65:16–23

Salvetti A, Rossi L, Deri P, Batistoni R (2000) An MCM2-related gene is expressed in proliferating cells of intact and regenerating planarians. Dev Dyn 218:603–614

Sánchez-Alvarado A, Newmark PA (1999) Double-stranded RNA specifically disrupts gene expression during planarian regeneration. Proc Natl Acad Sci U S A 96:5049–5054

Sharpless NE, Schatten G (2009) Stem cell aging. J Gerontol A Biol Sci Med Sci 64:202–204

Sherman MY, Gabai V, O'Callaghan C, Yaglom J (2007) Molecular chaperones regulate p53 and suppress senescence programs. FEBS Lett 581:3711–3715

Shibata N, Rouhana L, Agata K (2010) Cellular and molecular dissection of pluripotent adult somatic stem cells in planarians. Develop Growth Differ 52:27–41

Spradling A, Ganetsky B, Hieter P, Johnston M, Olson M, Orr-Weaver T, Rossant J, Sánchez Alvarado A, Waterston R (2006) New roles for model genetic organisms in understanding and treating human disease: Report from the 2006 Genetics Society of America Meeting Genetics 172:2025–2032

Wadhwa R, Kaul SC, Mitsui Y, Sugimoto Y (1993) Differential subcellular distribution of mortalin in mortal and immortal mouse and human fibroblasts. Exp Cell Res 207:442–448

Wadhwa R, Taira K, Kaul SC (2002) An HSP70 family chaperone, mortalin/mthsp70/PBP74/Grp75: what, when, and where? Cell Stress Chaperones 7:309–316

Wadhwa R, Takano S, Kaur K, Deocaris CC, Pereira-Smith OM, Reddel RR, Kaul SC (2006) Upregulation of mortalin/mthsp70/Grp75 contributes to human carcinogenesis. Int J Cancer 118:2973–2980

Walker C, Böttger S, Low B (2006) Mortalin-based cytoplasmic sequestration of p53 in a nonmammalian cancer model. Am J Pathol 168:1526–1530

Yokoyama K, Fukumoto K, Murakami T (2002) Extended longevity of Caenorhabditis elegans by knocking in extra copies of hsp70F, a homolog of mot-2 (mortalin/mthhsp70/Grp75). FEBS Lett 516:53–57

Part III
Mortalin in Health and Disease

Chapter 8
Mortalin in Cell Protection from Immune Attack

Moran Saar, Oren Moskovich and Zvi Fishelson

Abstract The complement system plays major roles in innate and adaptive immunity. Complement activation may proceed through the classical, lectin and alternative pathways that initiate assembly of the membranolytic C5b-9 complex, also known as the membrane attack complex or MAC. Insertion of the C5b-9 complex into target cells membrane is an important weapon the body uses to fight pathogenic microorganisms. However, in many diseases, the C5b-9 complexes are considered pathogenic and inflict damages to patients' organs. During antibody-based therapy, complement C5b-9 complexes penetrate the plasma membrane of antibody-coated cancer cells and facilitate cancer destruction and remission. Several resistance mechanisms protect cancer cells from complement-dependent cytotoxicity. We found that the mitochondrial hsp70 mortalin supports cancer evasion from cell death activated by C5b-9. The mechanism whereby mortalin is protecting cells is currently under investigation. Initial results indicate that mortalin promotes C5b-9 elimination from the cell surface. The plausible target proteins recognized by mortalin are the C8 and C9 components of the C5b-9 complex. Since mortalin expression is elevated in cancer, it is conceivable that mortalin poses an obstacle to antibody-based therapy and as such might be worth targeting.

Keywords Complement · Cytotoxicity · C5b-9 · Mortalin

8.1 Preface

Mortalin (mthsp70/grp75) is a major mitochondrial protein playing a major role in import of mitochondrial proteins (Bhattacharyya et al. 1995; Koehler 2004; Wiedemann et al. 2004). Studies in yeasts and mammalian cell lines showed that it is essential for cell growth and mitochondrial biogenesis (Craig et al. 1987; Kawai et al. 2001). Over expression of mortalin together with hTERT results in lifespan extension of fibroblasts (Kaul et al. 2003) and contributes to tumorigenesis (Wadhwa et al. 2006). Lowering mortalin level by RNA interference causes senescence-like growth arrest

Z. Fishelson (✉) · M. Saar · O. Moskovich
Department of Cell and Developmental Biology, Sackler School of Medicine,
Tel Aviv University, Tel Aviv 69978, Israel
e-mail: lifish@post.tau.ac.il

S. C. Kaul, R. Wadhwa (eds.), *Mortalin Biology: Life, Stress and Death*,
DOI 10.1007/978-94-007-3027-4_8, © Springer Science+Business Media B.V. 2012

in immortalized cells (Wadhwa et al. 2004). A role for mortalin in cell fate determination was also proposed (Rivolta and Holley 2002). Various stress conditions, including ionizing radiation, physical exercise and calorie restriction were shown to induce mortalin synthesis (Merrick et al. 1997; Sadekova et al. 1997; Mattson et al. 2000; Massa et al. 1995; Um et al. 2003; Tsuchiya et al. 2004). Up regulation of mortalin protects PC12 cells from glucose deprivation and ROS accumulation (Liu et al. 2005) and vascular smooth muscle cells from serum deprivation (Taurin et al. 2002).

Several years ago, when we accidentally encountered mortalin, the literature had already clearly indicated that mortalin is essential for life and has a role in cell defense from several insults. We have been investigating cell protective processes occurring at the plasma membrane level and finding the mitochondrial mortalin in that compartment came to us as a surprise. Our investigation of the mechanism through which cancer cells protect themselves from complement-dependent membrane damage led us to analyze proteins secreted by complement-attacked cells. A major protein band observed in the cell supernatant was identified as mortalin. Subsequent research, described in this chapter, revealed that mortalin protects cells from the complement membrane attack complex C5b-9 and supports C5b-9 elimination from the cell surface via membrane vesicles.

8.2 Complement-Dependent Cell Death (CDC)

The complement system is a family of soluble and membrane proteins found in blood plasma that serve as first line defense against infective pathogenic organisms (Walport 2001). Upon identification of the intruder, the system gets activated and reacts in 3 modes: (a) opsonization of the pathogen for phagocytosis, (b) direct destruction of the pathogen and (c) production of inflammatory signals. Complement genetic deficiencies are therefore associated with elevated incidence of infections (Tedesco 2008). Complement deficiencies and genetic variations as well as excessive pathogenic activation of complement are also predisposing to numerous diseases (Pettigrew et al. 2009; Sarma and Ward 2011). Complement is also a major effector mechanism in antibody-based cancer immunotherapy (Morris and Waldmann 2009).

Initiation of complement activation may occur via the classical, alternative and lectin pathways, leading first to deposition of complement C3 fragments on the pathogen surface and then activating a common terminal complement pathway. The latter activation steps generate the complement membrane attack complex (MAC), also known as the C5b-9 protein complex, comprised of the complement proteins C5b, C6, C7, C8 and C9 (Muller-Eberhard 1986). Complement-dependent cytotoxicity (CDC) is instigated upon insertion of the MAC into the target cell membrane.

Initiation of the classical pathway occurs when C1q together with C1r and C1s (C1 complex) bind to complement-fixing antibodies attached to pathogenic surfaces. Autocatalytic activation of C1r and C1s in turn leads to cleavage C4 and C2 into larger (C4b, C2a) and smaller (C4a, C2b) fragments. The larger fragments associate to form the C4b,2a C3 convertase on pathogenic surfaces, and gain the ability to cleave C3. Generation of the C3 convertase, which cleaves C3 into the anaphylatoxin C3a and the opsonin C3b, is the point at which the 3 initiation pathways converge. The lectin pathway is initiated when mannose-binding lectin (MBL) binds to pathogenic carbohydrate motifs. The MBL-associated serine protease is activated and cleaves C4 and C2 thus producing the C4b,2a C3 convertase which cleaves C3 into C3b and C3a. A second molecule of C3b can associate with C4b,2a to form the C5 convertase of the classical and lectin pathways, C4b,2a,3b (Rawal and Pangburn 2001). The alternative pathway (AP) is initiated when C3 is activated spontaneously and forms the initial AP C3 convertase, C3(H_2O),Bb, in the presence of Factors B and D, leading to additional C3 cleavage and eventual formation of the AP C3 convertase C3b, Bb and AP C5 convertase C3b,Bb,C3b (Fishelson et al. 1984). Properdin facilitates AP activation by stabilizing AP convertases (Kemper et al. 2010). The three initiation pathways form the convertases that generate the small protein fragments of C3, C4 and C5: C3a, C4a, C5a, respectively, that act as anaphylatoxins and chemotactic agents and the large protein fragments: C3b, C4b and C5b. C3b and C4b act as opsonins and C5b assembles the C5b-9 membrane attack complex.

8.3 Mechanisms of Resistance to Complement-Dependent Cytotoxicity

Cells, and in particular cancer cells, utilize multiple strategies to evade CDC. Membrane-associated complement regulatory proteins CD46, CD55 and CD59 play an important role in cancer immune evasion. CD46, known also as membrane cofactor protein (MCP) and CD55, known also as decay-accelerating factor (DAF), inhibit the initiation pathways of complement at the C3/C5 convertase stage (Seya et al. 1986; Nicholson-Weller et al. 1985), while CD59 inhibits the terminal complement pathway by preventing MAC formation and assembly via direct binding to C8 and C9 (Meri et al. 1990). Many primary tumors and tumor cell lines overexpress those membrane regulators (reviewed in Fishelson et al. 2003; Gelderman et al. 2004; Yan et al. 2008). In addition to protection by membrane regulators, tumor cells can evade CDC by secretion of soluble complement regulators such as C1 Inhibitor, Factor H, Factor I and soluble forms of membrane regulators (Bjorge et al. 2005; Okroj et al. 2009). Another strategy employed by tumor cell to evade CDC is modification of active complement components by proteases (Ollert et al. 1990; Jurianz et al. 1999; Rozanov et al. 2006) and ecto-protein kinases (Bohana-Kashtan et al. 2005). Presence of high levels of sialic acid on the surface of human prostate, breast and ovarian carcinoma cells may also contribute to protection against CDC (Gancz and Fishelson 2009). As described next, C5b-9 elimination from the cells is yet another major protective strategy.

8.4 Removal of the C5b-9 Complex by Ecto- and Endocytosis

Cell can rapidly eliminate the C5b-9 complex from their surface by outward or inward vesiculation. Complement-induced exo-vesiculation was demonstrated in several normal cell types and tumor cell lines (Carney et al. 1985; Scolding et al. 1989; Morgan et al. 1986; Stein and Luzio 1991). The presence of elevated levels of cholesterol and diacylglycerol in shed membrane vesicles containing the MAC suggests a selective membrane protein and lipid sorting process (Morgan et al. 1987). Elimination of MAC by endocytosis has also been demonstrated in several normal cell types and tumor cells (Kerjaschki et al. 1989; Morgan et al. 1986, 1987; Moskovich and Fishelson 2007). Using Alexa Fluor 488-tagged human C9 we demonstrated that bound MAC on the surface of K562 cells is either shed within 10 min attached to membrane vesicles or is endocytosed and accumulated in the endocytic recycling compartment before it undergoes exocytosis (Moskovich and Fishelson 2007). Ectosomes result from outward vesicle budding and fission from the cell surface. In contrast, exosomes are generated by a process involving endocytosis that is followed by exocytosis. Furthermore, ectosomes differ from exosomes also by size and composition, as reviewed in Pilzer et al. (2005). These shed vesicles probably have vital roles in tumor invasion, metastasis, inflammation, coagulation and stem cells renewal and expansion (Cocucci et al. 2009). Recent studies have suggested that ectosome shedding is a highly regulated process that occurs in a spectrum of cell types and, more frequently, in tumor cells, yet the mechanism behind direct outward plasma membrane vesiculation is still largely unknown.

8.5 Secretion of Mortalin from Complement-Attacked Cells

In an attempt to characterize the machinery involved in MAC elimination, we collected the proteins secreted along with the C5b-9 complex from complement-attacked K562 cells. A major protein band that showed up during SDS-PAGE analysis of the cell supernatant and was analyzed by Mass Spectroscopy was identified to be mortalin (Pilzer and Fishelson 2005). Mortalin is secreted to some extent also from non-treated cells but its secretion from complement-treated cells is markedly enhanced. The kinetics of the release of mortalin reveals a significant level of extracellular mortalin in supernatants collected 15 min after treatment with complement and the amount of mortalin keeps rising slightly up to 60 min. Concomitantly, the intracellular level of mortalin is gradually decreasing from 15–60 min. To rule out the possibility that mortalin leaks out of dead cells, a correlation between the amount of released mortalin and the level of cell damage was searched for. The most significant increase in mortalin release occurs under sub-lytic conditions (15–21% lysis). At higher, lytic conditions, there is no additional increase in the amount of extracellular mortalin. These findings indicate that mortalin is secreted from intact and not from damaged cells. Mortalin release depends on C5b-8 or C5b-9 formation. In the absence of C8, only a small amount of mortalin is secreted from the cells. C9-depleted serum that generates only C5b-8, causes a substantial release of mortalin, albeit only about half of the release induced by C5b-9. Interestingly, the effect of mortalin appears to be a C5b-9 selective phenomenon. Streptolysin O (SLO) and melittin are pore-forming proteins related to complement C9 (Pilzer and Fishelson 2005). However, neither SLO nor melittin induce secretion of mortalin.

Both mortalin and C9 that are released from complement-attacked K562 cells can be spun down at 100,000 g, conditions known to spin down small membrane vesicles. Detergent solubilization of the membrane vesicles bearing mortalin and C9 prior to the high-speed centrifugation, results in translocation of both mortalin and C9 from the pellet to the supernatant (Pilzer and Fishelson 2005). Mortalin secretion is interrupted by inhibition of the protein kinases PKC and ERK, suggesting that this process is regulated by phosphorylation (Pilzer and Fishelson 2005).

8.6 Indications that Mortalin Protects Cells from CDC

The fact that mortalin is released upon complement attack does not necessarily implicate mortalin in the MAC elimination process. However, following silencing of mortalin expression with siRNA, MAC elimination is reduced and cell sensitivity to lysis by antibody and complement increases (Pilzer et al. 2010). An inverse correlation between mortalin level in cells and their sensitivity to CDC is implicating mortalin as a protective stress protein. The relevance of mortalin to CDC was assessed also by blocking mortalin with MKT-077, a cationic rhodacyanine dye analogue that binds to mortalin and inhibits its activity (Wadhwa et al. 2000). Treatment of K562 and HCT116 cells with MKT-077 sensitizes them to cell death mediated by antibody and complement (Pilzer et al. 2010). Interestingly, MKT-077 is enhancing CDC of control cells but not of cells in which mortalin synthesis has been blocked with siRNA (Pilzer et al. 2010). This indicates that the sensitizing effect of MKT-077 on cell death by complement is indeed targeting mortalin. MKT-077 has no effect on SLO-mediated lysis of K562 cells demonstrating the selective effect of MKT-077 on mortalin response to C5b-9. Pre-treatment of cells with MKT-077 also reduces the extent of MAC-mortalin release by vesiculation following a sublytic complement attack.

The tumor suppressor protein p53 is a mortalin client protein. MKT-077 dissociates it from p53 and restores p53 transcriptional and pro-apoptotic activities (Wadhwa et al. 2000). The effect of MKT-077 on complement-mediated lysis of HCT116 p53+/+ and p53-/- cells was found to be independent on the presence of p53. This suggests that the protective effect of mortalin on CDC is p53-independent.

The presence of anti-mortalin antibodies outside the cells during complement attack also reduces release of MAC and mortalin and augments cell death (Pilzer and Fishelson 2005). Hence, apparently, mortalin emerges, during complement attack, on the cell surface and is still susceptible at this stage to inactivation by extracellular inhibitors.

8.7 Complement C8 and C9 are Mortalin Binding Proteins

Mortalin, being a chaperone supporting entry of proteins into mitochondria is expected to bind an array of proteins. Indeed, mortalin was shown to bind to p53 (Wadhwa et al. 1998), hsp60 (Mahlke et al. 1990; Langer and Neupert 1991; Hartl et al. 1992) and VDAC (Schwarzer et al. 2002) but also to IL-1 receptor type 1 (Sacht et al. 1999), GRP94 (Takano et al. 2001) and FGF-1 (Mizukoshi et al. 1999). How is mortalin regulating MAC removal from the plasma membrane by vesiculation? It may participate in the membrane vesiculation process and/or in the initial recognition and sorting of the C5b-9 complexes. Our results demonstrated that mortalin can bind directly to two components of the C5b-9 complex, to complement C8 and C9 (Pilzer and Fishelson 2005). Interestingly, exogenously added recombinant human mortalin significantly inhibits hemolysis of rabbit erythrocytes by complement in a dose-dependent manner (Pilzer et al. 2010). Furthermore, recombinant mortalin inhibits polymerization of purified C9 (accelerated by zinc chloride at 37°C) (Pilzer et al. 2010). Thus, the data strongly suggest that binding of mortalin to C8 and C9 during C5b-9 formation and membrane insertion may block MAC assembly. However, whether or not this direct binding occurs in cells and is indeed necessary for the mortalin pro-survival effect requires further experimentation. The fact that MKT-077 inhibits mortalin-C9 binding (Pilzer et al. 2010) and blocks the mortalin pro-survival activity only in cells containing mortalin further supports a requirement for mortalin-C5b-9 interaction for the complement inhibitory activity of mortalin. This does not rule out an additional role for mortalin in the exo-vesiculation process and awaits investigation.

8.8 Future Directions

Mortalin is a mitochondrial protein whereas complement C5b-9 complexes are assembled and act primarily at the plasma membrane level. Our yet unpublished results show that cytosolic mortalin is less effective in protection from C5b-9 than mitochondrial mortalin (Saar and Fishelson, manuscript in preparation). So apparently it is the mitochondrial fraction of mortalin that protects cells from C5b-9. It is a challenging objective now to determine how is a mitochondrial matrix protein performing a task at the plasma membrane level and which other proteins are involved in the process.

Mortalin expression is elevated in cancer cells (Takano et al. 1997). Thus, mortalin is highly expressed in human colorectal adenocarcinoma and its expression correlates with poor patients survival (Dundas et al. 2004). In addition, mortalin supports resistance of cancer cells to complement-dependent cytotoxicity (Pilzer and Fishelson 2005). Furthermore, mortalin inhibitors sensitize cancer cells to the lytic activity of complement C5b-9. This suggests that blocking mortalin in cancer cells would amplify the therapeutic capacity of cancer-targeted antibodies. Mortalin should thus be considered as a target for cancer adjuvant immunotherapy.

References

Bhattacharyya T, Karnezis AN, Murphy SP, Hoang T, Freeman BC, Phillips B, Morimoto RI (1995) Cloning and subcellular localization of human mitochondrial hsp70. J Biol Chem 270:1705–1710

Bjorge L, Hakulinen J, Vintermyr OK, Jarva H, Jensen TS, Iversen OE, Meri S (2005) Ascitic complement system in ovarian cancer. Br J Cancer 92:895–905

Bohana-Kashtan O, Pinna LA, Fishelson Z (2005) Extracellular phosphorylation of C9 by protein kinase CK2 regulates complement-mediated lysis. Eur J Immunol 35:1939–1948

Carney DF, Koski CL, Shin ML (1985) Elimination of terminal complement intermediates from the plasma membrane of nucleated cells: the rate of disappearance differs for cells carrying C5b-7 or C5b-8 or a mixture of C5b-8 with a limited number of C5b-9. J Immunol 134:1804–1809

Cocucci E, Racchetti G, Meldolesi J (2009) Shedding microvesicles: artefacts no more. Trends Cell Biol 19:43–51

Craig EA, Kramer J, Kosic-Smithers J (1987) SSC1, a member of the 70-kDa heat shock protein multigene family of Saccharomyces cerevisiae, is essential for growth. Proc Natl Acad Sci U S A 84:4156–4160

Dundas SR, Lawrie LC, Rooney PH, Murray GI (2004) Mortalin is over-expressed by colorectal adenocarcinomas and correlates with poor survival. J Pathol 205:74–81

Fishelson Z, Pangburn MK, Muller-Eberhard HJ (1984) Characterization of the initial C3 convertase of the alternative pathway of human complement. J Immunol 132:1430–1434

Fishelson Z, Donin N, Zell S, Schultz S, Kirschfink M (2003) Obstacles to cancer immunotherapy: expression of membrane complement regulatory proteins (mCRPs) in tumors. Mol Immunol 40:109–123

Gancz D, Fishelson Z (2009) Cancer resistance to complement-dependent cytotoxicity (CDC): Problem-oriented research and development. Mol Immunol 46:2794–2800

Gelderman KA, Tomlinson S, Ross GD, Gorter A (2004) Complement function in mAb-mediated cancer immunotherapy. Trends Immunol 25:158–164

Hartl FU, Martin J, Neupert W (1992) Protein folding in the cell: the role of molecular chaperones Hsp70 and Hsp60. Annu Rev Biophys Biomol Struct 21:293–322

Jurianz K, Ziegler S, Garcia-Schuler H, Kraus S, Bohana-Kashtan O, Fishelson Z, Kirschfink M (1999) Complement resistance of tumor cells: basal and induced mechanisms. Mol Immunol 36:929–939

Kaul SC, Yaguchi T, Taira K, Reddel RR, Wadhwa R (2003) Overexpressed mortalin (mot-2)/mthsp70/GRP75 and hTERT cooperate to extend the in vitro lifespan of human fibroblasts. Exp Cell Res 286:96–101

Kawai A, Nishikawa S, Hirata A, Endo T (2001) Loss of the mitochondrial Hsp70 functions causes aggregation of mitochondria in yeast cells. J Cell Sci 114:3565–3574

Kemper C, Atkinson JP, Hourcade DE (2010) Properdin: emerging roles of a pattern-recognition molecule. Ann Rev Immunol 28:131–155

Kerjaschki D, Schulze M, Binder S, Kain R, Ojha PP, Susani M, Horvat R, Baker PJ, Couser WG (1989) Transcellular transport and membrane insertion of the C5b-9 membrane attack complex of complement by glomerular epithelial cells in experimental membranous nephropathy. J Immunol 143:546–552

Koehler CM (2004) New developments in mitochondrial assembly. Ann Rev Cell Dev Biol 20:309–335

Langer T, Neupert W (1991) Heat shock proteins hsp60 and hsp70: their roles in folding, assembly and membrane translocation of proteins. Cur Top Microbiol Immunol 167:3–30

Liu Y, Liu W, Song XD, Zuo J (2005) Effect of GRP75/mthsp70/PBP74/mortalin overexpression on intracellular ATP level, mitochondrial membrane potential and ROS accumulation following glucose deprivation in PC12 cells. Mol Cell Biochem 268:45–51

Mahlke K, Pfanner N, Martin J, Horwich AL, Hartl FU, Neupert W (1990) Sorting pathways of mitochondrial inner membrane proteins. Eur J Biochem 192:551–555

Massa SM, Longo FM, Zuo J, Wang S, Chen J, Sharp FR (1995) Cloning of rat grp75, an hsp70-family member, and its expression in normal and ischemic brain. J Neurosci Res 40:807–819

Mattson JP, Ross CR, Kilgore JL, Musch TI (2000) Induction of mitochondrial stress proteins following treadmill running. Med Sci Sports Exerc 32:365–369

Meri S, Morgan BP, Davies A, Daniels RH, Olavesen MG, Waldmann H, Lachmann PJ (1990) Human protectin (CD59), an 18,000–20,000 MW complement lysis restricting factor, inhibits C5b-8 catalysed insertion of C9 into lipid bilayers. Immunology 71:1–9

Merrick BA, Walker VR, He C, Patterson RM, Selkirk JK (1997) Induction of novel Grp75 isoforms by 2-deoxyglucose in human and murine fibroblasts. Cancer Lett 119:185–190

Mizukoshi E, Suzuki M, Loupatov A, Uruno T, Hayashi H, Misono T, Kaul SC, Wadhwa R, Imamura T (1999) Fibroblast growth factor-1 interacts with the glucose-regulated protein GRP75/mortalin. Biochem J 343:461–466

Morgan BP, Imagawa DK, Dankert JR, Ramm LE (1986) Complement lysis of U937, a nucleated mammalian cell line in the absence of C9: effect of C9 on C5b-8 mediated cell lysis. J Immunol 136:3402–3406

Morgan BP, Dankert JR, Esser AF (1987) Recovery of human neutrophils from complement attack: removal of the membrane attack complex by endocytosis and exocytosis. J Immunol 138:246–253

Morris JC, Waldmann TA (2009) Antibody-based therapy of leukaemia. Exp Rev Molec Med 11:e29

Moskovich O, Fishelson Z (2007) Live cell imaging of outward and inward vesiculation induced by the complement c5b-9 complex. J Biol Chem 282:29977–29986

Muller-Eberhard HJ (1986) The membrane attack complex of complement. Annu Rev Immunol 4:503–528

Nicholson-Weller A, March JP, Rosen CE, Spicer DB, Austen KF (1985) Surface membrane expression by human blood leukocytes and platelets of decay-accelerating factor, a regulatory protein of the complement system. Blood 65:1237–1244

Okroj M, Corrales L, Stokowska A, Pio R, Blom AM (2009) Hypoxia increases susceptibility of non-small cell lung cancer cells to complement attack. Cancer Immunol Immunother 58:1771–1780

Ollert MW, Frade R, Fiandino A, Panneerselvam M, Petrella EC, Barel M, Pangburn MK, Bredehorst R, Vogel CW (1990) C3-cleaving membrane proteinase. A new complement regulatory protein of human melanoma cells. J Immunol 144:3862–3867

Pettigrew HD, Teuber SS, Gershwin ME (2009) Clinical significance of complement deficiencies. Ann N Y Acad Sci 1173:108–123

Pilzer D, Fishelson Z (2005) Mortalin/GRP75 promotes release of membrane vesicles from immune attacked cells and protection from complement-mediated lysis. Int Immunol 17:1239–1248

Pilzer D, Gasser O, Moskovich O, Schifferli JA, Fishelson Z (2005) Emission of membrane vesicles: roles in complement resistance, immunity and cancer. Springer Semin Immunopathol 27:375–387

Pilzer D, Saar M, Koya K, Fishelson Z (2010) Mortalin inhibitors sensitize K562 leukemia cells to complement-dependent cytotoxicity. Int J Cancer 126:1428–1435

Rawal N, Pangburn MK (2001) Structure/function of C5 convertases of complement. Int Immunopharmacol 1:415–422

Rivolta MN, Holley MC (2002) Asymmetric segregation of mitochondria and mortalin correlates with the multi-lineage potential of inner ear sensory cell progenitors in vitro. Brain Res 133:49–56

Rozanov DV, Savinov AY, Golubkov VS, Tomlinson S, Strongin AY (2006) Interference with the complement system by tumor cell membrane type-1 matrix metalloproteinase plays a significant role in promoting metastasis in mice. Cancer Res 66:6258–6263

Sacht G, Brigelius-Flohe R, Kiess M, Sztajer H, Flohe L (1999) ATP-sensitive association of mortalin with the IL-1 receptor type I. BioFactors 9:49–60

Sadekova S, Lehnert S, Chow TY (1997) Induction of PBP74/mortalin/Grp75, a member of the hsp70 family, by low doses of ionizing radiation: a possible role in induced radioresistance. Int J Rad Biol 72:653–660

Sarma JV, Ward PA (2011) The complement system. Cell Tissue Res 343:227–235

Schwarzer C, Barnikol-Watanabe S, Thinnes FP, Hilschmann N (2002) Voltage-dependent anion-selective channel (VDAC) interacts with the dynein light chain Tctex1 and the heat-shock protein PBP74. Int J Biochem Cell Biol 34:1059–1070

Scolding NJ, Morgan BP, Houston WA, Linington C, Campbell AK, Compston DA (1989) Vesicular removal by oligodendrocytes of membrane attack complexes formed by activated complement. Nature 339:620–622

Seya T, Turner JR, Atkinson JP (1986) Purification and characterization of a membrane protein (gp45–70) that is a cofactor for cleavage of C3b and C4b. J Exp Med 163:837–855

Stein JM, Luzio JP (1991) Ectocytosis caused by sublytic autologous complement attack on human neutrophils. The sorting of endogenous plasma-membrane proteins and lipids into shed vesicles. Biochem J 274:381–386

Takano S, Wadhwa R, Yoshii Y, Nose T, Kaul SC, Mitsui Y (1997) Elevated levels of mortalin expression in human brain tumors. Exp Cell Res 237:38–45

Takano S, Wadhwa R, Mitsui Y, Kaul SC (2001) Identification and characterization of molecular interactions between glucose-regulated proteins (GRPs) mortalin/GRP75/peptide-binding protein 74 (PBP74) and GRP94. Biochem J 357:393–398

Taurin S, Seyrantepe V, Orlov SN, Tremblay TL, Thibault P, Bennett MR, Hamet P, Pshezhetsky AV (2002) Proteome analysis and functional expression identify mortalin as an antiapoptotic gene induced by elevation of $[Na^{+}]i/[K^{+}]i$ ratio in cultured vascular smooth muscle cells. Circ Res 91:915–922

Tedesco F (2008) Inherited complement deficiencies and bacterial infections. Vaccine 26 (Suppl 8):I3–8

Tsuchiya T, Dhahbi JM, Cui X, Mote PL, Bartke A, Spindler SR (2004) Additive regulation of hepatic gene expression by dwarfism and caloric restriction. Physiol Genom 17:307–315

Um JH, Kim SJ, Kim DW, Ha MY, Jang JH, Chung BS, Kang CD, Kim SH (2003) Tissue-specific changes of DNA repair protein Ku and mtHSP70 in aging rats and their retardation by caloric restriction. Mech Age Develop 124:967–975

Wadhwa R, Takano S, Robert M, Yoshida A, Nomura H, Reddel RR, Mitsui Y, Kaul SC (1998) Inactivation of tumor suppressor p53 by mot-2, a hsp70 family member. J Biol Chem 273:29586–29591

Wadhwa R, Sugihara T, Yoshida A, Nomura H, Reddel RR, Simpson R, Maruta H, Kaul SC (2000) Selective toxicity of MKT-077 to cancer cells is mediated by its binding to the hsp70 family protein mot-2 and reactivation of p53 function. Cancer Res 60:6818–6821

Wadhwa R, Takano S, Taira K, Kaul SC (2004) Reduction in mortalin level by its antisense expression causes senescence-like growth arrest in human immortalized cells. J Gene Med 6:439–444

Wadhwa R, Takano S, Kaur K, Deocaris CC, Pereira-Smith OM, Reddel RR, Kaul SC (2006) Upregulation of mortalin/mthsp70/Grp75 contributes to human carcinogenesis. Int J Cancer 118:2973–2980

Walport MJ (2001) Complement. First of two parts. N Eng J Med 344:1058–1066

Wiedemann N, Frazier AE, Pfanner N (2004) The protein import machinery of mitochondria. J Biol Chem 279:14473–14476

Yan J, Allendorf DJ, Li B, Yan R, Hansen R, Donev R (2008) The role of membrane complement regulatory proteins in cancer immunotherapy. Adv Exp Med Biol 632:159–174

Chapter 9
Mortalin in Neurological Diseases

Jinghua Jin, Travis J. Cook, Jake G. Hoekstra and Jing Zhang

Abstract Mitochondria play critical roles in neuronal function and emerging evidence shows that mitochondrial dysfunction is essential in the pathogenesis of several major neurological diseases, including neurodegenerative disorders and cerebral ischemia. The present chapter centers on the roles of the functionally diverse mitochondrial protein, mortalin. Since its initial discovery in mouse embryonic fibroblasts and identification of its association with cellular mortality, subsequent research efforts have recognized mortalin as having potential roles in the maintenance of mitochondrial homeostasis, energy generation, mitochondrial import of nuclear-encoded proteins, and chaperoning of misfolded proteins. Dysfunction of mortalin has been implicated in a variety of neurological disorders, including Alzheimer's disease, Parkinson's disease, and brain tumors in addition to brain ischemia. This chapter will present the associative evidence implicating the protein as being involved in neurological diseases, as well as attempt to synthesize potential mechanisms by which mortalin participates in the processes of these conditions.

Keywords Alzheimer's disease · Glioma · Ischemia · Mortalin · Parkinson's disease

---

Jinghua Jin and Travis J. Cook have contributed to the manuscript equally.

J. Zhang (✉)
Department of Pathology, University of Washington School of Medicine,
RM 302B R&T; 325 9th Avenue, 359635, 98104-2499 Seattle, WA, USA

J. Jin
Department of Neurobiology, Zhejiang University School of Medicine,
Hangzhou, Zhejiang 310058, China
e-mail: jhjin@zju.edu.cn

T. J. Cook
Department of Environmental & Occupational Health Sciences,
University of Washington School of Public Health, Seattle, WA 98195, USA
e-mail: ttjc@u.washington.edu

J. G. Hoekstra
Department of Pathology, University of Washington School of Medicine,
Seattle, WA 98104, USA
e-mail: hoeke@u.washington.edu

S. C. Kaul, R. Wadhwa (eds.), *Mortalin Biology: Life, Stress and Death*,
DOI 10.1007/978-94-007-3027-4_9, © Springer Science+Business Media B.V. 2012

9.1 Introduction

The brain, one of the most complex structures in the body, accounts for 2% of the body weight while disproportionately consuming 20% of cardiac output and metabolic energy (Deocaris et al. 2008; Silver and Erecinska 1998). The energy requirements of the brain are heavily dependent upon adenosine triphosphate (ATP), with glucose as the main source of energy, although the storage of glucose within the brain is very limited. Mitochondria, the key players in ATP production fueled by products derived from glucose, are essential organelles for the continued viability of both neuronal and glial cells. Nonetheless, mitochondrial function goes beyond ATP production, with more recent data showing that mitochondria are also vital in the maintenance of ion homeostasis, steroid synthesis, generation of reactive oxygen species (ROS), and initiation of cell death (Kann and Kovács 2007; Schatz 1995). As a consequence, it is not surprising that mitochondrial dysfunction has been implicated in a variety of neurological diseases, including neurodegenerative disorders, brain tumors, and cerebral ischemia (Chen et al. 2011; Morais and De Strooper 2010; Ordys et al. 2010). While the contribution of mitochondrial dysfunction in cerebral ischemia is obvious since the loss of oxygen and glucose results in a rapid and severe onset of injury to neurons, many other pathophysiological states, e.g. Alzheimer's disease (AD) and Parkinson's disease (PD), can also be substantially influenced by mitochondrial dysfunction. The injury mechanisms secondary to mitochondrial dysfunction in these diseases typically manifest in a more slow, subtle, and complex manner than acute bioenergetic collapse. Additionally, injury may localize within distinct neuronal compartments, or potentially even within a subset of mitochondria inside a single neuron (Rintoul and Reynolds 2010).

Among the over 1,000 mitochondrial proteins that have been identified thus far (Calvo and Mootha 2010; Jin et al. 2007a), mortalin, localized on chromosome 5q31.1.1 (Kaul et al. 1995), is one of the most abundant proteins in the mitochondrial matrix, representing approximately 1% of the total protein content (Naylor et al. 1996). It is encoded by the nuclear gene HSPA9B (GeneID: 3313), and its designations also include: mitochondrial heat shock 70 kDa Protein (mtHsp70), 75 kDa glucose regulated protein (GRP75), heat shock 70 kD protein 9, heat shock 70 kD protein 9B, mortalin-2, heat shock 70 kDa protein 9B, perinuclear protein, p66-mortalin, peptide-binding protein 74 (PBP74), stress-70 protein and tumor necrosis factor receptor-associated protein 1 (TRAP-1) (Deocaris et al. 2009; Kaul et al. 2007; Wadhwa et al. 2002).

Functionally, mortalin is best known as a mitochondrial chaperone protein which plays a role in mitochondrial biogenesis, maintenance of protein integrity, and import of proteins into the mitochondria (Deocaris et al. 2008). However, the mitochondrion is not the only residence for mortalin. It can relocate outside of mitochondria and into the cytosol, where it has been found to localize with organelles such as the endoplasmic reticulum (ER) and the nucleus (Ran et al. 2000) and is believed to play a role in a wide range of cellular functions, including stress response (Takano et al. 2001), cell survival (Taurin et al. 2002), cell proliferation (Wadhwa et al. 2003b), and signal transduction (Mizukoshi et al. 1999, 2001). Altered expression

Table 9.1 The possible functions of mortalin in neurological diseases

Disease	Major findings	Implicated functions	Ref(s)
AD	Aβ inhibited mitochondrial import of mortalin in PC12 cells, which resulted in enhanced ROS production as well as loss of mitochondrial membrane potential (MMP) and altered mitochondrial morphology	Inhibition of ROS production, maintenance of MMP, and mitochondrial morphology	(Sirk et al. 2007)
	Mortalin overexpression was protective against Aβ induced cell death, while inhibition of mortalin expression sensitized Aβ induced neurotoxicity in SH-SY5Y cells	Maintainance of mitochondrial function and reduction of oxidative stress	(Qu et al. 2011)
	Oxidatively modified mortalin increased in APOE knockout mice	Oxidation of mortalin may play an important role in AD pathogenesis	(Choi et al. 2004)
	Mortalin isoforms were differentially expressed in both the hippocampus of *APOE4* mice compared to *APOE3* (control) mice, and in the hippocampi of patients with AD compared to controls	*APOE* genotype regulates mortalin modifications such as oxidation and phosphorylation	(Osorio et al. 2007)
PD	Mortalin decreased in mitochondrial-enriched fraction of SNpc in PD patients compared to controls. Manipulating mortalin level in a dopaminergic cell line altered the rotenone-induced neurotoxicity	Provided the first evidence that mortalin may be relevant in the pathogenesis of PD	(Jin et al. 2006)
	Mortalin interacted with α-synuclein and DJ-1 *in vitro*. Overexpression of WT but not A53T α-synuclein decreased mortalin level in mitochondrial fraction *in vitro*	Mortalin may be involved in the pathogenesis of PD by interacting with α-synuclein and DJ-1	(Pennington et al. 2010; Jin et al. 2007b; Li et al. 2005)
	Mortalin expression changed in association with LB spread from the brain stem to the limbic system and the isocortex	Mortalin may be involved in Lewy body progression of PD	(Shi et al. 2008)
	Three mutations of mortalin (R126W, P509S, A476T) have been identified and found to interfere with mitochondrial homeostasis	Mortalin variants found in PD patients may be correlative with the mitochondrial dysfunction in PD	(Burbulla et al. 2010; De Mena et al. 2009)

Table 9.1 (continued)

Disease	Major findings	Implicated functions	Ref(s)
	Mortalin decreased in rat mitochondria exposed to DA-quinones and later were found to be covalently modified by exposure to DA	Oxidative damage of mortalin may be detrimental to its function, which may lead to neuronal dysfunction and death	(Van Laar et al. 2008, 2009)
Brain tumors	Mortalin expression increased in human astrocytomas, with a trend of higher-grade malignancy tumors expressing increasingly more mortalin than lower-grade tumors	Overexpression of mortalin may be linked to tumorigenesis	(Takano et al. 1997)
Cerebral ischemia	Mortalin expression increased in areas of ischemic injury in rat brain. Overexpression of mortain *in vitro* or *in vivo* was shown to be protective from ischemic injury	Mortalin could protect mitochondrial function by reducing oxidative stress	(Massa et al. 1995; Xu et al. 2009; Voloboueva et al. 2008; Liu et al. 2005)
Schizo-phrenia	Mortalin splicing isoform CRP40 decreased in schizophrenic brain. CRP40 expression increased with increased use of antipsychotic drugs in schizophrenic patients. Mortalin knockdown in the rat medial prefrontal cortex resulted in prepulse inhibition deficits	Mortalin regulation of dopamine may be the critical mechanism linking its role to the pathogenesis of schizophrenia	(Gabriele et al. 2005, 2010a, b)
Friedreich's ataxia	Mortalin was identified as one of the mitochondrial frataxin-interacted proteins by bi-directional co-immunoprecipitation	Mortalin may be involved in iron–sulfur cluster biogenesis	(Shan et al. 2007)
Absence seizure	Mortalin overexpression was observed in the thalamus of the absence seizure mouse model stargazer	Mortalin overexpression may be a defense response initiated by increased basal cellular oxidative stress	(Ryu et al. 2008)

or modification of mortalin has been found in a variety of neurological disorders. Furthermore, manipulation of mortalin expression has been experimentally found to have diverse effects in the different disease conditions. A list of neurological diseases where mortalin has been implicated can be found in Table 9.1. In this chapter, we will discuss each of these diseases, with an emphasis on neurodegenerative disorders (AD and PD), gliomas, and brain ischemia, largely because the understanding of mortalin's roles in these diseases is quickly evolving. We also propose a hypothesis that might explain the major functions of mortalin in common neurological diseases.

9.2 Mortalin in Common Neurodegenerative Diseases

A unique feature of several neurodegenerative diseases is the accumulation of misfolded proteins that adversely affect neuronal function, plasticity and connectivity, ultimately triggering cell death (Meriin and Sherman 2005). This process is exemplified by AD and PD, with the presence of proteinaceous inclusions of senile plaques and neurofibrillary tangles in AD (Castellani et al. 2010) and Lewy bodies in PD (Castellani et al. 2010; Dauer and Przedborski 2003; Forno 1996). Most neurodegenerative diseases are associated with mitochondrial dysfunction, accompanied with an excessive generation of both reactive nitrogen species (RNS) and ROS, which contribute to neuronal cell injury and death (Patten et al. 2010). Given mortalin's chaperone function in mitochondrial biogenesis and protein regulation, it is unsurprising that mortalin is implicated to play a central role in maintaining mitochondrial homeostasis, reducing oxidative stress, and facilitating protein refolding. Therefore, alterations in the expression or modification of mortalin may produce deleterious effects contributing toward neurodegenerative diseases such as AD and PD (Jin et al. 2006; Osorio et al. 2007; Shi et al. 2008).

9.2.1 Mortalin in Alzheimer's Disease

AD, the most common geriatric dementia, affects approximately 13% of the population 65 years of age or older in the United States alone and many more worldwide (Association 2010). Clinical diagnosis of AD is made with memory loss in combination with decreased word comprehension, muscle coordination, object recognition and use, or ability to participate in normal daily activities (Castellani et al. 2010). As discussed, the defining pathologic features of AD include the presence of amyloid beta (Aβ)-containing senile plaques and tau-containing neurofibrillary tangles, primarily distributed in the cerebral cortex and the medial temporal lobes, depending on the stages of the disease (Castellani et al. 2010). The current belief is that synaptic dysfunction is a proximal event in the pathogenesis of AD and eventually leads to neuronal death, events which are thought to be contributed to initially by aggregated soluble Aβ (Castellani et al. 2010).

Several studies conducted *in vitro* and *in vivo* highlight mortalin's potential involvement in AD. When PC12 cells differentiated with nerve growth factor were treated with Aβ to simulate AD, the ability of mitochondria to import proteins was decreased (Sirk et al. 2007). This resulted in enhanced ROS production as well as loss of mitochondrial membrane potential and altered mitochondrial morphology (Sirk et al. 2007), indicating the common AD feature of mitochondrial dysfunction (Kish et al. 1992). Treatment with Aβ also caused a decrease in the expression of mortalin in the mitochondrial fraction of cells, suggesting that mortalin may be involved in the observed Aβ-induced mitochondrial defects (Sirk et al. 2007). A separate study looked at the potential role of mortalin in response to Aβ using the neuronal cell line SH-SY5Y and reported that mortalin overexpression was protective against Aβ induced cell death (Qu et al. 2011). It also protected against excessive ROS production and oxidative damage, as well as loss of mitochondrial membrane potential, complex IV dysfunction, and depletion of ATP in response to Aβ (Qu et al. 2011). When mortalin expression was decreased by siRNA, SH-SY5Y cells were rendered more vulnerable to Aβ-induced cell death (Qu et al. 2011). These results indicate that mortalin may serve a protective role in response to a neurotoxic insult by regulating mitochondrial function.

Mortalin has been further implicated in AD through studies of the *APOE* gene; specifically the ε4 allele, which confers increased risk to developing AD (Castellani et al. 2010). In one study, the *APOE* gene was knocked out in mice and protein oxidation in the hippocampus and cortex was compared to a wild type (WT) control at a young time-point and an old time-point using immunoblotting for protein carbonyls (Choi et al. 2004). Young knockout mice had levels of oxidized proteins which were higher than age-matched WT mice, but were comparable to old WT and knockout mice (Choi et al. 2004). Identification of oxidized proteins using mass spectrometry revealed mortalin to be at a higher oxidation level in *APOE* knockout mice for both time-points (Choi et al. 2004). Thus, oxidation of mortalin may play an important part in AD pathogenesis and progression. Using targeted replacement (TR), the human *APOE*-ε4 and *APOE*-ε3 alleles were put into separate mice, with the *APOE*-ε4 mouse serving as an AD model and the *APOE*-ε3 mouse as the control (Osorio et al. 2007). Fluorescent protein staining and two-dimensional gel electrophoresis of lysates from the hippocampi of each TR mouse, when coupled with mass spectrometry, identified different isoforms of mortalin in each genotype (Osorio et al. 2007). Interestingly, different isoforms of mortalin were found to contain different phosphorylation states. The same study also identified mortalin isoform alterations in AD compared to non-AD hippocampal tissue, as well as differences between AD cases from *APOE*-ε4 and *APOE*-ε3, indicating that mortalin is not only altered in AD, but that an AD-related allele may regulate mortalin (Osorio et al. 2007).

9.2.2 Mortalin in Parkinson's Disease

PD, most characteristically presented as a movement disorder, is initiated by the death of dopaminergic (DAergic) neurons in the substantia nigra pars compacta (SNpc) and the presence of Lewy bodies (LB), intracytoplasmic protein inclusions containing aggregates of, among other proteins, α-synuclein (Dauer and Przedborski 2003; Forno 1996). PD is the second most common serious neurodegenerative disease and its incidence increases with age, with populations over the age of 65 having an estimated disease prevalence of 3% (Orr et al. 2002). Clinical features of PD include bradykinesia, resting tremor, muscle rigidity, and postural instability which are responsive to treatment with levodopa (Jankovic 2008). There are also non-motor symptoms, such as depression and cognitive impairment (Merims and Freedman 2008; Starkstein et al. 1989) that are typically not responsive to dopamine replacement therapies.

Factors influencing the development of PD remain poorly understood despite intensive research efforts, although many investigators hypothesize contributing roles for both genes and environment. Leading mechanistic hypotheses include mitochondrial dysfunction, oxidative stress, and inflammation contributing to cell death (Dick et al. 2007; McCulloch et al. 2008; Warner and Schapira 2003).

Recent studies have implicated mortalin in the pathogenesis of PD. Using quantitative proteomics, it was discovered that mortalin expression is decreased in the mitochondrial-enriched fraction from the SNpc of PD patients compared to controls, a result which was validated using Western blot and immunohistochemistry (Jin et al. 2006). Mortalin is further implicated in PD pathogenesis as its expression was shown to change in association with LB spread from the brain stem to the limbic system and eventually to the isocortex (Shi et al. 2008), a spread which is associated with PD progression, where non-motor symptoms become more common (Braak et al. 2003). Genetic screens for mutations of mortalin in PD patients have added additional evidence supporting the notion that mortalin is involved in PD. In screening 17 exons of the mortalin gene in 330 Spanish PD patients and 250 controls, 3 novel DNA aberrations were found in separate PD patients (De Mena et al. 2009). Two were missense mutations, R126W and P509S, found in the ATPase and substrate-binding domains, respectively, while the third was an insertion in intron eight which was predicted to exert effects on RNA splicing (De Mena et al. 2009). A third mutation, A476T, in the substrate-binding domain of mortalin was also recently discovered in a study of German PD patients (Burbulla et al. 2010). Overexpression of all three mutations in SH-SY5Y cells showed elevated ROS production by the mitochondria, as well as altered mitochondrial membrane potential compared to SH-SY5Y cells overexpressing WT mortalin (Burbulla et al. 2010). Similarly, SH-SY5Y cells overexpressing mortalin had mitochondria which were branched and interconnected, while those overexpressing any of the three mortalin variants exhibited fewer connected mitochondria (Burbulla et al. 2010). These results represent the first functional evidence for a direct contribution of mortalin to impaired mitochondrial function and dynamics in PD.

Various studies have been performed in an attempt to elucidate a mechanism as to how mortalin may contribute to the DAergic neuronal death observed in PD. Treatment of the DAergic cell line MES23.5 (MES) with rotenone, a complex I inhibitor and toxicant commonly used to model PD *in vitro* (Miller et al. 2009), recapitulates the decreased mortalin expression observed in human tissue (Jin et al. 2006). In addition, manipulation of mortalin expression in MES cells resulted in altered sensitivity to rotenone-induced neurotoxicity, oxidative stress, and mitochondrial and proteasomal dysfunction (Jin et al. 2006). Each of these characteristics is implicated in PD pathogenesis (Dauer and Przedborski 2003), indicating that mortalin may play an integral role in neuronal death in PD (Jin et al. 2006). Arguments favoring involvement of mortalin in PD are strengthened by the fact that mortalin interacts with α-synuclein and DJ-1, two proteins involved in familial PD (Dauer and Przedborski 2003; Jin et al. 2007b). Mortalin's interaction with DJ-1 has been reported elsewhere and familial PD-associated mutations of DJ-1 strengthens this interaction (Burbulla et al. 2010; Li et al. 2005). Upon treatment with hydrogen peroxide, a greater amount of mortalin interacts with DJ-1 in mouse NIH3T3 and human 293T cells, indicating oxidative stress may influence the interaction of mortalin with DJ-1 (Li et al. 2005).

A potential link between mortalin and oxidative damage could be DA oxidation, a process thought to contribute to DA neuron death in PD due to the tendency of DA oxidation to produce DA-quinones, superoxide radicals, and hydrogen peroxide, which may cause oxidative damage, mitochondrial dysfunction, and enhanced ROS generation (Cardoso et al. 2005). When rat mitochondria were exposed to DA-quinones, mortalin was shown to be decreased (Van Laar et al. 2008). Mortalin was later found to be covalently modified by exposure to DA in both rat brain mitochondria and SH-SY5Y cells (Van Laar et al. 2009). If DA oxidation is indeed a contributing factor to neuronal death via mortalin modification, particularly through oxidative damage, mortalin's function may be detrimentally affected leading to neuronal dysfunction and death. These findings also elucidate the possibility that the rapid proteolytic degradation of oxidatively-modified mortalin may offer another contribution to the overall decline of the chaperone as mitochondria contain numerous proteases dedicated to degradation of misfolded, denatured, and oxidatively-modified proteins (Koppen and Langer 2007).

Another potential link of mortalin to PD pathogenesis might be through its interaction with α-synuclein (Jin et al. 2007b; Shi et al. 2008), a protein intricately involved in PD development and progression. α-Synuclein is not only a principal protein of which LBs are comprised of, but genetic screens have also found this gene to be mutated or overexpressed in familial PD, causing autosomal dominant forms of PD (Dauer and Przedborski 2003; Forno 1996; Levy et al. 2009). In the mitochondrial-enriched fractions of SH-SHY5Y cells, it was found that overexpression of A53T mutated α-synuclein (one of the three mutations reported in familial PD cases) did not alter mortalin expression levels when compared to cells overexpressing WT α-synuclein (Pennington et al. 2010). This experiment, though still preliminary, indicates that DAergic neuronal death in PD caused by the A53T mutation might act through blocking adaptive changes in mortalin expression in the mitochondria.

9.3 Mortalin in Brain Tumors

Mortalin has been implicated in a wide variety of cancers, including brain tumors, as its increased expression has been associated with not only tumor development, but tumor malignancy in a dose-dependent manner (Wadhwa et al. 2006). Though rare and representing only approximately 2% of all cancers, primary malignant brain tumors accounted for more than 22,000 new cancer diagnoses and over 13,000 deaths in the United States in 2010 (Jemal et al. 2010). Specific to human brain tumors, Takano et al. observed increased mortalin expression in human astrocytomas, with a trend of higher-grade malignancy tumors expressing increasingly more mortalin than lower-grade tumors (Takano et al. 1997). Furthermore, studies targeting the reduction of mortalin expression in immortalized tumor cells have found that its decreased expression results in reduced proliferative capability, phenotypical transformation toward cellular senescence, and ultimately cellular death (Wadhwa et al. 2003a, 2004).

Mechanistically, mortalin's involvement in carcinogenesis has been hypothesized to involve a variety of pathways. A primary finding implicating a direct role of mortalin in tumor development is its cytoplasmic interaction with the tumor-suppressing protein p53. The *mot-2* genetic form of mortalin has been observed to not only co-localize with p53, but also inhibit its expression (Wadhwa et al. 1998). This suppression of p53 has major implications in carcinogenesis, as it leaves the cell vulnerable to uncontrolled proliferation stemming from its inability to arrest the cell-division cycle. Further evidence linking mortalin to p53 expression has been demonstrated by mortalin's reduced expression upon treatment with the therapeutic agent MKT-077, a cationic rhodacyanine dye analogue that binds to mortalin, and subsequent increased p53 expression and cell-cycle arrest in cancer cells (Wadhwa et al. 2000). More recently, mortalin has been found to associate with duplicated centrosomes (Ma et al. 2006), a cellular property often characteristic of cancer cells and whose dysfunction has been proposed to be causally linked to the disease (Zyss and Gergely 2009). The study by Ma et al. further implicated an important role for mortalin in centrosomal duplication by demonstrating its overexpression resulted in a loss of the ability of p53 to suppress duplication of the centrosomes (Ma et al. 2006). A subsequent study identified mortalin interaction with Mps1 kinase, a protein that localizes to and is believed to play a role in the initiation of duplication of the centrosome (Kanai et al. 2007).

Compared to neurodegenerative diseases, investigations on the roles of mortalin in cancer are much more advanced, with several groups having already attempted to target mortalin therapeutically (Lu et al. 2011; Pilzer et al. 2010; Yoo et al. 2010).

9.4 Mortalin in Cerebral Ischemia

Cerebral ischemia results from the interruption of normal blood flow to the brain, often the result of traumatic injury or a cardiovascular disorder such as stroke. Due to their role in energy production, free radical generation, and regulation of apoptosis, mitochondria are known to be one of the primary targets of ischemic injury (Xu et al. 2009). A critical pathologic development following an ischemic event is the excess production of ROS and ensuing cellular oxidative stress resulting from acute mitochondrial dysfunction, leading to death of the affected cells (Saito et al. 2005; Siesjö et al. 1989). Ischemic conditions further place cells under conditions of glucose deprivation (GD) and/or oxygen-glucose deprivation (OGD) in the vicinity of the site of injury. Both *in vitro* and *in vivo* studies have indicated that mortalin has a protective effect against acute ischemic insult. Massa et al. first reported an association of mortalin with ischemia, observing that mortalin expression increased within the immediate injured area for smaller-scale ischemia, while expression additionally increased in regions surrounding the ischemic area following larger-scale ischemic events in the rat brain (Massa et al. 1995). In an *in vitro* study, mortalin overexpressing PC12 cells exhibited slowed and reduced ROS accumulation compared to control cells, suggesting that mortalin could inhibit ROS accumulation (Liu et al. 2005). Giffard et al. have investigated the protective effect of mortalin against ischemic injury in primary astrocytic cultures as well as in the rat brain (Voloboueva et al. 2008; Xu et al. 2009). They have shown that mortalin overexpression in astrocytes results in decreased ROS production and preserved mitochondrial membrane potential during GD, as well as preserved ATP levels and cell viability during OGD. Meanwhile, overexpression of mortalin in rat brain reduced infarct area and improved neurological outcome significantly with decreased ROS levels, and additionally reduced tissue oxidative damage, better preserved mitochondrial function, and attenuated ATP loss. These data suggest that protecting mitochondrial function via mortalin overexpression could also be a potential target for protection from cerebral ischemia.

9.5 Mortalin in Other Neurological Disorders

Besides common neurodegenerative diseases, gliomas, and cerebral ischemia, mortalin has been found to be involved in several other neurological disorders spanning a wide range of disease characteristics, including schizophrenia, Friedreich's ataxia, and absence seizure. Although knowledge of the underlying mechanisms involved in these less common neurological diseases remains quite limited, the potential roles mortalin may play in them are discussed below.

9.5.1 Mortalin in Schizophrenia

The rare mental disorder schizophrenia is characterized by symptoms of psychosis with the presence of delusions and/or hallucinations, loss of motivation, social withdrawal, neurocognitive deficits, depression and bipolarity (Van Os and Kapur 2009). Recent epidemiologic data indicate lifetime prevalence of the illness to be between 0.33 and 0.66%, and incidence to be approximately 10.2–22.0 per 100,000 person-years (McGrath et al. 2008). Multiple genome-wide analyses (GWAS) have indicated mutations in the chromosome and region in which mortalin is encoded (5q31) as a gene locus associated as a risk factor for disease development (Crowe and Vieland 1999; Hong et al. 2004; Lewis et al. 2003; Sklar et al. 2004). A study in human post-mortem ventral striatum isolated from schizophrenic patients and controls indicated a significant decrease in the mortalin-alternative gene product CRP40 in schizophrenic brain samples, and particularly reduced in patients who were not medicated with antipsychotic drugs (Gabriele et al. 2005). Furthermore, a similar study in human post-mortem dorsolateral prefrontal cortex samples isolated from bipolar disorder, schizophrenic, and control patients indicated that CRP40/mortalin expression increased with increased use of antipsychotic drugs in schizophrenic patients only (Gabriele et al. 2010a). These findings indicate mortalin expression may play a role in schizophrenia etiology, further substantiated by the recent observation that mortalin knockdown in the rat medial prefrontal cortex resulted in prepulse inhibition deficits similar to those characteristic of schizophrenic patients (Gabriele et al. 2010b). The authors suggest that mortalin regulation of dopamine may be the critical mechanism linking its role to the pathogenesis of schizophrenia.

9.5.2 *Mortalin in Friedreich's Ataxia*

Friedreich's ataxia is a genetic disorder caused by mutation of the frataxin gene, whose major clinical symptoms are characterized by progressive weakness of the limbs, unsteadiness of gait, and loss of coordination (Lodi et al. 2006; Shan et al. 2007). At a prevalence of approximately 1 in 50,000, it is the most common inherited ataxia (Harding 1981). Although reduced frataxin expression is clearly causal of Friedreich's ataxia, the protein's function remains unclear. Hypothesized roles include reduction of oxidative stress, iron transport and metabolism, and iron–sulfur cluster biogenesis and repair (Shan et al. 2007). In a study aimed at identifying proteins interacting with frataxin, mortalin was identified and confirmed as one of the mitochondrial frataxin-interacted proteins by bi-directional co-immunoprecipitation (Shan et al. 2007). The study also revealed frataxin interaction with ISD11, a protein found to be essential in the manufacture of iron–sulfur clusters (Adam et al. 2006). These results suggest that mortalin may also be involved in iron–sulfur cluster biogenesis.

9.5.3 *Mortalin in Absence Seizure*

Absence seizures account for approximately 10% of epileptic seizures in children, and are characterized by abrupt and repeated short periods of unconsciousness, typically lasting several seconds (Posner 2008). Disease prevalence has been estimated to range from 5 to 50 per 100,000 people based on epidemiologic data (Jallon and Latour 2005). A proteomic study of the absence seizure mouse model stargazer identified mortalin as being overexpressed in the thalamus, but not in the cortex or hippocampus, when compared to its wild type counterpart (Ryu et al. 2008). The authors speculate the observation may be a defense response initiated by increased basal cellular oxidative stress in the mutant mouse, as evidenced by an increased presence of carbonylated proteins.

9.6 Perspectives and Future Research Directions

The studies discussed in this review implicate mortalin as being critically involved in a wide array of neurological disorders. Altogether, evidence points towards the necessity of a delicate balance of its expression to be achieved in order to maintain

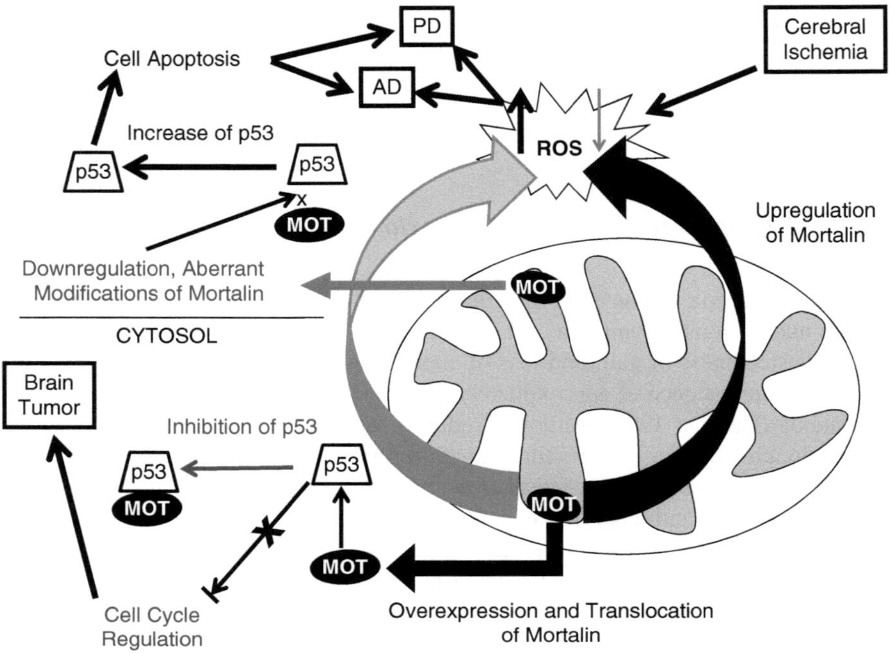

Fig. 9.1 Potential role of mortalin in neurodegenerative disease, ischemia, and brain tumor. Downregulation, underexpression, or aberrant modifications which compromise mortalin's (*mot*) functionality may result in increased cellular oxidative stress response and represents a risk factor for neurodegenerative diseases such as AD and PD (*gray*). Conversely, upregulation of mortalin is associated with neuroprotection following ischemic events due its ability to mitigate oxidative stress (*black*). However, mortalin overexpression may also be detrimental, as after translocation from the mitochondrial matrix to the cytosol it has been found to co-localize with p53 and inhibit its activity, potentially resulting in the loss of cell cycle regulation and the development of tumors. Mortalin's ability to interact with p53 is also hypothesized to play a role in AD & PD, as decreased mortalin expression or modifications which alter its function may result in increased p53 activity, ultimately triggering excessive cellular apoptosis

optimal neurological health. Examples cited above show that its overexpression may be protective, as is the case with cerebral ischemia, or may be deleterious, as is the case with brain carcinogenesis. Conversely, abnormal modifications or underexpression of mortalin may be equally detrimental, as indicated by its roles in neurodegenerative disorders such as AD and PD. One of the common processes involved in mortalin-related pathogenesis appears to be enhanced ROS production by dysfunctional mitochondria, and one principal mediator is p53 (Fig. 9.1). Mortalin's interaction with p53 has been well characterized, as described here with its overexpression shown to reduce p53 expression, and ultimately linked to tumorigenesis. It then stands to reason that its underexpression may result in increased p53 activity. Although a mortalin/p53 link to neurodegeneration has not been established experimentally (the link indicated in Fig. 9.1 is just a hypothesis), its mechanistic role in AD/PD may be inferred hypothetically via p53 regulation. Indeed, p53 has been implicated to play a role in neurodegenerative disease in multiple studies related to PD and AD, which will be discussed below.

Several PD-associated proteins have been shown by genetic studies to interact with p53, including definitively α-synuclein (Alves da Costa et al. 2002), DJ-1 (Bretaud et al. 2007; Shinbo et al. 2005), and parkin (da Costa et al. 2009). In addition, either hypothesized or indirect interactions with p53 have been discussed for LRRK2 (Gloeckner et al. 2009) and PINK-1 (Alves da Costa and Checler 2011; Gloeckner et al. 2009; Stambolic et al. 2001). Furthermore, p53 has been implicated in playing a role in PD based on post-mortem studies which revealed its increased expression in the caudate nucleus of autopsied tissues from parkinsonian subjects as compared to controls (Mogi et al. 2007), as well as an observed increase of phosphorylated p53 in the SNpc of tissue from PD patients (Nair et al. 2006). Additional evidence of a potential role of p53 modulation in PD is implicated by numerous post-mortem studies which have demonstrated the alteration of p53 pathway-related proteins in the disease, including: increased SN phospho-p38 in PD patients (Karunakaran et al. 2008), increased Bax in PD DAergic neurons (Hartmann et al. 2001), and increased GADPH in PD nigral melanized neurons (Tatton 2000).

Further support of p53 involvement in PD is evidenced by studies using *in vitro* and *in vivo* toxicant-induced models of the disease. Treatment with 6-hydroxydopamine (6-OHDA), which is selectively toxic to DAergic neurons, resulted in increased expression of p53 and Bax in PC12 cells (Blum et al. 1997), phosphorylation of p53 and subsequent activation of PUMA in PC12 cells and DAergic neurons derived from embryonic stem cells (Nair 2006), and increased p53 expression in DAergic neurons of rats (Liang et al. 2007). Treatment with 1-methyl-4-phenyl-1,2,3,6-tetrahydropyridine (MPTP), a toxicant commonly used to induce PD-like effects in model systems, resulted in increased p53 expression in SH-SY5Y cells (Kitamura et al. 1998), while DA neurons of transgenic p53-knockout mice were found to be more resistant to MPTP toxicity than their WT counterparts (Trimmer et al. 1996). Additionally, exposure to PC12 cells of MPTP's toxic metabolite MPP+ revealed upregulation of p53 utilizing a transcription-factor protein/DNA binding array approach (Xu et al. 2005). Taken together, post-mortem, *in vitro*, and *in vivo* studies strongly indicate p53 regulation as playing a potentially vital role in apoptotic cellular death associated with PD.

Similarly to PD, p53 regulation has been hypothesized to play a role in the etiology of AD. In a human post-mortem study of AD patients and controls, p53 was found to be increased in the superior temporal gyrus of the diseased brains (Hooper et al. 2007). Another post-mortem study identified elevated p53 in the inferior parietal lobule (IPL) of patients with both mild cognitive impairment (MCI) and AD, with only the increase in AD patients reaching statistical significance (Cenini et al. 2008).

Mechanistically, several *in vitro* and *in vivo* studies have demonstrated the potential involvement of p53 in AD. In the SKN-SH cell line, transfection and subsequent overexpression of Aβ42, whose accumulation in the brain is widely accepted as playing a role in AD pathology (Hardy and Selkoe 2002), resulted in a 6-fold increase in expression of p53 mRNA and induced cytotoxicity, implicating direct binding of

Aβ42 to p53's promoter in response to oxidative stress (Ohyagi et al. 2005). The same study also found evidence for a role of p53 in AD *in vivo*, as degenerating neurons of the Tg2572 mutant mouse model of AD were found to be co-reactive to both Aβ42 and p53 by immunohistochemistry, a finding that was corroborated in similar studies with post-mortem tissue of AD patients (Ohyagi et al. 2005).

Though familial forms of AD are very rare, a small number of genetic mutations have been found to be strongly associated with the disease and can lead to its early onset, including mutations of the presenilin 1 (*PS1*), presenilin 2 (*PS2*), and amyloid protein precursor (*APP*) genes (Rocchi et al. 2003). Some of these mutations have been implicated to alter p53 expression, perhaps explaining in part their deleterious nature. Both *PS1* and *PS2* knockout mice have been observed to express lower levels of p53 in their brains than wild type controls (Saura et al. 2004), while PC12 cells transfected with mutant *PS2* and transgenic mice mutant for the gene have been found to express increased levels of p53 mRNA as compared to their wild type counterparts (Nguyen et al. 2007). APP has also been tied to p53 activity, as its reduced expression was observed to result in a corresponding reduced amount of p53 mRNA and protein expression both *in vitro* and *in vivo* (Alves da Costa et al. 2006).

Finally, in addition to associative and mechanistic evidence implying its role in the disease, peripherally mutated p53 has been identified as a potential biomarker of AD, as AD patients were found to have increased levels of altered protein in blood cells as compared to controls (Lanni et al. 2008). Taken together, it seems that one of the productive lines of research in the near future could be investigating the potential interactions, either physically or physiologically, between mortalin and p53 in neurodegenerative disorders.

9.7 Concluding Remarks

The regulation of properly functioning mortalin has been linked to several neurological disorders with distinct and diverse pathologies. In the most thoroughly characterized ailments outlined above, brain tumor and the neurodegenerative diseases AD and PD, mortalin plays opposing roles, with its overexpression associated with the former and its underexpression associated with the latter. These observations argue for a critical role of mortalin expression neurologically in order to avoid these contrasting disease states. Mortalin's common connection with both ends of this broad spectrum may occur through its known capability to interact with and influence the expression and/or function of p53. As described above, the importance of p53 in both cancer and neurodegeneration has been well established. Furthermore, a link between mortalin overexpression and subsequent p53 decline has been clearly established in cancer etiology. Given mortalin's known reduced expression and/or loss of functionality in neurodegeneration and coincidental increase of p53 expression in these diseases, it is logical to hypothesize that mortalin's contribution to their pathogenesis is at least partially due to its reduced interaction with and subsequent upregulation of p53. Under this scenario, and considering p53's critical role in apoptotic cellular signaling, the dysregulation of p53 consequent of mortalin's reduced expression would be reasonably expected to result in neurodegenerative disease-associated apoptotic death.

Acknowledgments J.J. was supported by funding from National Natural Sciences Foundation of China (30770760). T.J.C., J.G.H., and J.Z. have been supported in part by grants of the NIH (AG033398, ES004696-5897, ES016873, NS062684-6221, NS057567, and ES007032).

References

Adam AC, Bornhövd C, Prokisch H, Neupert W, Hell K (2006) The Nfs1 interacting protein Isd11 has an essential role in Fe/S cluster biogenesis in mitochondria. EMBO J 25:174–183

Alves da Costa C, Checler F (2011) Apoptosis in Parkinson's disease: Is p53 the missing link between genetic and sporadic Parkinsonism? Cell Signal 23:963–968

Alves da Costa C, Paitel E, Vincent B, Checler F (2002) Alpha-synuclein lowers p53-dependent apoptotic response of neuronal cells. Abolishment by 6-hydroxydopamine and implication for Parkinson's disease. J Biol Chem 277:50980–50984

Alves da Costa C, Sunyach C, Pardossi-Piquard R, Sévalle J, Vincent B, Boyer N, Kawarai T, Girardot N, St George-Hyslop P, Checler F (2006) Presenilin-dependent gamma-secretase-mediated control of p53-associated cell death in Alzheimer's disease. J Neurosci 26:6377–6385

Association AS (2010) 2010 Alzheimer's disease facts and figures. Alzheimers Dement 6:158–194

Blum D, Wu Y, Nissou MF, Arnaud S, Alim-Louis-Benabid, Verna JM (1997) p53 and Bax activation in 6-hydroxydopamine-induced apoptosis in PC12 cells. Brain Res 751:139–142

Braak H, Del Tredici K, Rüb U, De Vos R, Jansen Steur E, Braak E (2003) Staging of brain pathology related to sporadic Parkinson's disease. Neurobiol Aging 24:197–211

Bretaud S, Allen C, Ingham PW, Bandmann O (2007) p53-dependent neuronal cell death in a DJ-1-deficient zebrafish model of Parkinson's disease. J Neurochem 100:1626–1635

Burbulla LF, Schelling C, Kato H, Rapaport D, Woitalla D, Schiesling C, Schulte C, Sharma M, Illig T, Bauer P, Jung S, Nordheim A, Schöls L, Riess O, Krüger R (2010) Dissecting the role of the mitochondrial chaperone mortalin in Parkinson's disease: functional impact of disease-related variants on mitochondrial homeostasis. Hum Mol Genet 19:4437–4452

Calvo SE, Mootha VK (2010) The mitochondrial proteome and human disease. Annu Rev Genomics Hum Genet 11:25–44

Cardoso S, Moreira P, Agostinho P, Pereira C, Oliveira C (2005) Neurodegenerative pathways in Parkinson's disease: therapeutic strategies. Curr Drug Targets CNS Neurol Disord 4:405–419

Castellani RJ, Rolston RK, Smith MA (2010) Alzheimer disease. Dis Mon 56:484–546

Cenini G, Sultana R, Memo M, Butterfield DA (2008) Elevated levels of pro-apoptotic p53 and its oxidative modification by the lipid peroxidation product, HNE, in brain from subjects with amnestic mild cognitive impairment and Alzheimer's disease. J Cell Mol Med 12:987–994

Chen H, Yoshioka H, Kim GS, Jung JE, Okami N, Sakata H, Maier CM, Narasimhan P, Goeders CE, Chan PH (2011) Oxidative stress in ischemic brain damage: Mechanisms of cell death and potential molecular targets for neuroprotection. Antioxid Redox Signal 14:1505–1517

Choi J, Forster MJ, Mcdonald SR, Weintraub ST, Carroll CA, Gracy RW (2004) Proteomic identification of specific oxidized proteins in ApoE-knockout mice: relevance to Alzheimer's disease. Free Radic Biol Med 36:1155–1162

Crowe RR, Vieland V (1999) Report of the Chromosome 5 Workshop of the Sixth World Congress on Psychiatric Genetics. Am J Med Genet 88:229–232

da Costa CA, Sunyach C, Giaime E, West A, Corti O, Brice A, Safe S, Abou-Sleiman PM, Wood NW, Takahashi H, Goldberg MS, Shen J, Checler F (2009) Transcriptional repression of p53 by parkin and impairment by mutations associated with autosomal recessive juvenile Parkinson's disease. Nat Cell Biol 11:1370–1375

Dauer W, Przedborski S (2003) Parkinson's disease: mechanisms and models. Neuron 39:889–909

De Mena L, Coto E, Sánchez-Ferrero E, Ribacoba R, Guisasola LM, Salvador C, Blázquez M, Alvarez V (2009) Mutational screening of the mortalin gene (HSPA9) in Parkinson's disease. J Neural Transm 116:1289–1293

Deocaris CC, Kaul SC, Wadhwa R (2008) From proliferative to neurological role of an hsp70 stress chaperone, mortalin. Biogerontology 9:391–403

Deocaris CC, Kaul SC, Wadhwa R (2009) The versatile stress protein mortalin as a chaperone therapeutic agent. Protein Pept Lett 16:517–529

Dick FD, De Palma G, Ahmadi A, Osborne A, Scott NW, Prescott GJ, Bennett J, Semple S, Dick S, Mozzoni P, Haites N, Wettinger SB, Mutti A, Otelea M, Seaton A, Soderkvist P, Felice A, Group GS (2007) Gene-environment interactions in parkinsonism and Parkinson's disease: the Geoparkinson study. Occup Environ Med 64:673–680

Forno LS (1996) Neuropathology of Parkinson's disease. J Neuropathol Exp Neurol 55: 259–272.

Gabriele JP, Chong VZ, Pontoriero GF, Mishra RK (2005) Decreased expression of a 40-kDa catecholamine-regulated protein in the ventral striatum of schizophrenic brain specimens from the Stanley Foundation Neuropathology Consortium. Schizophr Res 74:111–119

Gabriele JP, Pontoriero GF, Thomas N, Ferro MA, Mahadevan G, Maccrimmon DJ, Pristupa ZB, Mishra RK (2010a) Antipsychotic drug use is correlated with CRP40/mortalin mRNA expression in the dorsolateral prefrontal cortex of human postmortem brain specimens. Schizophr Res 119:228–231

Gabriele N, Pontoriero GF, Thomas N, Shethwala SK, Pristupa ZB, Gabriele JP (2010b) Knockdown of mortalin within the medial prefrontal cortex impairs normal sensorimotor gating. Synapse 64:808–813

Gloeckner CJ, Schumacher A, Boldt K, Ueffing M (2009) The Parkinson disease-associated protein kinase LRRK2 exhibits MAPKKK activity and phosphorylates MKK3/6 and MKK4/7, in vitro. J Neurochem 109:959–968

Harding AE (1981) Friedreich's ataxia: a clinical and genetic study of 90 families with an analysis of early diagnostic criteria and intrafamilial clustering of clinical features. Brain 104:589–620

Hardy J, Selkoe DJ (2002) The amyloid hypothesis of Alzheimer's disease: progress and problems on the road to therapeutics. Science 297:353–356

Hartmann A, Michel PP, Troadec JD, Mouatt-Prigent A, Faucheux BA, Ruberg M, Agid Y, Hirsch EC (2001) Is Bax a mitochondrial mediator in apoptotic death of dopaminergic neurons in Parkinson's disease? J Neurochem 76:1785–1793

Hong KS, Mcinnes LA, Service SK, Song T, Lucas J, Silva S, Fournier E, León P, Molina J, Reus VI, Sandkuijl LA, Freimer NB (2004) Genetic mapping using haplotype and model-free linkage analysis supports previous evidence for a locus predisposing to severe bipolar disorder at 5q31–33. Am J Med Genet B Neuropsychiatr Genet 125B:83–86

Hooper C, Meimaridou E, Tavassoli M, Melino G, Lovestone S, Killick R (2007) p53 is upregulated in Alzheimer's disease and induces tau phosphorylation in HEK293a cells. Neurosci Lett 418:34–37

Jallon P, Latour P (2005) Epidemiology of idiopathic generalized epilepsies. Epilepsia 46(Suppl 9):10–14

Jankovic J (2008) Parkinson's disease: clinical features and diagnosis. J Neurol Neurosurg Psychiatry 79:368–376

Jemal A, Siegel R, Xu J, Ward E (2010) Cancer statistics, 2010. CA Cancer J Clin 60:277–300

Jin J, Hulette C, Wang Y, Zhang T, Pan C, Wadhwa R, Zhang J (2006) Proteomic identification of a stress protein, mortalin/mthsp70/GRP75: relevance to Parkinson disease. Mol Cell Proteomics 5:1193–1204

Jin J, Davis J, Zhu D, Kashima DT, Leroueil M, Pan C, Montine KS, Zhang J (2007a) Identification of novel proteins affected by rotenone in mitochondria of dopaminergic cells. BMC Neurosci 8:67

Jin J, Li GJ, Davis J, Zhu D, Wang Y, Pan C, Zhang J (2007b) Identification of novel proteins associated with both alpha-synuclein and DJ-1. Mol Cell Proteomics 6:845–859

Kanai M, Ma Z, Izumi H, Kim SH, Mattison CP, Winey M, Fukasawa K (2007) Physical and functional interaction between mortalin and Mps1 kinase. Genes Cells 12:797–810

Kann O, Kovács R (2007) Mitochondria and neuronal activity. Am J Physiol Cell Physiol 292:C641–657

Karunakaran S, Saeed U, Mishra M, Valli RK, Joshi SD, Meka DP, Seth P, Ravindranath V (2008) Selective activation of p38 mitogen-activated protein kinase in dopaminergic neurons of substantia nigra leads to nuclear translocation of p53 in 1-methyl-4-phenyl-1,2,3,6-tetrahydropyridine-treated mice. J Neurosci 28:12500–12509

Kaul SC, Wadhwa R, Matsuda Y, Hensler PJ, Pereira-Smith OM, Komatsu Y, Mitsui Y (1995) Mouse and human chromosomal assignments of mortalin, a novel member of the murine hsp70 family of proteins. FEBS Lett 361:269–272

Kaul SC, Deocaris CC, Wadhwa R (2007) Three faces of mortalin: a housekeeper, guardian and killer. Exp Gerontol 42:263–274

Kish SJ, Bergeron C, Rajput A, Dozic S, Mastrogiacomo F, Chang LJ, Wilson JM, Distefano LM, Nobrega JN (1992) Brain cytochrome oxidase in Alzheimer's disease. J Neurochem 59:776–779

Kitamura Y, Kosaka T, Kakimura JI, Matsuoka Y, Kohno Y, Nomura Y, Taniguchi T (1998) Protective effects of the antiparkinsonian drugs talipexole and pramipexole against 1-methyl-4-phenylpyridinium-induced apoptotic death in human neuroblastoma SH-SY5Y cells. Mol Pharmacol 54:1046–1054

Koppen M, Langer T (2007) Protein degradation within mitochondria: versatile activities of AAA proteases and other peptidases. Crit Rev Biochem Mol Biol 42:221–242

Lanni C, Racchi M, Mazzini G, Ranzenigo A, Polotti R, Sinforiani E, Olivari L, Barcikowska M, Styczynska M, Kuznicki J, Szybinska A, Govoni S, Memo M, Uberti D (2008) Conformationally altered p53: a novel Alzheimer's disease marker? Mol Psychiatry 13:641–647

Levy O, Malagelada C, Greene L (2009) Cell death pathways in Parkinson's disease: proximal triggers, distal effectors, and final steps. Apoptosis 14:478–500

Lewis CM, Levinson DF, Wise LH, Delisi LE, Straub RE et al. (2003) Genome scan meta-analysis of schizophrenia and bipolar disorder, part II: Schizophrenia. Am J Hum Genet 73:34–48

Li HM, Niki T, Taira T, Iguchi-Ariga SM, Ariga H (2005) Association of DJ-1 with chaperones and enhanced association and colocalization with mitochondrial Hsp70 by oxidative stress. Free Radic Res 39:1091–1099

Liang ZQ, Li YL, Zhao XL, Han R, Wang XX, Wang Y, Chase TN, Bennett MC, Qin ZH (2007) NF-kappaB contributes to 6-hydroxydopamine-induced apoptosis of nigral dopaminergic neurons through p53. Brain Res 1145:190–203

Liu Y, Liu W, Song XD, Zuo J (2005) Effect of GRP75/mthsp70/PBP74/mortalin overexpression on intracellular ATP level, mitochondrial membrane potential and ROS accumulation following glucose deprivation in PC12 cells. Mol Cell Biochem 268:45–51

Lodi R, Tonon C, Calabrese V, Schapira AH (2006) Friedreich's ataxia: from disease mechanisms to therapeutic interventions. Antioxid Redox Signal 8:438–443

Lu WJ, Lee NP, Kaul SC, Lan F, Poon RT, Wadhwa R, Luk JM (2011) Induction of mutant p53-dependent apoptosis in human hepatocellular carcinoma by targeting stress protein mortalin. Int J Cancer (in press)

Ma Z, Izumi H, Kanai M, Kabuyama Y, Ahn NG, Fukasawa K (2006) Mortalin controls centrosome duplication via modulating centrosomal localization of p53. Oncogene 25:5377–5390

Massa SM, Longo FM, Zuo J, Wang S, Chen J, Sharp FR (1995) Cloning of rat grp75, an hsp70-family member, and its expression in normal and ischemic brain. J Neurosci Res 40:807–819

Mcculloch CC, Kay DM, Factor SA, Samii A, Nutt JG, Higgins DS, Griffith A, Roberts JW, Leis BC, Montimurro JS, Zabetia CP, Payami H (2008) Exploring gene-environment interactions in Parkinson's disease. Hum Genet 123:257–265

Mcgrath J, Saha S, Chant D, Welham J (2008) Schizophrenia: a concise overview of incidence, prevalence, and mortality. Epidemiol Rev 30:67–76

Meriin AB, Sherman MY (2005) Role of molecular chaperones in neurodegenerative disorders. Int J Hyperthermia 21:403–419
Merims D, Freedman M (2008) Cognitive and behavioural impairment in Parkinson's disease. Int Rev Psychiatry 20:364–373
Miller R, James-Kracke M, Sun G, Sun A (2009) Oxidative and inflammatory pathways in Parkinson's disease. Neurochem Res 34:55–65
Mizukoshi E, Suzuki M, Loupatov A, Uruno T, Hayashi H, Misono T, Kaul SC, Wadhwa R, Imamura T (1999) Fibroblast growth factor-1 interacts with the glucose-regulated protein GRP75/mortalin. Biochem J 343:461–466
Mizukoshi E, Suzuki M, Misono T, Loupatov A, Munekata E, Kaul SC, Wadhwa R, Imamura T (2001) Cell-cycle dependent tyrosine phosphorylation on mortalin regulates its interaction with fibroblast growth factor-1. Biochem Biophys Res Commun 280:1203–1209
Mogi M, Kondo T, Mizuno Y, Nagatsu T (2007) p53 protein, interferon-gamma, and NF-kappaB levels are elevated in the parkinsonian brain. Neurosci Lett 414:94–97
Morais VA, De Strooper B (2010) Mitochondria dysfunction and neurodegenerative disorders: cause or consequence. J Alzheimers Dis 20(Suppl 2):S255–263
Nair VD (2006) Activation of p53 signaling initiates apoptotic death in a cellular model of Parkinson's disease. Apoptosis 11:955–966
Nair VD, Mcnaught KS, González-Mạeso J, Sealfon SC, Olanow CW (2006) p53 mediates nontranscriptional cell death in dopaminergic cells in response to proteasome inhibition. J Biol Chem 281:39550–39560
Naylor DJ, Hoogenraad NJ, Høj PB (1996) Isolation and characterisation of a cDNA encoding rat mitochondrial GrpE, a stress-inducible nucleotide-exchange factor of ubiquitous appearance in mammalian organs. FEBS Lett 396:181–188
Nguyen HN, Lee MS, Hwang DY, Kim YK, Yoon DY, Lee JW, Yun YP, Lee MK, Oh KW, Hong JT (2007) Mutant presenilin 2 increased oxidative stress and p53 expression in neuronal cells. Biochem Biophys Res Commun 357:174–180
Ohyagi Y, Asahara H, Chui DH, Tsuruta Y, Sakae N, Miyoshi K, Yamada T, Kikuchi H, Taniwaki T, Murai H, Ikezoe K, Furuya H, Kawarabayashi T, Shoji M, Checler F, Iwaki T, Makifuchi T, Takeda K, Kira J, Tabira T (2005) Intracellular Abeta42 activates p53 promoter: a pathway to neurodegeneration in Alzheimer's disease. FASEB J 19:255–257
Ordys BB, Launay S, Deighton RF, Mcculloch J, Whittle IR (2010) The role of mitochondria in glioma pathophysiology. Mol Neurobiol 42:64–75
Orr CF, Rowe DB, Halliday GM (2002) An inflammatory review of Parkinson's disease. Prog Neurobiol 68:325–340
Osorio C, Sullivan PM, He DN, Mace BE, Ervin JF, Strittmatter WJ, Alzate O (2007) Mortalin is regulated by APOE in hippocampus of AD patients and by human APOE in TR mice. Neurobiol Aging 28:1853–1862
Patten DA, Germain M, Kelly MA, Slack RS (2010) Reactive oxygen species: stuck in the middle of neurodegeneration. J Alzheimers Dis 20(Suppl 2):S357–367
Pennington K, Peng J, Hung CC, Banks RE, Robinson PA (2010) Differential effects of wild-type and A53 T mutant isoform of alpha-synuclein on the mitochondrial proteome of differentiated SH-SY5Y cells. J Proteome Res 9:2390–2401
Pilzer D, Saar M, Koya K, Fishelson Z (2010) Mortalin inhibitors sensitize K562 leukemia cells to complement-dependent cytotoxicity. Int J Cancer 126:1428–1435
Posner E (2008) Absence seizures in children. Clin Evid (Online) 2008:0317
Qu M, Zhou Z, Xu S, Chen C, Yu Z, Wang D (2011) Mortalin overexpression attenuates beta-amyloid-induced neurotoxicity in SH-SY5Y cells. Brain Res 1368:336–345
Ran Q, Wadhwa R, Kawai R, Kaul SC, Sifers RN, Bick RJ, Smith JR, Pereira-Smith OM (2000) Extramitochondrial localization of mortalin/mthsp70/PBP74/GRP75. Biochem Biophys Res Commun 275:174–179
Rintoul GL, Reynolds IJ (2010) Mitochondrial trafficking and morphology in neuronal injury. Biochim Biophys Acta 1802:143–150

Rocchi A, Pellegrini S, Siciliano G, Murri L (2003) Causative and susceptibility genes for Alzheimer's disease: a review. Brain Res Bull 61:1–24
Ryu MJ, Lee C, Kim J, Shin HS, Yu MH (2008) Proteomic analysis of stargazer mutant mouse neuronal proteins involved in absence seizure. J Neurochem 104:1260–1270
Saito A, Maier CM, Narasimhan P, Nishi T, Song YS, Yu F, Liu J, Lee YS, Nito C, Kamada H, Dodd RL, Hsieh LB, Hassid B, Kim EE, González M, Chan PH (2005) Oxidative stress and neuronal death/survival signaling in cerebral ischemia. Mol Neurobiol 31:105–116
Saura CA, Choi SY, Beglopoulos V, Malkani S, Zhang D, Shankaranarayana Rao BS, Chattarji S, Kelleher RJ, Kandel ER, Duff K, Kirkwood A, Shen J (2004) Loss of presenilin function causes impairments of memory and synaptic plasticity followed by age-dependent neurodegeneration. Neuron 42:23–36
Schatz G (1995) Mitochondria: beyond oxidative phosphorylation. Biochim Biophys Acta 1271:123–126
Shan Y, Napoli E, Cortopassi G (2007) Mitochondrial frataxin interacts with ISD11 of the NFS1/ISCU complex and multiple mitochondrial chaperones. Hum Mol Genet 16:929–941
Shi M, Jin J, Wang Y, Beyer RP, Kitsou E, Albin RL, Gearing M, Pan C, Zhang J (2008) Mortalin: a protein associated with progression of Parkinson disease? J Neuropathol Exp Neurol 67:117–124
Shinbo Y, Taira T, Niki T, Iguchi-Ariga SM, Ariga H (2005) DJ-1 restores p53 transcription activity inhibited by Topors/p53BP3. Int J Oncol 26:641–648
Siesjö BK, Agardh CD, Bengtsson F (1989) Free radicals and brain damage. Cerebrovasc Brain Metab Rev 1:165–211
Silver I, Erecinska M (1998) Oxygen and ion concentrations in normoxic and hypoxic brain cells. Adv Exp Med Biol 454:7–16
Sirk D, Zhu Z, Wadia JS, Shulyakova N, Phan N, Fong J, Mills LR (2007) Chronic exposure to sub-lethal beta-amyloid (Abeta) inhibits the import of nuclear-encoded proteins to mitochondria in differentiated PC12 cells. J Neurochem 103:1989–2003
Sklar P, Pato MT, Kirby A, Petryshen TL, Medeiros H, Carvalho C, Macedo A, Dourado A, Coelho I, Valente J, Soares MJ, Ferreira CP, Lei M, Verner A, Hudson TJ, Morley CP, Kennedy JL, Azevedo MH, Lander E, Daly MJ, Pato CN (2004) Genome-wide scan in Portuguese Island families identifies 5q31–5q35 as a susceptibility locus for schizophrenia and psychosis. Mol Psychiatry 9:213–218
Stambolic V, Macpherson D, Sas D, Lin Y, Snow B, Jang Y, Benchimol S, Mak TW (2001) Regulation of PTEN transcription by p53. Mol Cell 8:317–325
Starkstein SE, Preziosi TJ, Berthier ML, Bolduc PL, Mayberg HS, Robinson RG (1989) Depression and cognitive impairment in Parkinson's disease. Brain 112:1141–1153
Takano S, Wadhwa R, Yoshii Y, Nose T, Kaul SC, Mitsui Y (1997) Elevated levels of mortalin expression in human brain tumors. Exp Cell Res 237:38–45
Takano S, Wadhwa R, Mitsui Y, Kaul SC (2001) Identification and characterization of molecular interactions between glucose-regulated proteins (GRPs) mortalin/GRP75/peptide-binding protein 74 (PBP74) and GRP94. Biochem J 357:393–398
Tatton NA (2000) Increased caspase 3 and Bax immunoreactivity accompany nuclear GAPDH translocation and neuronal apoptosis in Parkinson's disease. Exp Neurol 166:29–43
Taurin S, Seyrantepe V, Orlov SN, Tremblay TL, Thibault P, Bennett MR, Hamet P, Pshezhetsky AV (2002) Proteome analysis and functional expression identify mortalin as an antiapoptotic gene induced by elevation of [Na+]i/[K+]i ratio in cultured vascular smooth muscle cells. Circ Res 91:915–922
Trimmer PA, Smith TS, Jung AB, Bennett JP (1996) Dopamine neurons from transgenic mice with a knockout of the p53 gene resist MPTP neurotoxicity. Neurodegeneration 5:233–239
Van Laar VS, Dukes AA, Cascio M, Hastings TG (2008) Proteomic analysis of rat brain mitochondria following exposure to dopamine quinone: implications for Parkinson disease. Neurobiol Dis 29:477–489

Van Laar VS, Mishizen AJ, Cascio M, Hastings TG (2009) Proteomic identification of dopamine-conjugated proteins from isolated rat brain mitochondria and SH-SY5Y cells. Neurobiol Dis 34:487–500

Van Os J, Kapur S (2009) Schizophrenia. Lancet 374:635–645

Voloboueva LA, Duan M, Ouyang Y, Emery JF, Stoy C, Giffard RG (2008) Overexpression of mitochondrial Hsp70/Hsp75 protects astrocytes against ischemic injury in vitro. J Cereb Blood Flow Metab 28:1009–1016

Wadhwa R, Takano S, Robert M, Yoshida A, Nomura H, Reddel RR, Mitsui Y, Kaul SC (1998) Inactivation of tumor suppressor p53 by mot-2, a hsp70 family member. J Biol Chem 273:29586–29591

Wadhwa R, Sugihara T, Yoshida A, Nomura H, Reddel RR, Simpson R, Maruta H, Kaul SC (2000) Selective toxicity of MKT-077 to cancer cells is mediated by its binding to the hsp70 family protein mot-2 and reactivation of p53 function. Cancer Res 60:6818–6821

Wadhwa R, Taira K, Kaul SC (2002) An Hsp70 family chaperone, mortalin/mthsp70/PBP74/Grp75: what, when, and where? Cell Stress Chaperones 7:309–316

Wadhwa R, Ando H, Kawasaki H, Taira K, Kaul SC (2003a) Targeting mortalin using conventional and RNA-helicase-coupled hammerhead ribozymes. EMBO Rep 4:595–601

Wadhwa R, Yaguchi T, Hasan MK, Taira K, Kaul SC (2003b) Mortalin-MPD (mevalonate pyrophosphate decarboxylase) interactions and their role in control of cellular proliferation. Biochem Biophys Res Commun 302:735–742

Wadhwa R, Takano S, Taira K, Kaul SC (2004) Reduction in mortalin level by its antisense expression causes senescence-like growth arrest in human immortalized cells. J Gene Med 6:439–444

Wadhwa R, Takano S, Kaur K, Deocaris CC, Pereira-Smith OM, Reddel RR, Kaul SC (2006) Upregulation of mortalin/mthsp70/Grp75 contributes to human carcinogenesis. Int J Cancer 118:2973–2980

Warner TT, Schapira AH (2003) Genetic and environmental factors in the cause of Parkinson's disease. Ann Neurol 53(Suppl 3):S16–23 (discussion S23–25)

Xu L, Voloboueva LA, Ouyang Y, Emery JF, Giffard RG (2009) Overexpression of mitochondrial Hsp70/Hsp75 in rat brain protects mitochondria, reduces oxidative stress, and protects from focal ischemia. J Cereb Blood Flow Metab 29:365–374

Xu Z, Cawthon D, Mccastlain KA, Duhart HM, Newport GD, Fang H, Patterson TA, Slikker W, Ali SF (2005) Selective alterations of transcription factors in MPP+-induced neurotoxicity in PC12 cells. Neurotoxicol 26:729–737

Yoo JY, Ryu J, Gao R, Yaguchi T, Kaul SC, Wadhwa R, Yun CO (2010) Tumor suppression by apoptotic and anti-angiogenic effects of mortalin-targeting adeno-oncolytic virus. J Gene Med 12:586–595

Zyss D, Gergely F (2009) Centrosome function in cancer: guilty or innocent? Trends Cell Biol 19:334–346

Chapter 10
Loss of Mortalin Function in Parkinson's Disease-Supporting the Mitochondrial Pathway of Neurodegeneration

Lena F. Burbulla and Rejko Krüger

Abstract Dysfunctional mitochondria have been early implicated in the neurodegenerative process leading to idiopathic Parkinson's disease (PD) based on biochemical and immunohistochemical findings in affected brain regions of PD patients. The identification of genetic causes of PD during the last decade allowed for the first time to study disease-relevant molecular signaling cascades of neurodegeneration. Based on rare families with autosomal recessively inherited forms of early onset PD, first genes were identified that encode proteins that are critical for the maintenance of mitochondrial function and morphology and therefore contribute to the mitochondrial phenotype of PD. For instance, loss of function mutations in the *DJ-1* gene cause impaired mitochondrial respiration with accumulation of free oxygen species, disruption of the mitochondrial energy metabolism and characteristic morphological changes of these organelles. Interestingly, the mitochondrial chaperone protein mortalin was identified as a DJ-1-interacting protein by several groups. Subsequent genetic studies in large cohorts of PD patients revealed first loss of function mutations as risk factors for PD. These disease-associated mortalin variants showed effects on mitochondrial dynamics that paralleled the phenotype observed upon lack of DJ-1 in different *in vitro* and *ex vivo* models. These results provided first evidence for a mitochondrial matrix chaperone involved in neuronal cell death in PD and strengthen the relevance of a mitochondrial endophenotype in PD. Therefore, functional studies of mortalin allow to dissect the signaling pathways involved in mitochondrial causes of neurodegeneration in PD and may provide new targets for therapeutic approaches in PD.

Keywords DJ-1 · Mitochondrial dysfunction · Mutations · Oxidative stress · Parkinson's disease

---

R. Krüger (✉) · L. F. Burbulla
Laboratory of Functional Neurogenomics, Center of Neurology, Hertie-Institute for Clinical Brain Research and German Center for Neurodegenerative Diseases (DZNE), University of Tübingen, Hoppe-Seyler-Str. 3, 72076 Tübingen, Germany
e-mail: rejko.krueger@uni-tuebingen.de

S. C. Kaul, R. Wadhwa (eds.), *Mortalin Biology: Life, Stress and Death*, 159
DOI 10.1007/978-94-007-3027-4_10, © Springer Science+Business Media B.V. 2012

10.1 Current Concepts on Parkinson's Disease Pathogenesis

Parkinson's disease (PD) is the most common neurodegenerative movement disorder and affects more than 2.5% of the population over age 70 (de Rijk et al. 2000). As aging is commonly recognized as the most significant risk factor for the disease, the number of PD patients is supposed to further increase during the next decades in the aging populations of the Western hemisphere (Hindle 2010). Based on recent progress in molecular genetics, first insight into the molecular mechanisms leading to neurodegeneration was gained and subsequently revolutionized our understanding of PD as a hitherto 'idiopathic' and 'sporadic' disease. For up to 10% of all PD cases, genes were identified that lead to a Mendelian inheritance of the disease. Interestingly, large genetic studies in different populations worldwide have recently provided evidence that low penetrance variants in some of these genes also contribute susceptibility to the common sporadic form of PD (Maraganore et al. 2006; Simon-Sánchez et al. 2009).

10.1.1 Clinicopathological Characteristics of Parkinson's Disease

PD is typically characterized by its motor symptoms that include a slowing of movements, a tremor, muscular rigidity and postural instability. These symptoms are generally linked to impaired dopaminergic neurotransmission caused by degeneration of dopamine-producing neurons in the substantia nigra pars compacta (SNpc) in the brainstem [reviewed in (Schiesling et al. 2008)]. The pathological hallmark of PD is the presence of intracytoplasmic protein inclusions, called Lewy bodies, in affected brain regions. Therefore therapeutic strategies that aim to restore the deficit in dopamine by exogenous substitution of dopamine via its precursor levodopa are effective in compensating most of these motor symptoms.

However, as already mentioned in the first description of the disease by James Parkinson in 1817, a variety of non-motor symptoms including depression, obstipation, pain, REM sleep behavior disorder, dementia and impaired olfaction are typically observed in PD patients and hint towards a more general neurodegenerative process (Chaudhuri and Schapira 2009). Neuronal death in the central nervous system implies i.e. impaired serotonergic transmission in the raphe nucleus that is linked to depression, or loss of neurons in the cholinergic nucleus basalis Meynert that typically relates to cognitive impairment. Although knowledge about pathological alterations in PD brains, i.e. characteristic protein aggregation and oxidative damage of biological molecules, already existed for a long time, the pathways underlying the neurodegenerative process were still not well understood.

10.1.2 The Role of Mitochondria in Neurodegeneration in PD

Free oxygen radicals leading to oxidative damage of lipid membranes, proteins and DNA molecules have been early recognized as linked to neuronal cell death in PD brains. Therefore the mitochondria as the sites of energy generation by oxidative phosphorylation and major source of free radicals in the cell came into focus of interest (Gotz et al. 1990). First evidence for a direct involvement of impaired mitochondrial function in PD came from studies that defined a loss of mitochondrial complex I activity in the substantia nigra of PD patients compared to controls (Schapira et al. 1989). Indeed, subsequent studies confirmed this defect of the respiratory chain components also in other tissues like muscle and blood cells from PD patients (Schapira 1994; Mann et al. 1992). However, genetic studies found no primary mutations in the mitochondrial DNA (mtDNA) that could explain the complex I deficiency in PD patients and, vice versa, known mutations in the mtDNA were only rarely present with signs of parkinsonism (Morgan-Hughes et al. 1995).

Therefore mitochondrial damage was thought to primarily result from toxic insult caused by environmental factors. Indeed, first evidence for an exogenous toxin that could cause parkinsonism in humans was identified in drug addicts that accidentally incorporated a synthetic drug containing 1-methyl-4-tetrahydropyridine [MPTP; (Langston et al. 1983)]. Some of these individuals consecutively developed an akinetic-rigidic syndrome that was responsive to levodopa therapy. Subsequent *in vitro* and *in vivo* experiments revealed that MPP^+ as the active metabolite of MPTP, was selectively taken up into dopaminergic neurons by the dopamine transporter and caused mitochondrial damage by complex I inhibition (Nicklas et al. 1985). However, although rare additional cases of acute parkinsonism due to drug exposure or infection were reported, until now, no further relevant environmental factor causing PD could be defined (Matzler et al. 2007; Lopez-Alberola et al. 2009).

10.1.3 Genetic Contribution to Mitochondrial Dysfunction in Parkinson's Disease

Although genetic contributions to PD were initially discounted, the systematic genetic studies performed during the last 15 years in PD have profoundly changed our views on the etiology of the disease. Based on rare families with autosomal recessive inheritance of PD, several nuclear-encoded genes were identified that may directly affect mitochondrial function due to loss of function mutations, i.e. Parkin, PINK1 and DJ-1 (Kitada et al. 1998; Valente et al. 2004; Bonifati et al. 2003). These PD-associated genes encode proteins that are translocated to (Parkin, DJ-1) or transported into (PINK1, Omi/HtrA2) mitochondria to exert a protective function in terms of oxidative stress and/or mitochondrial dysfunction (Canet-Aviles et al. 2004; Strauss et al. 2005; Narendra et al. 2008; Dagda et al. 2009; Jin et al. 2010). Only recently, the interplay of PINK1 and Parkin in mitochondrial quality control was elucidated. This includes the recruitment of cytosolic Parkin to dysfunctional mitochondria after stabilization of PINK1 at the outer mitochondrial membrane. Upon interaction and potentially phosphorylation of Parkin by PINK1, this allows the effective clearance of dysfunctional mitochondria via ubiquitination of outer membrane proteins, i.e. voltage-dependent anion channel (VDAC) by Parkin, and subsequent lysosomal degradation of the organelles (Narendra et al. 2009; Jin et al. 2010; Gegg et al. 2010; Geisler et al. 2010).

Interestingly, DJ-1 acts in parallel to the PINK1/Parkin-mediated signaling pathway to protect mitochondria from oxidative insult and also influences autophagic clearance of mitochondria (Krebiehl et al. 2010; Thomas et al. 2011). Loss of DJ-1 causes dysfunction of mitochondrial complex I activity and increased levels of reactive oxygen species (ROS) within mitochondria that ultimately lead to the disruption of the mitochondrial membrane potential [MMP; (Krebiehl et al. 2010)]. Moreover, the DJ-1 protein that is mutated in only rare cases of familial PD turned out to be a consistent biomarker in cerebrospinal fluid for the common sporadic form of the disease (Hong et al. 2010). Interestingly, the chaperone mortalin (also known as GRP75, mtHsp70, PBP74) was consistently described as the only intramitochondrial protein interacting with DJ-1 in independent and unbiased proteomic approaches (Jin et al. 2007; Burbulla et al. 2010b). This was the first functional link of mortalin with a PD-associated protein and offered new perspectives for a role of mortalin in disease-related signaling cascades leading to neurodegeneration in PD.

10.2 Mortalin in Aging and Neurodegeneration

Aging is recognized as the most important risk factor for late-onset neurodegenerative disorders. Experimental data from mortalin indicate that it is involved in both processes, aging and neurodegeneration, thereby influencing the risk to develop neurodegenerative disorders such as PD and Alzheimer's disease (AD). In contrast to its pro-survival role that—upon strong overexpression—is crucially involved in the development of cancer (i.e brain tumors), mortalin is downregulated in conditions leading to neurodegeneration (Hsu et al. 2008; Takano et al. 1997). This indicates the need of a tight regulation of mortalin expression within the cell and points to a role of mortalin as an anti-aging protein with a physiological role in the continuous securing of cell survival. Indeed moderate overexpression of mortalin leads to increased life span *in vitro* (human cells) and *in vivo* (nematodes) (Kaul et al. 2003; Yokoyama et al. 2002).

10.2.1 Mortalin and Its Relation to Mitochondrial Pathology in PD

Neurodegeneration in PD is thought to be caused at least partially by oxidative stress and mitochondrial damage, which then initiate signaling cascades leading to cell death. Mitochondria use oxidative phosphorylation to provide energy (ATP), but as a by-product generate toxic ROS. These toxic species may damage many biological molecules (i.e. lipid membranes, DNA) including mitochondria and their DNA. In consequence, mitochondria accumulate mutations and damage over their lifetime, which results in impaired function during aging. As such, mitochondria represent a prime target for age-related insults (Wallace 2005). In particular, substantia nigra neurons are exposed to high levels of ROS that arise from the turnover of the neurotransmitter dopamine and may explain the relatively high proportion of mtDNA mutations observed in PD patients (Bender et al. 2006). This study provides a strong correlation between age-related accumulation of mtDNA deletions and the selective neuronal loss observed in neurodegeneration in PD.

Furthermore, impaired mitochondrial function due to genetic defects in mitochondrial proteins is frequently linked to impaired mitochondrial dynamics, i.e. imbalanced fusion and/or fission events or defective axonal transport and cause neurodegeneration (Mandemakers et al. 2007). For instance, mutations in the mitochondrially-targeted regulators of mitochondrial dynamics OPA1 (fusion), mitofusin 2 (fusion) and GDAP1 (fission) were identified as causes of autosomal dominant optic nerve atrophy or autosomal dominant peripheral neuropathy (Alexander et al. 2000; Züchner et al. 2004; Niemann et al. 2005). PD-associated proteins like Parkin, DJ-1 and PINK1 also modulate mitochondrial function and dynamics, when loss of function mutations occur (Krebiehl et al. 2010; Exner et al. 2007; Mortiboys et al. 2008).

Interestingly, mortalin has been identified as a novel interaction partner of DJ-1 by different methods including unbiased proteomic approaches (Li et al. 2005; Jin et al. 2007; Burbulla et al. 2010b). This mitochondrial chaperone was already known to play an important role for the maintenance of mitochondrial homeostasis upon stress (Liu et al. 2005; Yaguchi et al. 2007). The majority of mortalin is located in the mitochondrial matrix, where it acts as chaperone belonging to the heat shock protein 70 family (mtHsp70). The nuclear-encoded protein reaches its final destination after active import via the translocases of the mitochondrial outer and inner membranes (Webster et al. 1994; Rehling et al. 2004). Importantly, mortalin itself takes an active role in the import of other proteins via the translocases of the mitochondrial inner membrane and has been identified as the only ATPase component within this import complex. This indicates an essential role of mortalin in the effective import of nuclear encoded proteins into mitochondria. As part of the mtHsp70 system, mortalin also plays a critical function in protein quality control in the mitochondrial matrix compartment, thereby regulating mitochondrial biogenesis (Savel'ev et al. 1998; Bender et al. 2011).

10.2.2 Mortalin and Neurodegeneration in PD

Defects in the mitochondrial energy supply of the neurons in the SNpc of PD patients were an early finding from pathoanatomical and biochemical studies (Schapira et al. 1990). In this context, the dopaminergic neurons of the SNpc may be more susceptible to mitochondrial damage as they display a relative low mitochondrial mass compared to neurons in other brain regions as shown in mice (Liang et al. 2007). This could therefore contribute to the specific pattern of neurodegeneration observed in PD and count as responsible for the motor phenotype.

Interestingly, in accordance with the dopaminergic nerve cell loss in PD, recently, a substantial decrease of mortalin protein levels in affected brain areas of PD patients has been reported that was observed in the mitochondrial fraction prepared from SNpc (Jin et al. 2006). Subsequent studies revealed similar changes in the frontal cortex of PD patients and suggested disease-stage dependent loss of mortalin in this brain region (Shi et al. 2008). The observed quantitative decrease of mortalin in PD patients that correlated with more advanced disease stages suggested a novel marker for disease progression in PD. However, it still remains to be determined, whether reduced mortalin levels were a consequence of the disease process or rather the cause of progression of PD.

Additionally, it has been shown by the same group that mortalin is a modulator of rotenone-mediated toxicity via pathways involving oxidative stress, mitochondrial and proteasomal dysfunction (Jin et al. 2006). Rotenone is a potent inhibitor of mitochondrial complex I activity and linked to toxic models of PD pathogenesis in rodents (Betarbet et al. 2000). This further links mortalin to pathways of exogenic toxins related to mitochondrial impairments that recently gained further support by the reproduction of neuroanatomical and neurochemical features of PD in mice that intragastrically received chronic low doses of rotenone (Pan-Montojo et al. 2010). In correlation to the motor phenotype, these animals showed an accumulation of alpha-synuclein in the enteric nervous system that subsequently spread throughout the central nervous system as observed in the human disease condition (Braak et al. 2003).

Together these findings suggest a critical regulation of mortalin levels in relevant brain areas and may provide an answer to the question what determines the specific susceptibility of dopaminergic neurons to oxidative stress in PD.

10.2.3 Functional Implications of the Interaction Between Mortalin and DJ-1

First evidence for a functional relationship between mortalin and DJ-1 came from both, hypothesis-driven and unbiased proteomic approaches that revealed a direct interaction between these two proteins (Jin et al. 2006; Burbulla et al. 2010b; Li et al. 2005). There are conflicting findings on whether loss of function mutations related to DJ-1-mediated pathology display a stronger or weaker interaction with mortalin (Yaguchi et al. 2007; Li et al. 2005). We also observed interaction of mortalin with different variants of DJ-1 and found that this interaction was mediated by the substrate binding domain (Burbulla et al. 2010b; Burbulla and Krüger unpublished).

However the mitochondrial chaperone mortalin is not just simply a binding partner of DJ-1, their relation can truly be characterized as multidimensional, as mortalin and DJ-1 share subcellular localization (mitochondria), functional overlap (oxidative stress sensor, modulator of mitochondrial homeostasis) and disease-related mechanism [(loss of function; reviewed in (Burbulla et al. 2010a)].

Indeed both proteins act as indicators of the neurodegenerative process in the common idiopathic form of PD as the amount of mortalin is decreased in different brain regions of PD patients and DJ-1 levels are reduced in the CSF of PD patients compared to controls (Hong et al. 2010; Jin et al. 2006). Moreover, both proteins are involved in oxidative stress response, with DJ-1 being an important quencher of oxidative damage that is itself oxidatively damaged in brains of sporadic PD patients (Choi et al. 2006). Interestingly, the interaction and co-localization of mortalin and DJ-1 at the mitochondria is further increased during oxidative stress *in vitro* (Li et al. 2005). These findings imply that for both proteins, an important role in the neuronal defense mechanism against oxidative insults can be assumed. Neurons in the SNpc are especially exposed to dopamine oxidation. Interestingly, both, DJ-1 and mortalin, are a target of covalent modifications by dopamine as shown by a proteomic approach in isolated mitochondria from rat brain and human dopaminergic neuroblastoma cells (Van Laar et al. 2009). As the formation of reactive dopamine quinones is related to dysfunction of mitochondrial respiration, the respective modifications of the DJ-1 and mortalin protein could contribute to dysregulation and mitochondrial damage.

A critical role of DJ-1 and mortalin for mitochondrial homeostasis, dynamics and morphology has been recently recognized by different *in vitro* and *in vivo* approaches (Burbulla et al. 2010b; Krebiehl et al. 2010; Thomas et al. 2011; Irrcher et al. 2010). Loss of DJ-1 as well as loss of mortalin function in different human *in vitro* and *ex vivo* models cause a loss of mitochondrial interconnection that is paralleled by an energy depletion of the mitochondrial compartment (Burbulla et al. 2010b; Krebiehl et al. 2010). For DJ-1 additional effects on downstream lysosomal degradation pathways were observed that modulate the clearance of dysfunctional mitochondria from the cell (Krebiehl et al. 2010; Thomas et al. 2011). In this context it still remains to be determined whether or how mortalin may also take part in this function of DJ-1 and may modulate autophagic pathways.

10.2.4 Genetic Evidence for a Role of Mortalin in Parkinson's Disease

Whereas rare large families with Mendelian forms of PD are accessible to linkage studies in order to define the respective disease-causing gene, the more complex genetics of the common sporadic form of PD were a major challenge for geneticists during the last 15 years. Based on genome wide linkage studies in small families with PD and whole genome association studies in sporadic PD patients, an accumulating evidence for a risk locus for PD on the long arm of chromosome 5 was provided by recent studies (Scott et al. 2001; Hicks et al. 2002; Martinez et al. 2004; Maraganore et al. 2005; Sharma et al. 2009; Pankratz et al. 2003). Interestingly, the implicated chromosomal region harbored the *mortalin* gene located at 5q31.1. Thus, based on genetic and functional data, the human *mortalin* gene was a good candidate for a genetic mutation screening in PD patients to define genetic factors contributing to neurodegeneration in PD.

Thereby, the first coding variants in the human *mortalin* gene have been discovered in the Spanish and the German population (De Mena et al. 2009; Burbulla et al. 2010b) and subsequently functionally characterized concerning their role on mitochondrial homeostasis and morphology (Burbulla et al. 2010b). Cellular overexpression as well as knockdown models revealed a loss of the protective physiological mortalin function for all three disease-associated variants known to date. Both, human dopaminergic cell lines as well as fibroblasts from a heterozygous carrier of one of the mortalin variants showed an impaired mitochondrial integrity with impairments in morphology and dynamics (Burbulla et al. 2010b).

All disease-associated variants were observed in the heterozygous state in affected carriers. Two of the variants (p. R126 W, p. P509 S) were each identified in single sporadic PD patients (De Mena et al. 2009), however, the p. A476 T variant was observed in nine individuals after screening more than 2900 PD patients and controls. Most of the carriers fulfilled either the diagnostic criteria of PD or revealed extrapyramidal symptoms as assessed by a neurologist. Asymptomatic carriers of the p. A476 T variant were significantly younger than the typical age of onset (Burbulla et al. 2010b).

Interestingly, all currently identified disease-associated variants of the *mortalin* gene were observed in the heterozygous state. Although heterozygosity of these variants in the *mortalin* gene suggests an autosomal dominant effect, based on the lack of clear Mendelian inheritance in families of the identified affected carriers, it seems to be more likely a genetic trait with reduced penetrance and argues in favor of a genetic susceptibility factor for PD (Burbulla et al. 2010b). The complete loss of physiological protective mortalin function is not compatible with cell survival in various models from yeast to vertebrates (Slater and Craig 1987; Deocaris et al. 2008; Macario et al. 2010). Thus, heterozygous variants in the *mortalin* gene would be in line with the concept of a risk factor contributing to a late onset of neurodegenerative disease.

10.3 Parkinson's Disease-Associated Variants of Mortalin

10.3.1 Structure of the Mortalin Protein and Implications of Disease-Associated Variants

Mortalin is a 679 amino acid protein that has been found in multiple subcellular localizations such as the endoplasmic reticulum (ER), cytoplasmic vesicles, the cytosol and mitochondria (Kaul et al. 2002; Ran et al. 2000). The protein structure is subdivided in two major domains, i.e. ATPase and substrate binding domain, with individual characteristics. Mortalin is a nuclear encoded protein that is predominantly located in the mitochondria (Bhattacharyya et al. 1995) and therefore carries an N-terminal mitochondrial targeting sequence (amino acid position 1–47). The ATPase domain (amino acid position 57–434) is located just behind the aminoterminal start of the mature protein followed by the substrate binding domain (amino acid position 435–588). The oligomerization domain at the C-terminus (amino acid position 589–679) ensures stable structures leading to an oligomerized multivalent state of the other domains in the molecule, hence providing increased structure stabilization and combined functions of the different domains (Engel and Kammerer 2000).

To date only three PD-associated variants in the *mortalin* gene leading to changes in the amino acid sequence of the protein were described (Burbulla et al. 2010b; De Mena et al. 2009). One variant (R126 W) is located in the ATPase domain, whereas the other two variants (A476 T, P509 S) are located in the substrate binding domain. Interspecies comparison of the amino acid sequence of mortalin homologues showed complete conservation of the respective amino acid residues from vertebrates to yeast and therefore gave first evidence for the pathogenic relevance of the respective variants in PD (Burbulla et al. 2010b).

Indeed, known artificial mutations in different mortalin homologues allowed to speculate on structural and functional implications of the human variants. It has already been shown that mutations in the yeast homologue of mortalin, SSC1, revealed differential effects depending on the domain of the protein that was affected (Gambill et al. 1993). In this context, two temperature sensitive mutants were extensively studied and showed that an amino acid substitution in the ATPase domain (ssc1–3) caused more severe defects in translocation than the P442 S (ssc1–2) mutation that is located in the substrate binding domain (Gambill et al. 1993). While mutations located in the substrate binding domain affect the unfolding of the polypeptide chain for translocation across the mitochondrial membranes and the completion of the transport of matrix-targeted pre-proteins across the inner membrane, a mutation in the ATPase domain causes a more severe defect due to the lack of propensity to bind preproteins (Gambill et al. 1993). This supported the notion that ATP is needed for mtHsp70 to bind to the polypeptide chain and promote its translocation across the inner membrane (Gambill et al. 1993).

It has been shown in the E. coli homologue of mortalin, DnaK, that the arch mutants L450 A and A429 W, the latter corresponding to an A475 W exchange in the human mortalin peptide and therefore being located next to the identified PD-associated A476 T mortalin variant, affected the substrate binding of DnaK by impeding the formation of a hydrophobic arch spanning over the bound substrate (Rüdiger et al. 2000).

Concerning the identified human variants that were associated with PD, irrespective of the localization of these variants within the mortalin protein, our results in human cells argue in favor of a loss of protective mortalin function in the mitochondria as a cause of PD (Burbulla et al. 2010b). All disease-associated mortalin variants showed the same *in vitro* import kinetics into mitochondria, implicating that the different mutations do not affect the physiological subcellular localization of the mortalin protein itself. The observed functional differences (see 10.3.2) would be consequently not due to defective import of the mortalin variants into mitochondria, but may be rather due to a loss of functional mortalin within the mitochondrial matrix.

10.3.2 The Effect of Mortalin Variants on Mitochondrial Function

As mitochondria are the major source of energy and neurons in the brain are the major consumers of energy, impaired mitochondrial function directly results in the loss of neuronal integrity. Several PD-associated mitochondrial proteins are related to mitochondrial homeostasis. All known PD-associated mortalin variants lead to a loss of protective function of the respective protein (Fig. 10.1). Observed differences in the levels of ROS and the MMP in neuronal cell lines could be traced back to a loss of protective function of the wild type protein. Indeed, knockdown of *mortalin* in human cells caused an increase in ROS levels and a reduced MMP, which could not be restored by the reported variants, whereas the physiological protein was able to rescue this mitochondrial phenotype (Burbulla et al. 2010b). This study argues in favor of the idea that misfolding of mortalin and/or impaired interaction with intramitochondrial substrates might interfere with the protective role of mortalin in maintaining mitochondrial function and is therefore critical for the development of neurodegeneration in PD.

Similar to loss of mortalin function in human cells, decreased mortalin levels in zebrafish revealed a similar phenotype with alterations of mitochondrial function, producing oxidative stress and apoptosis (Craven et al. 2005). Also here a single point mutation leading to a glycine-to-glutamate substitution at position 492 (G492E) within the substrate binding domain of the zebrafish homologue of mortalin was sufficient to cause tissue-specific mitochondrial dysfunction. The presence of this substitution in zebrafish resulted in a loss of protein function showing a developmental phenotype with a myelodysplastic syndrome (Craven et al. 2005). Using *Saccharomyces cerevisae* in a combined study of human and yeast chaperone components, the mutation identified in zebrafish caused multiple chaperone-specific biochemical defects *in vitro* including interdomain communication defects and defective stimulation by other components of the mitochondrial chaperone machinery (Goswami et al. 2010). It has been speculated that the loss of mortalin activity in the myelodysplastic syndrome mutant impedes the import of many precursor proteins and their mortalin-dependent folding in the matrix.

10.3.3 The Effect of Mortalin Variants on Mitochondrial Morphology

Mitochondria are highly dynamic organelles that are engaged in repeated cycles of fusion and fission events to maintain its homeostasis. Fusion allows mitochondria to interact with each other, thereby serving as a protective mechanism by forming networks and diluting toxic intermediates. Fission events are essential for the sequestration of damaged parts of the mitochondria and allow mitochondrial renewal and redistribution. Therefore a well-balanced fusion-fission machinery is essential for maintaining the metabolic function of mitochondria and allowing to rapidly adapt to changes in cellular requirements [reviewed in (Burbulla et al. 2010a)]. Once this

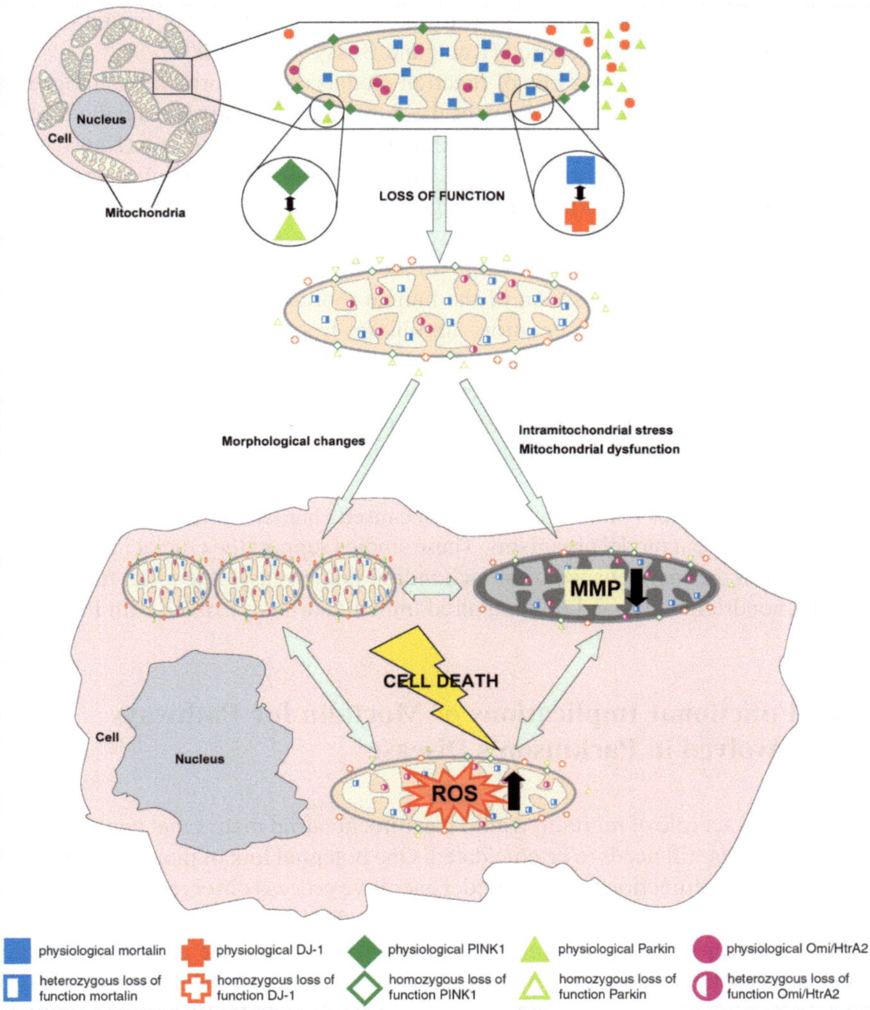

Fig. 10.1 The role of PD-associated proteins and the effect of their loss of function for the cell. The healthy cell contains a highly interconnected mitochondrial network. Single mitochondria contain physiological mortalin (*square, full blue*) within the matrix, physiological Omi/HtrA2 (*circle, full pink*) within the intermembrane space, physiological PINK1 (*diamond, full dark green*) within the outer membrane facing its kinase domain towards the cytosol, physiological DJ-1 (*cross, full orange*) and physiological Parkin (*triangle, full light green*) in the cytosol. To ensure mitochondrial function and dynamics in conditions of oxidative stress, Parkin and DJ-1 are recruited to the mitochondria and interact with their respective binding partners PINK1 and mortalin, respectively. If, as observed in PD, a heterozygous loss of function for mortalin (*square, half blue*) or Omi/HtrA2 (*circle, half pink*) or a homozygous loss of function for PINK1 (*diamond, unfilled dark green*), DJ-1 (*cross, unfilled orange*) or Parkin (*triangle, unfilled light green*) occurs, a disequilibrium of mitochondrial morphology towards increased fission or fusion events as well as intramitochondrial stress in terms of increased ROS production and an impaired MMP take place. This leads to severe mitochondrial dysfunction and finally to cell death observed in neurodegeneration

sensitive system is impaired and a shift towards increased fusion or fission occurs, cellular function is critically altered (Knott and Bossy-Wetzel 2008).

Indeed, early studies using loss of function models of the mortalin homologue in yeast (SSC1) revealed a disintegration of the mitochondrial network and the formation of aggregates due to severe effects within the mitochondrial matrix (Kawai et al. 2001). The functional inactivation of SSC1 directly affected the tubular network of mitochondria towards the formation of aggregates.

We showed for the first time that overexpression of wild type mortalin in human dopaminergic neuroblastoma cells led to a more interconnected mitochondrial network compared to overexpression of the human mortalin variants (Burbulla et al. 2010b). In line with this observation, fibroblasts from a heterozygous carrier of the A476 T variant showed a disrupted, fragmented mitochondrial arrangement, whereas control fibroblasts exhibited large interconnected mitochondria (Burbulla et al. 2010b). Herewith, it has been demonstrated that a loss of function mutation in the *mortalin* gene causes altered mitochondrial morphology even in a presymptomatic stage of PD.

In conclusion, a loss of mortalin function caused characteristic mitochondrial alterations in different model organisms. These studies support the crucial involvement of mortalin in the maintenance of mitochondrial homeostasis due to its function in the mitochondrial matrix and link impaired mitochondrial dynamics with PD.

10.4 Functional Implications of Mortalin for Pathways Involved in Parkinson's Disease

To date, the exact role of mortalin within the mitochondrial matrix that contributes to PD pathogenesis still needs to be elucidated. One potential link of mortalin-mediated mitochondrial dysfunction to the neurodegenerative process observed in PD patients could be proteolytic stress. All proteins, either encoded by the mitochondrial genome or by the nuclear genome (and subsequently transported into the mitochondria) are exposed to substantial oxidative stress within mitochondria (Bota and Davies 2001). The proper removal of damaged mitochondrial proteins includes lysosomal pathways and intramitochondrial proteolytic pathways, which rely on the concerted interplay of chaperones and mitochondrial proteases (Bota and Davies 2001; Pfeifer 1978). Indeed, proteolytic dysfunction is known to cause diminished resistance to oxidative stress and aging processes and a number of diseases were described related to mitochondrial proteolytic dysfunction (Casari et al. 1998; Lee et al. 1999). As mortalin is known to act as an intramitochondrial chaperone, it is conceivable that loss of function mutations in the *mortalin* gene lead to problems with proper import of mitochondrial proteins and/or subsequent misfolding of imported nuclear encoded mitochondrial proteins.

Indeed, an interaction and functional cooperation between mortalin (as a mitochondrial Hsp70 protein) and the mitochondrial chaperonin Hsp60 was described *in vitro* and *in vivo* that is responsible for proper folding of matrix proteins in intact mitochondria (Ostermann et al. 1989; Hong et al. 1990; Manning-Krieg et al. 1991; Wadhwa et al. 2005). It was shown that proteins containing a mitochondrial localization sequence interact with mortalin and Hsp60 as soon as they reach the mitochondrial matrix compartment in order to adopt their native conformation as a prerequisite to exert their physiological role (Hartl et al. 1992). Interestingly, mutations in the human *Hsp60* gene are sufficient to cause an autosomal dominant form of spastic paraplegia (SPG13), a neurodegenerative movement disorder related to progressive loss of motoneurons (Hansen et al. 2003). Based on data derived from the yeast homologue of mortalin (SSC1), it can be speculated that the chaperoning function is essential for the maintenance of the integrity of stress-sensitive proteins in the mitochondrial matrix. Indeed, the characteristic mitochondrial phenotype observed upon loss of SSC1 in yeast was not due to an impairment of protein import or an accumulation of mitochondrially-targeted proteins outside the organelles (Kawai et al. 2001). Thus, it can be speculated that changes observed in mitochondrial morphology upon loss of mitochondrial Hsp70 function in yeast and human cells are caused by functional inactivation of a yet unknown target protein in the matrix (Kawai et al. 2001; Burbulla et al. 2010b). In support to this notion, the loss of chaperone function due to a heterozygous G492E mutation in the substrate binding domain of mortalin in zebrafish caused impaired mitochondrial function and oxidative stress *in vivo* (Craven et al. 2005).

Although the Hsp70 protein mortalin is primarily localized in the mitochondrial matrix, a minor proportion of the protein can be found at the ER (Ran et al. 2000). To date, the exact role of mortalin within this compartment is not fully understood, however, as functional connections between the ER and the mitochondrial network are critical for the regulation of the intramitochondrial calcium levels and therefore directly affect mitochondrial function, another potential role of mortalin in the maintenance of mitochondrial homeostasis becomes evident (Levine and Rabouille 2005). In this context, mortalin was described as part of the connection between the mitochondrial VDAC, that is localized at the outer mitochondrial membrane, and the ER calcium-release channel inositol 1,4,5- trisphosphate receptor (Szabadkai et al. 2006). In this model, a loss of mortalin function severely affected the mitochondrial calcium uptake machinery supporting the central location of the chaperone mortalin in the interaction between ER and mitochondrial calcium channels. As intramitochondrial calcium levels are closely linked to mitochondrial function and cell death, mortalin exerts an important control function at the ER-mitochondria interface.

10.5 Outlook

Recent genetic and proteomic data indicate a role of the mitochondrial chaperone mortalin in the pathogenesis of PD and support the emerging concept of a genetically defined mitochondrial impairment leading to neurodegeneration of dopaminergic neurons in the substantia nigra (Burbulla et al. 2010b) (Fig. 10.1). How signaling pathways defined by Parkin and PINK1 or DJ-1 and mortalin may converge to impaired clearance of dysfunctional mitochondria still remains to be determined.

Interestingly, recent data indicate that mortalin may also be related to the most common neurodegenerative disorder, Alzheimer's disease. Using mouse models expressing different isoforms of the AD-associated protein Apolipoprotein E (APOE), a differential modulation of mortalin isoforms and their phosphorylation status could be shown (Osorio et al. 2007). The authors speculate that mortalin was modulated by the different APOE genotypes and contributed as a protective player to cellular homeostasis. Indeed, mortalin was shown to be protective in an independent cellular model of AD using amyloid-beta overexpression in human neuroblastoma cells (Qu et al. 2011). This link to AD pathogenesis was supported by the observation that *APOE* knockout mice exhibited an increased oxidation of mortalin (Choi et al. 2004) and may hint to mortalin as a potential biomarker in AD patients.

Thus, interfering with mitochondrial impairment caused by loss of the protective mortalin function may be a valid target for the treatment of the two most common neurodegenerative disorders.

Acknowledgments Work of RK is supported by research grants of the German Research Council (DFG; KR2119/3-1), the Michael J Fox Foundation, the Fritz Thyssen Foundation and the Federal Ministry for Education and Research [BMBF, NGFNplus; 01GS08134].

References

Alexander C, Votruba M, Pesch UE, Thiselton DL, Mayer S, Moore A, Rodriguez M, Kellner U, Leo-Kottler B, Auburger G, Bhattacharya SS, Wissinger B (2000) OPA1, encoding a dynamin-related GTPase, is mutated in autosomal dominant optic atrophy linked to chromosome 3q28. Nat Genet 26:211–215

Bender A, Krishnan KJ, Morris CM, Taylor GA, Reeve AK, Perry RH, Jaros E, Hersheson JS, Betts J, Klopstock T, Taylor RW, Turnbull DM (2006) High levels of mitochondrial DNA deletions in substantia nigra neurons in aging and Parkinson disease. Nat Genet 38:515–517

Bender T, Lewrenz I, Franken S, Baitzel C, Voos W (2011) Mitochondrial enzymes are protected from stress-induced aggregation by mitochondrial chaperones and the Pim1/LON protease. Mol Biol Cell 22:541–554

Betarbet R, Sherer TB, Mackenzie G, Garcia-Osuna M, Panov AV, Greenamyre JT (2000) Chronic systemic pesticide exposure reproduces features of Parkinson's disease. Nat Neurosci 3:1301–1306

Bhattacharyya T, Karnezis AN, Murphy SP, Hoang T, Freeman BC, Phillips B, Morimoto RI (1995) Cloning and subcellular localization of human mitochondrial hsp70. J Biol Chem 270:1705–1710

Bonifati V, Rizzu P, Squitieri F, Krieger E, Vanacore N, Van Swieten JC, Brice A, Van Duijn CM, Oostra B, Meco G, Heutink P (2003) DJ-1 (PARK7), a novel gene for autosomal recessive, early onset parkinsonism. Neurol Sci 24:159–160

Bota DA, Davies KJ (2001) Protein degradation in mitochondria: implications for oxidative stress, aging and disease: a novel etiological classification of mitochondrial proteolytic disorders. Mitochondrion 1:33–49

Braak H, Rub U, Gai WP, Del Tredici K (2003) Idiopathic Parkinson's disease: possible routes by which vulnerable neuronal types may be subject to neuroinvasion by an unknown pathogen. J Neural Transm 110;517–536

Burbulla LF, Krebiehl G, Kruger R (2010a) Balance is the challenge–the impact of mitochondrial dynamics in Parkinson's disease. Eur J Clin Invest 40:1048–1060

Burbulla LF, Schelling C, Kato H, Rapaport D, Woitalla D, Schiesling C, Schulte C, Sharma M, Illig T, Bauer P, Jung S, Nordheim A, Schols L, Riess O, Kruger R (2010b) Dissecting the role of the mitochondrial chaperone mortalin in Parkinson's disease: functional impact of disease-related variants on mitochondrial homeostasis. Hum Mol Genet 19:4437–4452

Canet-Aviles RM, Wilson MA, Miller DW, Ahmad R, Mclendon C, Bandyopadhyay S, Baptista MJ, Ringe D, Petsko GA, Cookson MR (2004) The Parkinson's disease protein DJ-1 is neuroprotective due to cysteine-sulfinic acid-driven mitochondrial localization. Proc Natl Acad Sci U S A 101:9103–9108

Casari G, De Fusco M, Ciarmatori S, Zeviani M, Mora M, Fernandez P, De Michele G, Filla A, Cocozza S, Marconi R, Durr A, Fontaine B, Ballabio A (1998) Spastic paraplegia and OXPHOS impairment caused by mutations in paraplegin, a nuclear-encoded mitochondrial metalloprotease. Cell 93:973–983

Chaudhuri KR, Schapira AH (2009) Non-motor symptoms of Parkinson's disease: dopaminergic pathophysiology and treatment. Lancet Neurol 8:464–474

Choi J, Forster MJ, Mcdonald SR, Weintraub ST, Carroll CA, Gracy RW (2004) Proteomic identification of specific oxidized proteins in ApoE-knockout mice: relevance to Alzheimer's disease. Free Radic Biol Med 36:1155–1162

Choi J, Sullards MC, Olzmann JA, Rees HD, Weintraub ST, Bostwick DE, Gearing M, Levey AI, Chin LS, Li L (2006) Oxidative damage of DJ-1 is linked to sporadic Parkinson and Alzheimer diseases. J Biol Chem 281:10816–10824

Craven SE, French D, Ye W, De Sauvage F, Rosenthal A (2005) Loss of Hspa9b in zebrafish recapitulates the ineffective hematopoiesis of the myelodysplastic syndrome. Blood 105:3528–3534

Dagda RK, Cherra SJ 3rd, Kulich SM, Tandon A, Park D, Chu CT (2009) Loss of PINK1 function promotes mitophagy through effects on oxidative stress and mitochondrial fission. J Biol Chem 284:13843–13855

De Mena L, Coto E, Sanchez-Ferrero E, Ribacoba R, Guisasola LM, Salvador C, Blazquez M, Alvarez V (2009) Mutational screening of the mortalin gene (HSPA9) in Parkinson's disease. J Neural Transm 116:1289–1293

De Rijk MC, Launer LJ, Berger K, Breteler MM, Dartigues JF, Baldereschi M, Fratiglioni L, Lobo A, Martinez-Lage J, Trenkwalder C, Hofman A (2000) Prevalence of Parkinson's disease in Europe: a collaborative study of population-based cohorts. Neurologic Diseases in the Elderly Research Group. Neurology 54:S21–23

Deocaris CC, Kaul SC, Wadhwa R (2008) From proliferative to neurological role of an hsp70 stress chaperone, mortalin. Biogerontology 9:391–403

Engel J, Kammerer RA (2000) What are oligomerization domains good for? Matrix Biol 19:283–288

Exner N, Treske B, Paquet D, Holmstrom K, Schiesling C, Gispert S, Carballo-Carbajal I, Berg D, Hoepken HH, Gasser T, Kruger R, Winklhofer KF, Vogel F, Reichert AS, Auburger G, Kahle PJ, Schmid B, Haass C (2007) Loss-of-function of human PINK1 results in mitochondrial pathology and can be rescued by parkin. J Neurosci 27:12413–12418

Gambill BD, Voos W, Kang PJ, Miao B, Langer T, Craig EA, Pfanner N (1993) A dual role for mitochondrial heat shock protein 70 in membrane translocation of preproteins. J Cell Biol 123:109–117

Gegg ME, Cooper JM, Chau KY, Rojo M, Schapira AH, Taanman JW (2010) Mitofusin 1 and mitofusin 2 are ubiquitinated in a PINK1/parkin-dependent manner upon induction of mitophagy. Hum Mol Genet 19:4861–4870

Geisler S, Holmstrom KM, Skujat D, Fiesel FC, Rothfuss OC, Kahle PJ, Springer W (2010) PINK1/Parkin-mediated mitophagy is dependent on VDAC1 and p62/SQSTM1. Nat Cell Biol 12:119–131

Goswami AV, Chittoor B, D'silva P (2010) Understanding the functional interplay between mammalian mitochondrial Hsp70 chaperone machine components. J Biol Chem 285:19472–19482

Gotz ME, Freyberger A, Riederer P (1990) Oxidative stress: a role in the pathogenesis of Parkinson's disease. J Neural Transm Suppl 29:241–249

Hansen JJ, Bross P, Westergaard M, Nielsen MN, Eiberg H, Borglum AD, Mogensen J, Kristiansen K, Bolund L, Gregersen N (2003) Genomic structure of the human mitochondrial chaperonin genes: HSP60 and HSP10 are localised head to head on chromosome 2 separated by a bidirectional promoter. Hum Genet 112:71–77

Hartl FU, Martin J, Neupert W (1992) Protein folding in the cell: the role of molecular chaperones Hsp70 and Hsp60. Annu Rev Biophys Biomol Struct 21:293–322

Hicks AA, Petursson H, Jonsson T, Stefansson H, Johannsdottir HS, Sainz J, Frigge ML, Kong A, Gulcher JR, Stefansson K, Sveinbjornsdottir S (2002) A susceptibility gene for late-onset idiopathic Parkinson's disease. Ann Neurol 52:549–555

Hindle JV (2010) Ageing, neurodegeneration and Parkinson's disease. Age Ageing 39:156–161

Hong WS, Kim CM, Lee JO, Kang TW, Yun TK, Kim CY (1990) Natural killer and lymphokine-activated killer activities in stomach cancer patients with special emphasis on the effect of 5-fluorouracil, adriamycin and mitomycin-C chemotherapy. Jpn J Clin Oncol 20:87–93

Hong Z, Shi M, Chung KA, Quinn JF, Peskind ER, Galasko D, Jankovic J, Zabetian CP, Leverenz JB, Baird G, Montine TJ, Hancock AM, Hwang H, Pan C, Bradner J, Kang UJ, Jensen PH, Zhang J (2010) DJ-1 and alpha-synuclein in human cerebrospinal fluid as biomarkers of Parkinson's disease. Brain 133:713–726

Hsu WM, Lee H, Juan HF, Shih YY, Wang BJ, Pan CY, Jeng YM, Chang HH, Lu MY, Lin KH, Lai HS, Chen WJ, Tsay YG, Liao YF, Hsieh FJ (2008) Identification of GRP75 as an independent favorable prognostic marker of neuroblastoma by a proteomics analysis. Clin Cancer Res 14:6237–6245

Irrcher I, Aleyasin H, Seifert EL, Hewitt SJ, Chhabra S, Phillips M, Lutz AK, Rousseaux MW, Bevilacqua L, Jahani-Asl A, Callaghan S, Maclaurin JG, Winklhofer KF, Rizzu P, Rippstein P, Kim RH, Chen CX, Fon EA, Slack RS, Harper ME, Mcbride HM, Mak TW, Park DS (2010) Loss of the Parkinson's disease-linked gene DJ-1 perturbs mitochondrial dynamics. Hum Mol Genet 19:3734–3746

Jin J, Hulette C, Wang Y, Zhang T, Pan C, Wadhwa R, Zhang J (2006) Proteomic identification of a stress protein, mortalin/mthsp70/GRP75: relevance to Parkinson disease. Mol Cell Proteomics 5:1193–1204

Jin J, Li GJ, Davis J, Zhu D, Wang Y, Pan C, Zhang J (2007) Identification of novel proteins associated with both alpha-synuclein and DJ-1. Mol Cell Proteomics 6:845–859

Jin SM, Lazarou M, Wang C, Kane LA, Narendra DP, Youle RJ (2010) Mitochondrial membrane potential regulates PINK1 import and proteolytic destabilization by PARL. J Cell Biol 191:933–942

Kaul SC, Taira K, Pereira-Smith OM, Wadhwa R (2002) Mortalin: present and prospective. Exp Gerontol 37:1157–1164

Kaul SC, Yaguchi T, Taira K, Reddel RR, Wadhwa R (2003) Overexpressed mortalin (mot-2)/mthsp70/GRP75 and hTERT cooperate to extend the in vitro lifespan of human fibroblasts. Exp Cell Res 286:96–101

Kawai A, Nishikawa S, Hirata A, Endo T (2001) Loss of the mitochondrial Hsp70 functions causes aggregation of mitochondria in yeast cells. J Cell Sci 114:3565–3574

Kitada T, Asakawa S, Hattori N, Matsumine H, Yamamura Y, Minoshima S, Yokochi M, Mizuno Y, Shimizu N (1998) Mutations in the parkin gene cause autosomal recessive juvenile parkinsonism. Nature 392:605–608

Knott AB, Bossy-Wetzel E (2008) Impairing the mitochondrial fission and fusion balance: a new mechanism of neurodegeneration. Ann N Y Acad Sci 1147:283–292

Krebiehl G, Ruckerbauer S, Burbulla LF, Kieper N, Maurer B, Waak J, Wolburg H, Gizatullina Z, Gellerich FN, Woitalla D, Riess O, Kahle PJ, Proikas-Cezanne T, Kruger R (2010) Reduced basal autophagy and impaired mitochondrial dynamics due to loss of Parkinson's disease-associated protein DJ-1. PLoS One 5:e9367

Langston JW, Ballard P, Tetrud JW, Irwin I (1983) Chronic Parkinsonism in humans due to a product of meperidine-analog synthesis. Science 219:979–980

Lee CK, Klopp RG, Weindruch R, Prolla TA (1999) Gene expression profile of aging and its retardation by caloric restriction. Science 285:1390–1393

Levine T, Rabouille C (2005) Endoplasmic reticulum: one continuous network compartmentalized by extrinsic cues. Curr Opin Cell Biol 17:362–368

Li HM, Niki T, Taira T, Iguchi-Ariga SM, Ariga H (2005) Association of DJ-1 with chaperones and enhanced association and colocalization with mitochondrial Hsp70 by oxidative stress. Free Radic Res 39:1091–1099

Liang CL, Wang TT, Luby-Phelps K, German DC (2007) Mitochondria mass is low in mouse substantia nigra dopamine neurons: implications for Parkinson's disease. Exp Neurol 203:370–380

Liu Y, Liu W, Song XD, Zuo J (2005) Effect of GRP75/mthsp70/PBP74/mortalin overexpression on intracellular ATP level, mitochondrial membrane potential and ROS accumulation following glucose deprivation in PC12 cells. Mol Cell Biochem 268:45–51

Lopez-Alberola R, Georgiou M, Sfakianakis GN, Singer C, Papapetropoulos S (2009) Contemporary Encephalitis Lethargica: phenotype, laboratory findings and treatment outcomes. J Neurol 256:396–404

Macario AJ, Cappello F, Zummo G, Conway De Macario E (2010) Chaperonopathies of senescence and the scrambling of interactions between the chaperoning and the immune systems. Ann N Y Acad Sci 1197:85–93

Mandemakers W, Morais VA, De Strooper B (2007) A cell biological perspective on mitochondrial dysfunction in Parkinson disease and other neurodegenerative diseases. J Cell Sci 120:1707–1716

Mann VM, Cooper JM, Krige D, Daniel SE, Schapira AH, Marsden CD (1992) Brain, skeletal muscle and platelet homogenate mitochondrial function in Parkinson's disease. Brain 115:333–342

Manning-Krieg UC, Scherer PE, Schatz G (1991) Sequential action of mitochondrial chaperones in protein import into the matrix. EMBO J 10:3273–3280

Maraganore DM, de Andrade M, Lesnick TG, Strain KJ, Farrer MJ, Rocca WA, Pant PV, Frazer KA, Cox DR, Ballinger DG (2005) High-resolution whole-genome association study of Parkinson disease. Am J Hum Genet 77:685–693

Maraganore DM, de Andrade M, Elbaz A, Farrer MJ, Ioannidis JP, Krüger R, Rocca WA, Schneider NK, Lesnick TG, Lincoln SJ, Hulihan MM, Aasly JO, Ashizawa T, Chartier-Harlin MC, Checkoway H, Ferrarese C, Hadjigeorgiou G, Hattori N, Kawakami H, Lambert JC, Lynch T, Mellick GD, Papapetropoulos S, Parsian A, Quattrone A, Riess O, Tan EK, Van Broeckhoven C; Genetic Epidemiology of Parkinson's Disease (GEO-PD) Consortium (2006) Collaborative analysis of alpha-synuclein gene promoter variability and Parkinson disease. JAMA 296:661–670

Martinez M, Brice A, Vaughan JR, Zimprich A, Breteler MM, Meco G, Filla A, Farrer MJ, Bétard C, Hardy J, De Michele G, Bonifati V, Oostra B, Gasser T, Wood NW, Dürr A (2004) Genome-wide scan linkage analysis for Parkinson's disease: the European genetic study of Parkinson's disease. J Med Genet 41:900–907

Matzler W, Nagele T, Gasser T, Kruger R (2007) Acute parkinsonism with corresponding lesions in the basal ganglia after heroin abuse. Neurology 68:414

Morgan-Hughes JA, Sweeney MG, Cooper JM, Hammans SR, Brockington M, Schapira AH, Harding AE, Clark JB (1995) Mitochondrial DNA (mtDNA) diseases: correlation of genotype to phenotype. Biochim Biophys Acta 1271:135–140

Mortiboys H, Thomas KJ, Koopman WJ, Klaffke S, Abou-Sleiman P, Olpin S, Wood NW, Willems PH, Smeitink JA, Cookson MR, Bandmann O (2008) Mitochondrial function and morphology are impaired in parkin-mutant fibroblasts. Ann Neurol 64:555–565

Narendra D, Tanaka A, Suen DF, Youle RJ (2008) Parkin is recruited selectively to impaired mitochondria and promotes their autophagy. J Cell Biol 183:795–803

Narendra D, Tanaka A, Suen DF, Youle RJ (2009) Parkin-induced mitophagy in the pathogenesis of Parkinson disease. Autophagy 5:706–708

Nicklas WJ, Vyas I, Heikkila RE (1985) Inhibition of NADH-linked oxidation in brain mitochondria by 1-methyl-4-phenyl-pyridine, a metabolite of the neurotoxin, 1-methyl-4-phenyl-1,2,5,6-tetrahydropyridine. Life Sci 36:2503–2508

Niemann A, Ruegg M, La Padula V, Schenone A, Suter U (2005) Ganglioside-induced differentiation associated protein 1 is a regulator of the mitochondrial network: new implications for Charcot-Marie-Tooth disease. J Cell Biol 170:1067–1078

Osorio C, Sullivan PM, He DN, Mace BE, Ervin JF, Strittmatter WJ, Alzate O (2007) Mortalin is regulated by APOE in hippocampus of AD patients and by human APOE in TR mice. Neurobiol Aging 28:1853–1862

Ostermann J, Horwich AL, Neupert W, Hartl FU (1989) Protein folding in mitochondria requires complex formation with hsp60 and ATP hydrolysis. Nature 341:125–130

Pan-Montojo F, Anichtchik O, Dening Y, Knels L, Pursche S, Jung R, Jackson S, Gille G, Spillantini MG, Reichmann H, Funk RH (2010) Progression of Parkinson's disease pathology is reproduced by intragastric administration of rotenone in mice. PLoS One 5:e8762

Pankratz N, Nichols WC, Uniacke SK, Halter C, Murrell J, Rudolph A, Shults CW, Conneally PM, Foroud T (2003) Genome-wide linkage analysis and evidence of gene-by-gene interactions in a sample of 362 multiplex Parkinson disease families. Hum Mol Genet 12:2599–2608

Pfeifer U (1978) Inhibition by insulin of the formation of autophagic vacuoles in rat liver. A morphometric approach to the kinetics of intracellular degradation by autophagy. J Cell Biol 78:152–167

Qu M, Zhou Z, Xu S, Chen C, Yu Z, Wang D (2011) Mortalin overexpression attenuates beta-amyloid-induced neurotoxicity in SH-SY5Y cells. Brain Res 1368:336–345

Ran Q, Wadhwa R, Kawai R, Kaul SC, Sifers RN, Bick RJ, Smith JR, Pereira-Smith OM (2000) Extramitochondrial localization of mortalin/mthsp70/PBP74/GRP75. Biochem Biophys Res Commun 275:174–179

Rehling P, Brandner K, Pfanner N (2004) Mitochondrial import and the twin-pore translocase. Nat Rev Mol Cell Biol 5:519–530

Rüdiger S, Mayer MP, Schneider-Mergener J, Bukau B (2000) Modulation of substrate specificity of the DnaK chaperone by alteration of a hydrophobic arch. J Mol Biol 304:245–251

Savel'ev AS, Novikova LA, Kovaleva IE, Luzikov VN, Neupert W, Langer T (1998) ATP-dependent proteolysis in mitochondria. m-AAA protease and PIM1 protease exert overlapping substrate specificities and cooperate with the mtHsp70 system. J Biol Chem 273:20596–20602

Schapira AH (1994) Evidence for mitochondrial dysfunction in Parkinson's disease–a critical appraisal. Mov Disord 9:125–138

Schapira AH, Cooper JM, Dexter D, Jenner P, Clark JB, Marsden CD (1989) Mitochondrial complex I deficiency in Parkinson's disease. Lancet 1:1269

Schapira AH, Cooper JM, Dexter D, Clark JB, Jenner P, Marsden CD (1990) Mitochondrial complex I deficiency in Parkinson's disease. J Neurochem 54:823–827

Schiesling C, Kieper N, Seidel K, Kruger R (2008) Review: familial Parkinson's disease–genetics, clinical phenotype and neuropathology in relation to the common sporadic form of the disease. Neuropathol Appl Neurobiol 34:255–271

Scott WK, Nance MA, Watts RL, Hubble JP, Koller WC, Lyons K, Pahwa R, Stern MB, Colcher A, Hiner BC, Jankovic J, Ondo WG, Allen FH Jr, Goetz CG, Small GW, Masterman D, Mastaglia F, Laing NG, Stajich JM, Slotterbeck B, Booze MW, Ribble RC, Rampersaud E, West SG, Gibson RA, Middleton LT, Roses AD, Haines JL, Scott BL, Vance JM, Pericak-Vance MA (2001) Complete genomic screen in Parkinson disease: evidence for multiple genes. JAMA 286:2239–2244

Sharma M, Lichtner P, Kruger R, Berg D, Schulte C, Illig T, Riess O, Gasser T (2009) Further delineation of the association signal on chromosome 5 from the first whole genome association study in Parkinson's disease. Neurobiol Aging 30:1706–1709

Shi M, Jin J, Wang Y, Beyer RP, Kitsou E, Albin RL, Gearing M, Pan C, Zhang J (2008) Mortalin: a protein associated with progression of Parkinson disease? J Neuropathol Exp Neurol 67:117–124

Simón-Sánchez J, Schulte C, Bras JM, Sharma M, Gibbs JR, Berg D, Paisan-Ruiz C, Lichtner P, Scholz SW, Hernandez DG, Krüger R, Federoff M, Klein C, Goate A, Perlmutter J, Bonin M, Nalls MA, Illig T, Gieger C, Houlden H, Steffens M, Okun MS, Racette BA, Cookson MR, Foote KD, Fernandez HH, Traynor BJ, Schreiber S, Arepalli S, Zonozi R, Gwinn K, Van Der Brug M, Lopez G, Chanock SJ, Schatzkin A, Park Y, Hollenbeck A, Gao J, Huang X, Wood NW, Lorenz D, Deuschl G, Chen H, Riess O, Hardy JA, Singleton AB, Gasser T (2009) Genome-wide association study reveals genetic risk underlying Parkinson's disease. Nat Genet 41:1308–1312

Slater MR, Craig EA (1987) Transcriptional regulation of an hsp70 heat shock gene in the yeast Saccharomyces cerevisiae. Mol Cell Biol 7:1906–1916

Strauss KM, Martins LM, Plun-Favreau H, Marx FP, Kautzmann S, Berg D, Gasser T, Wszolek Z, Müller T, Bornemann A, Wolburg H, Downward J, Riess O, Schulz JB, Krüger R (2005) Loss of function mutations in the gene encoding Omi/HtrA2 in Parkinson's disease. Hum Mol Genet 14:2099–2111

Szabadkai G, Bianchi K, Vàrnai P, De Stefani D, Wieckowski MR, Cavagna D, Nagy AI, Balla T, Rizzuto R (2006) Chaperone-mediated coupling of endoplasmic reticulum and mitochondrial Ca2 + channels. J Cell Biol 175:901–911

Takano S, Wadhwa R, Yoshii Y, Nose T, Kaul SC, Mitsui Y (1997) Elevated levels of mortalin expression in human brain tumors. Exp Cell Res 237:38–45

Thomas KJ, McCoy MK, Blackinton J, Beilina A, Van Der Brug M, Sandebring A, Miller D, Maric D, Cedazo-Minguez A, Cookson MR (2011) DJ-1 acts in parallel to the PINK1/parkin pathway to control mitochondrial function and autophagy. Hum Mol Genet 20:40–50

Valente EM, Salvi S, Ialongo T, Marongiu R, Elia AE, Caputo V, Romito L, Albanese A, Dallapiccola B, Bentivoglio AR (2004) PINK1 mutations are associated with sporadic early-onset parkinsonism. Ann Neurol 56:336–341

Van Laar VS, Mishizen AJ, Cascio M, Hastings TG (2009) Proteomic identification of dopamine-conjugated proteins from isolated rat brain mitochondria and SH-SY5Y cells. Neurobiol Dis 34:487–500

Wadhwa R, Takano S, Kaur K, Aida S, Yaguchi T, Kaul Z, Hirano T, Taira K, Kaul SC (2005) Identification and characterization of molecular interactions between mortalin/mtHsp70 and HSP60. Biochem J 391:185–190

Wallace DC (2005) Mitochondria and cancer: Warburg addressed. Cold Spring Harb Symp Quant Biol 70:363–374

Webster TJ, Naylor DJ, Hartman DJ, Høj PB, Hoogenraad NJ (1994) cDNA cloning and efficient mitochondrial import of pre-mtHSP70 from rat liver. DNA Cell Biol 13:1213–1220

Yaguchi T, Aida S, Kaul SC, Wadhwa R (2007) Involvement of mortalin in cellular senescence from the perspective of its mitochondrial import, chaperone, and oxidative stress management functions. Ann N Y Acad Sci 1100:306–311

Yokoyama K, Fukumoto K, Murakami T, Harada S, Hosono R, Wadhwa R, Mitsui Y, Ohkuma S (2002) Extended longevity of Caenorhabditis elegans by knocking in extra copies of hsp70F, a homolog of mot-2 (mortalin)/mthsp70/Grp75. FEBS Lett 516:53–57

Züchner S, Mersiyanova IV, Muglia M, Bissar-Tadmouri N, Rochelle J, Dadali EL, Zappia M, Nelis E, Patitucci A, Senderek J, Parman Y, Evgrafov O, Jonghe PD, Takahashi Y, Tsuji S, Pericak-Vance MA, Quattrone A, Battaloglu E, Polyakov AV, Timmerman V, Schröder JM, Vance JM (2004) Mutations in the mitochondrial GTPase mitofusin 2 cause Charcot-Marie-Tooth neuropathy type 2A. Nat Genet 36:449–451

Chapter 11
Hsp75/mortalin and Protection from Ischemic Brain Injury

Robin E. White, Yi-Bing Ouyang and Rona G. Giffard

Abstract Mortalin is a vital mitochondrial chaperone that is important to normal brain functioning. Several studies have shown that overexpression of mortalin reduces damage in both *in vitro* and *in vivo* models of ischemic stroke. This neuroprotection is likely mediated by mortalin's ability to preserve mitochondrial function and reduce reactive oxygen species following ischemic injury. In addition, by improving mitochondrial function after stroke, mortalin may contribute to recovery by additional beneficial actions, including improved neurogenesis. Thus, mortalin is a promising new candidate among chaperones as a potential target to reduce ischemic brain injury. Ways to increase expression of mortalin in the brain should be developed and tested for neuroprotection in additional models of ischemic brain injury and neurodegenerative disease.

Keywords Ischemia · Mitochondria · Reactive oxygen species · Stroke

11.1 Introduction

Stroke is currently the third leading cause of death and the leading cause of chronic disability in the United States (Roger et al. 2011). Stroke is caused by an interruption of cerebral blood flow that leads to a cascade of cellular events, including mitochondrial dysfunction, oxidative stress, and inflammation, that ultimately result in brain cell death and cognitive and physical disability (Doyle et al. 2008; Sims and Muyderman 2010). To date, 616 interventional clinical trials (Phase II, III, and IV) on stroke have been conducted in the United States, with 1273 worldwide (The U.S. National Institutes of Health 2011, http://www.clinicaltrials.gov). Despite these numerous clinical trials, the only currently approved effective therapy for stroke is treatment with the thrombolytic drug tissue plasminogen activator (t-PA) (The National Institute of Neurological Disorders and Stroke rt-PA Study Group 2011). The principal shortcoming of this treatment is that it is only effective if given within the first few

R. G. Giffard (✉) · R. E. White · Y.-B. Ouyang
Dept. of Anesthesia, Stanford University School of Medicine, 300 Pasteur Drive, S272 Grant Building, CA 94305-5117, USA
e-mail: rona.giffard@stanford.edu

S. C. Kaul, R. Wadhwa (eds.), *Mortalin Biology: Life, Stress and Death*,
DOI 10.1007/978-94-007-3027-4_11,

hours after stroke, and is contraindicated in hemorrhagic stroke, so less than 5% of patients receive this treatment.

Mortalin/mitochondrial heat shock protein 70/glucose regulated protein 75 is the mitochondrial member of the heat shock protein 70 (HSP70) family, an essential mitochondrial chaperone, and a vital component of the mitochondrial import machinery (Wadhwa et al. 2002). Although primarily located in the mitochondria, it has also been shown under certain circumstances to be located in the endoplasmic reticulum, plasma membrane, and cytoplasm (Wadhwa et al. 2002). Another approximately 70 kDa chaperone protein present in mitochondria is TRAP1, but it is part of the HSP90 family of proteins (Felts et al. 2000). Both our laboratory and others have demonstrated that overexpression of the inducible cytosolic member of the HSP70 family, Hsp72, is highly protective against both *in vitro* and *in vivo* ischemia (Giffard and Yenari 2004). Although promising, no clinical studies have explored the use of HSP70 family members as therapeutics, though drugs that induce a heat shock response and Hsp72 have been studied in animals and are shown to be protective (Lu et al. 2002). In addition, a few studies show positive correlations between Hsp72 and improved stroke outcome. For example, lymphocyte levels of Hsp72 correlate with neurological function in elderly stroke patients (Jin et al. 2004). Although less widely studied in the stroke field than Hsp72, several studies from our laboratory and others have shown promising results on the neuroprotective capabilities of mortalin after ischemia. Protection by mortalin is the main topic of this chapter. The potential neuroprotective mechanisms of mortalin will be discussed, and the possible use of mortalin to reduce injury from stroke will be explored.

11.2 Mortalin Expression in the Central Nervous System

Thus far, few studies have characterized the distribution of mortalin in the normal brain. Although mortalin mRNA is detected throughout the adult rat brain, several regions exhibit higher levels, including the diagonal band, ventral pallidum, supraoptic nucleus, globus pallidus, reticular nucleus, subthalamic nucleus, and amygdala (Massa et al. 1995). Mortalin expression is higher in neuronal cells than glial cells in the normal brain (Massa et al. 1995). Interestingly, mortalin cellular localization differs depending on the type of cell. Normal, non-immortalized cells have a cytoplasmic and mitochondrial distribution, while immortalized, cycling cell lines exhibit perinuclear distribution (Kaul et al. 1997). Mortalin is known to be upregulated in a variety of tumors (Takano et al. 1997), and reducing levels of mortalin has been proposed as a promising cancer treatment (Yoo et al. 2010).

Although mortalin is not upregulated in response to heat, it is increased in response to other forms of stress, including ischemia and glucose deprivation, as the name glucose regulated protein 75 would suggest. Indeed, mortalin mRNA increases in the ischemic region within 24 h of transient (30 min) focal brain ischemia. Interestingly, it is increased in the core of mild focal ischemia-induced injuries, but only in the penumbra, and not the core, in severe ischemic injuries (Massa et al. 1995). Glucose deprivation induces mortalin protein expression in murine fibroblasts (Merrick et al. 1997).

11.3 Mortalin and *in vitro* Ischemia

Glucose deprivation (GD) and combined oxygen-glucose deprivation (OGD) are common *in vitro* models of brain ischemia. Either cell cultures or slice cultures are subjected to medium lacking glucose, and in the case of OGD, also placed in a chamber with very low oxygen levels for a fixed period of time. This allows researchers maximum control over the cellular environment and allows assessment of the responses of specific cell types or brain regions. *In vitro* studies also allow assessment of specific aspects of injury, including oxidative stress, mitochondrial dysfunction and quantitation of cell death as a function of the insult or of the addition of a protective strategy more easily than in *in vivo* models .

One such model of ischemia uses primary astrocyte cultures isolated from early postnatal (P2) mice. Astrocytes cultured for approximately 3 weeks are exposed to GD or OGD, followed by 24 h of "reperfusion", or a return to glucose-containing media and normoxic conditions. To determine the effects of mortalin overexpression on these cultures, the HSPA9 gene (GenBank L15189) was cloned into the LXSN retroviral vector and subsequently packaged into a retrovirus. Infection with this vector resulted in a 2-fold increase in mortalin protein levels in transduced cells (Voloboueva et al. 2008). During GD, mortalin-overexpressing cells exhibited lower hydroethidine fluorescence (an indicator of the accumulation of reactive oxygen species (ROS)) and preserved mitochondrial membrane potential as shown using tetramethyl rhodamine staining, a dye whose sequestration by mitochondria depends on the mitochondrial membrane potential, compared to control cultures (Fig. 11.1; Voloboueva et al. 2008). Furthermore, mortalin overexpression prior to OGD resulted in preserved ATP levels and decreased cell death (Voloboueva et al. 2008).

Similar protective effects of mortalin overexpression have been demonstrated in neural cell lines. Following GD, PC12 cells, tumor cells from the adrenal medulla, exhibited increased survival and decreased ROS with overexpression of mortalin (Liu et al. 2005). Likewise, overexpression of mortalin in rat neonatal cardiac myocytes protected them from hypoxia-induced death, decreased ROS, and protected mitochondrial complex III and IV functioning (Williamson et al. 2008).

11.4 Mortalin and Focal Cerebral Ischemia

Middle cerebral artery occlusion (MCAO) in rats or mice is a rodent model of focal cerebral ischemia, or stroke. Using this method, transient ischemia is achieved by inserting a suture into the left middle cerebral artery, temporarily blocking blood flow to the middle cerebral artery territory, and removing the suture to allow reperfusion

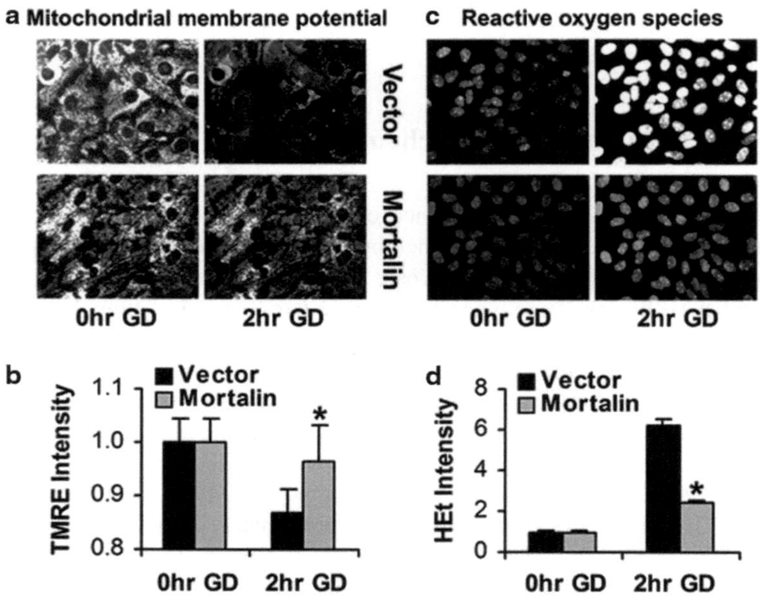

Fig. 11.1 Mortalin overexpression preserves mitochondrial membrane potential and reduces reactive oxygen species accumulation during glucose deprivation (GD). **a** Examples of tetramethyl rhodamine (TMRE) staining before (t = 0, *left column*) and after 2 hr of GD (*right column*). Control-transfected cells are depicted in the top row, and mortalin-transfected cells are in the bottom row. **b** Changes in TMRE mitochondrial fluorescence with GD in control vector-transduced, and mortalin-overexpressing astrocytes. Note the preserved fluorescence in mortalin-transfected cells compared to control-transfected cells following GD, indicative of preserved mitochondrial membrane potential. **c** Examples of changes in hydroethidine (HEt) fluorescence before (t = 0hr, *left column*) and after 2 hr of GD (*right column*). Control-transfected cells are depicted in the top row, and mortalin-transfected cells are in the bottom row. **d** ROS accumulation with GD in control vector-transduced and mortalin-overexpressing astrocytes. Note the reduced fluorescence in mortalin-transfected astrocytes compared to control after GD, indicative of reduced oxidative stress. For each cell, the (TMRE or HEt) fluorescence was normalized to the starting fluorescence for that cell. The data are representative of mean ± s.d. of three independent experiments, with at least 50 astrocytes per condition. *$p < 0.05$ compared to control vector-transduced under the same condition. Modified from Voloboueva et al. 2008

after a duration of 30–120 min depending on the specific study (Longa et al. 1989). Twenty-four hours following this procedure, animals are assessed for neurological function using a scale from 0–4, and then sacrificed at different reperfusion intervals. If histological or immunohistochemical evaluation is performed, the animals are transcardially perfused.

An effective way to increase protein expression in the brain is through intracerebroventricular infusion of a DNA plasmid encoding the gene of interest under control of a strong promoter in combination with a cationic lipid (DOTAP) to facilitate transfection. Using this method, our laboratory has demonstrated increased mortalin protein expression in both neurons and astrocytes 48 h following injection

(Xu et al. 2009). This treatment reduced infarct volume and improved neurological score 24 h after MCAO. Furthermore, mortalin overexpression reduced both reactive oxygen species and levels of oxidized lipids, suggesting less oxidative damage, as well as preserved mitochondrial complex IV function and ATP levels (Xu et al. 2009).

11.5 Mechanisms of Mortalin Protection Against Ischemia

As our laboratory has demonstrated, overexpression of mortalin leads to decreased neuronal cell death and reduced infarct size following ischemia. Because mortalin is a vital mitochondrial chaperone and performs a variety of functions, there are several neuroprotective mechanisms that could be responsible, several of which are discussed below and summarized in Fig. 11.2.

11.5.1 Mortalin Attenuates Oxidative Stress

We have found that reduction of reactive oxygen species and oxidative stress is one candidate mechanism of mortalin-induced neuroprotection after ischemia. In both our *in vitro* and *in vivo* models of ischemia, indicators of oxidative stress such as hydroethidine fluorescence and levels of hydroxynonenol are decreased with mortalin overexpression (Xu et al. 2009; Voloboueva et al. 2008). There are a few ways in which mortalin overexpression may be mediating this effect. First, mortalin increases production of superoxide dismutase 2 (SOD2) in cardiac myocytes (Williamson et al. 2008), a potent mitochondrial anti-oxidant enzyme important in all cells, including neurons and astrocytes. Astrocytes play an important role in brain antioxidant defense in part via support of glutathione levels and reduced ascorbate levels (Wilson 1997). As mortalin overexpression preserves astrocyte functioning in the face of ischemia *in vitro* (Voloboueva et al. 2008), it is likely that levels of SOD2 and glutathione may be enhanced following mortalin overexpression, contributing to the observed decrease in ROS. Finally, mitochondria are important producers of ROS during post-ischemia reperfusion (Piantadosi and Zhang 1996), and, as discussed below, mortalin overexpression has robust effects on preservation of mitochondrial function, likely contributing to attenuated production of ROS.

11.5.2 Mortalin Preserves Mitochondrial Function

As mentioned above, upregulation of the mitochondrial-specific SOD2 may be one way that mortalin influences mitochondrial function. Likewise, preservation of other mitochondrial components is also protective. Mitochondria are a primary target of ischemic injury (Soane et al. 2007). Iron-sulfur complexes, integral to proper electron transport chain and mitochondrial enzyme function (Atamna et al. 2002),

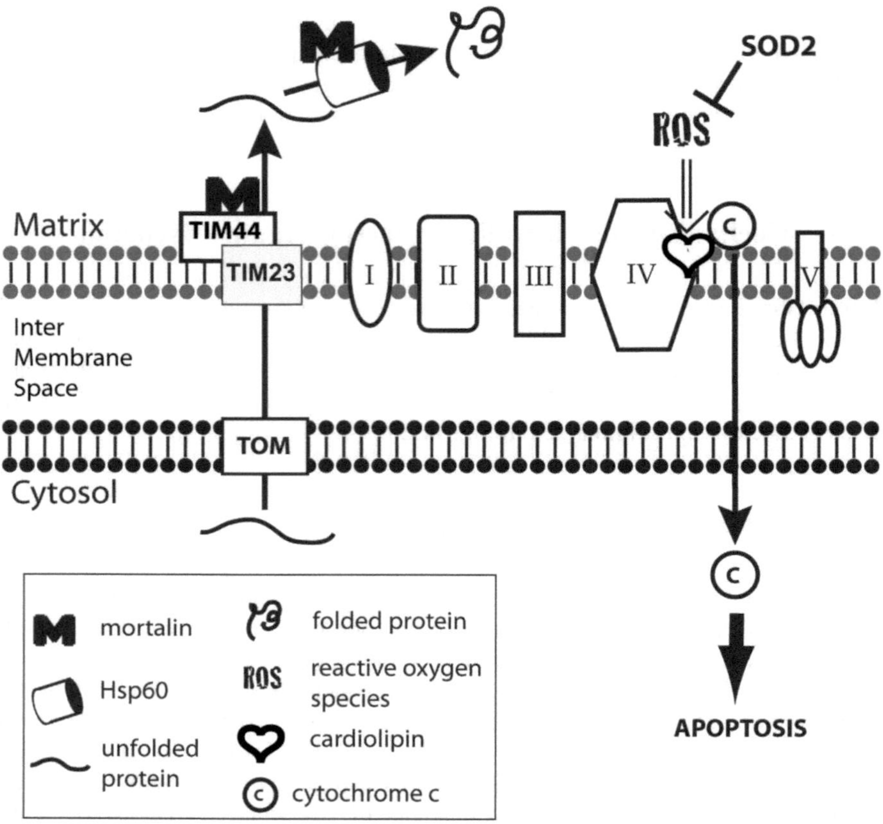

Fig. 11.2 Sites of interactions of mortalin in the mitochondrial inner and outer membranes, in addition to the electron transport chain. Nuclear encoded mitochondrial proteins are imported to the inner matrix of the mitochondria via the translocase of the outer membrane (TOM) and translocase of the inner membrane (TIM) complexes. Mortalin binds to Tim44 and is thus anchored to the inner mitochondrial membrane. The mortalin/TIM complex at the inner mitochondrial membrane moves unfolded proteins from the inter membrane space to the matrix. After arriving in the matrix, the protein is folded by the Hsp60/mortalin complex. Reactive oxygen species are increased in ischemia and oxidize cardiolipin, a protein associated with the electron transport chain Complex IV. The lipid peroxidation of cardiolipin induces release of cytochrome c from the mitochondria, leading to apoptosis. SOD2, an antioxidant that detoxifies superoxide, is increased following mortalin overexpression. In the figure, the thin arrows indicated translocation, the thick arrow indicates cytochrome c activation of apoptosis, and the double-lined arrow is representative of lipid peroxidation of cardiolipin by ROS

are particularly sensitive to oxidative stress (Beal 1995). Mitochondrial complex IV (cytochrome c oxidase) is a vital part of the electron transport chain necessary for ATP production (Moro et al. 2005) and is often compromised following ischemia (Canevari et al. 1997). Normal complex IV activity is dependent upon the phospholipid cardiolipin (Robinson 1993), which is particularly sensitive to free radical-induced damage (Paradies et al. 2000). Furthermore, peroxidation of cardiolipin by ROS is necessary for cytochrome c release from the mitochondria, a critical early step in the signaling cascade for apoptosis (Piantadosi and Zhang 1996). Deficits in complex IV may also lead to decreased ATP production, as the electron transport chain is the main producer of ATP in mammalian cells, and is especially critical in the brain, a highly oxidative organ responsible for 20% of whole body oxygen consumption at rest.

As our lab has demonstrated previously, overexpression of mortalin significantly preserves mitochondrial function following *in vitro* and *in vivo* ischemia (Voloboueva et al. 2008; Xu et al. 2009). Mortalin overexpression leads to preserved mitochondrial membrane potential, and maintained complex IV activity and ATP production in the face of ischemic stress. As discussed above, mortalin overexpression leads to decreased reactive oxygen species and increased SOD2, which may contribute to preservation of mitochondrial function.

In addition, mortalin plays a key role in mitochondrial ATP-dependant protein import. Mortalin forms complexes with mitochondrial inner membrane translocase subunits Tim44 and Tim23, among others (Wadhwa et al. 2002). Mortalin, the core of the mitochondrial import motor, binds with Tim44 at both the ATPase and peptide-binding domains (D'Silva et al. 2004) and thus is anchored to the internal mitochondrial membrane transporter, TIM. This multiprotein complex imports proteins that do not contain a mitochondrial localization sequence. In addition, mortalin helps to refold proteins as they enter the mitochondria by binding with Hsp60 (Deocaris et al. 2008; Wadhwa et al. 2005). Thus, mortalin may contribute to preservation of mitochondrial function by assuring continued import of essential mitochondrial proteins.

11.5.3 Mortalin Inhibits Apoptosis

One mechanism to trigger apoptosis is cytochrome c release from mitochondria. Release of cytochrome c enables Apaf-3 to form a complex with caspase-9, leading to activation of caspase-9, which activates caspase-3, one of the executioner caspases that carry out apoptosis (Nizuma et al. 2010). Mortalin may attenuate this process by suppressing cytochrome c release from mitochondria; mortalin overexpression decreases the number of Bax+ cells with the configuration needed for cytochorome c release (Yang et al. 2008).

Another common mechanism of cell death is through p53 activation, which leads to apoptosis by activating proapoptic members of the Bcl-2 family including Bax, or binding directly to mitochondria (Haupt et al. 2003). Mortalin and p53 can bind to each other in the cytoplasm (Wadhwa et al. 2002), inhibiting p53 translocation to the nucleus and subsequent activation.

11.5.4 Mortalin Enhances Neurogenesis

In addition to direct impacts on mitochondrial function, oxidative stress, and cell death, preservation of mitochondrial function may have additional effects in the acute response to ischemia and in the period of recovery following ischemia. Basal levels of neurogenesis occur in the adult brain in the dentate gyrus of the hippocampus and in the subventricular zone (Alvarez-Buylla and Lois 1995). Brain ischemia induces increased neurogenesis in both of these regions (Tonchev et al. 2005; Tureyen et al. 2004; Liu et al. 1998). However, this increase in neurogenesis does not lead to significant replacement of lost brain tissue. Despite this, enhancement of neurogenesis strongly correlates with lesion size and functional recovery after stroke (Tanaka et al. 2010). Immature neurons, defined as those expressing doublecortin, are highly dependent upon ATP for differentiation (Bernstein and Bamburg 2003). Impairment of mitochondrial function by transgenic deficiency of the alpha-ketoglutarate-dehydrogenase complex attenuates neurogenesis in the intact hippocampus (Calingasan et al. 2008), and protection of mitochondria by overexpression of Bcl-xl improves neurogenesis in the post-ischemic striatum (Chang et al. 2007). Oxidative stress in the brain such as that caused by irradiation, or exposure to inflammatory medium from activated microglia, has been shown to strongly suppress neurogenesis (Monje et al. 2003). In light of these findings our laboratory tested the effect of mortalin overexpression on neurogenesis *in vitro* using neurospheres. We found that neurospheres transfected with a mortalin overexpression plasmid exhibit increased survival of immature neurons. Neuronal survival was also significantly increased by mortalin overexpression in the presence of conditioned media from LPS-treated microglia, a treatment that drastically attenuates neurogenesis in association with oxidative stress and impaired mitochondrial function (Voloboueva et al. 2010). Thus mortalin likely improves neurogenesis by preserving mitochondrial function in immature neurons and may contribute to improved recovery after stroke via this mechanism.

11.5.5 Mortalin in Other Neurodegenerative Diseases

In addition to stroke, deficits in mortalin have also been implicated in other neurodegenerative diseases. Patients with Parkinson's diseases have significantly less mortalin in their substantia nigra mitochondria than control patients (Jin et al. 2006), and also exhibit polymorphisms in the mortalin gene (De Mena et al. 2009). Although not yet tested *in vivo*, mortalin may also have a potential role in Alzheimer's disease. Amyloid β protein, a hallmark of Alzheimer's disease, has harmful effects on mitochondrial functioning. Overexpression of mortalin *in vitro* prevents amyloid β-induced cell apoptosis and depolarization of mitochondrial membrane potential (Qu et al. 2011). Thus, development of pharmaceutical agents that selectively upregulate mortalin in the brain might provide an important new approach to the treatment of neurodegenerative disease. For a more extensive discussion of this area, see Chap. 12 by J. Zhang and Chap. 13 by R. Krueger in this volume.

11.6 Clinical Applications

It is increasingly apparent that overexpression of mortalin *in vitro* and in animal models provides protection against ischemic injury. Although not yet explored, mortalin may also provide protection from other neurodegenerative diseases that involve oxidative stress and mitochondrial impairment, including the selective hippocampal damage induced by forebrain, or global ischemia. This type of injury leads to selective astrocyte dysfunction and neuronal cell death in the CA1 region of the hippocampus, likely due to increased oxidative stress leading to decreased glutamate transporter function in CA1 astrocytes (Xu et al. 2010; Ouyang et al. 2007). As mortalin protects astrocytes from *in vitro* ischemic injury and reduces oxidative stress, it is likely that it could preserve astrocyte viability in this model of the type of brain injury seen following cardiac arrest and resuscitation, and thus prevent CA1 neuronal cell death.

Mortalin is a promising candidate for the treatment of stroke due to its multiple protective mechanisms. Although encouraging results have been found using genetic overexpression of mortalin, this is not an approach that lends itself to rapid clinical translation. Development of an inducer of mortalin would have applications to at least stroke, head and spinal cord trauma, which can be accompanied by ischemia, and Parkinson's and Alzheimer's disease, as well as other neurodegenerative conditions.

11.7 Conclusions

Mortalin, an essential mitochondrial chaperone, is vital for normal brain functioning and is a potential therapeutic target for human stroke. By reducing harmful reactive oxygen species, preserving mitochondrial function, decreasing apoptosis, and supporting neurogenesis, mortalin acts as a multifaceted protector from ischemic conditions and facilitator of recovery. An additional advantage of using mortalin as a pro-reparative tool is that it would likely be effective in hemorrhagic stroke, a setting in which tPA is contraindicated and no current treatment exists. By further exploring the mechanisms of mortalin's neuroprotective properties and possibly developing a small molecule-based approach to increase mortalin expression, researchers may be able to develop a strategy to utilize endogenous mortalin for neuroprotection.

Acknowledgments This work was supported in part by NIH grants NS053898 and GM49831. The authors thank Dr. Ludmila Voloboueva for feedback.

References

Alvarez-Buylla A, Lois C (1995) Neuronal stem cells in the brain of adult vertebrates. Stem Cells 13:263–272

Atamna H, Walter PB, Ames BN (2002) The role of heme and iron-sulfur clusters in mitochondrial biogenesis, maintenance, and decay with age. Arch Biochem Biophys 397:345–353

Beal MF (1995) Aging, energy, and oxidative stress in neurodegenerative diseases. AnnNeurol 38:357–366

Bernstein BW, Bamburg JR (2003) Actin-ATP hydrolysis is a major energy drain for neurons. J Neurosci 23:1–6

Calingasan NY, Ho DJ, Wille EJ. Campagna MV, Ruan J, Dumont M, Yang L, Shi Q, Gibson GE, Beal MF (2008) Influence of mitochondrial enzyme deficiency on adult neurogenesis in mouse models of neurodegenerative diseases. Neuroscience 153:986–996

Canevari L, Kuroda S, Bates TE, Clark JB, Siesjo BK (1997) Activity of mitochondrial respiratory chain enzymes after transient focal ischemia in the rat. J Cereb Blood Flow Metab 17:1166–1169

Chang MY, Sun W, Ochiai W, Nakashima K, Kim SY, Park CH, Kang JS, Shim JW, Jo AY, Kang CS, Lee YS, Kim JS, Lee SH (2007) Bcl-XL/Bax proteins direct the fate of embryonic cortical precursor cells. Mol Cell Biol 27:4293–4305

D'Silva P, Liu Q, Walter W, Craig EA (2004) Regulated interactions of mtHsp70 with Tim44 at the translocon in the mitochondrial inner membrane. Nat Struct Mol Biol 11:1084–1091

De Mena L, Coto E, Sanchez-Ferrero E, Ribacoba R, Guisasola LM, Salvador C, Blazquez M, Alvarez V (2009) Mutational screening of the mortalin gene (HSPA9) in Parkinson's disease. J Neural Transm 116:1289–1293

Deocaris CC, Kaul SC, Wadhwa R (2008) From proliferative to neurological role of an hsp70 stress chaperone, mortalin. Biogerontology 9:391–403

Doyle KP, Simon RP, Stenzel-Poore MP (2008) Mechanisms of ischemic brain damage. Neuropharmacol 55:310–318

Felts SJ, Owen BA, Ngyuen P, Trepel J, Donner DB, Toft DO (2000) The hsp90-related protein TRAP1 is a mitochondrial protein with distinct functional properties. J Biol Chem 275:3305–3312

Giffard RG, Yenari MA (2004) Many mechanisms for hsp70 protection from cerebral ischemia. J Neurosurg Anesthesiol 16:53–61

Haupt S, Berger M, Goldberg Z, Haupt Y (2003) Apoptosis—the p53 network. J Cell Sci 116:4077–4085

Jin J, Hulette C, Wang Y, Zhang T, Pan C, Wadwha R, Zhang J (2006) Proteomic identification of a stress protein, mortalin/mthsp70/GRP75: relevance to Parkinson disease. Mol Cell Proteomics 5:1193–1204

Jin X, Xiao C, Tanguay RM, Yang L, Wang F, Chen M, Fu X, Wang R, Deng J, Deng Z, Zheng Y, Wei Q, Wu T (2004) Correlation of lymphocyte heat shock protein 70 levels with neurologic deficits in elderly patients with cerebral infarction. Am J Med 117:406–411

Kaul SC, Matsui M, Takano S, Sugihara T, Mitsui Y, Wadwha R (1997) Expression analysis of mortalin, a unique member of the Hsp70 family of proteins, in rat tissues. Exp Cell Res 232:56–63

Liu J, Solway K, Messing RO, Sharp FR (1998) Increased neurogenesis in the dentate gyrus after transient global ischemia in gerbils. J Neurosci 18:7768–7778

Liu Y, Liu W, Song XD, Zuo J (2005) Effect of GRP75/mthsp70/PBP74/mortalin overexpression on intracellular ATP level, mitochondrial membrane potential and ROS accumulation following glucose deprivation in PC12 cells. Mol Cell Biochem 268:45–51

Longa EZ, Weinstein PR, Carlson S, Cummins R (1989) Reversible middle cerebral artery occlusion without craniectomy in rats. Stroke 20:84–91

Lu A, Ran R, Parmentier-Batteur S, Nee A, Sharp FR (2002) Geldanamycin induces heat shock proteins in brain and protects against focal cerebral ischemia. J Neurochem 81:355–364

Massa SM, Longo FM, Zuo J, Wang S, Chen J, Sharp FR (1995) Cloning of rat grp75, an hsp70-family member, and its expression in normal and ischemic brain. J Neurosci Res 40:807–819
Merrick BA, Walker VR, He C, Patterson RM, Selkirk JK (1997) Induction of novel Grp75 isoforms by 2-deoxyglucose in human and murine fibroblasts. Cancer Lett 119:185–190
Monje ML, Toda H, Palmer TD (2003) Inflammatory blockade restores adult hippocampal neurogenesis. Science 302:1760–1765
Moro MA, Almeida A, Bolanos JP, Lizasoain I (2005) Mitochondrial respiratory chain and free radical generation in stroke. Free Radic Biol Med 39:1291–1304
Nizuma K, Yoshioka H, Chen H, Kim GS, Jung JE, Katsu M, Okami N, Chan PH (2010) Mitochondrial and apoptotic neuronal death signaling pathways in cerebral ischemia. Biochem Biophys Acta 1802:92–99
Ouyang YB, Voloboueva LA, Xu LJ, Giffard RG (2007) Selective dysfunction of hippocampal CA1 astrocytes contributes to delayed neuronal damage after transient forebrain ischemia. J Neurosci 27:4253–4260
Paradies G, Petrosillo G, Pistolese M, Ruggiero FM (2000) The effect of reactive oxygen species generated from the mitochondrial electron transport chain on the cytochrome c oxidase activity and on the cardiolipin content in bovine heart submitochondrial particles. FEBS Lett 466:323–326
Piantadosi CA, Zhang J (1996) Mitochondrial generation of reactive oxygen species after brain ischemia in the rat. Stroke 27:327–332
Qu M, Zhou Z, Xu S, Chen C, Yu Z, Wang D (2011) Mortalin overexpression attenuates beta-amyloid-induced neurotoxicity in SH-SY5Y cells. Brain Res 1368:336–345
Robinson NC (1993) Functional binding of cardiolipin to cytochrome c oxidase. J Bioenerg Biomembr 25:153–163
Roger VL, Go AS, Lloyd-Jones DM, Adams RJ, Berry JD, Brown TM, Carnethon MR, Dai S, De Simone G, Ford ES, Fox CS, Fullerton HJ, Gillespie C, Greenlund KJ, Hailpern SM, Heit JA, Ho PM, Howard VJ, Kissela BM, Kittner SJ, Lackland DT, Lichtman JH, Lisabeth LD, Makuc DM, Marcus GM, Marelli A, Matchar DB, McDermott MM, Meigs JB, Moy CS, Mozaffarian D, Mussolino ME, Nichol G, Paynter NP, Rosamond WD, Sorlie PD, Stafford RS, Turan TN, Turner MB, Wong ND, Wylie-Rosett J, American Heart Association Statistics Committee and Stroke Statistics Subcommittee (2011) Heart disease and stroke statistics–2011 update: a report from the American Heart Association. Circulation 123:e18–209
Sims NR, Muyderman H (2010) Mitochondria, oxidative metabolism and cell death in stroke. Biochem Biophys Acta 1802:80–91
Soane L, Kahraman S, Kristian T, Fiskum G (2007) Mechanisms of impaired mitochondrial energy metabolism in acute and chronic neurodegenerative disorders. J Neurosci Res 85:3407–3415
Takano S, Wadhwa R, Yoshii Y, Nose T, Kaul SC, Mitsui Y (1997) Elevated levels of mortalin expression in human brain tumors. Exp Cell Res 237:38–45
Tanaka Y, Tanaka R, Liu M, Hattori N, Urabe T (2010) Cilostazol attenuates ischemic brain injury and enhances neurogenesis in the subventricular zone of adult mice after transient focal cerebral ischemia. Neuroscience 171:1367–1376
The National Institute of Neurological Disorders and Stroke rt-PA Study Group (2000) Effect of intravenous recombinant tissue plasminogen activator on ischemic stroke lesion size measured by computed tomography. Stroke 31:2912–2919
The U.S. National Institutes of Health (2011) http://www.clinicaltrials.gov. Accessed 13 Jan 2011
Tonchev AB, Yamashima T, Sawamoto K, Okano H (2005) Enhanced proliferation of progenitor cells in the subventricular zone and limited neuronal production in the striatum and neocortex of adult macaque monkeys after global cerebral ischemia. J Neurosci Res 81:776–788
Tureyen K, Vemuganti R, Sailor KA, Bowen KK, Dempsey RJ (2004) Transient focal cerebral ischemia-induced neurogenesis in the dentate gyrus of the adult mouse. J Neurosurg 101:799–805

Voloboueva LA, Duan M, Ouyang Y, Emery JF, Stoy C, Giffard RG (2008) Overexpression of mitochondrial Hsp70/Hsp75 protects astrocytes against ischemic injury in vitro. J Cereb Blood Flow Metab 28:1009–1016

Voloboueva LA, Lee SW, Emery JF, Palmer TD, Giffard RG (2010) Mitochondrial protection attenuates inflammation-induced impairment of neurogenesis in vitro and in vivo. J Neurosci 30:12242–12251

Wadhwa R, Taira K, Kaul SC (2002) An Hsp70 family chaperone, mortalin/mthsp70/PBP74/Grp75: what, when, and where? Cell Stress Chaperones 7:309–316

Wadhwa R, Takano S, Kaur K, Aida S, Yaguchi T, Kaul Z, Hirano T, Taira K, Kaul SC (2005) Identification and characterization of molecular interactions between mortalin/mtHsp70 and HSP60. Biochem J 391:185–190

Williamson CL, Dabkowski ER, Dillmann WH, Hollander JM (2008) Mitochondria protection from hypoxia/reoxygenation injury with mitochondria heat shock protein 70 overexpression. Am J Physiol Heart Circ Physiol 294:H249–256

Wilson JX (1997) Antioxidant defense of the brain: a role for astrocytes. Can J Physiol Pharmacol 75:1149–1163

Xu L, Emery JF, Ouyang YB, Voloboueva LA, Giffard RG (2010) Astrocyte targeted overexpression of Hsp72 or SOD2 reduces neuronal vulnerability to forebrain ischemia. Glia 58:1042–1049

Xu L, Voloboueva LA, Ouyang Y, Emery JF, Giffard RG (2009) Overexpression of mitochondrial Hsp70/Hsp75 in rat brain protects mitochondria, reduces oxidative stress, and protects from focal ischemia. J Cereb Blood Flow Metab 29:365–374

Yang L, Liu X, Hao J, Yang Y, Zhao M, Zuo J, Liu W (2008) Glucose-regulated protein 75 suppresses apoptosis induced by glucose deprivation in PC12 cells through inhibition of Bax conformational change. Acta Biochim Biophys Sin 40:339–348

Yoo JY, Ryu J, Gao R, Yaguchi T, Kaul SC, Wadhwa R, Yun CO (2010) Tumor suppression by apoptotic and anti-angiogenic effects of mortalin-targeting adeno-oncolytic virus. J Gene Med 12:586–595

Chapter 12
Catecholamine Regulated Protein (CRP40), A Splice Variant of Mortalin-2: Functional Role in CNS Disorders

Joseph P. Gabriele, Sarah E. Groleau, Ritesh P. Daya, Zdenek B. Pristupa and Ram K. Mishra

Abstract Heat shock proteins are constitutively expressed, but are also induced by heat and other environmental stressors. These proteins play an important role in maintaining correct folding of cellular proteins; protecting the cell from deleterious stresses by inhibiting the aggregation of partially denatured proteins and refolding them to correct confirmations. The Catecholamine Regulated Protein of size 40-kDa (CRP40) is a heat shock-like splice variant of Mortalin-2; its peptide sequence is identical to the carboxyl terminal of Mortalin-2. This protein is found co-localized with tyrosine hydroxylase in the dopaminergic pathways of the midbrain. CRP40 is differentially expressed by dopamine D1 and D2 receptor antagonists; when dopamine concentrations increase, CRP40 is upregulated. Since this increased CRP40 protein expression at high levels of DA is inhibited by treatment of cells with antioxidants, and since the natural oxidation of excess dopamine forms a potent toxic species of oxidant, CRP40 is implicated in the response pathways associated with oxidative stress. Oxidative stress and mitochondrial dysfunction have been implicated in dopaminergic disorders of the brain. CRP40 and Mortalin-2, therefore, are of particularly interest when conducting novel research in the realm of Parkinson's disease and Schizophrenia. Further, as a splice variant of Mortalin-2, CRP40 could play a central role in research involving other Mortalin-related diseases such as stroke and ischemia, carcinogenesis, Alzheimer's disease, and Huntington's disease. With so many unexplored avenues, CRP40 could boast an exciting future in health research.

Keywords CNS disorder · Schizophrenia · Parkinson's · Heat Shock Proteins · Dopamine

---

J. P. Gabriele (✉) · S. E. Groleau · R. P. Daya · Z. B. Pristupa · R. K. Mishra
Department of Psychiatry, McMaster University, 1200 Main St. West, HSC 4N81, Hamilton, L8N 3Z5, Canada
e-mail: gabriejp@mcmaster.ca

S. C. Kaul, R. Wadhwa (eds.), *Mortalin Biology: Life, Stress and Death*,
DOI 10.1007/978-94-007-3027-4_12, © Springer Science+Business Media B.V. 2012

12.1 Introduction: Heat Shock Proteins, and Catecholamine Regulated Proteins

12.1.1 Heat Shock Proteins

Heat shock proteins (HSPs) such as HSP90, HSP70, HSP60 and HSP28 are constitutively expressed, and play an important role in maintaining correct folding of cellular proteins (Ashok et al. 2001; Frydman 2001; Kalivendi et al. 2001; Kiang and Tsokos 1998; Liu and Thiele 1999; Nanasi and Jednakovits 2001; Sheth et al. 2001; Sorger 1991; Zinsmaier and Bronk 2001). HSPs are mainly induced by heat, although other environmental stressors can also induce their expression. Most HSPs act as molecular chaperones, protecting the cell from deleterious stresses. These proteins are able to inhibit the aggregation of partially denatured proteins and refold them to correct confirmations (Frydman 2001; Kiang and Tsokos 1998; Sorger 1991; Zinsmaier and Bronk 2001).

There are over 15,000 references on HSPs. Many of these proteins are multifunctional. Protein B23 (Hingorani et al. 2000), as an example, acts as both a chaperone and a ribosomal assembly factor with a nucleic acid binding site. It has the ability to shuttle between nucleus and cytoplasm (Borer et al. 1989; Hingorani et al. 2000), bind nuclear localization signal containing proteins (Borer et al. 1989; Hingorani et al. 2000; Szebeni and Olson 1999) and stimulate import of proteins into the nucleus (Borer et al. 1989; Hingorani et al. 2000; Szebeni and Olson 1999). As such, B23 acts as a molecular chaperone and displays multifunctionality. Similarly, HSP90 displays chaperone activity; it has steroid binding sites and a site for P23 protein binding. Together, HSP90 and P23 help assemble inactivated steroid receptors into HSP90 heterocomplexes, allowing the receptor to bind the hormone (Dittmar and Pratt 1997; Pratt et al. 1999). Further, the inactive glucocorticoids receptor first binds to HSP70/HSC70, then complexes with HSP90. Together this complex opens the receptor pocket for ligand binding. In some cases, complexes of HSP60 and HSP70 can also open ligand binding pockets. The activated receptor then moves to the nucleus and allows for DNA binding (Cato and Mink 2001). Another protein, clusterin, also has multifunctional properties (Humphreys et al. 1999). It can protect catalase from heat-induced aggregation, act as a protein-binding partner, remodel membranes, and transport lipids. Many HSPs (HSP90, HSP70 and HSP27) have also been shown to display anti-apoptotic activities. These proteins have the collective ability to regulate the activities of signalling molecules involved in programmed cell death (Beere 2001).

12.1.2 Catecholamine Regulated Proteins

Previous studies from our laboratory have described a new class of central nervous system (CNS) proteins (Catecholamine Regulated Proteins; CRPs; 47-, 40- and

26-kDa) capable of binding dopamine (DA) and related catecholamines when incubated *in vitro* with synaptosomal membrane preparations (Ross et al. 1993, 1995). The CRP of size 40-kDa (CRP40) is a heat shock-like splice variant of Mortalin-2. Specifically, CRP40 is identical to the carboxyl terminal of Mortalin-2, and completely lacks the p53-binding region found at the N-terminus of Mortalin-2 (Nair and Mishra 2001) . Recently, we reported molecular cloning of bovine CRP40 (Nair and Mishra 2001). We discovered that the bovine CRP40 is inducible in SH-SY5Y cells incubated with excess DA. This increased CRP40 protein expression at high levels of DA is inhibited by treatment of cells with antioxidants, implicating CRP40 in the response pathways associated with oxidative stress due to DA auto-oxidation. We also demonstrated differential protein expression of CRP40 by DA D1 and DA D2 receptor antagonists, and its colocalization with a DA-neuron marker, tyrosine hydroxylase (TH), in the nigrostriatal DA pathway (Gabriele et al. 2003; Goto et al. 2001). More recently, a human CRP40 has been cloned and characterized in our laboratory that shows significant homology with the bovine strain. This human CRP40 protein functions similarly to the bovine protein.

CRP40 (Fig. 12.1) belongs to a family of HSPs known as HSP70, as evidenced by the following characteristics. CRP40 exhibits the characteristic and conserved heat shock motif and its expression is inducible by elevated temperature. Further, the organellar localization motif, which allows CRP40 into different cellular organelles; as well as the chaperone motif, which allows CRP40 to protect other proteins from thermal denaturation, is conserved. In fact, CRP40 is known to protect the model substrate, luciferase, against thermal aggregation, in a similar fashion to that of HSP70 (Fig. 12.2). However, unlike HSP70, CRP40 contains the tyrosine and aspartate residues necessary for DA binding and is CNS and blood specific.

Unlike CRP40, HSP70 (or HSC70) are not differentially expressed by DA-ergic drugs under identical experimental conditions. This result suggests a highly specific function of CRP40 in DA-ergic neurotransmission. Considering the novel features of CRP40 (presence of nuclear localization signals, substrate binding domains, heat sensitive motif (Fig. 12.1) and regulation by DA-ergic drugs), this protein is expected to display multifunctionality similar to HSPs. CRP40 may chaperone an inactive DA synthesis/inhibition transcription factor, activate it, and translocate it to the DNA site for regulation of DA. Further, the distribution and localization of CRP40 suggests that it may act at several sites in order to maintain DA homeostasis. For example, CRP40 may bind DA to buffer the intracellular pool of DA; may act as a transcription factor and inhibit DA synthesis at the level of TH; and may, by virtue of its chaperone properties, protect against cellular stress in the substantia nigra (SN; Fig. 12.3). CRP40 can also play a protective role during DA-induced cell death. It has been shown that transcription factor, κB (NF κB) is activated and translocated to the nucleus where it binds to its cognate DNA sequences during DA-induced oxidative stress (Hunot et al. 1997). Binding of NF κB to DNA leads to activation of signalling pathways that are involved in apoptosis. CRP40 (alone or in conjunction with another CRP, such as CRP47; or HSP's, such as HSP70) may prevent this transcription factor from binding to DNA, thereby suppressing cell death.

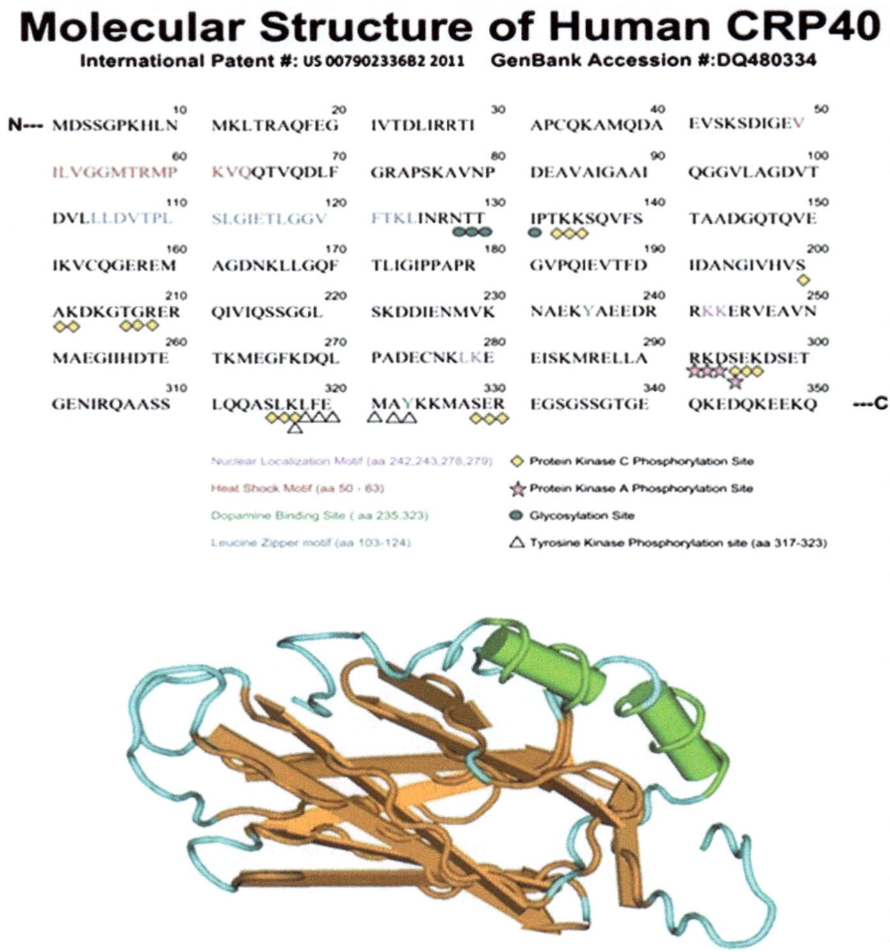

Fig. 12.1 Amino acid sequence and conformational structure of human CRP40

12.2 Characterization of CRP40

12.2.1 Pharmacological Studies

Although CRP40 belongs to the HSP70 family, unlike HSPs, it is specifically and differentially modulated by DA D_1 and D_2 receptor agonists and antagonists such as L-DOPA, apomorphine, DA and haloperidol in various brain regions (Modi et al. 1996). Chronic treatment with cocaine has been demonstrated to increase the levels of CRP40 in the nucleus accumbens. Meanwhile, chronic treatment with

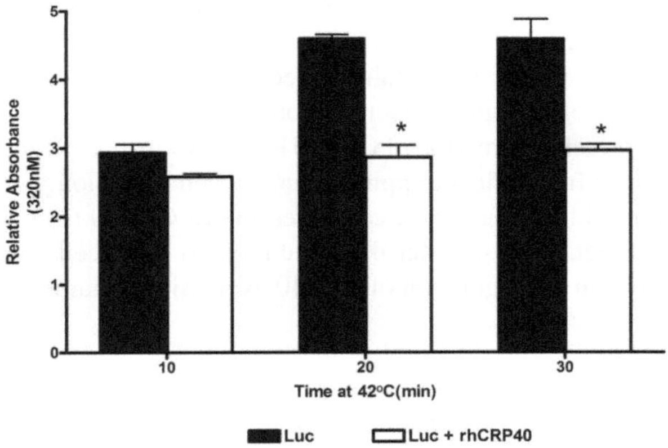

Fig. 12.2 Human CRP40 displays chaperone functions. Following 20 and 30 min of thermal denaturation at 42°C, incubation with recombinant human CRP40 (rhCRP40) significantly inhibited firefly luciferase (Luc) aggregation ($*p < 0.05$). Data are presented as the mean relative absorbance ± SEM from five separate experiments. This figure was reproduced with permission from Cell stress and Chaperones (Springer; License # 2738291317449)

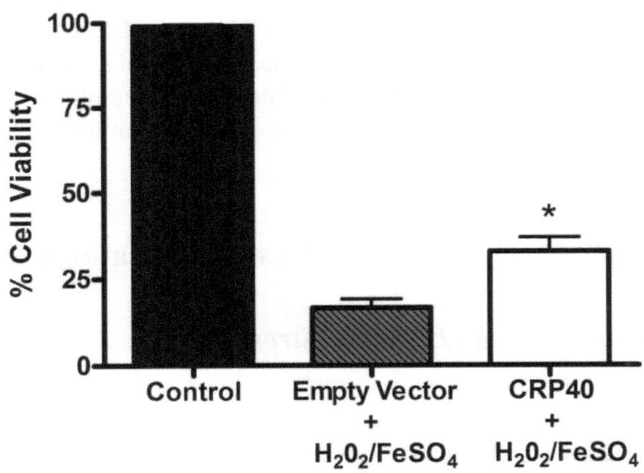

Fig. 12.3 Bar graph displaying results from an MTT assay of cell viability after incubation with $H_2O_2/FeSO_4$ (16 h) and transfection with human CRP40 or the empty vector. Overexpression of CRP40 significantly improved the viability of cells, as compared with empty vector-transfected cells ($*p < 0.05$) following $H_2O_2/FeSO_4$ treatment. Data are presented in triplicate as the mean percentage of viable cells ± SEM. This figure was reproduced with permission from Cell stress and Chaperones (Springer; License # 2738291317449)

d-amphetamine has shown to markedly increase CRP40 expression in both the striatum and the nucleus accumbens. Conversely, chronic treatment with a high affinity DA D2 receptor agonist, quinpirole, selectively decreases the CRP40 in the prefrontal cortex, ventral tegmental area, and striatum (Gabriele et al. 2002). Collectively, these findings support the expression of CRP40 in the nigrostriatal and mesocorticolimbic regions. Furthermore, our experiments support the hypothesis that anti-psychotic drugs and psychostimulants like cocaine, affect CRP40 protein levels in the mesocorticolimbic brain regions and further reinforce its role in these disorders (Sharan et al. 2003). The protein expression of CRP40 levels by DA-ergic drugs is not simply circumstantial since free radical trapping agents, when used along with haloperidol or cocaine, did not block the increased expression of CRP40 (Fig. 12.4). In other words, DA or its metabolites, rather than free radicals produced through oxidative stress, are involved in the regulation of CRP40 expression (Sharan et al. 2003).

12.2.2 Cloning and Pharmacological Characterization of the Human CRP40

As discussed earlier, studies on bovine CRP40 (Nair and Mishra 2001) revealed a human sequence (Genbank # BQ 224193) with similar protein homology and identical amino acid gap regions to bovine CRP40 cDNA. Using this information, specific primers were designed and the human CRP40 sequence was elucidated (Fig. 12.1) via RNA ligase-mediated rapid amplification of cDNA ends (RLM-RACE). As shown in (Fig. 12.5), the entire human CRP40 sequence is identical to the carboxy-terminal end of Mortalin-2 (Wadhwa et al. 1998). Specifically, the homology spans from exon 10 to exon 17 with an independent promoter region found at intron 9.

12.3 Dopamine and Oxidative Stress: Implications for CRP40

12.3.1 Dopamine and Oxidative Stress

DA is a catecholamine neurotransmitter that plays a fundamental and critical role in movement, learning, memory, motivation and cognition. DA is implicated in disorders such as schizophrenia, Parkinson's disease (PD) and drug addiction (Abi-Dargham et al. 2000; Carlsson et al. 2001; Seeman and Kapur 2000). DA undergoes oxidative reactions that lead to formation of several oxidative metabolites. The consequent oxidative stress and free radical species production result in an impairment of DA neurotransmission and DA related processes (Abraham et al. 1997).

Increased oxidative stress and free radical production have frequently been suggested as potential candidates or contributing factors for the pathogenesis of PD (Friedman and Galazka-Friedman 2001; Kikuchi et al. 2002; Migliore et al. 2002; Mouradian 2002; Nair and Mishra 2001; Olanow and Tatton 1999; Serra et al. 2001; Stern 1997). The current and most common hypothesis suggests that nigral neuronal death in PD is due to excessive oxidative stress resulting from auto-oxidation.

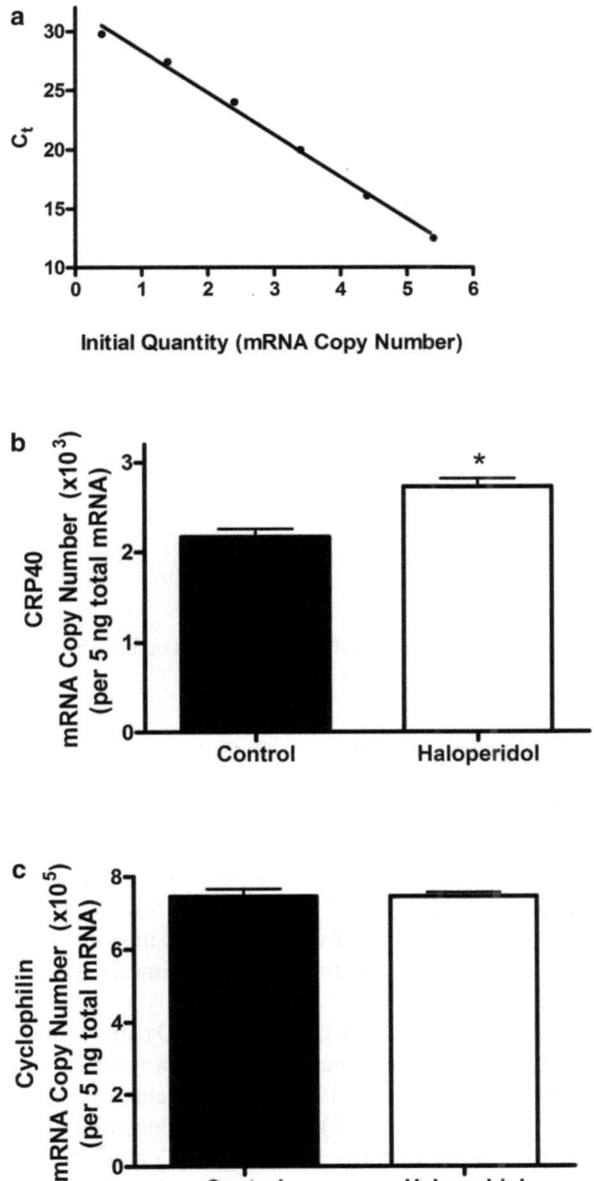

Fig. 12.4 Real-time PCR analysis of CRP40 mRNA expression. **a** Standard curve by real-time PCR using specific primers and varying concentrations of RNA. **b** Treatment of DA D_{2L}-transfected SH-SY5Y cells ($n = 3$), with haloperidol, resulted in a statistically significant ($^*p < 0.05$) increase in human CRP40 mRNA expression as compared with untreated (PBS) control cells ($n = 3$). **c** Treatment of DA D_{2L} -transfected SHSY5Y cells ($n = 3$), with haloperidol, failed to result in any significant alteration in human cyclophilin mRNA expression, when compared with untreated (PBS) control cells ($n = 3$). Data are presented as mRNA copy number per 5 ng of total RNA $\pm$ SEM. This figure was reproduced with permission from Cell stress and Chaperones (Springer; License # 2738291317449)

It is proposed that free radicals, formed as a result of catecholamine metabolism, play an important role in the pathogenesis of neurodegenerative disorders (Cohen 2000; Dexter et al. 1994; Jenner 1994; Olanow 1992, 1993; Olanow and Tatton 1999; Youdim 1988). These findings can be summarized as follows: (1) DA turnover is increased in nigrostriatal surviving neurons which consequently results in the formation

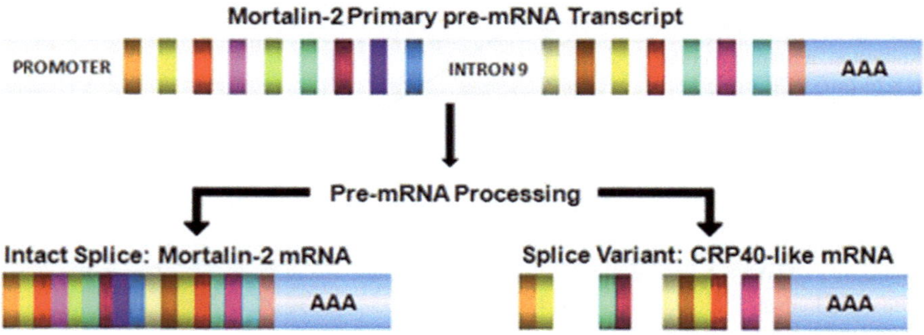

Fig. 12.5 Proposed model of CRP40 mRNA splice variant formation

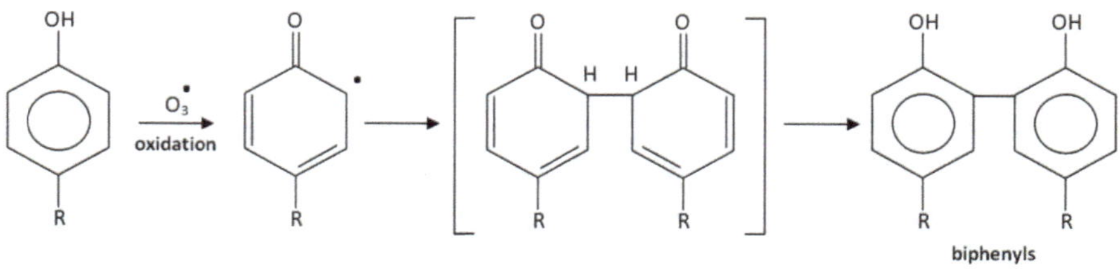

Fig. 12.6 Diagram of phenolic coupling which occurs when two phenols dimerize oxidatively and form biphenyls

of hydrogen peroxide (H_2O_2); (2) the levels of total and reduced forms of glutathione are decreased in the SN, suggesting decreased protection against hydroxyl radical ($OH^\bullet$) resulting from DA oxidation; (3) iron concentration is increased in the SN thus increasing the interaction with H_2O_2 and subsequent formation of $OH^\bullet$; and (4) lipid peroxidation is subsequently increased. This chain of reactions has been suggested to account for cell death. Furthermore, biochemical data indicate that the oxidative products of catecholamines may involve "phenolic coupling". Phenolic coupling occurs when two phenols dimerize oxidatively and form biphenyls as shown in (Fig. 12.6).

This reaction has been implicated in PD patients receiving large doses of L-DOPA, and has been demonstrated to occur with tyrosine units in peptides (Lal and de la Vega 1975; Prutz et al. 1983). DA concentrations in synaptic vesicles range from 0.1–1 mM (Jonnson 1971). These high concentrations can be toxic to neurons in the absence of cytoprotective mechanisms (Cohen 2000). DA and other catecholamines have also been shown to form cysteinyl-DA and cysteinyl-DOPAC adducts in striatal tissue (Hastings et al. 1996; Rotman et al. 1976; Saner and Thoenen 1971; Scheulen et al. 1975). In addition, DA quinones formed during DA oxidation have been demonstrated to modify DA transporters without causing any loss in the membrane integrity (Berman et al. 1996, Berman and Hastings 1997). Similarly, the glutamate transporter can also be inhibited by DA oxidation products and reactive oxygen species (ROS) (Berman and Hastings 1997; Bernheimer et al. 1973; Terland et al. 1997). Many scientists have reported that high synaptic catecholamine concentrations (0.1–1 mM) can be toxic in the absence of protective mechanisms (Beaudet and Descarries 1984; Chiueh and Rauhala 1998; Jonnson 1971; Zigmond and Hastings 1998).

Mitochondrial dysfunction in PD occurs as a significant decrease in activity levels of Complex I proteins in the pars compacta of the substantia nigra (PC-SN) (Lang and Lozano 1998; Moore et al. 2005). As mitochondria are the power centre for the cell, a defect of this magnitude could be responsible for energy failure leading to apoptosis, or a weakening of the cell against such insults as toxins, oxidative stress, or mutagens. ROS are a natural by-product of the electron transport chain (ETC) that creates cell energy in the mitochondria. When the cell and mitochondria are healthy, these natural toxins are disposed of by various, highly conserved processes (Lang and Lozano 1998). However, as discussed earlier, if these ROS are not properly controlled, they can react with proteins, lipids, and other cellular molecules, altering their structures and leading to cell damage or cell death. In fact, the natural oxidation of excess DA forms DA quinones, a potent toxic species, making this pathway particularly interesting when conducting research in the realm of PD.

12.4 Role of CRP40 in Parkinson's Disease

PD is a universal disease, affecting people of both sexes and of all ethnicities. Approximately 3% of people over the age of 65 and about 0.3% of the general population suffer from PD (Lang and Lozano 1998). Major symptoms include muscle tone rigidity, akinesia, and resting tremor (Lang and Lozano 1998). Currently, there is no definitive biomarker to confirm diagnosis of PD and, according to postmortem studies by Rajput and colleagues; approximately 24% of patients with PD are misdiagnosed (Rajput et al. 1991).

PD is characterized by progressive cell death specific to the DA-ergic neurons in the PC-SN. The pattern of deterioration begins at the ventro-lateral PC-SN and progresses to the medial-ventral PC-SN. This trend is highly specific to PD, and is the opposite of trends observed in normal aging (Lang and Lozano 1998).

Several possible causes have been implicated for PD, including excitotoxins, neurotrophic deficiencies, genetics, and immune responses (Lang and Lozano 1998). Most relevant to this chapter is the possibility that mitochondrial dysfunction and oxidative stress play a major role in the pathology of PD.

CRP40 is expressed in the nigrostriatal and ventral tegmental DA-ergic neurons, as evidenced by its co-localization with TH (Goto et al. 2001). In PD, there is a degeneration of nigrostriatal DA-ergic neurons. Thus, under-expression of CRP40 in this region may contribute to imbalance of DA. In fact, it is possible that excessive DA auto-oxidation in the nigrostriatal DA pathways lead to the neural degeneration that is characteristic of PD.

In a study using rats with intrastriatal injections of 6-hydroxydopamine and reserpine treatment (used as models of PD), results showed decreased levels of CRP40 in the striatum (Modi et al. 1996). Similarly, preliminary experiments carried out in

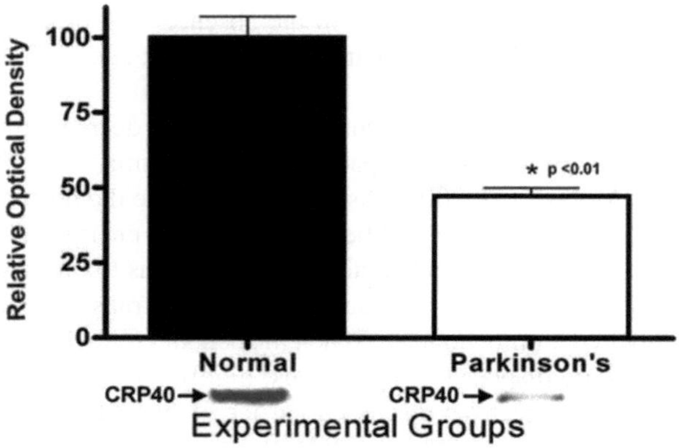

Fig. 12.7 Reduced levels of CRP40 at the protein level in post mortem Parkinson's disease specimens as compared to healthy aged matched controls ($n = 6$). Representative western immunoblot values were normalized with β-actin housekeeping gene. All samples were matched for gender, age, and post mortem time. Samples were supplied by National Neurological Research Specimen Bank, Los Angeles, USA

post mortem human striatal samples (obtained from controls and PD patients) show significantly decreased expression of CRP40 (Fig. 12.7). It is not known whether similar changes also occur in other DA-ergic brain regions such as medial prefrontal cortex, nucleus accumbens, olfactory tubercle, and SN. From these experiments, it is also unclear whether the decrease in CRP40 results from to the degeneration of neurons, or if low expression of CRP40 leads to degeneration of nigrostriatal DA-ergic neurons. Further studies are necessary to establish the "cause or consequence" phenomenon for the role of CRP40 in this degenerative disorder.

12.5 Role of CRP40 in Schizophrenia

Schizophrenia, like PD, involves a dysregulation of the DA-ergic system that manifests as positive, negative and cognitive symptoms. Positive symptoms include hallucinations, delusions and disorganised behaviour, while negative symptoms are characterized by emotional flattening, social withdrawal and attention impairment. The expression of CRP40 in healthy (n = 16, age 70 ± 18 years (mean ± SD), post mortem interval (PMI) 16 ± 4 h) and schizophrenic (n = 16, age 53 ± 22 years (mean ± SD), PMI 26 ± 22 h) patients was examined in human post mortem samples (Gabriele et al. 2005). Western blot analysis revealed a significant decrease in the protein expression of CRP40 in the ventral striatum of schizophrenic patients compared to that of healthy subjects (Fig. 12.8). Non-medicated schizophrenic patients showed an even greater reduction in CRP40 expression than medicated schizophrenics when compared to controls (Fig. 12.9) (Gabriele et al. 2005).

In a separate study, investigation of human schizophrenic dorsolateral prefrontal cortex brain samples revealed human CRP40 mRNA expression is in fact related

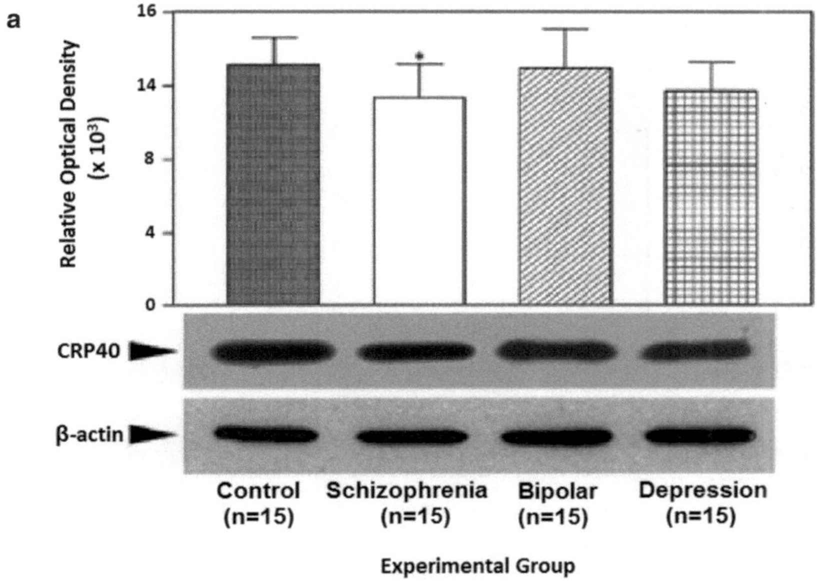

b

CATEGORY	n	MEAN	SD
Normal Control	15	131412	14694
*Schizophrenia	15	113017	18769
Bipolar	15	129461	21348
Depression	15	116926	15794

INDIVIDUAL 95% CONFIDENCE INTERVALS FOR MEANS

```
--------+---------+---------+--------
                      (--------*---------)
(--------*--------)
                    (--------*---------)
    (--------*--------)
--------+---------+---------+--------
     110000    120000    130000
```

*Significant effect [F=3.91, df(3,59), p=0.013]

Fig. 12.8 **a** Graph showing relative optical densities of ventral striatal CRP40 expression among normal control subjects (Normal control), schizophrenic patients (Schizophrenia), bipolar disorder patients (Bipolar) and depression patients (Depression). Data are presented as Mean ± S.D. Representative immunoblots showing relative ventral striatal CRP40 expressions and corresponding β-actin levels are displayed beneath the graph. **b** Post hoc analysis of CRP40 expression among these disease groups. Non-overlapping confidence intervals represent significant differences (*$p < 0.02$). This figure was reproduced with permission from Schizophrenia Research (Elsevier; License # 2738871259784)

to antipsychotic drug use (Gabriele et al. 2010a). Human CRP40 is differentially expressed in the presence of DA and DA metabolites; therefore the effect of antipsychotic drug use on CRP40 mRNA levels in schizophrenic post-mortem samples was investigated. Post-mortem dorsolateral prefrontal cortex RNA samples were extracted from 35 control patients, bipolar disorder patients, and schizophrenic patients. CRP40/mortalin mRNA expression was assessed via real-time PCR and backward,

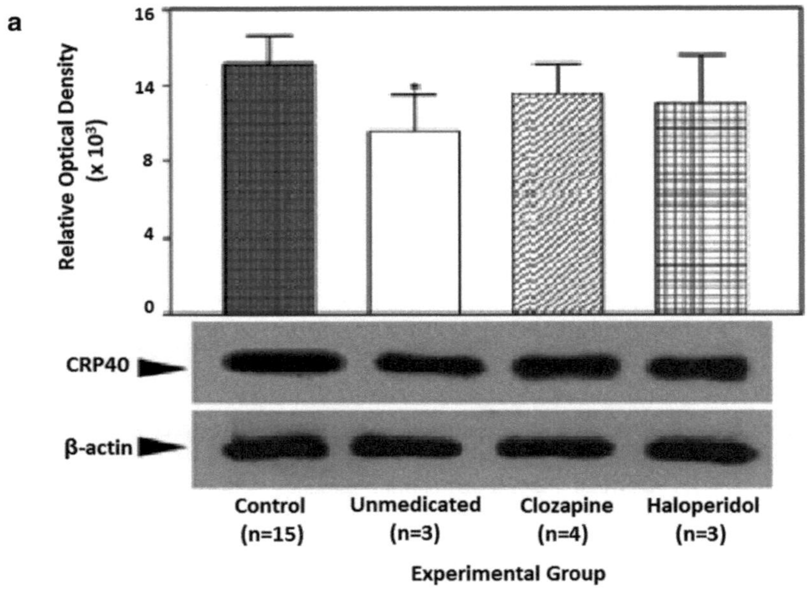

b

CATEGORY	n	MEAN	SD
Normal Control	15	131412	14694
*Unmedicated	3	96194	18955
Clozapine	4	115828	15399
Haloperidol	3	109602	25988

INDIVIDUAL 95% CONFIDENCE INTERVALS FOR MEANS

```
--+---------+---------+---------+----
                            (----*---)
   (---------*---------)
              (--------*--------)
         (---------*---------)
--+---------+---------+---------+----
80000   100000   120000   140000
```

*Significant effect [F=4.75, df(3,24), p=0.011]

Fig. 12.9 **a** Graph showing relative optical densities of ventral striatal CRP40 expression among normal control subjects (Normal control), unmedicated schizophrenic patients (Unmedicated), clozapine-treated patients (Clozapine) and haloperidol-treated patients (Haloperidol). Data are presented as mean ± S.D. Representative immunoblots showing relative ventral striatal CRP40 expressions and corresponding β-actin levels are displayed beneath the graph. **b** Post hoc analysis of CRP40 expression among these groups. Non-overlapping confidence intervals represent significant differences (*$p < 0.02$). This figure was reproduced with permission from Schizophrenia Research (Elsevier; License # 2738871259784)

stepwise multiple regression. Controlling for important confounding factors, multiple regression analysis revealed a positive correlation between lifetime antipsychotic drug use and increased CRP40/mortalin mRNA expression in schizophrenic patients (Table 12.1). Conversely, multiple regression analysis revealed no evidence to suggest a correlation between antipsychotic drug use and mRNA expression in bipolar patients (Gabriele et al. 2010a).

Table 12.1 Descriptive subject statistics. This figure was reproduced with permission from Schizophrenia Research (Elsevier; License # 2738871259784)

Variable	Normal control (n = 35)	Bipolar disorder (n = 35)	Schizophrenia (n = 35)
Age (mean ± SD, years)	44.2 ± 7.6 (Range 31–60)	4.5 ± 10.5 (Range 19–64)	42.6 ± 8.5 (Range 19–59)
Sex (M/F)	29/9	17/18	26/9
Race (Caucasian/other)	35/0	33/2	34/1
Age of onset (mean ± SD, years)	N/A	25.1 ± 9.1 (Range 14–48)	21.3 ± 6.1 (Range 9–34)
Duration of illness (mean ± SD, years)	N/A	20.1 ± 9.5 (Range 2–45)	21.3 ± 10.2 (Range 1–45)
Time in hospital (mean ± SD, years)	N/A	0.5 ± 1.4 (Range 0–8)	1.2 ± 2.3 (Range 0–12)
Alcohol abuse at TOD (n)	2	11	12
Drug abuse at TOD (n)	1	10	9
Smoking at TOD (yes/no/unknown)	9/9/17	16/6/13	23/4/8
Psychotic feature (yes/no/unknown)	0/35/0	21/12/2	35/0/0
Lifetime antipsychotic use (FE ± SD, mg)	N/A	10.035 ± 22.896 (Range 0–130.000)	85.004 ± 100.335 (Range 50–400.000)
Relative brain mass (mean ± SD, g)	1444 ± 148.4 (Range 1120–1900)	1394 ± 139.1 (Range 1120–1670)	1442 ± 107.5 (Range 1170–1630)
Right brain (n)	19	15	18
Left brain (n)	16	20	17
PMI (mean ± SD, h)	29.4 ± 12.9 (Range 9–58)	37.9 ± 18.4 (Range 12–81)	31.4 ± 15.5 (Range 9–80)
Refrigerator interval (mean ± SD, h)	3.6 ± 2.6 (Range 0–14)	10.1 ± 10.4 (Range 1–54)	6.0 ± 4.2 (Range 1–19)
Brain pH (mean ± SD)	6.61 ± 0.27 (Range 6.00–7.03)	6.43 ± 0.30 (Range 5.76–6.97)	6.48 ± 0.24 (Range 5.90 ± 6.93)
28S:18S rRNA ratio (mean ± SD)	2.18 ± 0.50 (Range 0.90–3.74)	2.21 ± 0.75 (Range 0.49–4.18)	2.13 ± 0.55 (Range 1.18–3.81)

M male; *F* female; *N/A* not applicable; *TOD* time of death; *PMI* postmartem interval; *FE* fluphenazine equivalents

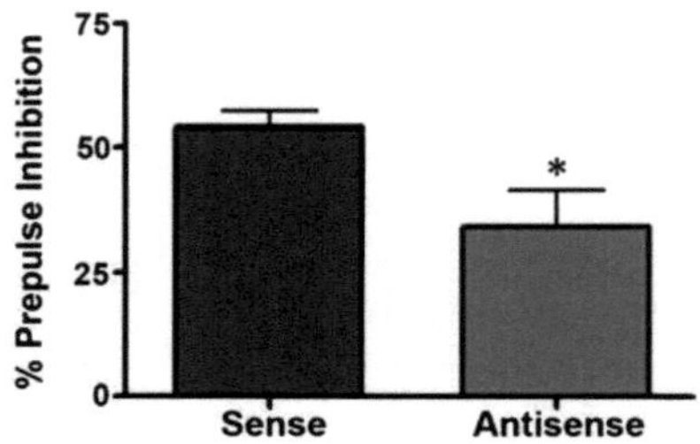

Fig. 12.10 PPI deficits in mortalin antisense ODN-treated rats. Antisense and sense ODN-infused rats were subjected to analyses to measure sensorimotor gating deficits using the SR-Lab Startle Response System. Antisense ODN-treated rats (n = 6) exhibited significant PPI deficits (*$p < 0.045$) when compared to sense ODN-treated control rats (n = 6). Data were expressed as mean % prepulse inhibition (±SEM). This figure was reproduced with permission from Synapse (Wiley; License # 2738880147214)

Mortalin, the 70-kDa mitochondrial alternate gene product of CRP40, was misexpressed through a gene knockdown in the medial prefrontal cortex of male Sprague Dawley rats (Gabriele et al. 2010b). The misexpression of mortalin is hypothesized to contribute to the development of schizophrenic symptoms. Mortalin protein knockdown was achieved through the infusion of antisense oligodeoxynucleotide molecules in the rat medial prefrontal cortex. Sensorimotor gating and social interaction were measured in sense and antisense groups one week following surgery and continuous infusion of oligodeoxynucleotide sequences. Sensorimotor gating was measured though prepulse inhibition (PPI), which assesses the startle reflex response to acoustic stimuli. Deficits in PPI are typically observed in patients diagnosed with SZ. Significant deficits in PPI were found in the antisense oligodeoxynucleotide treated group when compared to the control group (Fig. 12.10). However, no changes were observed in social interaction. Social interaction included nonaggressive sniffing, following, crawling under or over, grooming, and aggressive behaviour (Gabriele et al. 2010b).

12.6 Future Directions for CRP40 Research

CRP40 is expressed at both the protein and mRNA levels in white blood cells and platelets, but not in other peripheral tissues. These characteristics highlight CRP40's potential as a biomarker for the diagnosis of PD or schizophrenia.

Further, as a splice variant of Mortalin-2, CRP40 could play a central role in research involving other Mortalin-related diseases. For example, CRP40 research may hold answers for questions with regards to stroke and ischemia, carcinogenesis, Alzheimer's disease, and Huntington's disease. With so many unexplored avenues, CRP40 is a unique heat shock protein with an exciting future in health research.

References

Abi-Dargham A, Rodenhiser J, Printz D, Zea-Ponce Y, Gil R, Kegeles LS, Weiss R, Cooper TB, Mann JJ, Van Heertum RL, Gorman JM, Laruelle M (2000) From the cover: increased baseline occupancy of D2 receptors by dopamine in schizophrenia. Proc Natl Acad Sci U S A 97:8104–8109

Abraham WC, Logan B, Thompson VL, Williams JM, Tate WP (1997) Sequence-independent effects of phosphorothiolated oligonucleotides on synaptic transmission and excitability in the hippocampus in vivo. Neuropharmacol 36:345–352

Ashok BT, Kim E, Mittelman A, Tiwari RK (2001) Proteasome inhibitors differentially affect heat shock protein response in cancer cells. Int J Mol Med 8:385–390

Beaudet A, Descarries L (1984) Fine structures of monoamine axon terminals in cerebral cortex. Neurol Neurobiol 10:77–93

Beere HM (2001) Stressed to death: regulation of apoptotic signaling pathways by the heat shock proteins. Sci STKE 2001:re1

Berman SB, Hastings TG (1997) Inhibition of glutamate transport in synaptosomes by dopamine oxidation and reactive oxygen species. J Neurochem 69:1185–1195

Berman SB, Zigmond MJ, Hastings TG (1996) Modification of dopamine transporter function: effect of reactive oxygen species and dopamine. J Neurochem 67:593–600

Bernheimer H, Birkmayer W, Hornykiewicz O, Jellinger K, Seitelberger F (1973) Brain dopamine and the syndromes of Parkinson and Huntington. Clinical, morphological and neurochemical correlations. J Neurol Sci 20:415–455

Borer RA, Lehner CF, Eppenberger HM, Nigg EA (1989) Major nucleolar proteins shuttle between nucleus and cytoplasm. Cell 56:379–390

Carlsson A, Waters N, Holm-Waters S, Tedroff J, Nilsson M, Carlsson ML (2001) Interactions between monoamines, glutamate, and GABA in schizophrenia: new evidence. Annu Rev Pharmacol Toxicol 41:237–260

Cato AC, Mink S (2001) BAG-1 family of cochaperones in the modulation of nuclear receptor action. J Steroid Biochem Mol Biol 78:379–388

Chiueh CC, Rauhala P (1998) Free radicals and MPTP-induced selective destruction of substantia nigra compacta neurons. Adv Pharmacol 42:796–800

Cohen G (2000) Oxidative stress, mitochondrial respiration, and Parkinson's disease. Ann N Y Acad Sci 899:112–120

Dexter DT, Holley AE, Flitter WD, Slater TF, Wells FR, Daniel SE, Lees AJ, Jenner P, Marsden CD (1994) Increased levels of lipid hydroperoxides in the parkinsonian substantia nigra: an HPLC and ESR study. Mov Disord 9:92–97

Dittmar KD, Pratt WB (1997) Folding of the glucocorticoid receptor by the reconstituted Hsp90-based chaperone machinery. The initial hsp90.p60.hsp70-dependent step is sufficient for creating the steroid binding conformation. J Biol Chem 272:13047–13054

Friedman A, Galazka-Friedman J (2001) The current state of free radicals in Parkinson's disease. Nigral iron as a trigger of oxidative stress. Adv Neurol 86:137–142

Frydman J (2001) Folding of newly translated proteins in vivo: the role of molecular chaperones. Annu Rev Biochem 70:603–647

Gabriele J, Culver K, Sharma S, Zhang B, Szechtman H, Mishra R (2003) Asymmetric modulation of a catecholamine-regulated protein in the rat brain, following quinpirole administration. Synapse 49:261–269

Gabriele J, Rajaram M, Zhang BJ, Sharma S, Mishra RK (2002) Modulation of a 40-kDa catecholamine-regulated protein following D-amphetamine treatment in discrete brain regions. Eur J Pharmacol 453:13–19

Gabriele JP, Chong VZ, Pontoriero GF, Mishra RK (2005) Decreased expression of a 40-kDa catecholamine-regulated protein in the ventral striatum of schizophrenic brain specimens from the Stanley Foundation Neuropathology Consortium. Schizophr Res 74:111–119

Gabriele JP, Pontoriero GF, Thomas N, Ferro MA, Mahadevan G, MacCrimmon DJ, Pristupa ZB, Mishra RK (2010a) Antipsychotic drug use is correlated with CRP40/mortalin mRNA expression in the dorsolateral prefrontal cortex of human postmortem brain specimens. Schizophr Res 119:228–231
Gabriele N, Pontoriero GF, Thomas N, Shethwala SK, Pristupa ZB, Gabriele JP (2010b) Knockdown of mortalin within the medial prefrontal cortex impairs normal sensorimotor gating. Synapse 64:808–813
Goto A, Doering L, Nair VD, Mishra RK (2001) Immunohistochemical localization of a 40-kDa catecholamine regulated protein in the nigrostriatal pathway. Brain Res 900:314–319
Hastings TG, Lewis DA, Zigmond MJ (1996) Role of oxidation in the neurotoxic effects of intrastriatal dopamine injections. Proc Natl Acad Sci U S A 93:1956–1961
Hingorani K, Szebeni A, Olson MO (2000) Mapping the functional domains of nucleolar protein B23. J Biol Chem 275:24451–24457
Humphreys DT, Carver JA, Easterbrook-Smith SB, Wilson MR (1999) Clusterin has chaperone-like activity similar to that of small heat shock proteins. J Biol Chem 274:6875–6881
Hunot S, Brugg B, Ricard D, Michel PP, Muriel MP, Ruberg M, Faucheux BA, Agid Y, Hirsch EC (1997) Nuclear translocation of NF-kappaB is increased in dopaminergic neurons of patients with parkinson disease. Proc Natl Acad Sci U S A 94:7531–7536
Jenner P (1994) Oxidative damage in neurodegenerative disease. Lancet 344:796–798
Jonnson G (1971) Quantitation of fluorescence of biogenic monoamines. Prog Histochem Cytochem 2:299–344
Kalivendi SV, Kotamraju S, Zhao H, Joseph J, Kalyanaraman B (2001) Doxorubicin-induced apoptosis is associated with increased transcription of endothelial nitric-oxide synthase. Effect of antiapoptotic antioxidants and calcium. J Biol Chem 276:47266–47276
Kiang JG, Tsokos GC (1998) Heat shock protein 70 kDa: molecular biology, biochemistry, and physiology. Pharmacol Ther 80:183–201
Kikuchi A, Takeda A, Onodera H, Kimpara T, Hisanaga K, Sato N, Nunomura A, Castellani RJ, Perry G, Smith MA, Itoyama Y (2002) Systemic increase of oxidative nucleic acid damage in Parkinson's disease and multiple system atrophy. Neurobiol Dis 9:244–248
Lal S, de la Vega CE (1975) Apomorphine and psychopathology. J Neurol Neurosurg Psychiatry 38:722–726
Lang AE, Lozano AM (1998) Parkinson's disease. First of two parts. N Engl J Med 339:1044–1053
Liu PC, Thiele DJ (1999) Modulation of human heat shock factor trimerization by the linker domain. J Biol Chem 274:17219–17225
Migliore L, Petrozzi L, Lucetti C, Gambaccini G, Bernardini S, Scarpato R, Trippi F, Barale R, Frenzilli G, Rodilla V, Bonuccelli U (2002) Oxidative damage and cytogenetic analysis in leukocytes of Parkinson's disease patients. Neurology 58:1809–1815
Modi PI, Kashyap A, Nair VD, Ross GM, Fu M, Savelli JE, Marcotte ER, Barlas C, Mishra RK (1996) Modulation of brain catecholamine absorbing proteins by dopaminergic agents. Eur J Pharmacol 299:213–220
Moore DJ, West AB, Dawson VL, Dawson TM (2005) Molecular pathophysiology of Parkinson's disease. Annu Rev Neurosci 28:57–87
Mouradian MM (2002) Recent advances in the genetics and pathogenesis of Parkinson disease. Neurology 58:179–185
Nair VD, Mishra RK (2001) Molecular cloning, localization and characterization of a 40-kDa catecholamine-regulated protein. J Neurochem 76:1142–1152
Nanasi PP, Jednakovits A (2001) Multilateral in vivo and in vitro protective effects of the novel heat shock protein coinducer, bimoclomol: results of preclinical studies. Cardiovasc Drug Rev 19:133–151
Olanow CW (1992) An introduction to the free radical hypothesis in Parkinson's disease. Ann Neurol 32(Suppl):S2–S9
Olanow CW (1993) A radical hypothesis for neurodegeneration. Trends Neurosci 16:439–444

Olanow CW, Tatton WG (1999) Etiology and pathogenesis of Parkinson's disease. Annu Rev Neurosci 22:123–144

Pratt WB, Silverstein AM, Galigniana MD (1999) A model for the cytoplasmic trafficking of signalling proteins involving the hsp90-binding immunophilins and p50cdc37. Cell Signal 11:839–851

Prutz WA, Butler J, Land EJ (1983) Phenol coupling initiated by one-electron oxidation of tyrosine units in peptides and histone. Int J Radiat Biol Relat Stud Phys Chem Med 44:183–196

Rajput AH, Rozdilsky B, Rajput A (1991) Accuracy of Clinical-Diagnosis in Parkinsonism—A Prospective-Study. Can J Neurol Sci 18:275–278

Ross GM, McCarry BE, Thakur S, Mishra RK (1993) Identification of novel catecholamine absorbing proteins in the central nervous system. J Mol Neurosci 4:141–148

Ross GM, McCarry BE, Mishra RK (1995) Covalent affinity labeling of brain catecholamine-absorbing proteins using a high-specific-activity substituted tetrahydronaphthalene. J Neurochem 65:2783–2789

Rotman A, Daly JW, Creveling CR (1976) Oxygen-dependent reaction of 6-hydroxydopamine, 5,6-dihydroxytryptamine, and related compounds with proteins in vitro: a model for cytotoxicity. Mol Pharmacol 12:887–899

Saner A, Thoenen H (1971) Model experiments on the molecular mechanism of action of 6-hydroxydopamine. Mol Pharmacol 7:147–154

Scheulen M, Wollenberg P, Bolt HM, Kappus H, Remmer H (1975) Irreversible binding of DOPA and dopamine metabolites to protein by rat liver microsomes. Biochem Biophys Res Commun 66:1396–1400

Seeman P, Kapur S (2000) Schizophrenia: more dopamine, more D2 receptors. Proc Natl Acad Sci U S A 97:7673–7675

Serra JA, Dominguez RO, de Lustig ES, Guareschi EM, Famulari AL, Bartolome EL, Marschoff ER (2001) Parkinson's disease is associated with oxidative stress: comparison of peripheral antioxidant profiles in living Parkinson's, Alzheimer's and vascular dementia patients. J Neural Transm 108:1135–1148

Sharan N, Chong VZ, Nair VD, Mishra RK, Hayes RJ, Gardner EL (2003) Cocaine treatment increases expression of a 40 kDa catecholamine-regulated protein in discrete brain regions. Synapse 47:33–44

Sheth K, De A, Nolan B, Friel J, Duffy A, Ricciardi R, Miller-Graziano C, Bankey P (2001) Heat shock protein 27 inhibits apoptosis in human neutrophils. J Surg Res 99:129–133

Sorger PK (1991) Heat shock factor and the heat shock response. Cell 65:363–366

Stern MB (1997) Contemporary approaches to the pharmacotherapeutic management of Parkinson's disease: an overview. Neurology 49:S2–S9

Szebeni A, Olson MO (1999) Nucleolar protein B23 has molecular chaperone activities. Protein Sci 8: 905–912

Terland O, Flatmark T, Tangeras A, Gronberg M (1997) Dopamine oxidation generates an oxidative stress mediated by dopamine semiquinone and unrelated to reactive oxygen species. J Mol Cell Cardiol 29:1731–1738

Wadhwa R, Takano S, Robert M, Yoshida A, Nomura H, Reddel RR, Mitsui Y, Kaul SC (1998) Inactivation of tumor suppressor p53 by Mot-2, a hsp70 family member. J Biol Chem 273:29586–29591

Youdim MB (1988) Iron in the brain: implications for Parkinson's and Alzheimer's diseases. Mt Sinai J Med 55:97–101

Zigmond MJ, Hastings TG (1998) Neurochemical responses to lesions of dopaminergic neurons: implications for compensation and neuropathology. Adv Pharmacol 42:788–792

Zinsmaier KE, Bronk P (2001) Molecular chaperones and the regulation of neurotransmitter exocytosis. Biochem Pharmacol 62:1–11

Chapter 13
Chaperonopathies: Diseases in Which Mortalin and Other Hsp-Chaperones Play a Role in Etiology and Pathogenesis

Alberto J. L. Macario, Francesco Cappello and Everly Conway de Macario

Abstract Human mortalin is presented against the background of the Hsp70 family to show its distinctive properties and disease-causing potential. Seventeen genes of the Hsp70 family have recently been characterized applying the chaperonomics protocol to the human genome. One of these genes is HSPA9B, which encodes mortalin, identified in the early 1990s. Mortalin also called mtHsp70, PBP74, GRP75, and HSPA9B, resides inside the mitochondria but can also occur elsewhere. Mortalin is unique within the family because it is more closely related to bacterial than to eukaryotic orthologs, indicating distinctive evolution and functions. Its canonical role pertains to protein folding inside mitochondria in association with Hsp60 (Cpn60) and other molecules. However, it is involved also in other processes distinct from protein folding inside and outside mitochondria. It can be predicted that mortalin structure-function defects, inherited or acquired, will have a serious impact on key cellular events, particularly when mitochondria play a role, and in aging. Chaperonopathies due to mortalin malfunction will surely be identified, soon. Existing data indicate that mortalin can be pathogenic, particularly in some types of cancer: mortalin is normal but helps cancer cells to grow, exemplifying the chaperonopathies by mistake, in which a normal chaperone contributes to disease rather than to protection, as expected from a chaperone. Future research offers a multifaceted perspective for mortalin as etiologic factor (chaperonopathies due to chaperone malfunction or mistaken allegiance), disease biomarker, therapeutic target for anti-chaperone compounds (when mortalin is pathogenic), and therapeutic agent in replacement chaperonotherapy (when mortalin is absent or defective).

Keywords Hsp70 family · HSPA9B · Mortalin · Chaperonopathies by defect · Chaperonopathies by mistake

---

A. J. L. Macario (✉) · E. Conway de Macario
Department of Microbiology and Immunology, School of Medicine, University of Maryland at Baltimore; and IMET, Columbus Center, 701 East Pratt Street, Baltimore, Maryland 21202, USA
e-mail: Ajlmacario@som.umaryland.edu

A. J. L. Macario · F. Cappello
IEMEST, Istituto Euro-Mediterraneo di Scienza e Tecnologia, Palermo, Italy

F. Cappello
Department of Experimental Biomedicine and Clinical Neurosciences, Section of Human Anatomy, University of Palermo, Palermo, Italy

S. C. Kaul, R. Wadhwa (eds.), *Mortalin Biology: Life, Stress and Death*,
DOI 10.1007/978-94-007-3027-4_13, © Springer Science+Business Media B.V. 2012

13.1 Introduction

Molecular chaperones assist in protein folding, re-folding, and translocation as well as participate in a number of cellular processes unrelated to protein maturation (Henderson 2009; Henderson et al. 2010; Macario et al. 2010; Mayer 2010; Turbpaiboon and Wilairat 2010). Since chaperones are present in every cell, tissue, and organ and have diverse functions, it is expected that their failure will have important and widespread consequences. However, not much is known yet about the impact of chaperone deficiency. One reason for this scarcity of knowledge and for the fact that most practitioners of health sciences and services are unaware of the existence of chaperone deficiencies is that they are not studied or taught in a systematic and organized way as, for example, many other disorders like those of the immune and hematopoietic systems. To remedy this situation, we have started several years ago to organize pathologic conditions presumably, or certainly, due to chaperone malfunction in a unified category (Macario 1995; Macario and Conway de Macario 2000), under the name of chaperonopathies (Macario and Conway de Macario 2004, 2005). These pathologic conditions represent a section of a new scientific discipline, chaperonology, which encompasses the study of chaperones in all their aspects, normal and pathological (Macario and Conway de Macario 2007a; Macario et al. 2010). Likewise, the chaperonopathies constitute a new area of medicine and pathology and is the subject matter of this article.

13.2 Scope

There are many chaperonopathies already classified as such but here we will deal only with those affecting Hsp70 family members, briefly, and specifically in more detail with those pertaining to one of the members of this family, HSPA9B or mortalin (Wadhwa et al. 2002).

The genomic characteristics of the gene encoding mortalin and its evolutionary features will be discussed in comparison with the other members of the extended Hsp70 family to bring up to light some distinctive properties of mortalin. Likewise, the structure of the protein mortalin will be presented within the framework of structures of all the family members to reveal its typical attributes. In the same manner and for the same reasons the chaperonopathies, already described or predicted, involving mortalin, will be discussed within the context of the Hsp70 chaperonopathies in general, i.e., non-mitochondrial Hsp70 chaperonopathies.

13.3 Chaperonopathies

Chaperonopathies are diseases in which the malfunction of a chaperone plays an etiologic role (Macario and Conway de Macario 2004, 2005; Macario et al. 2005). Typically, chaperones assist in the correct folding of nascent polypeptides, thus, if chaperones do not function properly misfolding can occur (Fig. 13.1). There are,

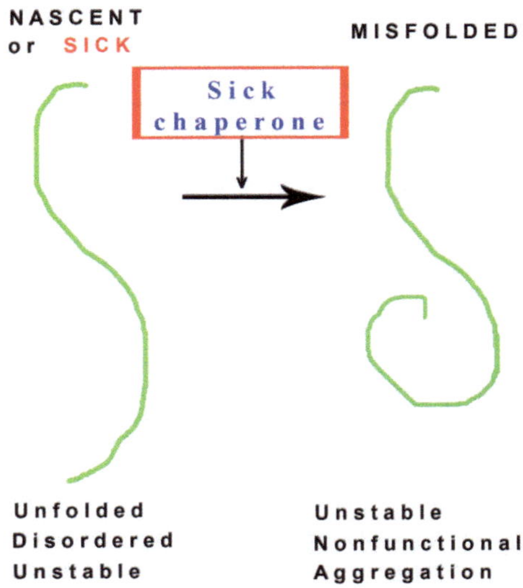

Fig. 13.1 Chaperone assisted folding of polypeptides may be seriously impaired if the chaperone is defective and malfunctions. Nascent or abnormal polypeptides are unfolded, disordered, and unstable (and tend to aggregate and form precipitates) whereas folded proteins are ordered, stable, and functional, i.e., they are in the native conformation. Passage from the unfolded state to the native conformation is assisted by chaperones. If chaperones are defective in structure-function (chaperonopathy) correct folding may not occur, which is followed by misfolding and aggregation

however, many other molecular and cellular processes that can be affected more or less seriously by the malfunctioning of one or more chaperones (Macario et al. 2010).

Chaperone malfunctioning can be due to structural alteration of the chaperone molecule, alteration that can be genetic or acquired (Macario and Conway de Macario 2002, 2004, 2005, 2007b). Both these types of conditions are qualitative chaperonopathies in which the structural and functional properties of the chaperone molecule are impaired, i.e., the chaperone malfunctions. In addition, there are other chaperone abnormalities that can have an impact on function. For instance, quantitative chaperonopathies are characterized by an increase or decrease of the concentration of chaperone molecules in any given cell or tissue. One mechanism of this kind of chaperonopathies can be gene dysregulation: a given chaperone gene is over or under expressed due to changes in the gene or other factors that regulate the chaperone gene itself. In addition, polymorphisms in the promoter or other regulatory regions can cause quantitative chaperonopathies.

13.4 Hsp and Chaperones

In order to understand chaperonopathies and easily remember their basic features it is necessary first to clarify certain issues of nomenclature and classification. For example, many chaperones are heat-shock proteins (Hsp), i.e., they are the products

Table 13.1 Subpopulations of Hsp-chaperones. (Reproduced (modified) with permission from references Macario (1995) and Macario et al. (2010)

Name	Other names; examples	MW (kDa)
Super-Heavy	Sacsin	200 or higher
Heavy	High MW; Hsp100	100–199
Hsp90	HSP86; HSP89A; HSP90A; HSP90N; HSPC1; HSPCA; LAP2; FLJ31884	81–99
Hsp70	Chaperones; DnaK	65–80
Hsp60	Chaperonins (Groups I and II); Cpn60 and CCT	55–64
Hsp40	DnaJ	35–54
Small Hsp	sHsp; alpha-crystallins; Hsp10	34 or less
Other	Proteases; isomerases; AAA + proteins (e.g., paraplegin [SPG7]; spastin [SPG4]; torsin A); Alpha Hemoglobin-Stabilizing Protein (AHSP); clusterin; PPI (cyclophilin); etc	Various

of genes inducible by stressors, most typically heat, thereby the generic name of Hsp. However, many other chaperones are not Hsp, and vice versa, many Hsp are not chaperones. Unfortunately, these distinctions have not been made in the great majority of published work on these molecules and at present it is convenient, in fact necessary, to use the names Hsp and chaperones interchangeably.

Another important issue is that there are several groups of Hsp-chaperones and one classification of practical value in research and practice is that based on molecular weight, Table 13.1. Some chaperones within any given molecular weight range are evolutionarily related and form families, like the Hsp70 family which is the one that will be discussed in this article, and within this family we will focus on HSPA9B or mortalin.

13.5 The Hsp70 Family

This family includes the products of at least 17 genes, Table 13.2, which were characterized by applying the chaperonomics protocol to the study of the human genome (Brocchieri et al. 2007, 2008). The members of this family are all evolutionarily related (Fig. 13.2) and share structural features, which reveal their common ancestry and similarity of function (Fig. 13.3). In this regard, it is pertinent to keep in mind that a chaperone molecule is composed of domains with specialized functions, which can be conserved in two or more family members and that structural alterations in any one or more of these domains are at the basis of many chaperonopathies (Macario and Conway de Macario 2007c; Macario et al. 2010).

13.6 Mortalin

Mortalin is the mitochondrial Hsp70 (Wadhwa et al. 2002), also called HSPA9B, mtHsp70, CSA, GRP75, MGC4500, MOT, MOT2, mot-2, MTHSP75, and PBP74 (Czarnecka et al. 2006). Our recent analysis of the human genome (Brocchieri

Table 13.2 *hsp70* genes in the human genome: Main characteristics (NCBI Build 36.1), showing mortalin (HSPA9B) within the context of all the members of the Hsp70 family. (Reproduced (modified) with permission from reference Brocchieri et al. (2008))[a]

Name/ID	Location	S	Start/End	nt	aa	Ex	Is
Hsp70 kDa 6 (HSP70B′)/HSPA6	1q23.3	+	159,761,073/159,763,001	1,929	643	1	1
Hsp70 kDa 7 (HSP70B)/HSPA7	1q23.3	+	159,842,705/159,844,628[b]	1,924	641	1	1
Hsp70 kDa 4-like/HSPA4L	4q28.1	+	128,923,156/128,973,476	50,321	839	19	1
Hsp70 kDa 9B/HSPA9B	*5q31.2*	–	*137,919,628/137,938,906*	*19,279*	*679*	*17*	*1*
Hsp70 kDa 4/HSPA4	5q31.1	+	132,415,842/132,468,024	52,183	840	19	a
			132,415,842/132,468,024	52,183	148	5	b
Hsp70 kDa 1-like/HSPA1L (HSP70-Hom)	6p21.33	–	31,885,806/31,887,728	1,923	641	1	1
Hsp70 kDa 1A/HSPA1A (HSP70-1)	6p21.33	+	31,891,513/31,893,435	1,923	641	1	1
Hsp70 kDa 1B/HSPA1B (HSP70-2)	6p21.32	+	31,903,707/31,905,629	1,923	641	1	1
Hsp70 kDa 5 (Grp78)/HSPA5	9q33.3	–	127,038,695/127,043,226	4532	654	8	1
Hsp70 kDa 12A/HSPA12A	10q25.3	–	118,424,285/118,456,787	32,503	675	12	1
Hsp70 kDa 14/HSPA14	10p13	+	14,920,408/14,953,608	33,201	509	14	1
			14,920,408/14,924,181	3,774	88	4	2
Hsp70 kDa 8/HSPA8	11q24.1	–	122,433,655/122,437,242	3,588	646	8	1
			122,433,655/122,437,242	3,588	493	7	2
150 kDa oxygen-regulated protein/HYOU1	11q23.3	–	118,421,518/118,432,082	10,565	999	25	1
			118,421,518/118,432,082	10,565	999	25	1
			118,421,518/118,431,751	10,234	964	24	2
			118,424,940/118,432,082	7,143	687	16	3
Hsp105 kDa/HSPH1	13q12.3	–	30,609,458/30,633,719	24,262	858	18	α
			30,609,458/30,633,719	24,262	814	17	β
Hsp70 kDa 2/HSPA2	14q23.3	+	64,077,321/64,079,237	1,917	639	1	1
Hsp70 kDa 12B/HSPA12B	20p13	+	3,667,322/3,680,810	13,489	686	12	1
Stress 70 protein chaperone/STCH	21q11.2	–	14,667,812/14,677,311	9,500	471	5	1

[a]Abbreviations and symbols: *ID* identifying designation used in reference BMC Evol Biol (2008); *S* DNA strand; *Start/End* chromosome positions comprising the coding region; *nt* number of nucleotides between Start and End; *aa* length in amino acids of the protein product(s); *Ex* exons (non coding exons excluded); *Is* isoform names (for the 150-kDa oxygen-regulated protein/HYOU1 there are four mRNA variants but only three protein isoforms because two variants have the same translation initiation site)

[b]Bypassing an internal frame-shift (see text). The putative coding region would otherwise end at position 159,843,805 and encode for a protein 367 aa long

Mortalin data are in italics

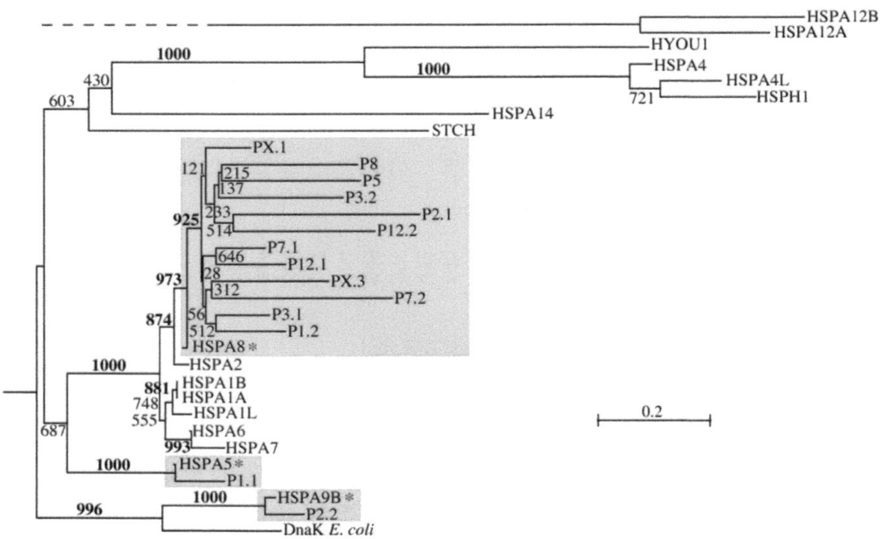

Fig. 13.2 Phylogenetic tree of the 17 human Hsp70 proteins and the proteins that would be encoded in related pseudogenes, showing mortalin. Phylogenetic analysis of the most conserved pseudogenes and detailed comparison of the most corrupted ones with the closest sequence among the 17 human *hsp70* genes, revealed that 27 of the 30 pseudogenes found originated from HSPA8, two from HSPA9B (i.e., mortalin), and one from HSPA5, indicated with asterisks. We included in the construction of the tree shown the 16 less corrupted sequences out of the 30 Hsp70-related pseudogenes, but for the sake of clarity only 14 are shown in this figure, including only one of the two related to mortalin. Alignments were done with Clustal W and the evolutionary tree was constructed using the neighbor-joining algorithm with the distance transformation method of Kimura, as implemented in Clustal W; a similar tree (not shown) was obtained using the maximum likelihood approach implemented in the program PHYML. Reproduced (modified) with permission from reference Brocchieri et al. (2008)

et al. 2008) showed that the HSPA9B or mortalin gene is located on chromosome 5 (as originally described, Kaul et al. 1995), and encodes 679 amino acids in 17 exons, Table 13.2. In the human genome there are 30 pseudogenes related to the Hsp70 family (Brocchieri et al. 2008). Phylogenetic analysis of the most conserved pseudogenes and detailed comparison of the most corrupted ones with the closest sequence among the 17 *hsp70* genes revealed that 27 of the 30 pseudogenes originated from the HSPA8, two from HSPA9B (mortalin), and one from HSPA5, Fig. 13.2. We included in the construction of the tree shown in Fig. 13.2 the 16 less corrupted sequences out of the 30 pseudogenes, but for the sake of clarity only 14 are shown. The pseudogenes related to mortalin are located on chromosomes 2 and 18, Table 13.3.

The product of the HSPA9B (mortalin) gene, typically an intramitochondrial protein, is evolutionarily related to the bacterial Hsp70/DnaK protein, Figs. 13.2 and 13.4, a fact that is taken to support the notion that the mitochondria and their components are of bacterial origin. Like the other mitochondrial chaperones Hsp60

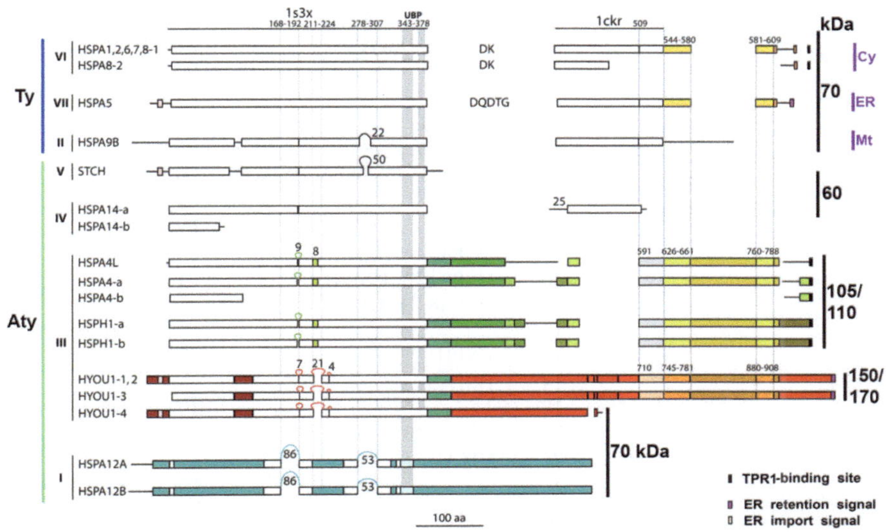

Fig. 13.3 Structural features of human mortalin (HSPA9B) shown within the context of the other members of the Hsp70 proteins referred to in Table 13.2 and Fig. 13.2. Typical Hsp70 molecules comprise two major regions: ATP-binding domain (ABD) and substrate-binding domain (SBD). The regions comprising the crystal structure of the ABD (1s3x, human HSPA1A, positions 1–382) and the NMR structure of the SBD (1ckr, rat HSPA8, corresponding to human HSPA8 and HSPA1A homologous positions 385–543) are shown as lines at the top of the figure. Domains conserved in human Hsp70 sequences within these regions are shown as aligned white boxes. Other domains conserved among groups of sequences are shown as aligned boxes of matching colors. Roman numeral to the left indicate the Hsp70 evolutionary groups. To the right molecular weights (kDa) are indicated, as well as locale of residence of the proteins (*Cy* cytosol; *ER* endoplasmic reticulum; *Mt* mitochondria). Abbreviations at bottom right: *UBP* ubiquitin-binding peptide; *TPR1* tetratrico peptide repeat 1; *ER* endoplasmic reticulum. It can be seen that mortalin forms its own phylogenetic group within the Hsp70 family (*see also* Fig. 13.2) and, considering size and structural features, is placed among the typical members of this family. Reproduced (modified) with permission from reference Brocchieri et al. (2008) (*Ty* typical; *Aty* atypical)

and Hsp10, mortalin is produced in the cytoplasm and translocated into mitochondria in which it functions, interacting with a host of other molecules (Wadhwa et al. 2002; Cappello et al. 2008; Corrao et al. 2010). In addition, similarly to the other mitochondrial chaperones, mortalin also occurs in extramitochondrial sites, such as cytosol and perinuclear region, especially when it is overexpressed. In the perinuclear region, mortalin has been found co-localizing with p53 in several cancer cell lines, in which the chaperone sequesters p53 and, thus, inhibits the expression of p53-target genes (e.g., p 21) (Wadhwa et al. 1998, 2002). Consequently, mortalin overexpression and cytosolic localization have been associated with an increase of malignancy and tumoral transformation (Wadhwa et al. 2006). These tumors, in which mortalin is overexpressed and enhances malignancy, are examples of chaperonopathies by mistake, as discussed in more detail later.

Table 13.3 Human chromosomes with *hsp70* genes and *hsp70*-related pseudogenes, including mortalin (HSPA9B) and its corresponding pseudogenes P2.2 and F18.1. (Reproduced (modified) with permission from reference Brocchieri et al. (2008))

Chr[a]	Protein-coding gene	Pseudogene
1	HSPA6, HSPA7	P1.1, P1.2
2		P2.1, *P2.2*, P2.3
3		P3.1, P3.2
4	HSPA4L	P4
5	*HSPA9B*, HSPA4	P5, F5
6	HSPA1L, HSPA1A, HSPA1B	F6.1, F6.2, F6.3, F6.4
7		P7.1, P7.2, F7
8		P8, 8.1, F8.2
9	HSPA5	P9
10	HSPA12A, HSPA14	P10
11	HSPA8, HYOU1	
12		P12.1, P12.2
13	HSPH1	F13
14	HSPA2	
18		*F18.1*, F18.2
20	HSPA12B	
21	STCH	
X		PX.1, PX.2, PX.3

[a]Chr, chromosome

13.7 Chaperonopathies Involving Members of the Hsp70 Family

A list of pathological conditions in which abnormalities of Hsp70 are the primary or secondary etiological factors is displayed in Table 13.4. Genetic and acquired abnormalities of one or more members of the Hsp70 family have been implicated in a variety of diseases, including tumors. Hsp70 has been found increased in various types of cancer, such as urothelial carcinoma of the urinary bladder, Fig. 13.5 (Cappello et al. 2006). In this regard, there are examples in which Hsp70 is essential for tumor cell growth (see details in Table 13.4), which makes the Hsp70 a potential target for chaperonotherapy aiming at blocking the chaperone action or at eliminating or suppressing its gene (Macario and Conway de Macario 2007a, b).

13.8 Chaperonopathies Involving Mortalin

Some types of cancer may have increased resistance to complement-dependent cytotoxicity, in part because they can eliminate the membrane attack complex (MAC) from their surface, a mechanism in which mortalin seems to be involved (Pilzer et al. 2010). Mortalin functions in favor of cancer cells and, in this type of situations, the pathological conditions can be considered chaperonopathies by mistake or collaborationism (Macario and Conway de Macario 2007b). These are conditions in which a normal chaperone promotes disease rather than the contrary as it would be expected from a molecule classically involved in defense mechanisms: in these situations the chaperone defends cancer cells instead of protecting the organism (Cappello et al. 2011). For instance, elimination of mortalin with siRNA reduced MAC action and

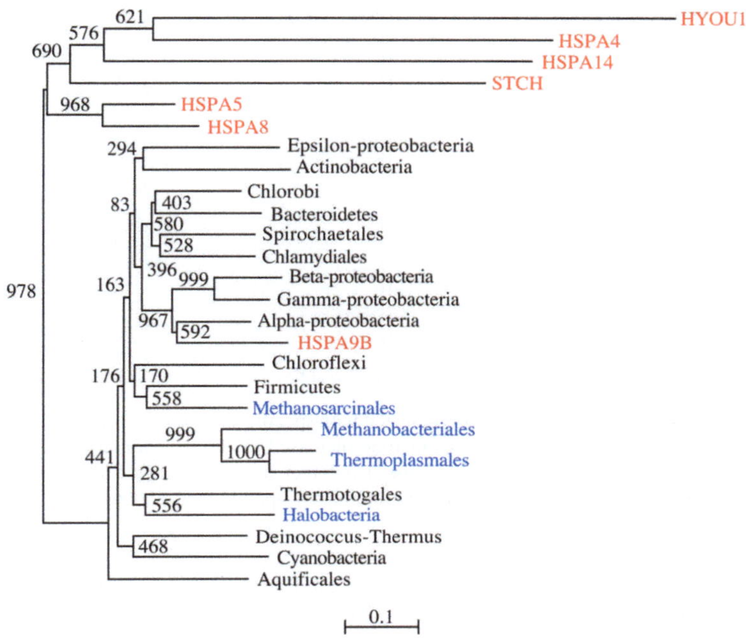

Fig. 13.4 Phylogenetic tree of human Hsp70 proteins and prokaryotic Hp70/DnaK proteins, showing mortalin (HSPAB9) clustering with the Alpha-proteobacteria orthologs. Included in this tree are all human Hsp70 proteins from the genes shown in Table 13.2 except the highly diverged HSPA12A and HSPA12B, Fig. 13.2. One sequence was chosen for each prokaryotic (bacterial in *black*, and archaeal in *blue*) group. Human proteins are in red. It can be seen that HSPA9B or mortalin clusters with bacterial orthologs. Reproduced (modified) with permission from reference Brocchieri et al. (2008)

increased the tumor cell sensitivity to MAC-induced cell death (Pilzer et al. 2010). Similar results were obtained with the cationic rhodacyanine dye MKT-077, a mortalin inhibitor. This effect was observed using human erythroleukemia K562 and colorectal carcinoma HCT116 cells. Furthermore, mortalin inhibited hemolysis of rabbit erythrocytes and zinc-induced C9 polymerization.

That mortalin may be involved in the initiation and/or progression of certain cancers is also suggested by the finding of it being increased in tumor tissues. For instance, mortalin was found increased in colorectal adenocarcinomas in correlation with poor prognosis (Dundas et al. 2005) and in human brain tumors (Takano et al. 1997). In this regard it is pertinent to note that mortalin has been found elevated in a number of other tumors and tumor-cell lines and to be associated with the capacity of tumor cells to become anchorage-independent and to form tumors in nude mice (Wadhwa et al. 2006). For all these reasons, mortalin may be considered a candidate for a chaperone-targeted therapy similarly to other instances in which a chaperone contributes to disease (Cappello et al. 2008).

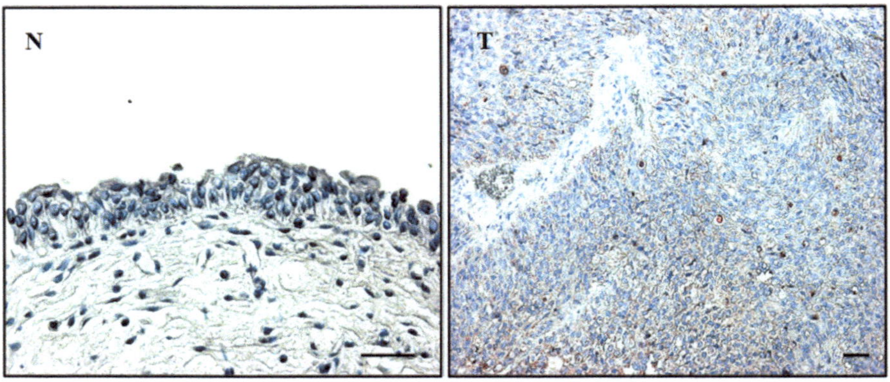

Fig. 13.5 Increase in the levels of Hsp70 as detected with specific antibody in urothelial carcinomas of the urinary bladder. The levels of the chaperone increased to closely parallel tumor stage and grade: the more advanced the tumor the larger the Hsp70 increase. *N* normal bladder; *T* tumor. Bar: 100 microns. Reproduced with permission from reference Cappello et al. (2006)

Table 13.4 Chaperonopathies involving members of the Hsp70 family (References in Cappello et al. (2006); Macario and Conway de Macario (2005, 2007a); Macario et al. (2010); Xu et al. (2010). For mortalin see text)

Gene/protein affected	Disease/syndrome
Structural hereditary chaperonopathies	
STCH (del223V-226L in ATP-binding domain)	Stomach cancer
Chaperonopathies due to gene dysregulation	
hsp70	Constitutive and stress-induced levels: Low in the aged Stress-induced expression: Decreased in cardiopathy of the aged
hsp70RY; hsc70; grp78	Constitutive and stress-induced expression: Decreased or increased (e.g., some neurodegenerative diseases)
Chaperone-gene polymorphisms	
Genetic polymorphism	*Abnormality/disease*
hsp70-1 promoter region, allele (A)-110	Does not favor longevity in women
hsp70-Hom, T247C, methionine/493/threonine	Does not favor longevity
hsp70-1 promoter region, allele (C)-110	Associates with Parkinson's disease in Taiwanese
hsp70-1 5′UTR region, +190CC genotype and haplotypes including +190C	Associate with higher risk of coronary heart disease vs. +190GG and +190G
Chaperonopathies by mistake: Cancer	
Increased in cancer of:	*Required for cancer:*
Breast; bladder urothelial carcinoma (Hsp70-2/HSPA2); laryngeal squamous cell carcinoma (Hsp70)	*Cell growth:* Breast (Hsp70/HSP70 A; Hsp70–2/HSPA2; Hsc70/HSPA8); bladder (Hsp70–2/HSPA2); *Radiation resistance:* laryngeal squamous cell carcinoma

13.9 Genetic Defects in Mortalin That Could Cause Disease: Genetic Chaperonopathies

Two mutations in the mortalin gene, one in the portion encoding the ATPase domain and the other in that encoding the substrate-binding domain (see Fig. 13.4) that result in the amino acids changes R126W and P509S, respectively, have been reported in Spanish patients with Parkinson's disease (PD; De Mena et al. 2009). The seventeen HSPA9B (mortalin) coding exons (Table 13.2) were examined in 330 PD patients and 250 controls. In addition to the two mutations mentioned above, a 17 bp insertion in intron 8 (predicted to affect RNA splicing) was also found only in PD patients. It was concluded that genetic changes in the mortalin molecule could be involved in the pathogenesis of some PD patients. This hypothesis is further encouraged by the finding of another mutation (A476T) in the substrate-binding domain of mortalin from German PD patients (Burbulla et al. 2010). All three mutant molecules were assayed in cellular models to determine the impact of the mutations on the functions of mortalin. The mutant proteins caused in neuronal and non-neuronal human cell lines a mitochondrial phenotype characterized by increased reactive oxygen species (ROS) and decreased membrane potential. No doubt these instances represent good examples of what a genetic chaperonopathy affecting mortalin would manifest and offer for detection by the clinician and the pathologist.

13.10 Conclusion

Mortalin is involved in a variety of cellular processes and resides in various places, not just mitochondria (Deocaris et al. 2007; Gabriele et al. 2010; Ran et al. 2000; Wadhwa et al. 2002, 2005). Consequently, mortalin defects of structure and function, inherited or acquired, are bound to have an impact on diverse cellular and tissue functions: these would be chaperonopathies by defect candidates to replacement chaperonotherapy (Macario and Conway de Macario 2007a) with mortalin molecules or with the mortalin gene. In addition, mortalin can be an etiologic factor in chaperonopathies by mistake or collaborationism, as discussed earlier, since it promotes growth and survival of malignant cells. In this case, blocking or eliminating mortalin would be a promising therapeutic approach (Yoo et al. 2010). Lastly, considering the important role that mortalin has in mitochondria and the impact of mitochondrial malfunction on cell senescence (Burbulla et al. 2010; Deocaris et al. 2007; Wadhwa et al. 1995, 2005), it is quite likely that chaperonin defects might have a key role in the progression of the age-related decline of vigor and age-related diseases (Macario and Conway de Macario 2002).

The study of mortalin and its direct role in disease, particularly the latter, is an emerging field that holds great promise to advance the understanding not only of mitochondrial physiopathology but also of diverse conditions, including those accompanying senescence. Mortalin can be etiologic factor if defective, biomarker of disease with diagnostic-prognostic value if overexpressed for example, target for treatment of conditions in which although normal mortalin contributes to pathogenesis, and therapeutic agent in replacement chaperonotherapy to deal with mortalin deficiency.

Acknowledgments This study was supported by funds from the University of Palermo (MIUR ex 60%, FC), Palermo, Italy, and from IEMEST, Istituto Euro-Mediterraneo di Scienza e Tecnologia, Palermo (FC and AJLM), Italy.

References

Brocchieri L, Conway de Macario E, Macario AJL (2007) Chaperonomics, a new tool to study ageing and associated diseases. Mechan Ageing Develop 128:125–136

Brocchieri L, Conway de Macario E, Macario AJL (2008) Hsp70 genes in the human genome: Conservation and differentiation patterns predict a wide array of overlapping and specialized functions. BMC Evol Biol 8:19

Burbulla LF, Schelling C, Kato H, Rapaport D, Woitalla D, Schiesling C, Schulte C, Sharma M, Illig T, Bauer P, Jung S, Nordheim A, Schöls L, Riess O, Krüger R (2010) Dissecting the role of the mitochondrial chaperone mortalin in Parkinson's disease: functional impact of disease-related variants on mitochondrial homeostasis. Hum Mol Genet 19:4437–4452

Cappello F, David S, Ardizzone N, Rappa F, Marasà L, Bucchieri F, Zummo G (2006) Expression of heat shock proteins HSP10, HSP27, HSP60, HSP70, and HSP90 in urothelial carcinoma of urinary bladder. J Can Mol 2:1–5

Cappello F, Conway de Macario E, Marasà L, Zummo G, Macario AJL (2008) Hsp60 expression, new locations, functions and perspectives for cancer diagnosis and therapy. Cancer Biol Ther 7:801–809

Cappello F, David S, Peri G, Farina F, Conway de Macario E, Macario AJL, Zummo G (2011) The human chaperonin Hsp60: molecular anatomy, role in carcinogenesis and potential for diagnosis and treatment of colorectal cancer. Front Biosci 3:341–351

Corrao S, Campanella C, Anzalone R, Farina F, Zummo G, Conway de Macario E, Macario AJL, Cappello F, La Rocca G (2010) Human Hsp10 and Early Pregnancy Factor (EPF) and their relationship and involvement in cancer and immunity: current knowledge and perspectives. Life Sci 86:145–152

Czarnecka AM, Campanella C, Zummo G, Cappello F (2006) Mitochondrial chaperones in cancer. Cancer Biol Ther 5:714–720

De Mena L, Coto E, Sánchez-Ferrero E, Ribacoba R, Guisasola LM, Salvador C, Blázquez M, Alvarez V (2009) Mutational screening of the mortalin gene (HSPA9) in Parkinson's disease. J Neural Transm 116:1289–193

Deocaris CC, Widodo N, Ishii T, Kaul SC, Wadhwa R (2007) Functional significance of minor structural and expression changes in stress chaperone mortalin. Ann NY Acad Sci 1119:165–175

Dundas SR, Lawrie LC, Rooney PH, Murray GI (2005) Mortalin is over-expressed by colorectal adenocarcinomas and correlates with poor survival. J Pathol 205:74–81

Gabriele N, Pontoriero GF, Thomas N, Shethwala SK, Pristupa ZB, Gabriele JP (2010) Knockdown of mortalin within the medial prefrontal cortex impairs normal sensorimotor gating. Synapse 64:808–813

Henderson B (2009) Integrating the cell stress response: a new view of molecular chaperones as immunological and physiological homeostatic regulators. Cell Biochem Funct 28:1–14

Henderson B, Calderwood SK, Coates AR, Cohen I, van Eden W, Lehner T, Pockley AG (2010) Caught with their PAMPs down? The extracellular signalling actions of molecular chaperones are not due to microbial contaminants. Cell Stress Chaperones 15:123–141

Kaul SC, Wadhwa R, Matsuda Y, Hensler PJ, Pereira-Smith OM, Komatsu Y, Mitsui Y (1995) Mouse and human chromosomal assignments of mortalin, a novel member of the murine hsp70 family of proteins. FEBS Lett 361:269–272

Macario AJL (1995) Heat-shock proteins and molecular chaperones: Implications for pathogenesis, diagnostics, and therapeutics. Intl J Clin Lab Res 25:59–70
Macario AJL, Conway de Macario E (2000) Stress and molecular chaperones in disease. Intl J Clin Lab Res 30:49–66
Macario AJL, Conway de Macario E (2002) Sick chaperones and ageing: A perspective. Ageing Res Rev 1:295–311
Macario AJL, Conway de Macario E (2004) The pathology of anti-stress mechanisms: A new frontier. Stress 7:243–249
Macario AJL, Conway de Macario E (2005) Sick chaperones, cellular stress and disease. New Eng J Med 353:1489–1501
Macario AJL, Conway de Macario E (2007a) Chaperonopathies and chaperonotherapy. FEBS Letters 581:3681–3688
Macario AJL, Conway de Macario E (2007b) Chaperonopathies by defect, excess, or mistake. Ann New York Acad Sci 1113:178–191
Macario AJL, Conway de Macario E (2007c) Molecular chaperones: Multiple functions, pathologies, and potential applications. Front Biosci 12:2588–2600
Macario AJL, Grippo TM, Conway de Macario E (2005) Genetic disorders involving molecular-chaperone genes: A perspective. Genet Med 7:3–12
Macario AJL, Cappello F, Zummo G, Conway de Macario E (2010) Chaperonopathies of senescence and the scrambling of the interactions between the chaperoning and the immune systems. Ann New York Acad Sci 1197:85–93
Mayer MP (2010) Gymnastics of molecular chaperones. Mol Cell 39:321–331
Pilzer D, Saar M, Koya K, Fishelson Z (2010) Mortalin inhibitors sensitize K562 leukemia cells to complement-dependent cytotoxicity. Int J Cancer 126:1428–1435
Ran Q, Wadhwa R, Kawai R, Kaul SC, Sifers RN, Bick RJ, Smith JR, Pereira-Smith OM (2000) Extramitochondrial localization of mortalin/mthsp70/PBP74/GRP75. Biochem Biophys Res Commun 275:174–179
Takano S, Wadhwa R, Yoshii Y, Nose T, Kaul SC, Mitsui Y (1997) Elevated levels of mortalin expression in human brain tumors. Exp Cell Res 237:38–45
Turbpaiboon C, Wilairat P (2010) Alpha hemoglobin stabilizing protein: molecular function and clinical correlation. Front Biosci 15:1–11
Wadhwa R, Pereira-Smith OM, Reddel RR, Sugimoto Y, Mitsui Y, Kaul SC (1995) Correlation between complementation group for immortality and the cellular distribution of mortalin. Exp Cell Res 216:101–106
Wadhwa R, Takano S, Robert M, Yoshida A, Nomura H, Reddel RR, Mitsui Y, Kaul SC (1998) Inactivation of tumor suppressor p53 by mot-2, a hsp70 family member. J Biol Chem 273:29586–29591
Wadhwa R, Taira K, Kaul SC (2002) An Hsp70 family chaperone, mortalin/mthsp70/PBP74/Grp75: what, when, and where? Cell Stress Chaperones 7:309–316
Wadhwa R, Takano S, Kaur K, Aida S, Yaguchi T, Kaul Z, Hirano T, Taira K, Kaul SC (2005) Identification and characterization of molecular interactions between mortalin/mtHsp70 and HSP60. Biochem J 391:185–190
Wadhwa R, Takano S, Kaur K, Deocaris CC, Pereira-Smith OM, Reddel RR, Kaul SC (2006) Upregulation of mortalin/mthsp70/Grp75 contributes to human carcinogenesis. Int J Cancer 118:2973–2980
Xu J, Wang K, Zhang X, Qiu Y, Huang D, Li W, Xiao X, Tian Y (2010) HSP70: a promising target for laryngeal carcinoma radiotherapy by inhibiting cleavage and degradation of nucleolin. J Exp Clin Cancer Res 29:106
Yoo JY, Ryu J, Gao R, Yaguchi T, Kaul SC, Wadhwa R, Yun CO (2010) Tumor suppression by apoptotic and anti-angiogenic effects of mortalin-targeting adeno-oncolytic virus. J Gene Med 12:586–595

Part IV
Mortalin and Cancer

Chapter 14
Many Faces of Mortalin and Tid1

Ohad Iosefson and Abdussalam Azem

Abstract The human mitochondrial Hsp70, mortalin, is dually distributed between the mitochondria and the cytosol. Inside the mitochondrial matrix, it plays crucial roles in mediating the import of nuclear-encoded precursor proteins and in the folding of newly imported and stress-denatured client-proteins. Outside the mitochondria it interacts with a wide-variety of proteins, which are mainly involved in cell-cycle regulation and the immune response. Similar to mortalin, its partner Tid1, the human mitochondrial DnaJ homologue, is also dually distributed and is found associated with numerous cytosolic proteins. The structural and functional significance of a selected number of interactions is discussed.

Keywords Chaperone · Import · Folding · Mitochondria · MtHsp70 · Protein–protein interactions

14.1 Introduction

Molecular chaperones are at the hub of the cell's protein quality control network. They are divided into families according to their molecular weight: Hsp100, Hsp90, Hsp70, Hsp60, Hsp40 and small Hsps. Their common denominator, apart from the small Hsps, is the ability to utilize ATP as a fuel in executing their folding functions. The ubiquitously expressed Hsp70 proteins are extremely conserved throughout evolution, both structurally and functionally. Representatives of the family are found from prokaryotes to plants to humans. They carry out their diverse functions in a broad spectrum of scenarios, including: folding of nascent polypeptide chains, translocation of proteins across membranes, assembly, disassembly and maintenance of macromolecular complexes and in the chaperoning and regulation of various signal-transduction molecules (Mayer and Bukau 2005). In accordance with their cytoprotective role, Hsp70 proteins are also involved in inhibition of stress-induced apoptosis (Gabai et al. 1997). Therefore, it is not surprising that over expression of these chaperones was detected in a wide variety of malignant tumors. The human

A. Azem (✉) · O. Iosefson
Department of Biochemistry and Molecular Biology, George S. Wise Faculty of Life Sciences, Tel Aviv University, Tel Aviv 69978, Israel
e-mail: azema@tauex.tau.ac.il

S. C. Kaul, R. Wadhwa (eds.), *Mortalin Biology: Life, Stress and Death*,
DOI 10.1007/978-94-007-3027-4_14, © Springer Science+Business Media B.V. 2012

Hsp70 family consists of eight known members: six are mainly localized to the cytosol, one member is found inside the ER, and mortalin, which resides in the mitochondria (Daugaard et al. 2007). Apart from the different localization and sequence variability, the various Hsp70 functional specificities are imposed by their J-protein co-chaperones. All the proteins within this family contain a conserved N-terminal J-domain, whose role is to stimulate the low basal ATPase activity of the Hsp70 chaperones. A number of J-domain containing proteins also recognize and bind to a misfolded polypeptide, followed by the recruitment of Hsp70 for mutual collaboration in substrate unfolding and disaggregation. There are 41 putative J-proteins in humans that differ significantly in their additional domains (Kampinga and Craig 2010). The diversity within this family enables precise regulation of individual Hsp70 counterparts while executing their different functions. For example, both Tim14/Pam18 and Tid1 are mitochondrial J-proteins; however, they cooperate with the same mitochondrial Hsp70 in two distinct functions: protein import into mitochondria and folding of proteins within the matrix, respectively. In addition, J-proteins also play a role in targeting client proteins for degradation, as well as in some Hsp70-independent functions, e.g., prevention of PolyQ protein aggregation and spliceosome disassembly (reviewed in Kampinga and Craig 2010). J-domain containing proteins can be sorted into three groups according to their domain composition:

1. Type I proteins consist of three additional domains that follow the N-terminal J-domain as illustrated in Fig. 14.1.
2. Type II lacks the cysteine-rich domain
3. Type III consists of the J-domain only (Walsh et al. 2004).

Mortalin plays key roles in the biogenesis of mitochondrial proteins as well as several extra mitochondrial cellular functions. These disparate functions are the consequence of the ability of this protein to form distinct hetero-oligomeric complexes with many protein components. In this chapter we review the interactions of mortalin, and its J-protein co-chaperone, Tid1, both inside and outside the mitochondria. Within the mitochondria, we focus on the function of mortalin as the core component of the mitochondrial protein import machinery and on the cooperation with Tid1 in folding of misfolded substrates. In the cytosol, we spotlight on the p53-mortalin interaction, as well as, the emerging role of Tid-1 as a regulator of p53 translocation to the mitochondria.

14.2 Mitochondrial Functions of Mortalin

14.2.1 Indispensable Role of Mortalin in Mediating Protein Import

It is estimated that yeast and mammalian mitochondria harbor 800 and ~1500 proteins, respectively (Anderson 1981; Lopez et al. 2000). The vast majority of these proteins are nuclear-encoded and must be obtained from the cytosol into the

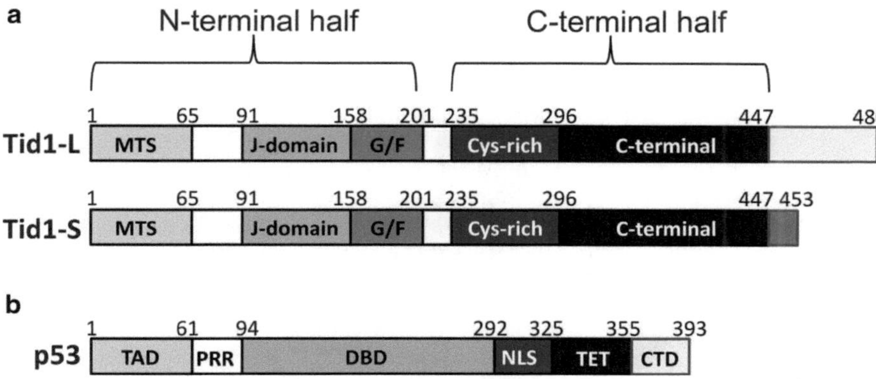

Fig. 14.1 The structural domains of Tid1 isoforms and p53. The numbers above each diagram represent the position of the relevant amino-acid in the protein sequence. **a** Both Tid1 isoforms consist of an N-terminal mitochondrial targeting sequence (MTS) followed by a J-domain, glycine/phenylalanie-rich domain (G/F), cysteine-rich domain (Cys-rich) and a C-terminal domain. The difference between the isoforms lies at their unique C-terminal tails. Tid1-L and Tid1-S, each has a distinctive sequence extension, as shown. Various deletion mutants' analyses, which are presented in Table 14.1, dissected the protein into two halves, as indicated. **b** p53 domains. The transactivation domain constitutes the extreme N-terminal followed by the proline-rich region (PRR), DNA-binding domain (DBD), nuclear localization signal (NLS), tetramerization domain (TET) and the C-terminal domain

mitochondria, passing through the organelle membranes. Most of our knowledge about the mechanism of mitochondrial protein import was obtained from studies carried-out with yeast systems. Thus, we will first provide a review of these studies. A protein destined for the mitochondria contains a mitochondrial targeting signal (MTS). This signal also encrypts the information needed to sort the protein to its correct mitochondrial compartment, *i.e.* the outer or inner membrane, the intermembrane space or the matrix. In order to import a protein to each of these specific sections, cells have evolved special translocases (Reviewed in Neupert and Herrmann 2007; Chacinska et al. 2009). Relevant to mortalin is the function of the TIM23 complex (TIM—*T*ranslocase of the *I*nner *M*embrane), which is responsible for the import of precursor proteins containing N-terminal-cleavable targeting signals. This group includes matrix targeted proteins and some of the inner membrane and IMS proteins. Like all the mitochondrial proteins, those are also primarily recognized and translocated by the TOM complex (TOM—*T*ranslocase of the *O*uter *M*embrane) across the outer membrane. The initial force that drives the translocation of a preprotein to the matrix is the membrane potential across the mitochondrial inner membrane. As a result of this force, the pre-sequence N-terminal segment is pushed through the TIM23 complex and starts emerging on the matrix side. In this state, The C-terminal region of the precursor is still located in the cytosol, and thus, the TOM and the TIM23 complexes are bound simultaneously to the polypeptide in transit (Schleyer and Neupert 1985) (Fig. 14.2).

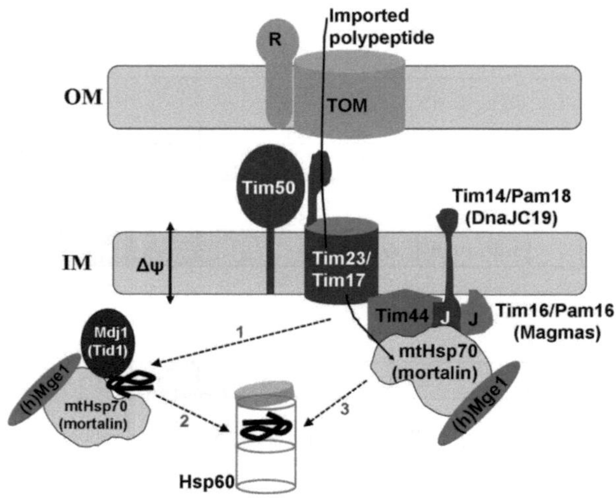

Fig. 14.2 Protein translocation from the cytosol to the matrix. The main entry gate for mitochondrial targeted preproteins is the TOM complex, embedded in the outer membrane (OM). The import selectivity is facilitated by receptors within this complex that recognize the mitochondrial targeting signal on the N-terminal part of the preprotein. The primary force that causes simultaneous threading of the preprotein through both the TOM and TIM23 complexes is the membrane potential ($\Delta\Psi$) across the inner membrane (IM). As soon as the N-terminal end of the translocating polypeptide emerges at the matrix side, it is subjected to the second pulling force—the import motor complex. The motor is composed of Tim44, mtHsp70, Tim14/Pam18, Tim16/Pam16 and Mge1 (The nomenclature for the human constituents is presented parenthetically). Tim44 recruits mtHsp70 to the incoming precursor. ATP hydrolysis by mtHsp70 induces locking of its binding-pocket on exposed substrate hydrophobic patches. Tim16/Pam16 regulates the stimulatory effect that Tim14/Pam18 exerts on mtHsp70 ATPase activity. Mge1 induces ADP release from mtHsp70, which unlocks the substrate. The chaperone is now ready to initiate under the cycle of substrate binding and release. A part of the freshly imported proteins need the assistance of molecular chaperones in order to gain their native conformation. This can be accomplished in several ways. They can be folded by the matrix folding complex consisting of mtHsp70, Mdj1 and Mge1 (*Dashed arrow 1*). A portion of the proteins folded in this pathway will require additional folding assistance by Hsp60 (*Dashed arrow* 2). Several imported proteins need the help of merely Hsp60 to acquire their native conformation (*Dashed arrow 3*)

A precursor protein in transit through the TIM23 channel requires additional help in order to be imported completely into the mitochondrial matrix. This assistance is provided by the translocation motor, whose key constituent is mtHsp70. In order to mediate protein import, mtHsp70 must anchor to the TIM23 import channel at a certain stage of the import process. This is facilitated by another component of the translocation motor, Tim44, a peripheral membrane protein that binds simultaneously to mtHsp70 and to the core of the TIM23 complex (Bomer et al. 1997; Moro et al. 1999; Truscott et al. 2003). Binding and release of precursor proteins by mtHsp70 is controlled by nucleotides. Additional vital components play a regulatory role on the ATPase function of mtHsp70 (Stojanovski et al. 2006). J-domain containing proteins enhance the ATPase activity of Hsp70 chaperones. Such enhancement of the ATPase activity is required for promoting the tight binding of unfolded substrate proteins to the peptide-binding pocket of mtHsp70. Another component Mge1 acts as a nucleotide-exchange factor and promotes the release of imported precursor proteins from mtHsp70.

As mentioned above, the translocation motor must be anchored to the TIM23 complex in order to mediate protein import into the mitochondrial matrix. This mission is achieved by Tim44, which forms a platform for tethering the components of the motor to the TIM23 complex. As such, Tim44 interacts directly with the Tim23 protein and with negatively charged phospholipids (Weiss et al. 1999; Marom et al. 2009). *In vivo* and *in vitro* studies showed that Tim44 also binds directly to several additional components: First, it binds to mtHsp70, thereby recruiting it to the Tim23 channel, in a nucleotide dependent manner. Second, Tim44 is found in close vicinity to a preprotein in transit and probably binds it (Krayl et al. 2007). Third, in a search for novel import-motor constituents that regulate mtHsp70 function, two type III J-domain proteins, Tim14/Pam18 and Tim16/Pam16, were discovered, as components that are recruited via Tim44 (D'Silva et al. 2003; Frazier et al. 2004; Kozany et al. 2004; Mokranjac et al. 2003; Truscott et al. 2003). Both proteins are essential and were shown to contain a hydrophobic N-terminal segment, which attaches them to the inner membrane, so that their C-terminal J-domain faces the matrix (Fig. 14.2). The typical J-domain of Tim14/Pam18 contains the three amino-acid signature motif, HPD, which is crucial for stimulating the ATPase activity of Hsp70 proteins. Tim16/Pam16, on the other hand, lacks this characterizing sequence, and thus is considered as a J-like protein. It was demonstrated that Tim16/Pam16 inhibits the ATPase stimulatory effect that Tim14/Pam18 exerts on mtHsp70 (Li et al. 2004; Mokranjac et al. 2006). Tim14/Pam18 and Tim16/Pam16 were shown to form an extremely stable complex both *in vitro* and *in vivo* (Iosefson et al. 2007; Kozany et al. 2004).

Studies carried out in yeast led to the following general outline for the molecular function of the translocation motor. A pre-sequence emerging from the matrix side of the TIM23 complex will bind to the ATP-bound form of mtHsp70, which itself is part of a preformed translocation motor. Substrate binding together with Tim14/Pam18 will accelerate ATP hydrolysis by mtHsp70, thereby forming the conformation of the protein that is stably bound to the precursor protein. Substrate binding also triggers the dissociation of mtHsp70 from Tim44 allowing it to diffuse together with the bound protein far from the import channel. The dissociation of mtHsp70 from its precursor protein is promoted by another component of the translocation motor, Mge1. This protein accelerates the exchange of ADP from mtHsp70 by ATP, a form that binds weakly to precursor proteins. Complete translocation into the matrix is achieved by several rounds of mtHsp70 to the imported precursor protein.

14.2.1.1 The Human Mitochondrial Import Motor

In recent years, considerable data has accumulated regarding the structural and functional conservation between the components of the yeast and human import motor:

1. The Human mtHsp70, mortalin, was shown to physically interact with Tim23 and Tim44 in human cells (Yaguchi et al. 2007).
2. Similar to their yeast homologues, human Tim14/Pam18 and Tim16/Pam16 form a heterodimer, both *in vitro* and *in vivo* (Elsner et al. 2009; Sinha et al. 2010). Furthermore, their ability to interact in the same manner with their yeast counterparts was also demonstrated (Elsner et al. 2009; Sinha et al. 2010).
3. Human Tim14/Pam18 stimulates the ATPase activity of both Ssc1 and mortalin. In the presence of human Tim16/Pam16 this activity was inhibited (Elsner et al. 2009; Sinha et al. 2010).
4. Human Tim16/Pam16 complements the deletion of yeast Tim16/Pam16 and is essential for the import of matrix-targeted precursors (Sinha et al. 2010).

This data demonstrates that there is a structural and functional evolutionary conservation from yeast to human. Recently, a number of human disorders were associated with the malfunction of import motor proteins (Davey et al. 2006; Jubinsky et al. 2001; Jubinsky et al. 2005). However, a direct link to import defects has not yet been established. This reinforces the importance of expanding our research efforts to the human translocation machinery.

14.2.2 The Mitochondrial Folding Complex of Mortalin

Proteins pass through the TIM23 channel in an unfolded conformation. Newly imported proteins reach their folded conformation with the help of two chaperone systems: the hsp60/Hsp10 team and the folding complex of mtHsp70. This second complex of mtHsp70 does not associate with the membrane and is soluble in the matrix. Consequently, the folding function of mtHsp70 is assisted by a second soluble J-domain containing J-protein, Mdj1. Notably, nucleotide exchange in this complex is also assisted by the nucleotide-exchange factor, Mge1. The role of this triad is to assist a freshly imported protein to acquire its native conformation (Fig. 14.2). The folding complex, thus, recognizes the precursor only after it is entirely introduced into the matrix by translocases. This sequence of actions was illustrated by the ability of Mdj1 to distinguish between the various folding states of a precursor in transit. Mdj1 can only be coimunnoprecipitated with a newly imported substrate, which requires folding assistance, and not with an arrested unfolded translocation-intermediate (Horst et al. 1997; Westermann et al. 1996). Indeed, the mitochondrial protein import is not impaired in Mdj1 deletion yeast strains (Prip-Buus et al. 1996). In addition, mtHsp70, with the aid of his co-chaperones, orchestrate other folding tasks in the matrix. These include prevention of protein aggregation under stress conditions, reactivation of aggregated substrates and supporting the assembly of proteins encoded by the mitochondrial DNA (Herrmann et al. 1994; Prip-Buus et al. 1996).

The human homologues of Mdj1 and Mge1 are Tid1 and hMge1, respectively. Tid1 is also a homologue of the *drosophila* protein, Tid56. The latter was identified as a tumor-suppressor, who functions in the regulation of the larvae development and differentiation (Kurzik-Dumke et al. 1992, 1995). A null mutation of the gene that encodes for Tid56 results in a lethal overgrowth of the imaginal discs (Hence the name Tid, initials for *T*umorous *I*maginal *D*iscs). These are unique structures of the larvae's epithelial tissue that normally differentiate into a variety of exterior organs in an adult fly. In human cells, there are two splice variants of Tid1, Tid1-long and Tid1-short (Henceforth Tid1-L and Tid1-S, respectively) (Syken et al. 1999). Generally, both proteins localize to the mitochondrial matrix and share identical amino acid sequences up to their 447 residues. From there on, Tid1-S and Tid1-L, each contain an extension of 6 and 33 unique amino-acid sequences, respectively (Fig. 14.1). Several lines of evidence indicate that both Tid1 isoforms are functional homologues of Mdj1:

1. They contain the characteristic J-domain of the Hsp40 family;
2. Both isoforms were coimmunoprecipitated with mortalin (Syken et al. 1999);
3. Tid1-L and Tid1-S were both able to compensate for an Mdj1 deletion in yeast. This capacity was dependent on the presence of an intact J-domain and a mitochondrial targeting sequence (Lu et al. 2006).

The functionality of the human folding complex was also evaluated *in vitro*, using purified components. This was achievable owing to the recent development of efficient purification protocols for mortalin and Tid1 (Goswami et al. 2010; Iosefson and Azem 2010; Luo et al. 2010; Zhai et al. 2008). Interestingly, even though both Tid1 isoforms could equally complement growth defects induced by a yeast temperature-sensitive mutant form of Mdj1, they differ in three functions:

1. Mortalin's ATPase activity was stimulated to a greater extent by Tid1-S (Goswami et al. 2010);
2. Several J-domain proteins have the ability to bind to a client protein and prevent it from aggregation under stress conditions. This capacity was also demonstrated for both Tid1 isoforms (Goswami et al. 2010). However, Tid1-S was more effective in protecting a denatured model substrate from aggregation;
3. Mitochondrial key metabolic enzymes are prone to aggregation under elevated temperatures, a process with deleterious consequences to the eukaryotic cell (Bender et al. 2011).

It is known that the bacterial chaperone system has the ability to restore proteins from an aggregated state back to their active conformation (Diamant et al. 2000; Mogk et al. 1999). Importantly, mortalin, in cooperation with its Tid1 co-chaperones, is also capable of effectively reactivating an aggregated protein (unpublished data). In this function as well, the short isoform was more active. Similar to mortalin, Tid1 isoforms also encompass extra-mitochondrial functions. In the cytosol, Tid1-L is considerably more stable than its shorter counterpart (Lu et al. 2006). It is possible that these differences in the splice-variants stability are responsible for their disparate ability to mediate certain mortalin functions.

14.2.3 Hep1, a Chaperone that Stabilizes Mortalin

Chaperones are considered as very stable proteins. However, this is not the case for mtHsp70s chaperones including mortalin. A first indication for the fact that mtHsp70 chaperones are prone to aggregation was provided by the fact that they are recovered in the inclusion bodies upon their expression in bacteria (Azem et al. 1997). Hep1 (*H*uman *E*scort *P*rotein 1, also termed Zim17 or Tim15) is an L-shaped, zinc finger-containing protein, which localizes to the mitochondrial matrix (Burri et al. 2004; Momose et al. 2007; Sanjuan Szklarz et al. 2005; Sichting et al. 2005; Yamamoto et al. 2005). Hep1 homologues exist in higher eukaryotes and were found in algae and higher plants, as well. The protein interacts with Ssc1 and mortalin in a nucleotide-dependent manner, both *in vivo* and *in vitro*. The most stable interaction was established in the absence of nucleotides, while ATP and ADP were shown to diminish the complex formation (Goswami et al. 2010; Sichting et al. 2005). Deletion of Hep1 in yeast mitochondria impairs the protein import via the TIM23 complex, the biogenesis of proteins that contain Fe–S clusters and the mitochondrial morphology (Sanjuan Szklarz et al. 2005; Sichting et al. 2005). These effects are attributed to Ssc1 aggregation in the absence of Hep1 (Blamowska et al. 2010; Sanjuan Szklarz et al. 2005; Sichting et al. 2005). Therefore, Hep1 is suggested to protect mtHsp70 proteins from self-association and to maintain them in a functional conformation. Intriguingly, in contrast to the full-length protein, the substrate binding domain and the ATPase domain are each soluble in yeast mitochondria lacking Hep1. However, when the construct which consists of the ATPase domain flanked by the short interdomain linker is expressed under the same conditions, it aggregates (Blamowska et al. 2010). Thus, it is tempting to speculate that interdomain communication leads to aggregation-prone mtHsp70 conformation, which is prevented upon interaction with Hep1. In summary, studies carried out both *in vivo* and *in vitro* demonstrated that Hep1 is a chaperone that is dedicated to the protection of mtHsp70 s from mitochondria and chloroplasts (Goswami et al. 2010; Iosefson and Azem 2010; Willmund et al. 2008; Zhai et al. 2008).

14.3 Extra-mitochondrial Functions of Mortalin

Elevated levels of molecular chaperones were observed in a variety of cancer cells (Helmbrecht et al. 2000; Jaattela 1999). Tumors can benefit from over expression of chaperones in several ways. For example, it was shown that Hsp90 stabilizes a mutant conformation of p53, both *in vivo* and *in vitro* (Blagosklonny et al. 1996). This type of buffering capacity, provided by the chaperone, enables the tumor cell to tolerate otherwise lethal mutations. In addition, Hsp70 proteins are well-known to function as anti-apoptotic agents (Beere and Green 2001). In this context, the ability of cytosolic Hsp70 to inhibit JNK, a kinase that prevents p53 degradation, was reported (Gabai et al. 1997; Mosser et al. 1997). An over expression of Hsp70 also conferred resistance to chemotherapeutic treatments, while silencing its gene

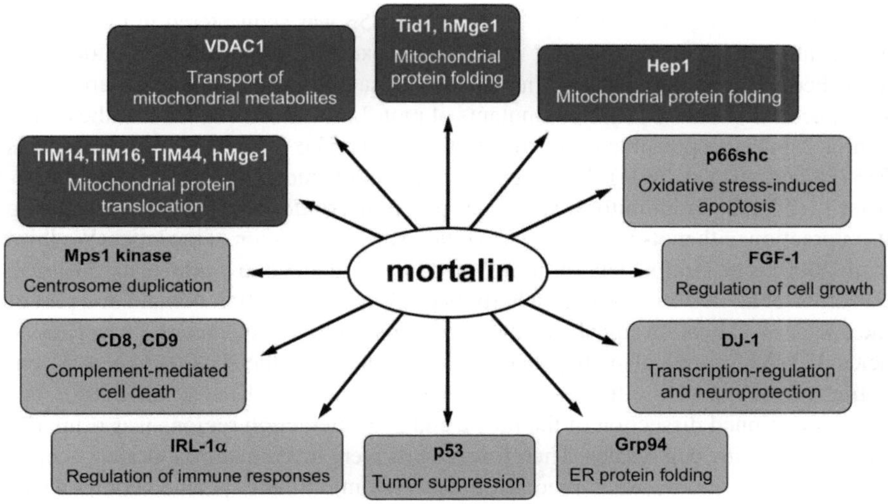

Fig. 14.3 The multiplicity of mortalin's binding partners. Each frame indicates a known mortalin binding partner and its typical function. Intra- and extra-mitochondrial interactions are represented by *dark* and *light grey*, respectively

expression, by means of shRNA, induced apoptosis in certain types of cancer (reviewed in Evans et al. 2010). Overall, it appears that the stress-protective role that heat-shock proteins play in healthy cells is subverted by the tumor, in a way that allows it to evade apoptosis and maintain homeostasis, even under hostile conditions such as hypoxia, heat or accumulation of mutations. The described association with key cell-cycle regulators was studied for the most part with cytosolic chaperones. However, there is a growing body of evidence concerning the interaction of the mitochondrial chaperones, mortalin and Tid1, with p53 and other cancer-related proteins. Mortalin and Tid1-L/S are nuclear encoded proteins that harbor amino terminal mitochondrial targeting signals. However, as will be discussed below, mounting evidence suggests that these two proteins perform extra-mitochondrial functions. Moreover, both proteins were detected physically inside and outside the mitochondria. Thus, mortalin and Tid1 belong to the large family of dually distributed proteins (Yogev and Pines 2011). What is the mechanism that allows dual distributions of these proteins? Future studies should address this question.

14.3.1 Mortalin-p53 Interaction

The multi-functionality of mortalin is reflected in its broad spectrum of cytosolic interactions. As shown in Fig. 14.3, these interactions are mainly associated with cell-growth regulation and immune responses. The interaction with p53, in particular, is the subject of intensive research. The complex was first identified in the cytosol of tumor cells, in which transfected murine mot-2, but not mot-1, co-immunoprecipitated with p53 (Wadhwa et al. 1998). Consequently, p53 was sequestered in the cytosol, and its ability to translocate to the nucleus, and execute its transcription duties, was abolished. Using binding assays, in which cell lysates containing wt p53 were mixed with bacterially-derived deletion mutants of mot-2, it was established that the region in mot-2 that is responsible for the interaction with p53 is located within amino acids 253–282 (Kaul et al. 2001). This segment is part of the mot-2 ATPase domain. Moreover, MKT-077, an inhibitor that rescues p53 transcriptional activity by abrogating its interaction with mot-2, was shown to bind to the same region of the latter (Wadhwa et al. 2000). The binding site for mot-2 on p53 was mapped to amino acids 312–352 (Wadhwa et al. 2002). This region harbors p53's tetramerization domain and part of the nuclear localization signal. A peptide composed of the sequences between amino acids 323–337 of p53 also abolished the interaction with the Hsp70 chaperone, enabling relocation of the tumor-suppressor to the nucleus (Kaul et al. 2005). The above mentioned dissection of the mot-2 and p53 interaction regions was primarily carried out, using cell lysates. Therefore, efforts were made in our laboratory to map these binding regions, using a reconstitution system of highly purified components. Human mortalin was used instead of its murine homologue. Surprisingly, the substrate binding domain of mortalin, and not its ATPase domain, was shown to mediate the interaction (Iosefson and Azem 2010). Moreover, a mutation of a conserved residue in mortalin's substrate-binding pocket impaired its ability to complex with p53. This implies that mortalin recognizes p53 as a substrate client protein. In addition, the presence of mortalin enabled p53 to form a stable complex with its consensus-DNA recognition motif. This result was also observed upon binding of the bacterial homologue, DnaK, to p53 (Hupp et al. 1992). It was confirmed that DnaK exerts this DNA-binding enhancement on p53 by masking the latter's extreme C-terminus, a region that was implicated in negative regulation of specific DNA binding (Hansen et al. 1996). Indeed, p53's extreme C-terminus (a.a. 356–393) was shown to interact with mortalin, as well as the aforementioned tetramerization domain (Iosefson and Azem 2010).

What is the underlying basis for the observed differences in the regions that mediate p53-mortalin interaction? It should be taken into consideration that molecular chaperones usually function as part of large multiprotein complexes rather than acting alone. They are assisted and regulated by specific co-chaperones and other partner-proteins that also assemble these complexes. This raises the possibility that other proteins are also involved in the regulation of p53-mortalin complex formation. An *in vitro* study carried out with purified proteins showed that cytosolic Hsp70 alone could not form a stable complex neither with wildtype nor with a conformational mutant p53 (King et al. 2001). However, a stable complex with both p53 forms was detected in the presence of Hsp40 and ATP. An interaction with Hsp40 alone was also observed. It was further demonstrated that Hsp90, the GrpE homologue, Bag-1, and Hop co-chaperone are all strongly involved in the regulation of the chaperones-p53 assembly (King et al. 2001). Moreover, these complexes were composed of different chaperone constituents, depending on whether wt or mutant p53 was present. The authors of this study suggested that Hsp40 plays a key regulatory role in recruiting other chaperones to interact with p53, following its own binding to the tumor-suppressor. This function of Hsp40 was recently reinforced as it was shown that the co-chaperone first binds to misfolded α-synuclein oligomers and then directs Hsp70 to interact with the aggregates (Hinault et al. 2010). Interestingly, a recent study has shown that Tid1, mortalin's mitochondrial Hsp40-co-chaperone, also physically interacts with p53 outside of the mitochondria (Ahn et al. 2010). Deletion of the Tid1 J-domain, as opposed to its other domains, abrogates p53 binding (Trinh et al. 2010). The J-domain is known to interact with the ATPase domain of Hsp70 proteins and stimulates their hydrolyzing activity. Therefore, it is tempting to speculate, that Tid1 or other proteins from the Hsp40 family, mediate the interaction of mortalin with p53. The existence of such a ternary complex, *i.e.,* p53-Tid-mortalin, might explain the contradictory results regarding which domain of mortalin is responsible for the interaction with p53.

14.4 Extra-mitochondrial Functions of Tid1

Similar to mortalin, Tid1 is also implicated in a wide variety of extra-mitochondrial interactions. The list of the Tid1 binding partners and their functional classification is listed in Table 14.1. The vast majority of these interactions were identified in two-hybrid screen systems. The emerging picture points to certain functional groups with which Tid1 mainly associates. Among these groups are transcription factors, receptor tyrosine kinases, viral proteins and tumor-suppressors (Table 14.1). Some proteins preferentially bind to one of the isoforms, but it is difficult to distinguish between the splice-variants on the basis of their role in apoptosis, as was previously suggested (Syken et al. 1999). In addition, there are interactions in which the J-domain was shown to physically bind to a client protein, despite the fact that the C-terminal part was suggested to mediate substrate binding (Banecki et al. 1996; Szabo et al. 1996). Surprisingly, some of the interactions were shown to involve cytosolic Hsp70's, rather than mortalin, the Tid1 mitochondrial partner. The functional significance of the interplay of chaperones in general and Tid1 in particular with the indicated binding partners is the subject of some interesting reviews (Gaestel 2006; Mosser and Morimoto 2004; Sehgal 2003; Song and Balice-Gordon 2008; Xiao et al. 2010).

14.4.1 Tid1 as a p53 Regulator

p53-dependent apoptosis can be triggered either by transcription of nuclear-encoded proapoptotic genes, or by releasing cytochrome c and other factors from the mitochondria (reviewed in Schuler and Green 2001). It was reported that irradiation of thymocytes induces the translocation of p53 to the mitochondria, where it forms a complex with proteins from the Bcl-2 family (Mihara et al. 2003). This interaction activates a rapid release of cytochrome c from intact mitochondria. However the mechanism by which p53 relocates from the cytosol to the mitochondria was not

Table 14.1 Tid1 binding-partners

	Protein	Function	Preferential interacting isoform	Interacting domain	Functional role of the interaction	Ref.
Transcriptions factors	Stat5b	Promotion of cell proliferation and survival	Both	Cys-rich	Inhibition of Stat5b-mediated transcriptional activity	(Dhennin-Duthille et al. 2011)
	NF-κB	Promotion of immune, inflammation and antiapoptotic responses	Both	J	Induction of growth arrest and cell-death	(Cheng et al. 2005, 2002)
	Stat1/3	Regulation of cell growth and apoptosis	Tid1-L	Tid1-L unique tail	N.D	(Lu et al. 2006)
Receptor Tyrosine Kinases	c-Met	Hepatocyte growth factor receptor	Tid1-S	J	Increased receptor activation	(Copeland et al. 2011; Schaaf et al. 2005)
	ErbB-2	Epidermal growth factor Receptor	Both	C-terminal	Induction of ubiquitin-dependent ErbB-2 down-regulation	(Kim et al. 2004; Kurzik-Dumke et al. 2010)
	Trk (tropomyosin-receptor kinase)	Regulation of neural survival and differentiation	Both	C-terminal half	positive regulation of nerve growth factor-induced outgrowth	(Liu et al. 2005)
	MuSK (muscle-specific receptor tyrosine kinase)	Regulation of nerve-induced acetylcholine clustering in muscle cells	Tid1-S	C-terminal half	Maintenance of normal synaptic structure and function	(Linnoila et al. 2008)
Viral proteins	UL9 protein	Replication initiator of herpes simplex virus type 1 (HSV-1)	N.D	N.D	Enhancement of the binding of UL9 to the origins of HSV-1 DNA	(Eom and Lehman 2002)
	BARF1	Epstein-Barr virus secretory protein	Both	C-terminal half	Promotion of the maturation and secretion of BARF1	(Wang et al. 2006)

Table 14.1 (continued)

	Protein	Function	Preferential interacting isoform	Interacting domain	Functional role of the interaction	Ref.
	Human papillomavirus E7 oncoprotein	Inactivation of host tumor suppressor proteins	N.D	C-terminal half and N-terminal half independently	N.D	(Schilling et al. 1998)
	HTLV-1 (Human T cell leukemia virus type 1) Tax protein	Promotion of viral survival and proliferation	N.D	Cys-rich	Inhibition of NF-κB activity induced by Tax	(Cheng et al. 2001, 2002)
Tumor suppressors	p53	Promotion of anti-cancer functions	Both	J	Direction of p53 to the mitochondria under stress conditions	(Ahn et al. 2010)
	pVHL (von Hippel-Lindau protein)	Promotion of hypoxia-inducible-factor degradation	Tid1-L	N.D	Suppression of HIF-1α-mediated angiogenesis	(Bae et al. 2005)
	APC (adenomatous polyposis cell)	Regulation of cell proliferation	Tid1-S	N.D	Inhibition of the proapoptotic function of APC	(Kurzik-Dumke and Czaja, 2007; Qian et al. 2010)
Chaperones	Hsc70	constitutively-expressed cytosolic Hsp	Tid1-L	Tid1-L unique tail	N.D	(Lu et al. 2006)
	Mortalin	Mitochondrial Hsp70 chaperone	Both	N.D	Folding of mitochondrial proteins	(Goswami et al. 2010; Lu et al. 2006; Syken et al. 1999)
Other	RasGAP (Ras GTPase-activating protein)	Negative regulator of Ras	Both	N.D	N.D	(Trentin et al. 2001)
	Jak2—interferon-γ receptor complex	Regulation of immune and anti-tumor functions	Both	N.D	Inhibition of interferon-γ activity	(Sarkar et al. 2001)

The indicated interacting domains are depicted in Fig. 14.1. Abbreviations: *J*—J-domain, *Cys-rich*—cysteine-rich domain, *N.D*—not-determined elucidated. Since it lacks any known mitochondrial targeting signal, p53 should be directed to the organelle by other means. Ahn et al. identified Tid1 as a regulator of p53 sub cellular localization that chaperones the latter to the mitochondria under stress conditions (Ahn et al. 2010). Under these circumstances, suppression of Tid1 expression resulted in the loss of a p53 mitochondrial targeting and reduced levels of apoptosis. Both the crucial HPD J-domain motif and the mitochondrial targeting signal of Tid1 were necessary for its shuttling function. Several types of cancer cells were unable to activate the transcription-dependent pathway as a result of a mutation in p53, which abolishes its DNA-binding capacity. Over expression of Tid1 in these cells induced apoptosis through the intrinsic mitochondrial pathway. Therefore, Tid1 can be considered as a new therapeutic target. It is not yet clear whether mortalin or other proteins from the Hsp70 family are involved in Tid1-dependent mitochondrial translocation of p53. However, the evidence that an intact HPD sequence is required indicates that a larger chaperone complex might form. The mechanism by which Tid1 directs p53 to the mitochondria awaits further elucidation.

14.5 Future Prospective

Despite extensive studies carried out to elucidate the molecular function of mortalin and its partner proteins, the following central aspects of the function of this system is still obscure:

1. The large variety of proteins that interact with mortalin and Tid1 reflect the central role that molecular chaperones play in the regulation of key cell signaling pathways. Although both are considered as mitochondrial proteins, the majority of the interactions were reported to reside outside the organelle. This emphasizes the need to elucidate the mechanism that enables these chaperones to dually distribute between different subcellular locations. It is not clear whether mortalin and Tid1 first enter the mitochondria, are then cleaved by a matrix peptidase and only then are exported outside, or whether a fraction of the translated chaperones are partially imported (Lu et al. 2006; Yogev and Pines 2011). The cellular circumstances that could trigger each of these options are not known either.
2. As described above, Ssc1 is the yeast mtHsp70 member that is in charge for the import of precursor proteins into the matrix and their folding thereafter. Two additional Hsp70 chaperones can also be found in yeast mitochondria: Ssq1 and Ecm10. Ssq1 is required for the biogenesis of proteins that contain iron-sulfur clusters while the function of Ecm10 has yet to be established (Lill and Kispal 2000; Craig and Marszalek 2002). On the other hand, in human mitochondria, mortalin is the sole Hsp70 constituent. As such, it is expected to take over the functions of its three yeast relatives. However, whether mortalin is also involved in the assembly of iron-sulfur clusters or other mitochondrial tasks is yet to be elucidated.
3. Bacteria harbor ClpB, which is involved in resolubilization of protein aggregates (Diamant et al. 2000; Goloubinoff et al. 1999; Mogk et al. 1999). Yeast

mitochondria contain a ClpB homologue known as Hsp78. Similar to its bacterial counterpart, the ability of Hsp78 to dissolve protein aggregates was also demonstrated (Krzewska et al. 2001; von Janowsky et al. 2006). Surprisingly, an Hsp78 homologue is absent from human mitochondria (von Janowsky et al. 2006). Hence, it will be interesting to explore whether mortalin is able to reactivate various populations of protein aggregates and whether it cooperates with, yet, unidentified partners in executing such a function.

Acknowledgments This work was supported by the German-Israeli Foundation for Scientific Research and Development (GIF- 1012/08) and the Israel Science Foundation (452/09).

References

Ahn BY, Trinh DL, Zajchowski LD, Lee B, Elwi AN, Kim SW (2010) Tid1 is a new regulator of p53 mitochondrial translocation and apoptosis in cancer. Oncogene 29:1155–1166

Anderson L (1981) Identification of mitochondrial proteins and some of their precursors in two-dimensional electrophoretic maps of human cells. Proc Natl Acad Sci U S A 78:2407–2411

Azem A, Oppliger W, Lustig A, Jeno P, Feifel B, Schatz G, Horst M (1997) The mitochondrial hsp70 chaperone system. Effect of adenine nucleotides, peptide substrate, and mGrpE on the oligomeric state of mhsp70. J Biol Chem 272:20901–20906

Bae MK, Jeong JW, Kim SH, Kim SY, Kang HJ, Kim DM, Bae SK, Yun I, Trentin GA, Rozakis-Adcock M, Kim KW (2005) Tid-1 interacts with the von Hippel-Lindau protein and modulates angiogenesis by destabilization of HIF-1alpha. Cancer Res 65:2520–2525

Banecki B, Liberek K, Wall D, Wawrzynow A, Georgopoulos C, Bertoli E, Tanfani F, Zylicz M (1996) Structure-function analysis of the zinc finger region of the DnaJ molecular chaperone. J Biol Chem 271:14840–14848

Beere HM, Green DR (2001) Stress management—heat shock protein-70 and the regulation of apoptosis. Trends Cell Biol 11:6–10

Bender T, Lewrenz I, Franken S, Baitzel C, Voos W (2011) Mitochondrial enzymes are protected from stress-induced aggregation by mitochondrial chaperones and the Pim1/LON protease. Mol Biol Cell 22:541–554

Blagosklonny MV, Toretsky J, Bohen S, Neckers L (1996) Mutant conformation of p53 translated in vitro or in vivo requires functional HSP90. Proc Natl Acad Sci U S A 93:8379–8383

Blamowska M, Sichting M, Mapa K, Mokranjac D, Neupert W, Hell K (2010) ATPase domain and interdomain linker play a key role in aggregation of mitochondrial Hsp70 chaperone Ssc1. J Biol Chem 285:4423–4431

Bomer U, Meijer M, Maarse AC, Honlinger A, Dekker PJ, Pfanner N, Rassow J (1997) Multiple interactions of components mediating preprotein translocation across the inner mitochondrial membrane. EMBO J 16:2205–2216

Burri L, Vascotto K, Fredersdorf S, Tiedt R, Hall MN, Lithgow T (2004) Zim17, a novel zinc finger protein essential for protein import into mitochondria. J Biol Chem 279:50243–50249

Chacinska A, Koehler CM, Milenkovic D, Lithgow T, Pfanner N (2009) Importing mitochondrial proteins: machineries and mechanisms. Cell 138:628–644

Cheng H, Cenciarelli C, Shao Z, Vidal M, Parks WP, Pagano M, Cheng-Mayer C (2001) Human T cell leukemia virus type 1 Tax associates with a molecular chaperone complex containing hTid-1 and Hsp70. Curr Biol 11:1771–1775

Cheng H, Cenciarelli C, Tao M, Parks WP, Cheng-Mayer C (2002) HTLV-1 Tax-associated hTid-1, a human DnaJ protein, is a repressor of Ikappa B kinase beta subunit. J Biol Chem 277:20605–20610

Cheng H, Cenciarelli C, Nelkin G, Tsan R, Fan D, Cheng-Mayer C, Fidler IJ (2005) Molecular mechanism of hTid-1, the human homolog of Drosophila tumor suppressor l(2)Tid, in the regulation of NF-kappaB activity and suppression of tumor growth. Mol Cell Biol 25:44–59

Copeland E, Balgobin S, Lee CM, Rozakis-Adcock M (2011) hTID-1 defines a novel regulator of c-Met Receptor signaling in renal cell carcinomas. Oncogene 30:2252–2263

Craig EA, Marszalek J (2002) A specialized mitochondrial molecular chaperone system: a role in formation of Fe/S centers. Cell Mol Life Sci 59:1658–1665

D'silva PD, Schilke B, Walter W, Andrew A, Craig EA (2003) J protein co-chaperone of the mitochondrial inner membrane required for protein import into the mitochondrial matrix. Proc Natl Acad Sci U S A 100:13839–13844

Daugaard M, Rohde M, Jaattela M (2007) The heat shock protein 70 family: highly homologous proteins with overlapping and distinct functions. FEBS Lett 581:3702–3710

Davey KM, Parboosingh JS, Mcleod DR, Chan A, Casey R, Ferreira P, Snyder FF, Bridge PJ, Bernier FP (2006) Mutation of DNAJC19, a human homologue of yeast inner mitochondrial membrane co-chaperones, causes DCMA syndrome, a novel autosomal recessive Barth syndrome-like condition. J Med Genet 43:385–393

Dhennin-Duthille I, Nyga R, Yahiaoui S, Gouilleux-Gruart V, Regnier A, Lassoued K, Gouilleux F (2011) The tumor suppressor hTid1 inhibits STAT5b activity via functional interaction. J Biol Chem 286:5034–5042

Diamant S, Ben-Zvi AP, Bukau B, Goloubinoff P (2000) Size-dependent disaggregation of stable protein aggregates by the DnaK chaperone machinery. J Biol Chem 275:21107–21113

Elsner S, Simian D, Iosefson O, Marom M, Azem A (2009) The mitochondrial protein translocation motor: structural conservation between the human and yeast Tim14/Pam18-Tim16/Pam16 co-chaperones. Int J Mol Sci 10:2041–2053

Eom CY, Lehman IR (2002) The human DnaJ protein, hTid-1, enhances binding of a multimer of the herpes simplex virus type 1 UL9 protein to oris, an origin of viral DNA replication. Proc Natl Acad Sci U S A 99:1894–1898

Evans CG, Chang L, Gestwicki JE (2010) Heat shock protein 70 (hsp70) as an emerging drug target. J Med Chem 53:4585–4602

Frazier AE, Dudek J, Guiard B, Voos W, Li Y, Lind M, Meisinger C, Geissler A, Sickmann A, Meyer HE, Bilanchone V, Cumsky MG, Truscott KN, Pfanner N, Rehling P (2004) Pam16 has an essential role in the mitochondrial protein import motor. Nat Struct Mol Biol 11:226–233

Gabai VL, Meriin AB, Mosser DD, Caron AW, Rits S, Shifrin VI, Sherman MY (1997) Hsp70 prevents activation of stress kinases. A novel pathway of cellular thermotolerance. J Biol Chem 272:18033–18037

Gaestel M (2006) Molecular chaperones in signal transduction. Handb Exp Pharmacol 172:93–109

Goloubinoff P, Mogk A, Zvi AP, Tomoyasu T, Bukau B (1999) Sequential mechanism of solubilization and refolding of stable protein aggregates by a bichaperone network. Proc Natl Acad Sci U S A 96:13732–13737

Goswami AV, Chittoor B, D'silva P (2010) Understanding the functional interplay between mammalian mitochondrial Hsp70 chaperone machine components. J Biol Chem 285:19472–19482

Hansen S, Midgley CA, Lane DP, Freeman BC, Morimoto RI, Hupp TR (1996) Modification of two distinct COOH-terminal domains is required for murine p53 activation by bacterial Hsp70. J Biol Chem 271:30922–30928

Helmbrecht K, Zeise E, Rensing L (2000) Chaperones in cell cycle regulation and mitogenic signal transduction: a review. Cell Prolif 33:341–365

Herrmann JM, Stuart RA, Craig EA, Neupert W (1994) Mitochondrial heat shock protein 70, a molecular chaperone for proteins encoded by mitochondrial DNA. J Cell Biol 127:893–902

Hinault MP, Farina-Henriquez-Cuendet A, Mattoo RU, Mensi M, Dietler G, Lashuel HA, Goloubinoff P (2010) Stable alpha-synuclein oligomers strongly inhibit chaperone activity of the Hsp70 system by weak interactions with J-domain co-chaperones. J Biol Chem 285:38173–38182

Horst M, Oppliger W, Rospert S, Schonfeld HJ, Schatz G, Azem A (1997) Sequential action of two hsp70 complexes during protein import into mitochondria. EMBO J 16:1842–1849

Hupp TR, Meek DW, Midgley CA, Lane DP (1992) Regulation of the specific DNA binding function of p53. Cell 71:875–886

Iosefson O, Azem A (2010) Reconstitution of the mitochondrial Hsp70 (mortalin)-p53 interaction using purified proteins—Identification of additional interacting regions. FEBS Lett 584:1080–1084

Iosefson O, Levy R, Marom M, Slutsky-Leiderman O, Azem A (2007) The Pam18/Tim14-Pam16/Tim16 complex of the mitochondrial translocation motor: the formation of a stable complex from marginally stable proteins. Protein Sci 16:316–322

Jaattela M (1999) Escaping cell death: survival proteins in cancer. Exp Cell Res 248:30–43

Jubinsky PT, Messer A, Bender J, Morris RE, Ciraolo GM, Witte DP, Hawley RG, Short MK (2001) Identification and characterization of Magmas, a novel mitochondria-associated protein involved in granulocyte-macrophage colony-stimulating factor signal transduction. Exp Hematol 29:1392–1402

Jubinsky PT, Short MK, Mutema G, Morris RE, Ciraolo GM, Li M (2005) Magmas expression in neoplastic human prostate. J Mol Histol 36:69–75

Kampinga HH, Craig EA (2010) The HSP70 chaperone machinery: J proteins as drivers of functional specificity. Nat Rev Mol Cell Biol 11:579–592

Kaul SC, Reddel RR, Mitsui Y, Wadhwa R (2001) An N-terminal region of mot-2 binds to p53 in vitro. Neoplasia 3:110–114

Kaul SC, Aida S, Yaguchi T, Kaur K, Wadhwa R (2005) Activation of wild type p53 function by its mortalin-binding, cytoplasmically localizing carboxyl terminus peptides. J Biol Chem 280:39373–29379

Kim SW, Chao TH, Xiang R, Lo JF, Campbell MJ, Fearns C, Lee JD (2004) Tid1, the human homologue of a Drosophila tumor suppressor, reduces the malignant activity of ErbB-2 in carcinoma cells. Cancer Res 64:7732–7739

King FW, Wawrzynow A, Hohfeld J, Zylicz M (2001) Co-chaperones Bag-1, Hop and Hsp40 regulate Hsc70 and Hsp90 interactions with wild-type or mutant p53. EMBO J 20:6297–6305

Kozany C, Mokranjac D, Sichting M, Neupert W, Hell K (2004) The J domain-related co-chaperone Tim16 is a constituent of the mitochondrial TIM23 preprotein translocase. Nat Struct Mol Biol 11:234–241

Krayl M, Lim JH, Martin F, Guiard B, Voos W (2007) A cooperative action of the ATP-dependent import motor complex and the inner membrane potential drives mitochondrial preprotein import. Mol Cell Biol 27:411–425

Krzewska J, Langer T, Liberek K (2001) Mitochondrial Hsp78, a member of the Clp/Hsp100 family in Saccharomyces cerevisiae, cooperates with Hsp70 in protein refolding. FEBS Lett 489:92–96

Kurzik-Dumke U, Czaja J (2007) Htid-1, the human homolog of the Drosophila melanogaster l(2)tid tumor suppressor, defines a novel physiological role of APC. Cell Signal 19:1973–1985

Kurzik-Dumke U, Phannavong B, Gundacker D, Gateff E (1992) Genetic, cytogenetic and developmental analysis of the Drosophila melanogaster tumor suppressor gene lethal(2)tumorous imaginal discs (1(2)tid). Differentiation 51:91–104

Kurzik-Dumke U, Gundacker D, Renthrop M, Gateff E (1995) Tumor suppression in Drosophila is causally related to the function of the lethal(2) tumorous imaginal discs gene, a dnaJ homolog. Dev Genet 16:64–76

Kurzik-Dumke U, Horner M, Nicotra MR, Koslowski M, Natali PG (2010) In vivo evidence of htid suppressive activity on ErbB-2 in breast cancers over expressing the receptor. J Transl Med 8:58

Li Y, Dudek J, Guiard B, Pfanner N, Rehling P, Voos W (2004) The presequence translocase-associated protein import motor of mitochondria. Pam16 functions in an antagonistic manner to Pam18. J Biol Chem 279:38047–38054

Lill R, Kispal G (2000) Maturation of cellular Fe–S proteins: an essential function of mitochondria. Trends Biochem Sci 25:352–356

Linnoila J, Wang Y, Yao Y, Wang ZZ (2008) A mammalian homolog of Drosophila tumorous imaginal discs, Tid1, mediates agrin signaling at the neuromuscular junction. Neuron 60:625–641

Liu HY, Macdonald JI, Hryciw T, Li C, Meakin SO (2005) Human tumorous imaginal disc 1 (TID1) associates with Trk receptor tyrosine kinases and regulates neurite outgrowth in nnr5-TrkA cells. J Biol Chem 280:19461–19471

Lopez MF, Kristal BS, Chernokalskaya E, Lazarev A, Shestopalov AI, Bogdanova A, Robinson M (2000) High-throughput profiling of the mitochondrial proteome using affinity fractionation and automation. Electrophoresis 21:3427–3440

Lu B, Garrido N, Spelbrink JN, Suzuki CK (2006) Tid1 isoforms are mitochondrial DnaJ-like chaperones with unique carboxyl termini that determine cytosolic fate. J Biol Chem 281:13150–13158

Luo WI, Dizin E, Yoon T, Cowan JA (2010) Kinetic and structural characterization of human mortalin. Protein Expr Purif 72:75–81

Marom M, Safonov R, Amram S, Avneon Y, Nachliel E, Gutman M, Zohary K, Azem A, Tsfadia Y (2009) Interaction of the Tim44C-terminal domain with negatively charged phospholipids. Biochemistry 48:11185–11195

Mayer MP, Bukau B (2005) Hsp70 chaperones: cellular functions and molecular mechanism. Cell Mol Life Sci 62:670–684

Mihara M, Erster S, Zaika A, Petrenko O, Chittenden T, Pancoska P, Moll UM (2003) p53 has a direct apoptogenic role at the mitochondria. Mol Cell 11:577–590

Mogk A, Tomoyasu T, Goloubinoff P, Rudiger S, Roder D, Langen H, Bukau B (1999) Identification of thermolabile Escherichia coli proteins: prevention and reversion of aggregation by DnaK and ClpB. EMBO J 18:6934–6949

Mokranjac D, Sichting M, Neupert W, Hell K (2003) Tim14, a novel key component of the import motor of the TIM23 protein translocase of mitochondria. EMBO J 22:4945–4956

Mokranjac D, Bourenkov G, Hell K, Neupert W, Groll M (2006) Structure and function of Tim14 and Tim16, the J and J-like components of the mitochondrial protein import motor. EMBO J 25:4675–4685

Momose T, Ohshima C, Maeda M, Endo T (2007) Structural basis of functional cooperation of Tim15/Zim17 with yeast mitochondrial Hsp70. EMBO Rep 8:664–670

Moro F, Sirrenberg C, Schneider HC, Neupert W, Brunner M (1999) The TIM17.23 preprotein translocase of mitochondria: composition and function in protein transport into the matrix. EMBO J 18:3667–3675

Mosser DD, Morimoto RI (2004) Molecular chaperones and the stress of oncogenesis. Oncogene 23:2907–2918

Mosser DD, Caron AW, Bourget L, Denis-Larose C, Massie B (1997) Role of the human heat shock protein hsp70 in protection against stress-induced apoptosis. Mol Cell Biol 17:5317–5327

Neupert W, Herrmann JM (2007) Translocation of proteins into mitochondria. Annu Rev Biochem 76:723–749

Prip-Buus C, Westerman B, Schmitt M, Langer T, Neupert W, Schwarz E (1996) Role of the mitochondrial DnaJ homologue, Mdj1p, in the prevention of heat-induced protein aggregation. FEBS Lett 380:142–146

Qian J, Perchiniak EM, Sun K, Groden J (2010) The mitochondrial protein hTID-1 partners with the caspase-cleaved adenomatous polyposis cell tumor suppressor to facilitate apoptosis. Gastroenterol 138:1418–1428

Sanjuan Szklarz LK, Guiard B, Rissler M, Wiedemann N, Kozjak V, Van Der Laan M, Lohaus C, Marcus K, Meyer HE, Chacinska A, Pfanner N, Meisinger C (2005) Inactivation of the mitochondrial heat shock protein zim17 leads to aggregation of matrix hsp70s followed by pleiotropic effects on morphology and protein biogenesis. J Mol Biol 351:206–218

Sarkar S, Pollack BP, Lin KT, Kotenko SV, Cook JR, Lewis A, Pestka S (2001) hTid-1, a human DnaJ protein, modulates the interferon signaling pathway. J Biol Chem 276:49034–49042

Schaaf CP, Benzing J, Schmitt T, Erz DH, Tewes M, Bartram CR, Janssen JW (2005) Novel interaction partners of the TPR/MET tyrosine kinase. FASEB J 19:267–269

Schilling B, De-Medina T, Syken J, Vidal M, Munger K (1998) A novel human DnaJ protein, hTid-1, a homolog of the Drosophila tumor suppressor protein Tid56, can interact with the human papillomavirus type 16 E7 oncoprotein. Virology 247:74–85

Schleyer M, Neupert W (1985) Transport of proteins into mitochondria: translocational intermediates spanning contact sites between outer and inner membranes. Cell 43:339–350

Schuler M, Green DR (2001) Mechanisms of p53-dependent apoptosis. Biochem Soc Trans 29:684–688

Sehgal PB (2003) Plasma membrane rafts and chaperones in cytokine/STAT signaling. Acta Biochim Pol 50:583–594

Sichting M, Mokranjac D, Azem A, Neupert W, Hell K (2005) Maintenance of structure and function of mitochondrial Hsp70 chaperones requires the chaperone Hep1. EMBO J 24:1046–1056

Sinha D, Joshi N, Chittoor B, Samji P, D'silva P (2010) Role of Magmas in protein transport and human mitochondria biogenesis. Hum Mol Genet 19:1248–1262

Song Y, Balice-Gordon R (2008) New dogs in the dogma: Lrp4 and Tid1 in neuromuscular synapse formation. Neuron 60:526–528

Stojanovski D, Rissler M, Pfanner N, Meisinger C (2006) Mitochondrial morphology and protein import—a tight connection? Biochim Biophys Acta 1763:414–421

Syken J, De-Medina T, Munger K (1999) TID1, a human homolog of the Drosophila tumor suppressor l(2)tid, encodes two mitochondrial modulators of apoptosis with opposing functions. Proc Natl Acad Sci U S A 96:8499–8504

Szabo A, Korszun R, Hartl FU, Flanagan J (1996) A zinc finger-like domain of the molecular chaperone DnaJ is involved in binding to denatured protein substrates. EMBO J 15:408–417

Trentin GA, Yin X, Tahir S, Lhotak S, Farhang-Fallah J, LI Y, Rozakis-Adcock M (2001) A mouse homologue of the Drosophila tumor suppressor l(2)tid gene defines a novel Ras GTPase-activating protein (RasGAP)-binding protein. J Biol Chem 276:13087–13095

Trinh DL, Elwi AN, Kim SW (2010) Direct interaction between p53 and Tid1 proteins affects p53 mitochondrial localization and apoptosisy as a target for cancer therapy? Oncotarget 1:396–404

Truscott KN, Voos W, Frazier AE, Lind M, Li Y, Geissler A, Dudek J, Muller H, Sickmann A, Meyer HE, Meisinger C, Guiard B, Rehling P, Pfanner N (2003) A J-protein is an essential subunit of the presequence translocase-associated protein import motor of mitochondria. J Cell Biol 163:707–713

Von Janowsky B, Major T, Knapp K, Voos W (2006) The disaggregation activity of the mitochondrial ClpB homolog Hsp78 maintains Hsp70 function during heat stress. J Mol Biol 357:793–807

Wadhwa R, Takano S, Robert M, Yoshida A, Nomura H, Reddel RR, Mitsui Y, Kaul SC (1998) Inactivation of tumor suppressor p53 by mot-2, a hsp70 family member. J Biol Chem 273:29586–29591

Wadhwa R, Sugihara T, Yoshida A, Nomura H, Reddel RR, Simpson R, Maruta H, Kaul SC (2000) Selective toxicity of MKT-077 to cancer cells is mediated by its binding to the hsp70 family protein mot-2 and reactivation of p53 function. Cancer Res 60:6818–6821

Wadhwa R, Yaguchi T, Hasan MK, Mitsui Y, Reddel RR, Kaul SC (2002) Hsp70 family member, mot-2/mthsp70/GRP75, binds to the cytoplasmic sequestration domain of the p53 protein. Exp Cell Res 274:246–253

Walsh P, Bursac D, Law YC, Cyr D, Lithgow T (2004) The J-protein family: modulating protein assembly, disassembly and translocation. EMBO Rep 5:567–571

Wang L, Tam JP, Liu DX (2006) Biochemical and functional characterization of Epstein-Barr virus-encoded BARF1 protein: interaction with human hTid1 protein facilitates its maturation and secretion. Oncogene 25:4320–4331

Weiss C, Oppliger W, Vergeres G, Demel R, Jeno P, Horst M, De Kruijff B, Schatz G, Azem A (1999) Domain structure and lipid interaction of recombinant yeast Tim44. Proc Natl Acad Sci U S A 96:8890–8894

Westermann B, Gaume B, Herrmann JM, Neupert W, Schwarz E (1996) Role of the mitochondrial DnaJ homolog Mdj1p as a chaperone for mitochondrially synthesized and imported proteins. Mol Cell Biol 16:7063–7071

Willmund F, Hinnenberger M, Nick S, Schulz-Raffelt M, Muhlhaus T, Schroda M (2008) Assistance for a chaperone: Chlamydomonas HEP2 activates plastidic HSP70B for co-chaperone binding. J Biol Chem 283:16363–16373

Xiao A, Wong J, Luo H (2010) Viral interaction with molecular chaperones: role in regulating viral infection. Arch Virol 155:1021–1031

Yaguchi T, Aida S, Kaul SC, Wadhwa R (2007) Involvement of mortalin in cellular senescence from the perspective of its mitochondrial import, chaperone, and oxidative stress management functions. Ann N Y Acad Sci 1100:306–311

Yamamoto H, Momose T, Yatsukawa Y, Ohshima C, Ishikawa D, Sato T, Tamura Y, Ohwa Y, Endo T (2005) Identification of a novel member of yeast mitochondrial Hsp70-associated motor and chaperone proteins that facilitates protein translocation across the inner membrane. FEBS Lett 579:507–511

Yogev O, Pines O (2011) Dual targeting of mitochondrial proteins: mechanism, regulation and function. Biochim Biophys Acta 1808:1012–1020

Zhai P, Stanworth C, Liu S, Silberg JJ (2008) The human escort protein Hep binds to the ATPase domain of mitochondrial hsp70 and regulates ATP hydrolysis. J Biol Chem 283:26098–26106

Chapter 15
Mortalin: A Positive Regulator of Centrosome Duplication and Amplification

Masayuki Kanai and Kenji Fukasawa

Abstract Numeral abnormality of centrosomes (presence of more than two centrosomes) leads to mitotic defects and high frequencies of chromosome segregation errors. Centrosome amplification occurs frequently in various types of human cancers, and is believed to be a major cause of chromosomal instability. Centrosome amplification in many cases results from over-duplication (more than one duplication in a single cell cycle) caused by aberrant activities of the positive and negative regulators of centrosome duplication. Centrosome duplication is triggered by the cyclin-dependent kinase 2 (CDK2)-cyclin E kinase complex through targeting several centrosomally localized proteins, including Mps1 kinase. In contrast, p53 tumor suppressor protein suppresses centrosome duplication partly through transactivating p21 CDK inhibitor, and partly through direct regulation at centrosomes. Mortalin localizes to centrosomes, and when overexpressed, promotes centrosome duplication and induces centrosome amplification via targeting both Mps1 and p53. Here, we discuss the involvement of mortalin in the regulation of centrosome duplication and re-duplication, revealing a new and important aspect of oncogenic activity of mortalin, namely induction of numeral abnormality of centrosomes and consequent chromosome instability.

Keywords Mortalin · Centrosome · Centrosome amplification · Chromosome instability · p53 · Mps1 · CDK2 · Cyclin E

15.1 Overview of Centrosome Biology

The centrosome is a small non-membranous organelle (1∼2 mm in diameter), normally found at the periphery of the nucleus. The centrosome consists of a pair of centrioles and surrounding materials consisting of many different proteins, which is often referred to as pericentriolar material (PCM) as a whole (Fig. 15.1a). The primary function of centrosome is to nucleate and anchor microtubules. In interphase

K. Fukasawa (✉) · M. Kanai
Molecular Oncology Program, H. Lee Moffitt Cancer Center & Research Institute, Tampa, FL 33612, USA
e-mail: kenji.fukasawa@moffitt.org

S. C. Kaul, R. Wadhwa (eds.), *Mortalin Biology: Life, Stress and Death*,
DOI 10.1007/978-94-007-3027-4_15, © Springer Science+Business Media B.V. 2012

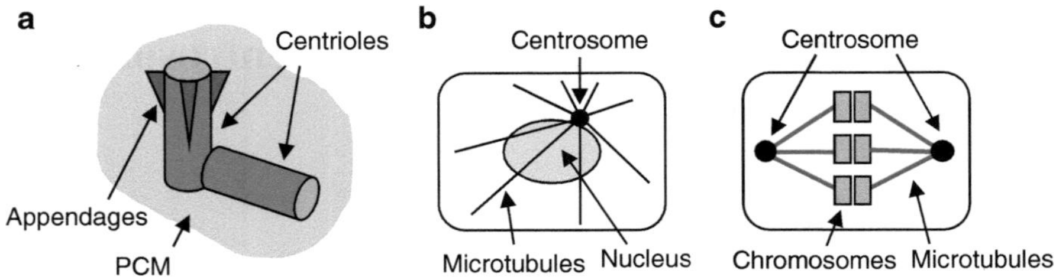

Fig. 15.1 Structure and function of the centrosome. (**a**) The centrosome consists of a pair of centrioles and surrounding protein aggregates (PCM) consisting of many different proteins. The paired centrioles are structurally different: one with a set of appendages at the distal ends (mother centriole) and another without them (daughter centriole). During interphase, centrosomes organize cytoplasmic microtubule networks (**b**), and in mitosis, centrosomes direct the formation of bipolar mitotic spindles (**c**)

cells, centrosomes play a key role in establishment of cytoplasmic microtubule network (Fig. 15.1b), which is essential for vesicle transport, intracellular distribution of small organelles, and determination of cell shape and polarity. In mitosis, two centrosomes, as central components of spindle poles, direct the formation of bipolar mitotic spindles (Fig. 15.1c), which is critical for accurate segregation of chromosomes to daughter cells during cytokinesis. Because each daughter cell receives only one centrosome upon cytokinesis, the centrosome must duplicate once prior to the following mitosis. Centrosome duplication is controlled both positively and negatively to ensure the one and only one duplication—timely initiation of centrosome duplication during the cell cycle and suppression of re-duplication of already duplicated centrosomes. Aberrant activities of the positive and negative regulators lead to more than one duplication of centrosomes in a single cell cycle, resulting in generation of more than two centrosomes, which is often referred to as centrosome amplification (Fukasawa 2005, 2007). Centrosome amplification leads to an increased frequency of defective mitoses, and consequential chromosome segregation errors. Numerous studies have shown the frequent occurrence of centrosome amplification in almost all types of solid tumors and certain leukemia (Neben et al. 2004) and lymphoma (Krämer et al. 2003), including breast (Pihan et al. 1998; Carroll et al. 1998; Lingle et al. 2002), lung (Pihan et al. 1998; Koutsami et al. 2006), bladder (Kawamura et al. 2004; Jiang et al. 2003), colon (Pihan et al. 1998), brain (Weber et al. 1998), liver (Nakajima et al. 2004), pancreas (Sato et al. 1999), prostate (Pihan et al. 2001), ovary (Hsu et al. 2005), to name a few. Moreover, aberrant mitotic spindles organized by abnormally amplified centrosomes are frequently observed in those tumors, and centrosome amplification is believed to be a major cause of chromosome instability in human cancers (D'Assoro et al. 2002; Fukasawa 2007) (discussed in more details below). Recent studies have shown that mortalin is a positive regulator of centrosome duplication (Ma et al 2006; Kanai et al. 2007). This section focuses on the mechanisms of how mortalin controls centrosome duplication at a molecular level, and how aberrant expression of mortalin leads to centrosome amplification.

15.2 Centrosome Amplification and Chromosomal Instability

Amplified centrosomes often lead to formation of multipolar (>2) mitotic spindles. Cells with tripolar spindles (Fig. 15.2a) normally undergo cytokinesis. Some daughter cells from tripolar divisions undergo cell death due to detrimental karyotypic alterations, while some cells survive, but suffer severe aneuploidy. Cells with the >3 pole-spindles (Fig. 15.2b) usually fail to undergo cytokinesis, and either become bi-nucleated or large mono-nucleated cells with doubling of genome or die due to mitotic catastrophe. Because polyploidy itself is known to destabilize chromosomes, those cells that have escaped the mitotic catastrophe will suffer chromosome segregation errors due to polyploidy as well as the presence of amplified centrosomes (cytokinesis failure results in doubling of centrosome number). Cells with amplified centrosomes also frequently form "pseudo-bipolar" spindles (Fig. 15.2c)—all amplified centrosomes position on a bipolar axis, forming spindles that structurally resemble the "true" bipolar spindles organized by two centrosomes. Although cells with pseudo-bipolar spindles usually undergo normal cytokinesis without any chromosome segregation errors, some cells with the pseudo-bipolar spindles suffer chromosome segregation errors (Fig. 15.2d): one or a few amplified centrosomes fail to position on the bipolar axis, yet they are functionally intact, nucleating microtubules which capture chromosomes. Depending on which daughter cell receives those mal-positioned chromosomes, aneuploid cells can be generated. More recently, it has been shown that the pseudo-bipolar spindles are often incapable of producing equal tension/force of microtubules for pulling chromosomes toward each daughter cell, resulting in chromosome segregation errors (Ganem et al. 2009).

15.3 Mechanism of Centrosome Duplication and Amplification

There are several mechanisms for generation of amplified centrosomes, including over-duplication of centrosomes, cytokinesis failure, physical disintegration of centrosomes and defects in centriole pairing (Fukasawa 2007). Among these, centrosome over-duplication is probably the most common cause of centrosome amplification. As mentioned earlier, centrosome duplication is controlled positively and negatively, and over-activation/over-expression of positive regulators as well as inactivation/depletion of negative regulators leads to centrosome over-duplication. As centrosome and DNA are the only two organelles that duplicate once in a single cell cycle, initiation of centrosome duplication and DNA replication in animal cells is coupled likely to ensure these two organelles to duplicate only once in a single cell cycle (Mazia 1987). The coupling of these two events are at least in part achieved by cyclin-dependent kinase 2 (CDK2)-cyclin E. Cyclin E is expressed specifically in late G1, and binds to CDK2, leading to the activation of the CDK2-cyclin E kinase complex. The activated CDK2-cyclin E then phosphorylates several target proteins such as the retinoblastoma (Rb) protein, leading to initiation of DNA replication (reviewed in Nevins 1992). CDK2-cyclin E was also found to be a key initiator of

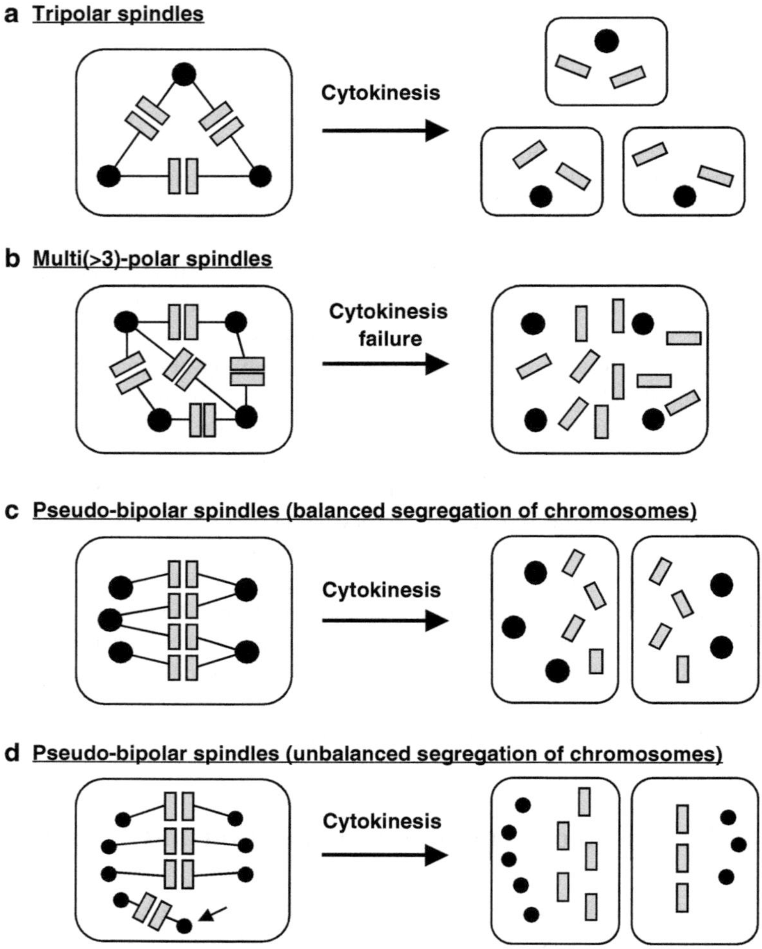

Fig. 15.2 Mitotic defects associated with amplified centrosomes and the consequential destabilization of chromosomes. Amplified centrosomes frequently form multiple (>2) spindle poles. (**a**) Cells with tripolar spindles can undergo cytokinesis. Some daughter cells are viable, yet suffer severe aneuploidy, while some daughter cells die due to detrimental karyotypic alterations. (**b**) Cells with spindles with >3 poles fail to undergo cytokinesis, becoming either bi-nucleated or large mono-nucleated cells, while some cells undergo cell death due to mitotic catastrophe. The surviving cells experience doubling of genomes as well as centrosomes. (**c**) Amplified centrosomes frequently form pseudo-bipolar spindles through positioning on a bipolar axis, resulting in mitotic spindles that structurally resemble the true bipolar spindles organized by two centrosomes. Cells with pseudo-bipolar spindles in many cases undergo normal cytokinesis without any chromosome segregation errors. (**d**) Upon formation of pseudo-bipolar spindles, there is a risk of chromosome segregation errors. One or a few centrosomes often fail to position on the bipolar axis (pointed by an *arrow*). Those misaligned centrosomes are still functionally intact, nucleating microtubules, which capture chromosomes. Depending on whether those particular chromosomes are segregated into one or the other daughter cell, aneuploid daughter cells are generated

centrosome duplication (Hinchcliffe et al. 1999; Lacey et al. 1999; Tarapore et al. 2002; Hanashiro et al. 2008). Thus, the late G1-specific activation of CDK2-cyclin E triggers initiation of both centrosome duplication and DNA replication. Several proteins were identified as the potential targets of CDK2-cyclin E for the initiation of centrosome duplication, including Mps1, CP110 and NPM/B23 (nucleophosmin) (Okuda et al. 2000; Fisk and Winey 2001; Chen et al. 2002).

Recent studies have unveiled how centrosomes duplicate more than once in a single cell cycle, in which the status of the p53 tumor suppressor protein plays a critical role (reviewed in Fukasawa 2005, 2007). Proliferating cells, even under optimal growth conditions, often suffer physiological stress that temporarily halts cell cycling irrespective of the p53 status (i.e. imbalance/deprivation of critical molecules such as dNTPs). Cells inflicted with DNA damage (i.e. irradiation, exposure to genotoxic drugs) also become arrested in S and late G2 phase irrespective of the p53 status (Taylor and Stark 2001). The cell cycle arrest provides time needed for the duplicated centrosomes to re-duplicate if active CDK2 is available. However, in normal cells, p53 is upregulated in response to the physiological stress associated with the prolonged arrest by the ARF-mediated inhibition of Mdm2 that promotes p53 degradation (Sherr 2006), and in response to DNA damage by the ATM/ATR- and Chk1/Chk2-mediated phosphorylation (Sancar et al. 2004). Up-regulated p53 then transactivates $p21^{Waf1/Cip1}$ CDK inhibitor, which in turn effectively inhibits CDK2 (reviewed in Harper 1997). Because initiation of centrosome duplication requires the CDK2 activity, re-duplication of the duplicated centrosomes is blocked in those arrested cells. In contrast, in cells lacking functional p53, the CDK2 activity is unchecked, leading to fortuitous re-duplication of centrosomes. In addition to the p21-mediated suppression of centrosome duplication, p53 appears to suppress centrosome duplication more directly at centrosomes independently from its transactivation function (see below for details). Once the stress causing problems are solved, cells resume cell cycling in the presence of abnormally amplified centrosomes.

15.4 Regulation of Centrosome Duplication by Mortalin in Association with p53

The involvement of mortalin in the regulation of centrosome duplication was initially speculated from the finding that mortalin localized to centrosomes, specifically to the centrosomes that have initiated duplication as well as to those that have already duplicated, but not to unduplicated centrosomes (Ma et al. 2006). Subsequently, it was found that overexpression of mortalin promotes centrosome duplication and re-duplication (note: in this section, "centrosome re-duplication" is used as an exchangeable term with "centrosome amplification") (Ma et al. 2006). Importantly, mortalin overexpression can override the p53-mediated suppression of centrosome duplication and re-duplication (Ma et al. 2006), suggesting the functional association between mortalin and p53 in the regulation of centrosome duplication. It has previously been shown that mortalin binds and sequesters p53 in cytoplasm, preventing p53 from exerting its transactivation function in nucleus (Wadhwa et al. 1998). As mentioned earlier, p53 plays a critical role in suppressing centrosome duplication and restricting centrosomes to duplicate once in the cell cycle in part through transactivating p21 CDK inhibitor (Tarapore et al. 2001a). Thus, promotion of centrosome duplication and re-duplication by mortalin is likely by inhibiting the p53's transactivation function. However, the effectiveness of mortalin-mediated inhibition of the transactivation function of p53 appears to differ from cell lines to cell lines (and probably from cell types to cell types), and it also appears to be influenced by the level of overexpressed mortalin, especially when p53 is up-regulated in response to the stress—a very high level of mortalin expression appears to be needed for sequestering p53 accumulated in response to the stress (Ma et al. 2006). Thus, in some cell lines, when challenged by physiological stress or DNA damage, overexpressed mortalin fails to inhibit the p53's transactivation function, leading to up-regulation of p21 at a similar level with the control cells. However, in those cells, mortalin is still able to at least partly override the p53-associated suppression of centrosome duplication and re-duplication (Ma et al. 2006). The answer to this puzzling observation was provided from the finding that mortalin also acts on the p53's transactivation activity-independent function to suppress centrosome duplication. A number of studies have shown that p53 localizes to centrosomes (Blair Zajdel et al. 1988; Brown et al. 1994; Morris et al. 2000; Tarapore et al. 2001b; Tritarelli et al. 2004; Shinmura et al. 2007; Mikule et al. 2007). Moreover, comparative analysis of the p53 mutants in respect to the transactivation activity and centrosome-localization activity have shown that the transactivation(+)/centrosome-localization(−) and transactivation (−)/centrosome-localization(+) p53 mutants can partially suppress centrosome duplication, while the transactivation(−)/centrosome-localization(−) mutant fails to do so, indicating that p53 also suppresses centrosome duplication directly at centrosomes (Shinmura et al. 2007). It was found that mortalin overexpression promotes dissociation of p53 from centrosomes, which depends on mortalin's ability to bind to p53; the mortalin mutant that lacks the p53 binding domain can no longer induce dissociation of p53 from centrosomes (Ma et al. 2006). At present, it is not known how physical interaction between mortalin and p53 leads to dissociation of p53 from centrosomes. Mortalin may prevent p53 from localizing to centrosomes via sequestering p53 to other cellular compartments. Alternatively, p53 may lose the ability to localize to centrosomes when bound by mortalin. Nonetheless, mortalin binds to p53, which blocks p53 to localize to centrosome, and thus preventing p53 from suppressing centrosome duplication and re-duplication at the centrosome. Another critical question remained to be answered is the molecular mechanism of how p53 suppresses centrosome duplication directly at centrosomes. Identification of the centrosomal protein(s) that physically interact with p53 will certainly shed a light on this issue.

15.5 The Role of Mortalin-Mps1 Interaction in the Regulation of Centrosome Duplication

Through the analysis of centrosomally localized kinases, Mps1 kinase was found to physically interact with mortalin (Kanai et al. 2007). Mps1 (Mps1p in yeast) was first identified in yeast as a kinase essential for duplication of spindle pole body (equivalent to centrosomes) as well as mitotic spindle checkpoint (Winey et al. 1991). It has later been found that both human and mouse Mps1 proteins localizes to centrosomes, and play a critical role in the initiation of centrosome duplication (Fisk and Winey 2001; Fisk et al. 2003). Moreover, Mps1 is phosphorylated by CDK2-Cyclin E, which results in stabilization and accumulation of Mps1 at centrosomes (Fisk and Winey 2001).

Physical association between mortalin and Mps1 results in super-activation of the Mps1 kinase activity. Moreover, mortalin acts as a phosphorylation target of Mps1: Mps1 phosphorylates mortalin on Thr62 and Ser65 (Kanai et al. 2007). The Mps1-mediated phosphorylation of mortalin has no effect on the association between mortalin and Mps1 *per se*, but it is found to be critical for super-activation of Mps1. For instance, both phospho-mimetic and phosphorylation dead mortalin mutants binds to Mps1 with similar efficiencies, but the phosphorylation-dead mutant fails to super-activate Mps1 (Kanai et al. 2007). Moreover, overexpression of wild-type mortalin as well as phospho-mimetic mutant promotes centrosome duplication and re-duplication, while overexpression of the phosphorylation-dead mutant fails to do so, indicating that Mps1-mediated phosphorylation is critical for mortalin to promote centrosome duplication and re-duplication. Thus, Mps1 and mortalin mutually depends on their activities for the regulation of centrosome duplication. Indeed, mortalin fails to promote centrosome duplication and re-duplication in cells depleted for Mps1, and likewise Mps1 fails to promote centrosome duplication and re-duplication in cells depleted for mortalin (Kanai et al. 2007). Based on these observations, one interesting scenario can be drawn for the mortalin-Mps1-associated regulation of centrosome duplication. In late G1 phase, Mps1, which is highly unstable in its nascent form, becomes stabilized and accumulates at centrosomes by the action of CDK2-cyclin E. The stabilized Mps1 then physically interacts with mortalin, and phosphorylates mortalin. Phosphorylated mortalin then super-activates Mps1 in a feedback manner. The super-activated Mps1 then acts on the centrosomal target(s), leading to initiation of centrosome duplication. The series of these events are perhaps important for the rapid initiation of centrosome duplication during a narrow window of the cell cycle, namely the G1-S phase transition. However, the drawback is, when mortalin is overexpressed or Mps1 acquires mutations that render the mortalin-independency for super-activation, induction of centrosome amplification and consequential destabilization of chromosomes.

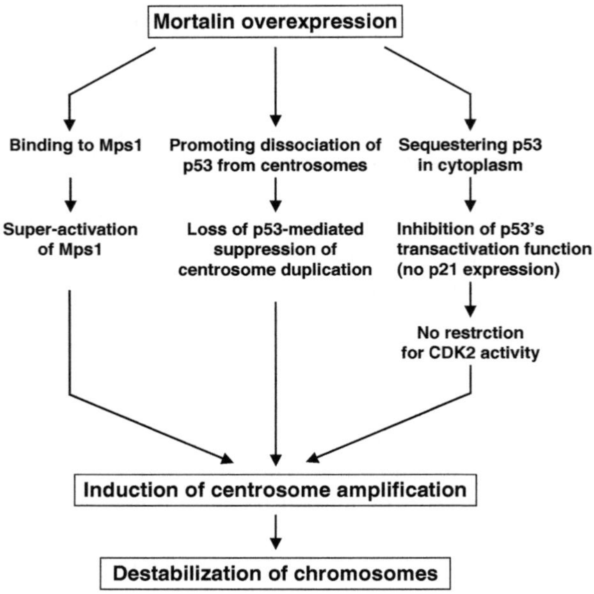

Fig. 15.3 Mortalin overexpression promotes centrosome amplification and chromosome instability through targeting multiple proteins/pathways. Overexpression of mortalin results in sequestration of p53 in the cytoplasm, preventing p53 from transactivating target genes including p21 CDK inhibitor. Without sufficient levels of p21, CDK2 activity is less controlled, allowing fortuitous activation of CDK2 and initiation of centrosome re-duplication. Mortalin overexpression also results in promotion of dissociation of p53 from centrosomes. Mortalin does so through directly binding to p53. Dissociation of p53 from centrosomes results in loss of the negative regulation of centrosome duplication by p53, favoring centrosome re-duplication. Mortalin also binds to and super-activates centrosomally localized Mps1 kinase, which is known to drive centrosome duplication. Thus, overexpression of mortalin results in continual super-activation of Mps1, which leads to over-duplication of centrosomes. Through inactivation of the p53-associated suppression of centrosome duplication and constitutive super-activation of Mps1, mortalin overexpression results in centrosome amplification, which in turn leads to chromosome instability

15.6 Summary

Mortalin has an activity to promote centrosome duplication via targeting multiple proteins/pathways, and when overexpressed, mortalin induces centrosome amplification (Fig. 15.3). Overexpressed mortalin sequesters p53 in cytoplasm, preventing p53 from transactivating the target genes including p21 CDK inhibitor. Without sufficient levels of p21, the activity of CDK2 is less controlled, leading to uncontrolled triggering of centrosome duplication. Overexpressed mortalin also blocks p53 to localize to centrosomes through physical association, depriving the p53's direct (transactivation function-independent) negative regulation of centrosome duplication at centrosomes. Mortalin also super-activates Mps1 kinase via physical association, and thus overexpression of mortalin can result in continual super-activation of Mps1, leading to uncontrolled initiation of centrosome duplication. Previous studies have implicated mortalin overexpression in carcinogenesis (Yi et al. 2008). Also, mortalin overexpression was found to correlate with poor survival (Dundas et al. 2005). Because chromosome instability is an important factor for carcinogenesis and survival rates of the patients, induction of centrosome amplification is clearly an important oncogenic activity of mortalin.

References

Blair Zajdel ME, Blair GE (1988) The intracellular distribution of the transformation-associated protein p53 in adenovirus-transformed rodent cells. Oncogene 2:579–584

Brown CR, Doxsey SJ, White E, Welch WJ (1994) Both viral (adenovirus E1B) and cellular (hsp 70, p53) components interact with centrosomes. J Cell Physiol 160:47–60

Carroll PE, Okuda M, Horn HF, Biddinger P, Stambrook PJ, Gleich LL, Li YQ, Fukasawa K (1998) Centrosome hyperamplification in human cancer: chromosome instability induced by p53 mutation and/or Mdm2 overexpression. Oncogene 18:1935–1944

Chen Z, Indjeian VB, McManus M, Wang L, Dynlacht BD (2002) CP110, a cell cycle-dependent CDK substrate, regulates centrosome duplication in human cells. Dev Cell 3:339–350

D'Assoro AB, Lingle WL, Salisbury JL (2002) Centrosome amplification and the development of cancer. Oncogene 21:6146–6153

Dundas SR, Lawrie LC, Rooney PH, Murray GI (2005) Mortalin is over-expressed by colorectal adenocarcinomas and correlates with poor survival. J Pathol 205:74–81

Fisk HA, Winey M (2001) The mouse Mps1p-like kinase regulates centrosome duplication. Cell 106:95–104

Fisk HA, Mattison CP, Winey M (2003) Human Mps1 protein kinase is required for centrosome duplication and normal mitotic progression. Proc Natl Acad Sci U S A 100:14875–14880

Fukasawa K (2005) Centrosome amplification, chromosome instability and cancer development. Cancer Lett 230:6–19

Fukasawa K (2007) Oncogenes and tumour suppressors take on centrosomes. Nat Rev Cancer 7:911–924

Ganem NJ, Godinho SA, Pellman D (2009) A mechanism linking extra centrosomes to chromosomal instability. Nature 460:278–282

Hanashiro K, Kanai M, Geng Y, Sicinski P, Fukasawa K (2008) Roles of cyclins A and E in induction of centrosome amplification in p53-compromised cells. Oncogene 27:5288–5302

Harper JW (1997) Cyclin dependent kinase inhibitors. Cancer Surv 29:91–107

Hinchcliffe HE, Li C, Thompson EA, Maller JL, Sluder G (1999) Requirement of Cdk2–cyclin E activity for repeated centrosome reproduction in Xenopus egg extracts. Science 283:851–854

Hsu LC, Kapali M, Deloia JA, Gallion HH (2005) Centrosome abnormalities in ovarian cancer. Int J Cancer 113:746–751

Jiang F, Caraway NP, Sabichi AL, Zhang HZ, Ruitrok A, Grossman HB, Gu J, Lerner SP, Lippman S, Katz RL (2003) Centrosomal abnormality is common in and a potential biomarker for bladder cancer. Int J Cancer 106:661–665

Kanai M, Ma Z, Izumi H, Kim SH, Mattison CP, Winey M, Fukasawa K (2007) Physical and functional interaction between mortalin and Mps1 kinase. Genes Cells 12:797–810

Kawamura K, Izumi H, Ma Z, Ikeda R, Moriyama M, Tanaka T, Nojima T, Levin LS, Fujikawa-Yamamoto K, Suzuki K, Fukasawa K (2004) Induction of centrosome amplification and chromosome instability in human bladder cancer cells by p53 mutation and cyclin E overexpression. Cancer Res 64:4800–4809

Koutsami MK, Tsantoulis PK, Kouloukoussa M, Apostolopoulou K, Pateras IS, Spartinou Z, Drougou A, Evangelou K, Kittas C, Bartkova J, Bartek J, Gorgoulis VG (2006) Centrosome

abnormalities are frequently observed in non-small-cell lung cancer and are associated with aneuploidy and cyclin E overexpression. J Pathol 209:512–521
Krämer A, Schweizer S, Neben K, Giesecke C, Kalla J, Katzenberger T, Benner A, Müller-Hermelink HK, Ho AD, Ott G (2003) Centrosome aberrations as a possible mechanism for chromosomal instability in non-Hodgkin's lymphoma. Leukemia 17:2207–2213
Lacey KR, Jackson PK, Stearns T (1999) Cyclin-dependent kinase control of centrosome duplication. Proc Natl Acad Sci U S A 96:2817–2822
Lingle WL, Barrett SL, Negron VC, D'Assoro AB, Boeneman K, Liu W, Whitehead CM, Reynolds C, Salisbury JL (2002) Centrosome amplification drives chromosomal instability in breast tumor development. Proc Natl Acad Sci U S A 99:1978–1983
Ma Z, Izumi H, Kanai M, Kabuyama Y, Ahn NG, Fukasawa K (2006) Mortalin controls centrosome duplication via modulating centrosomal localization of p53. Oncogene 25:5377–5390
Mazia D (1987) The chromosome cycle and the centrosome cycle in the mitotic cycle. Int Rev Cytology 100:49–92
Mikule K, Delaval B, Kaldis P, Jurcyzk A, Hergert P, Doxsey S (2007) Loss of centrosome integrity induces p38-p53-p21-dependent G1-S arrest. Nat Cell Biol 9:160–170
Morris VB, Brammall J, Noble J, Reddel R (2000) p53 localizes to the centrosomes and spindles of mitotic cells in the embryonic chick epiblast, human cell lines, and a human primary culture: an immunofluorescence study. Exp Cell Res 256:122–130
Nakajima T, Moriguchi M, Mitsumoto Y, Sekoguchi S, Nishikawa T, Takashima H, Watanabe T, Katagishi T, Kimura H, Okanoue T, Kagawa K (2004) Centrosome aberration accompanied with p53 mutation can induce genetic instability in hepatocellular carcinoma. Mod Pathol 17:722–727
Neben K, Tews B, Wrobel G, Hahn M, Kokocinski F, Giesecke C, Krause U, Ho AD, Krämer A, Lichter P (2004) Gene expression patterns in acute myeloid leukemia correlate with centrosome aberrations and numerical chromosome changes. Oncogene 23:2379–2384
Nevins JR (1992) E2F: a link between the Rb tumor suppressor protein and viral oncoproteins. Science 258:424–429
Okuda M, Horn HF, Tarapore P, Tokuyama Y, Smulian AG, Chan PK, Knudsen ES, Hofmann IA, Snyder JD, Bove KE, Fukasawa K (2000) Nucleophosmin/B23 is a target of CDK2–cyclin E in centrosome duplication. Cell 103:127–140
Pihan GA, Purohit A, Wallace J, Knecht H, Woda B, Quesenberry P, Doxsey SJ (1998) Centrosome defects and genetic instability in malignant tumors. Cancer Res 58:3974–3985
Pihan GA, Purohit A, Wallace J, Malhotra R, Liotta L, Doxsey SJ (2001) Centrosome defects can account for cellular and genetic changes that characterize prostate cancer progression. Cancer Res 61:2212–2219
Sancar A, Lindsey-Boltz LA, Unsal-Kacmaz K, Linn S (2004) Molecular mechanisms of mammalian DNA repair and the DNA damage checkpoints. Annu Rev Biochem 73:39–85
Sato N, Mizumoto K, Nakamura M, Nakamura K, Kusumoto M, Niiyama H, Ogawa T, Tanaka M (1999) Centrosome abnormalities in pancreatic ductal carcinoma. Clin Cancer Res 5:963–970
Sherr CJ (2006) Divorcing ARF and p53: an unsettled case. Nat Rev Cancer 6:663–673
Shinmura K, Bennett RA, Tarapore P, Fukasawa K (2007) Direct evidence for the role of centrosomally localized p53 in the regulation of centrosome duplication. Oncogene 26:2939–2944
Tarapore P, Horn HF, Tokuyama Y, Fukasawa K (2001a) Direct regulation of the centrosome duplication cycle by the p53-p21Waf1/Cip1 pathway. Oncogene 20:3173–3184
Tarapore P, Tokuyama Y, Horn HF, Fukasawa K (2001b) Difference in the centrosome duplication regulatory activity among p53 'hot spot' mutants: potential role of Ser 315 phosphorylation-dependent centrosome binding of p53. Oncogene 20:6851–6863
Tarapore P, Okuda M, Fukasawa K (2002) A mammalian in vitro centriole duplication system: evidence for involvement of CDK2/cyclin E and Nucleophosmin/B23 in centrosome duplication. Cell Cycle 1:75–81
Taylor WR, Stark GR (2001) Regulation of the G2/M transition by p53. Oncogene 20:1803–1815

Tritarelli A, Oricchio E, Ciciarello M, Mangiacasale R, Palena A, Lavia P, Soddu S, Cundari E (2004) p53 localization at centrosomes during mitosis and postmitotic checkpoint are ATM-dependent and require serine 15 phosphorylation. Mol Biol Cell 15:3751–3757

Wadhwa R, Takano S, Robert M, Yoshida A, Nomura H, Reddel RR, Mitsui Y, Kaul SC (1998) Inactivation of tumor suppressor p53 by mot-2, a hsp70 family member. J Biol Chem 273:29586–29591

Weber RG, Bridger JM, Benner A, Weisenberger D, Ehemann V, Reifenberger G, Lichter P (1998) Centrosome amplification as a possible mechanism for numerical chromosome aberrations in cerebral primitive neuroectodermal tumors with TP53 mutations. Cytogenet. Cell Genet 83:266–269

Winey M, Goetsch L, Baum P, Byers B (1991) MPS1 and MPS2: novel yeast genes defining distinct steps of spindle pole body duplication. J Cell Biol 114:745–754

Yi X, Luk JM, Lee NP, Peng J, Leng X, Guan XY, Lau GK, Beretta L, Fan ST (2008) Association of mortalin (HSPA9) with liver cancer metastasis and prediction for early tumor recurrence. Mol Cell Proteomics 7:315–325

Chapter 16
Mortalin Expression in Normal and Neoplastic Tissues

Angheliki Nomikos, Sinclair R. Dundas and Graeme I. Murray

Abstract The purpose of this chapter is to discuss the distribution of mortalin in different normal tissues and tumour types highlighting its role in neoplasia. Correlating mortalin expression with tumour grade and stage and its influence on cancer metastasis will also be reviewed along with the emerging role mortalin has as a clinically relevant therapeutic target and prognostic marker.

Keywords Mortalin · Neoplasia · Metastases · Prognosis · Targeted therapy

16.1 Introduction

Mortalin, a member of the hsp70 family of proteins, is an essential protein that performs various functions related to cell proliferation, functional maintenance and stress response as well as being implicated in cellular immortalisation and tumourigenesis (Wadhwa et al. 1991, 1993a; Kaul et al. 1998, 2002). Mortalin is localised at different subcellular sites, predominantly in mitochondria, but also endoplasmic reticulum, cytoplasmic vesicles and the cytosol (Wadhwa et al. 1993b; Ran et al. 2000).

In murine cells, mortalin is a 66-kDa protein that exists in two forms, designated mot-1 and mot-2, that are coded for by separate genes (Kaul et al. 2000a; Wadhwa et al. 1996). In contrast, in human cells mortalin is present as a single 74 kDa protein. Only one transcript for human mortalin (hmot-2) has been identified and its gene is located on chromosome 5q31.1 (Kaul et al. 1995, 1998).

In terms of its involvement in cellular transformation, the intracellular distribution of mortalin, rather than its absence or presence, is important in determining the mortality or immortality of a cell. Whereas both murine and human normal cells have widely distributed protein in their cytoplasm (pancytosolic), immortal murine cells are seen to have a perinuclear concentration of the protein and human transformed cells show four types of non-pancytoplasmic staining patterns: granular-juxtanuclear cap, granular gradient, granular-juxtanuclear arch and fibrous-perinuclear (Wadhwa

S. R. Dundas (✉) · A. Nomikos · G. I. Murray
Department of Pathology, University Medical Buildings, Foresterhill, Aberdeen, AB25 2ZD, Scotland, UK
e-mail: s.r.dundas@abdn.ac.uk

S. C. Kaul, R. Wadhwa (eds.), *Mortalin Biology: Life, Stress and Death*,
DOI 10.1007/978-94-007-3027-4_16, © Springer Science+Business Media B.V. 2012

et al. 1993b, 1995). Indeed, transfecting expression plasmids coding for mot-1 or mot-2 into murine fibroblasts showed that mot-1 protein localised to the cytoplasm induced cellular senescence (Wadhwa et al. 1993b), whereas the mot-2 form demonstrated a restricted perinuclear distribution inducing malignant transformation (Kaul et al. 1998). It has also been shown that induction of senescence in transformed cells was accompanied by reversion of subcellular distribution of mortalin from non-pancytoplasmic to the pancytoplasmic type and therefore growth arrest (Wadhwa et al. 1993c).

In contrast to the murine system, normal and transformed human cells showing differential staining patterns of mortalin possess identical sequences (Kaul et al. 1998). This suggests that the differential mortalin distribution observed in human cell lines is modulated by post-translational modification or binding to other factors and does not originate from distinct cDNAs as in murine cells (Kaul et al. 1998).

In support of its multiple subcellular sites, mortalin has multiple binding partners, including tumour suppression protein p53 (Kaul et al. 2002). A mechanism for the transforming capacity of mortalin is suggested by its interaction with wild type p53 in the cytoplasm of transformed cells (Wadhwa et al. 1998, 1999, 2002a; Kaul et al. 2000b). This results in cytoplasmic retention and transcriptional inactivation of the tumour suppressor activity of p53 (Kaul et al. 2000b), hence protecting cells from ATP depletion and energy deprivation associated with cell death by preventing the rapid rise in mitochondrial reactive oxygen species (Liu et al. 2005). Furthermore, mortalin influences the Ras-Raf-MAPK pathway by modulating the prenylation and stability of Ras, the function of which is to transduce signals from the extracellular milieu to the cell nucleus where specific genes are activated for cell growth, division and differentiation (Wadhwa et al. 2003). Thus, signalling through the Ras-Raf-MAPK regulates a variety of cellular functions that are important in tumourigenesis (Stacey 2003; Cary et al. 1999; Kranenburgm et al. 2004).

16.2 Mortalin Expression in Normal Tissues

Viable mammalian cells that lack the expression of mortalin are neither known nor have been so far created by antisense or knockout technologies (Kaul et al. 2002). Indeed, mortalin is expressed in all cell types and tissues examined to date (Wadhwa et al. 2002b) although levels of expression differ considerably. An obvious implication is that mortalin is vital for essential cellular functions. As a molecular chaperone, mortalin is actively involved in the stabilisation, import of nuclear gene products and refolding of precursor proteins into mitochondria (Deocaris et al. 2006). This role is fundamental for mitochondrial structure, function and biogenesis. In addition, mortalin protein has been localised in extra-mitochondrial sites assuming, in conjunction with other multiple roles, a function in trafficking non-mitochondrial proteins (Wadhwa et al. 2002b). The mechanism of translocation of mortalin in other organelles remains unknown so far.

16.3 Anti-Proliferative Function

Histochemical analysis of stained sections coordinated with RNA *in situ* hybridisation results and Northern blot analysis results (Kaul et al. 1997) for mortalin in normal rat tissues revealed that quiescent tissues and non-cycling cells displayed higher levels of expression compared to cells with division potential. Brain, heart and skeletal muscle are associated with high levels of expression. Intermediate levels of detectable mortalin were found in lung, liver and kidney. The lowest levels of expression were noted in actively proliferating tissues such as testis and spleen. Levels of mortalin expression are also cell type-specific. For example, squamous cells in the non-dividing upper layers of the epidermis express significantly higher levels than proliferating basal layer keratinocytes. In brain, neurons and nerve fibres express more mortalin than glial cells, whilst in the heart cardiomyocytes are associated with higher mortalin levels than non-myocytes. Ovarian tissue also displays differential expression with interstitial secretory cells than in germinating ovarian follicles. Furthermore, in the liver, the medium level of expression related to the endothelial cells rather than the hepatocytes.

Mortalin is also widely expressed in various human tissues. Using a Northern blot approach, Xie et al. (2000), demonstrated high levels of mortalin mRNA in heart with intermediate levels of expression in skeletal muscle, kidney, liver and brain although the specific histological cell types were not determined. Only low levels are detectable in lung, placenta and pancreas.

Tissue-specific mortalin expression supports the hypothesis that it plays a central anti-proliferative role in normal tissues and it is interesting to speculate that this may be due to its effect in modulating apoptosis, cell cycle regulation and cellular senescence. During p53-induced apoptosis, mortalin-p53 complexes are detectable in mitochondria but the precise function and mechanism mortalin performs in apoptotic pathways requires further investigations (Marchenko et al. 2000).

16.4 Muscle Activity, Mitochondrial Activity and Biogenesis

Mortalin levels correlate with skeletal muscle activity, mitochondrial activity and biogenesis. Contractile activity is a physiological stress which stimulates mitochondrial biogenesis in skeletal muscle and this is achieved by modifying extra- and intra-mitochondrial micro-environments, in particular accelerating the import of precursor proteins into the organelle matrix (Takahashi et al. 1998). Contractile activity increases the levels of the intra-mitochondrial mortalin (Takahashi et al. 1998).

In another study (Ornatsky et al. 1995), under steady state conditions, the tissue content of mortalin is proportional to muscle oxidative capacity and mitochondrial enzyme activity. However, chronic stimulation results in an apparent overproduction of mortalin compared to typical mitochondrial enzyme activity. Ornatsky et al. (1995) hypothesised that over-induction of mortalin may maintain mitochondrial biogenesis and ensure that the rate of protein import into growing mitochondria of stimulated muscle is not a rate limiting process in organelle synthesis. These studies also demonstrated that during chronic contractile activity mortalin protein levels increase but that there is no commensurate increase in mortalin-encoding mRNA. This presumably reflects post-transcriptional regulation, either at the level of translation or perhaps increased stability of the mortalin protein molecule itself.

16.5 Mortalin and Neoplasia

As discussed earlier, mortalin performs essential functions within the mitochondrial matrix acting as a chaperone and as a key component in the trafficking of mitochondrial-targeted proteins. Alterations in other mortalin-associated activities, such as inactivation of wild-type tumour suppressor protein p53, modulation of the Ras-Raf-MAPK pathway, as well as intracellular trafficking function, seem to confer proliferative and tumourigenic properties (Wadhwa et al. 2006). A pivotal finding indicating the neoplastic capacity of mortalin was its up-regulation in cells that had undergone spontaneous malignant transformation *in vivo*. Subclones of immortalised human embryonic fibroblasts with spontaneously increased mortalin expression levels display anchorage-independent growth and hence acquire malignant properties with the ability to form tumours in nude mice (Wadhwa et al. 2006). In the same paper, other work established that over-expression of mortalin increased the aggressive behaviour of cancer cells *in vivo*. Up-regulation of mortalin in the human MCF7 breast tumour-derived cell increased their malignant growth potential. Although MCF7 cells do not form tumours in nude mice, mortalin over-expressing derivatives injected into nude mice developed palpable tumours of variable size. This finding has important implications in understanding tumour biology and will be highlighted below.

16.6 Mortalin and Dysplasia

The currently accepted concept of carcinogenesis, particularly in relation to colonic carcinogenesis, states that normal cells progressively accumulate multiple genetic and epigenetic abnormalities leading to hyperplasia, dysplasia and ultimately acquisition of invasive and metastatic phenotypes. The study published by Dundas et al. (2005), remains one of the few to demonstrate that mortalin is over-expressed even at morphologically early stages of neoplasia. Small 4 mm low-grade tubular adenomas of colon over-express mortalin indicating that aberrations in mortalin levels may well be implicated in the earliest stages of neoplasia.

16.7 Mortalin Expression in Cancers

Mortalin levels are increased in numerous cancerous tissues and tumour-derived cell lines further substantiating the emerging principle that mortalin over-expression is fundamentally associated with malignancy.

More than fifty malignant human cells lines were studied in ground-breaking immunofluorescence experiments comparing the intra-cellular distribution of mortalin in normal and transformed cells (Wadhwa et al. 1995; Pereira-Smith and Smith 1988). All neoplastic cells demonstrated differential non-pancytoplasmic mortalin staining patterns. These studies were extended further by examining the level of mortalin expression in a large variety of tumour-derived tissue extracts and their patient-matched normal tissues using total RNA dot blots. Sixteen of twenty-four tumour types showed higher levels of mortalin expression than normal controls from the same individual. This included brain, duodenum, colon, liver, gallbladder, pancreas, parotid, kidney, ureter, bladder, prostate, breast, fallopian tube, uterus, thyroid, thymus neoplasms. Eight tumour tissues (lung, oesophagus, stomach, rectum, ovary, adrenal gland, testis and lymphoma) did not show any significant change in mortalin expression level in this initial study (Wadhwa et al. 2006).

Tumour-derived cells lines showing elevated levels of mortalin expression compared to counterpart normal primary cells encompass epithelial, mesenchymal and neural crest-derived malignancies. These include ovarian adenocarcinoma, glioblastoma, lung carcinoma, breast adenocarcinoma, osteosarcoma, colon adenocarcinoma, renal cell carcinoma, cervical carcinoma, squamous cell carcinoma of lung, prostatic carcinoma, malignant melanoma and fibrosarcoma (Wadhwa et al. 2006). This indicates that mortalin function is implicated in molecular aberrations responsible for the broad spectrum of neoplastic diseases afflicting humans.

Several studies demonstrate that mortalin up-regulation in breast, brain and colon tumours is a consistent abnormality (Dundas et al. 2005; Takano et al. 1997; Bini et al. 1997; Wadhwa et al. 2006). Compared to normal breast tissue mortalin expression is up-regulated in human breast ductal carcinoma (Bini et al. 1997). Additionally, Blanc et al. (2005) reported a two-fold up-regulation of mortalin protein in hepatocellular carcinoma arising in patients with chronic viral hepatitis C. Furthermore, northern blot evaluation of RNA derived from a panel of 13 leukaemia cell lines showed that mortalin was expressed in all cell lineages, although the level of expression varied considerably (Xie et al. 2000).

Renal carcinomas are associated least frequently with mortalin up-regulation. The fact that tumour-derived tissues, including renal cell carcinoma, up-regulate mortalin whilst others do not may relate to clinical stage or size of the primary cancers, variables which were unknown or, at least, not documented in many of the studies. Aside from these issues, the reason for differences in mortalin expression in various cancers remains unclear but it is not simply due to variations in cellular growth rate or stage of cell cycle progression. In fact there is no direct association between up-regulation of mortalin and cell growth rate in transformed cells (Wadhwa et al. 2006).

16.8 Mortalin Expression as a Prognostic Marker

Analysis of colorectal cancers indicates that as well as a role in the molecular pathways of tumourigenesis, mortalin expression directly correlates with clinically important prognostic parameters. Utilising subtraction proteomics Dundas et al. (2005) demonstrated that mortalin is consistently differentially over-expressed in colonic adenocarcinomas and is a key molecule in colonic carcinogenesis and tumour behaviour. Subsequent immunocytochemical analysis of 264 colorectal adenocarcinomas in a tissue microarray confirmed adenocarcinomas over-expressed mortalin compared with patient-matched normal colorectal mucosa. Additionally, not only did colonic adenocarcinomas over-express mortalin, but levels of expression correlated directly with pathological stage of disease. Cancers showing moderate, strong or very strong immunopositivity, scored in a blinded fashion, stratified into Duke stage A, B or C respectively. This correlation is independent of tumour differentiation, site, sex and age. Furthermore, multivariate analysis demonstrated that mortalin expression in colorectal adenocarcinomas correlates significantly with overall patient survival with strongly positive tumours being associated with poorer outcome irrespective of other standard variables.

Other studies substantiate the potential utility of mortalin as a clinically useful marker. Hepatocellular carcinoma represents a tumour type with aggressive biological and clinical behaviour. With the aim of finding biomarkers of early recurrence, Yi et al. (2008) analysed hepatitis B-related hepatocellular carcinomas and showed that mortalin expression levels differed between neoplastic and normal liver and, more specifically, detected early recurrent hepatocellular carcinoma. Mortalin protein levels effectively discriminated between recurrence-free patients and those afflicted by early recurrence. For early recurrences (ER subgroup), defined as tumour appearing clinically between four and twelve months after curative surgery, mortalin emerged with a high detection sensitivity of 90.0% on proteomic profiling analysis of a total of 36 potential biomarkers that were significantly up-regulated in the ER subgroup. Again mortalin over-expression was found to correlate with important clinical parameters and, for hepatocellular carcinomas, closely associates with advanced tumour stage and venous infiltration, two critical prognostic factors indicative of poor outcomes and risk of recurrence in this disease type. This study showed no apparent correlation with other standard prognostic variables for this disease, including serum AFP, tumour size and Edmonson grade and, additionally, mortalin level appeared to be a superior indicator of disease risk than these clinicopathological parameters.

In normal brain mortalin is confined to neurons and is undetectable in astrocytes. However, Takano et al. (1997) demonstrated mortalin expression in the malignant brain tumour astrocytoma and that this correlated with increasing tumour grade. An increasing number of tumour cells are immunocytochemically mortalin-positive in astrocytoma as tumours progress from low to high grade glioblastoma and also in intra-cerebral metastases. This study also demonstrated that other brain lesions such as meningiomas stained mortalin-positive more frequently than pituitary adenomas and neuromas.

The proliferative potential of brain tumours is an important prognostic feature. In diagnostic practice this is determined immunocytochemically using the proliferative marker MIB-1 recognising Ki-67 protein. Interestingly, mortalin expression mirrors that of increasing Ki-67 protein as astrocytomas become a more aggressive phenotype. This correlation between mortalin and MIB-1 positivity is statistically significant.

All of these studies emphasise the critical role mortalin plays in malignant transformation and that the function of this molecule is reflected in biological and clinically relevant tumour behaviour.

16.9 Mortalin and Metastasis

The development of secondary metastatic tumours at a distance from the primary site is a defining feature of malignancy. The fact that mortalin function is responsible at some level for an aggressive tumour phenotype with metastatic ability is proved by several studies. Yi et al. (2008) demonstrated that metastatic hepatocellular carcinoma cell lines exhibited higher levels of mortalin protein compared to parental clones with intrinsically lower metastatic potential. Brain metastasis from four different primary sites including lung, renal, bladder and thyroid carcinomas are highly positive compared to more indolent neoplasms (Takano et al. 1997). The mechanism by which mortalin modulates metastasis is undoubtedly complex and may be a consequence of extra-mitochondrial activity. Shin et al. (2003) found by comparative proteome profiling an overabundance of mortalin detectable on the cell surface of various cancer cells and reasonably concluded that some of the enhancing effect on tumourigenesis by mortalin over-expression may be attributed to its extra-mitochondrial activity. This hypothesis is further substantiated by the fact that cell membrane bound mortalin confers altered biophysical properties, which may contribute to the metastatic behaviour of neoplastic cells, by virtue of changes in membrane fluidity and deformability (Wadhwa et al. 2006). Furthermore, an enhanced chaperone loading may improve cell survival during the stresses of the metastatic process, which include loss of cellular adhesions, cell migration, intravasation, extravasation and survival and proliferation in a distant site.

16.10 Conclusion

It is clear that mortalin, with its multiple intra- and extra-mitochondrial functions and numerous binding partners form a complex pathway playing a critical role in normal cellular function. Mortalin is detectable in many normal cell types and tissues with higher levels of expression in functional non-dividing cells than in dividing populations. This points towards a central role in modulating both the cell cycle and apoptosis. However, when this function is disrupted mortalin is a key component in the earliest stages of neoplasia and both the development and progression of cancers. Although the precise factors responsible for mortalin over-expression during tumourigenesis remain to be fully elucidated, its proven correlation with tumour grade, stage and aggressiveness, including metastatic potential, indicates that mortalin may be a candidate biomarker for predicting clinically relevant tumour behaviour.

References

Bini L, Magi B, Marzocchi B, Arcuri F, Tripodi S, Cintorino M, Sanchez JC, Frutiger S, Hughes G, Pallini V, Hochstrasser DF, Tosi P (1997) Protein expression profiles in human breast ductal carcinoma and histologically normal tissue. Electrophoresis 18:2832–2841

Blanc JF, Lalanne C, Plomion C, Schmitter JM, Bathany K, Gion JM, Bioulac-Sage P, Balabaud C, Bonneu M, Rosenbaum J (2005) Proteomic analysis of differentially expressed proteins in hepatocellular carcinoma developed in patients with chronic viral hepatitis C. Proteomics 5:3778–3789

Cary LA, Han DC, Guan JL (1999) Integrin-mediated signal transduction pathways. Histol Histopathol 14:1001–1009

Deocaris CC, Kaul SC, Wadhwa R (2006) On the brotherhood of the mitochondrial chaperones mortalin and heat shock protein 60. Cell Stress Chaperones 11:116–128

Dundas SR, Lawrie LC, Rooney PH, Murray GI (2005) Mortalin is over-expressed by colorectal adenocarcinomas and correlates with poor survival. J Pathol 205:74–81

Kaul SC, Wadhwa R, Matsuda Y, Hensler PJ, Pereira-Smith OM, Komatsu Y, Mitsui Y (1995) Mouse and human chromosomal assignments of mortalin, a novel member of the murine hsp70 family of proteins. FEBS Lett 361:269–272

Kaul S, Matsui M, Takano S, Sugihara T, Mitsui Y, Wadhwa R (1997) Expression analysis of mortalin, a unique member of the HSP70 family of proteins, in rat tissues. Exp Cell Res 232:56–63

Kaul SC, Duncan EL, Englezou A, Takano S, Reddel RR, Mitsui Y, Wadhwa R (1998) Malignant transformation of NIH3T3 cells by overexpression of mot-2 protein. Oncogene 17:907–911

Kaul SC, Duncan E, Sugihara T, Reddel RR, Mitsui Y, Wadhwa R (2000a) Structurally and functionally distinct mouse hsp70 family members Mot-1 and Mot-2 proteins are encoded by two alleles. DNA Res 7:229–231

Kaul SC, Reddel RR, Sugihara T, Mitsui Y, Wadhwa R (2000b) Inactivation of p53 and life span extension of human diploid fibroblasts by mot-2. FEBS Lett 474:159–164

Kaul SC, Taira K, Pereira-Smith OM, Wadhwa R (2002) Mortalin: present and prospective. Exp Gerontol 37:1157–1164

Kranenburgm O, Gebbink MF, Voest EE (2004) Stimulation of angiogenesis by Ras proteins. Biochim Biophys Acta 1654:23–37

Liu Y, Liu W, Song XD, Zuo J (2005) Effect of GRP75/mthsp70/PBP74/mortalin overexpression on intracellular ATP level, mitochondrial membrane potential and ROS accumulation following glucose deprivation in PC12 cells. Mol Cell Biochem 268:45–51

Marchenko ND, Zaika A, Moll UM (2000) Death signal-induced localization of p53 protein to mitochondria. A potential role in apoptotic signalling. J Biol Chem 275:16202–16212

Pereira-Smith OM, Smith JR (1988) Genetic analysis of indefinite division in human cells-Identification of 4 complementation groups. Proc Natl Acad Sci U S A 85:6042–6046

Ornatsky OI, Connor MK, Hood DA (1995) Expression of stress proteins and mitochondrial chaperonins in chronically stimulated skeletal muscle. Biochem J 311:119–123

Ran Q, Wadhwa R, Kawai R, Kaul SC, Sifers RN, Bick RJ, Smith JR, Pereira-Smith OM (2000) Extramitochondrial localization of mortalin/mthsp70/PBP74/GRP75. Biochem Biophys Res Commun 275:174–179

Shin BK, Wang H, Yim AM, Le Naour F, Brichory F, Jang JH, Zhao R, Puravs E, Tra J, Michael CW, Misek DE, Hanash SM (2003) Global profiling of the cell surface proteome of cancer cells uncovers an abundance of proteins with chaperone function. J Biol Chem 278:7607–7616

Stacey DW (2003) Cyclin D1 serves as a cell cycle regulatory switch in actively proliferating cells. Curr Opin Cell Biol 15:158–163

Takahashi M, Chesley A, Freyssenet D, Hood DA (1998) Contractile activity-induced adaptations in the mitochondrial protein import system. Am J Physiol 274:C1380–C1387

Takano S, Wadhwa R, Toshii Y, Nose T, Kaul S, Mitsui Y (1997) Elevated levels of mortalin expression in human brain tumours. Exp Cell Res 237:38–45

Wadhwa R, Kaul SC, Ikawa Y, Sugimoto Y (1991) Protein markers for cellular mortality and immortality. Mutat Res 256:243–254

Wadhwa R, Kaul SC, Ikawa Y, Sugimoto Y (1993a) Identification of a novel member of mouse hsp70 family: its association with cellular mortal phenotypes. J Biol Chem 268:6615–6621

Wadhwa R, Kaul SC, Mitsui Y, Sugimoto Y (1993b) Differential subcellular distribution of mortalin in mortal and immortal mouse and human fibroblasts. Exp Cell Res 207:442–448

Wadhwa R, Kaul SC, Sugimoto Y, Mitsui Y (1993c) Induction of cellular senescence by transfection of cytosolic mortalin cDNA in NIH 3T3 cells. J Biol Chem 268:22239–22242

Wadhwa R, Pereira-Smith OM, Reddel RR, Sugimoto Y, Mitsui Y, Kaul SC (1995) Correlation between complementation group for immortality and the cellular distribution of mortalin. Exp Cell Res 216:101–106

Wadhwa R, Akiyama S, Sugihara T, Reddel RR, Mitsui Y, Kaul SC (1996) Genetic differences between the pancytosolic and perinuclear forms of murine mortalin. Exp Cell Res 226:381–386

Wadhwa R, Takano S, Robert M, Yoshida A, Nomura H, Reddel RR, Mitsui Y, Kaul SC (1998) Inactivation of tumour suppressor p53 by mot-2, a hsp70 family member. J Biol Chem 273:29586–29591

Wadhwa R, Takano S, Mitsui Y, Kaul SC (1999) NIH 3T3 cells malignantly transformed by mot-2 show inactivation and cytoplasmic sequestration of the p53 protein. Cell Res 9:261–269

Wadhwa R, Yaguchi T, Hasan MK, Mitsui Y, Reddel RR, Kaul SC (2002a) Hsp70 family member, mot-2/mthsp70/GRP75, binds to the cytoplasmic sequestration domain of the p53 protein. Exp Cell Res 274:246–253

Wadhwa R, Taira K, Kaul SC (2002b) An Hsp70 family chaperone, mortalin/mthsp70/PBP74/Grp75: what, when, and where? Cell Stress Chaperones 7:309–316

Wadhwa R, Yaguchi T, Hasan MK, Taira K, Kaul SC (2003) Mortalin-MDP (mevalonate pyrophosphate decarboxylase) interactions and their role in control of cellular proliferation. Biochem Biophys Res Commun 302:735–742

Wadhwa R, Takano S, Kaur K, Deocaris CC, Pereira-Smith OM, Reddel RR, Kaul SC (2006) Upregulation of mortalin/mthsp70/Grp75 contributes to human carcinogenesis. Int J Cancer 118:2973–2980

Xie H, Hu B, Chyna B, Horrigan SK, Westbrook CA (2000) Human mortalin (HSPA9): a candidate for the myeloid leukemia tumor suppressor gene on 5q31. Leukemia 14:2128–2134

Yi X, Luk JM, Lee NP, Peng J, Leng X, Guan XY, Lau GK, Beretta L, Fan ST (2008) Association of mortalin (HSPA9) with liver cancer metastasis and prediction for early tumour recurrence. Mol Cell Proteomics 7:315–325

Chapter 17
Mortalin-p53 Interaction as a Target for Liver Cancer Therapy

Wen-Jing Lu, Nikki P. Lee, Renu Wadhwa and John M. Luk

Abstract Mortalin is a member of Hsp70 family that exhibits a tight correlation with carcinogenesis. It has been shown to inactivate tumor suppressor protein p53. The present chapter reviews the history and presents information of mortalin, p53 and how their interaction contributes to carcinogenesis. It is demonstrated that mortalin-p53 interaction that is a cancer-driving event depends on the stress state of cells. The small molecules and short peptides that were able to block this interaction could induce apoptosis in cancer cells that harbour mutant p53 suggesting that mortalin-p53 is a good target for cancer therapy. Understanding of the molecular regulation of mortalin-p53 interaction and their role in highly malignant cells will help designing better strategies for cancer treatment.

Keywords Hepatocellular carcinoma · Targeted therapy · Mortalin-p53 interaction · Cellular stress

17.1 Mortalin and p53

HSP family is composed of at least 40 members in humans. Based on their molecular weight, they are grouped into 6 major subfamilies named as HSP100, HSP90, HSP70, HSP60, HSP40 and small HSPs (Powers and Workman 2007). HSPs of one family share similar domain structure, however may have unique pattern of expression and cellular localization (Lindquist and Craig 1988). HSPs are known for their ability in response to stress; they are expressed at low steady-state levels under normal conditions and are induced under stress conditions like high temperature,

J. M. Luk (✉)
Department of Oncology, Roche R&D Center (China) Ltd., 720 Cai Lun Road,
Shanghai 201203, China
e-mail: john.luk@roche.com

W.-J. Lu · N. P. Lee
Department of Surgery, Li Ka Shing Faculty of Medicine, The University of Hong Kong,
Hong Kong, China

R. Wadhwa
National Institute of Advanced Industrial Science and Technology (AIST),
Higashi, Tsukuba Ibaraki 305-8562, Japan

S. C. Kaul, R. Wadhwa (eds.), *Mortalin Biology: Life, Stress and Death*,
DOI 10.1007/978-94-007-3027-4_17, © Springer Science+Business Media B.V. 2012

chemical and physical stress (Sherman and Multhoff 2007). This induced expression of HSPs in response to various stresses leads to an augmentation of the biological functions for sustaining cell survival. They also bear housekeeping functions in maintaining the protein structure, gene transcription, signal transduction and immunity (Helmbrecht et al. 2000) suggesting their multi-functional characteristics. Among all functions, HSPs are known for their distinct ability in chaperoning protein folding and maturation. However their functionality goes well beyond their chaperoning powers. This chapter focuses mainly on constitutively expressed HSP70 also known as mortalin/mthsp70/PBP75/Grp75 and its role in inactivation of tumor suppressor protein p53.

p53 is a tumor suppressor gene that codes for a 53-kDa protein, also called guardian of the genome. It is a multifunctional protein that regulates genomic stability, cell cycle, stress, cell survival and death pathways. It is widely established that in a large variety of cancers, p53 is either lost or mutated or is functionally inactivated by a wide variety of mechanisms of which its inactivation by its binding partners plays a significant role. Lately, the pro-apoptotic function of p53 was appreciated in a number of studies such as re-introduction of p53 into p53-deficient myeloid leukemia cells induced apoptosis (Arakawa 2005; Vogelstein et al. 2000; Vousden and Lu 2002; Yonish-Rouach et al. 1991). It has been shown that pro-apoptotic function of p53 is mediated by transcription–dependent and transcription–independent pathways. For transcription-dependent pathway, p53 has been shown to induce transactivation of a panel of pro-apoptotic genes of several death pathways, including those related to death receptors, mitochondria and endoplasmic reticulums (Bourdon et al. 2002). For instance, p53 induces the expression of death receptors like Fas and DR5 and is involved in tumor necrosis factor-related apoptosis-inducing ligand (TRAIL)-induced apoptosis (Takimoto and El-Deiry 2000). In intrinsic apoptotic pathway, p53 promotes the transcription of pro-apoptotic molecules of Bcl-2 family, including Bax (Miyashita et al. 1994), Bid (Sax et al. 2002), Puma (Nakano and Vousden 2001) and Noxa (Oda et al. 2000). On the contrary, p53 represses the transcription of anti-apoptotic molecule like Bcl-xL (Sugars et al. 2001) and survivin (Hoffman et al. 2002). Therefore, p53 acts as a regulator for transcription-dependent apoptosis by modulating the expression of pro-apoptotic genes and anti-apoptotic genes. For transcription-independent pathway, p53 exerts apoptotic function independent of its transcription-related activities. It was demonstrated that the cytoplasmic p53 lacking the DNA-binding domain and nuclear localization signal could induce apoptosis by following two ways: (1) p53 could directly activate Bax in the absence of other proteins evidenced by increased mitochondrial membrane permeabilizing activity measured using lipid vesicle system model; (2) p53 could bind to anti-apoptotic proteins Bcl-2 and Bcl-xL (Petros et al. 2004), and release the Bak and BH-3 only domain proteins (i.e. Bid and Bim) from these apoptotic inhibitor proteins (Chipuk et al. 2004). Furthermore, Bid and Bim act on induction of the oligomerization and conformational changes of Bax (Chipuk et al. 2004).

The interaction between HSPs and p53 was discovered more than 20 years ago (Pinhasi-Kimhi et al. 1986). Many HSPs like Hsp40, Hsp70 and Hsp90 are known to interact with p53 and this interaction is more obvious with mutant p53 (Selkirk et al.

1994; Sugito et al. 1995). This type of interaction can be competitive, cooperative and versatile. For instance, the presence of wild-type p53-Hsp90 protein complex can inhibit the interaction between Hsp40 and Hsc70 with wild-type p53 (King et al. 2001). In another situation, the presence of Hsc70 and Hsp40 is needed for the formation of Hsp90 and mutant p53 (R175H) complex (King et al. 2001). One functional implication for this type of interaction is based on the ability of HSPs in binding to p53, which in turns controls the cellular location of p53. Hsp70 and Hsp90 bind to p53 and sequestrate p53 in the cytoplasm, thereby preventing the translocation of p53 into the nucleus and blocking p53-mediated transcription (Muller et al. 2004). If such binding is abrogated, p53 can be translocated into the nucleus (Selkirk et al. 1994). Other studies have shown that the treatment of mutant p53-harboring cells with antibiotics geldanamycin (an Hsp90-binding agent) released mutant p53 from Hsp90 and induced ubiquitination and nuclear translocation of the mutant p53 (Whitesell et al. 1997). The underlying mechanism of p53 sequestration in the cytoplasm is a subject of current research.

17.2 Mortalin-p53 Interaction in Human Cancers

Mortalin and p53 interaction was first detected in human cancer cells by co-immunolocalization and co-immunoprecipitation of the two proteins; normal cells lacked this interaction (Kaul et al. 1995; Wadhwa et al. 1998, 2000). Functional significance of these interactions was resolved by wild-type p53 dependent reporter assays in which mortalin was found to inactivate p53 function. It was then shown that p53 is sequestered in the cytoplasm by its interaction with mortalin. However, in contrast to the binding of p53 to other proteins by its N-terminal region that masks its nuclear translocation signal (Liu et al. 2007; Nikolaev et al. 2003; Zhao and Liao 2003), its binding to mortalin involves its C-terminal domain (312–352 amino acid residues) (Kaul et al. 1995). Tumor cells with cytoplasmic accumulation of p53 have impaired G1 checkpoint (Moll et al. 1996) and are resistant to MDM-2-mediated degradation (Zaika et al. 2000). Patients with cytoplasmic accumulation of p53 in tumors usually correlate with poor prognosis (Bosari et al. 1995). The involvement of mortalin in cancers was first underscored by its upregulation in cancer cell lines. Furthermore, transduction of non-tumorigenic MCF7 cells with mortalin-expressing retrovirus induced their tumor formation in nude mice (Wadhwa et al. 2006) suggesting that mortalin overexpression promotes carcinogenesis. Ma et al. (2006) by using comparative mass spectrometric analysis of the duplicated and unduplicated centrosomes demonstrated that mortalin-p53 interaction also caused deregulation of centrosome duplication (Ma et al. 2006), a hallmark of cancer cells. Various human malignancies including colon, breast, gastric and liver carcinoma (Dundas et al. 2005; Takano et al. 1997; Wadhwa et al. 2006; Yi et al. 2008, Oki et al. 2011) showed upregulation of mortalin expression. By a high-resolution proteomic approach, mortalin was found as enriched in hepatocellular carcinoma (HCC) when compared to the adjacent non-tumor tissues. Clinico-pathological analysis further illustrated high tumor expression of mortalin associates with early recurrence of HCC in patients subjected to curative treatments (Yi et al. 2008). A second independent study by immunohistochemistry in which 100 pairs of hepatic cancer tissues were examined revealed that mortalin was overexpressed in 61% of the tumor tissues. It correlated with the tumor aggressiveness; there was a marked increase in mortalin level from the early (stage I and stage II) to late (stage III and stage IV) stages. The clinical-pathological parameter analysis indicated that mortalin up-regulation was significantly associated with HCC recurrence ($p = 0.005$) and microsatellite formation ($p = 0.028$), a consistent feature of intra-hepatic metastasis (Lu et al. 2011b). In contrast to the thymus and spleen derived cells that undergo p53-mediated apoptosis in response to genotoxic stress, liver cancer cells tend to be resistant suggesting that the HCC may have mechanism(s) leading to inactivation of p53-mediated apoptosis. Based on the mortalin-p53 interaction and its upregulation in liver cancers, we speculated that mortalin may deregulate apoptotic function of p53 resulting in continued survival of cancer cells.

17.3 Mortalin-p53 Interaction is Stress Dependent and Constitutes a Selective Target for Cancer Therapy

p53 is a stress response factor. Various cellular stresses (intrinsic and extrinsic) have been shown to evoke p53 signaling (Murray-Zmijewski et al. 2008) through its modifications including phosphorylation (at serines and/or threonines) (Bode and Dong 2004), acetylation (Luo et al. 2000), sumoylation (Melchior and Hengst 2002), glycosylation (Shaw et al. 1996), ribosylation (Wesierska-Gadek et al. 1996) or ubiquitylation (Fuchs et al. 1998). It has been shown to interact with several stress proteins *in vivo* (Selkirk et al. 1994; Sugito et al. 1995) that modulate its functions. Contribution of such interactions in cancer, physiologically a stressed condition is a subject of current research. In our study on the investigation of mortalin-p53 interaction in liver cancer cell line (MHCC97H/97H, MHCC97L/97L, H2P, H2M, PLC/PRF/5, HepG2, Hep3B), we found that unlike most cancer cells, HepG2 hepatoma lacked mortalin-p53 interaction (Fig. 17.1a). Similar to the normal cells and in contrast to the liver tumor tissues, normal liver tissue did not show interaction of mortalin and p53 (Lu et al. 2011b). This finding initiated further studies to understand the regulation of mortalin-p53 interaction and their significance in cancer cells. First of all, we examined if mortalin-p53 interaction was regulated by wild type or mutant status of the p53 protein. Exogenous expression of wild-type or mutant (249ser) p53 into the p53 null HCC cells, Hep3B followed by immunoprecipitation of mortalin revealed that both the wild-type and the mutant p53 interact with mortalin (Fig. 17.1b). This had led us to suspect that mortalin-p53 interaction may be a stress dependent event. Therefore, we examined the stress status of cells by detecting the stress-induced phosphorylation of p53 in these HCC lines. We found phosphorylation of p53 at three (Ser 15, Ser 37, Ser 392) of the seven sites in five HCC cell lines (97H, 97L,

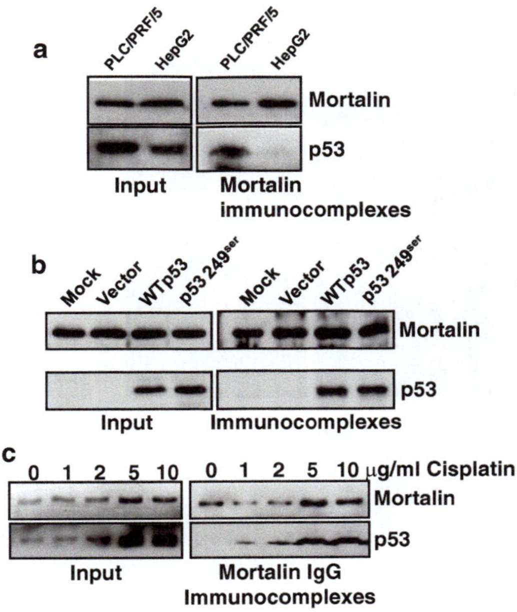

Fig. 17.1 Mortalin-p53 interaction in human liver cancer cells with mutant p53 (PLC/PRF/5) and with wild-type p53 (HepG2). PLC cells, but not HepG2 showed p53 in mortalin complexes (**a**). Interaction of wild type and mutant (249 Ser) p53 expressed from an exogenous promoter with mortalin showing that both the wild-type and mutant p53 interact with mortalin (**b**). Mortalin-p53 interaction in HepG2 cells stressed with cisplatin showing induction of interaction with stress (**c**). Knockdown of mortalin (KD) resulted in caspase dependent apoptosis in PLC (cancer cells), but not in MIHA (normal immortalized cells) (**d** and **e**). Modified from Lu et al. (2011a, b)

H2P, H2M, PLC) that showed mortalin-p53 interaction (Lu et al. 2011a). HepG2 that did not show mortalin-p53 interaction also did not have ser/thr phosphorylation suggesting that this interaction may be stress-regulated and HepG2 cells served as a good model to study the regulation of this interaction. These cells were subjected to stress by treatment with low doses of cisplatin, hydrogen peroxide and doxorubicin. These treatments induced mortalin-p53 interaction (Fig. 17.1c). Based on the finding that the normal cells lack mortalin-p53 interaction and hence inactivation of p53 by mortalin, we suspected that mortalin-p53 interaction might offer a selective target for cancer therapy. Indeed, we found that knockdown of mortalin by shRNA plasmid caused apoptosis of PLC/PRF/5 liver carcinoma in which mortalin-p53 interaction was detected, but not in the normal liver cells and HepG2 cancer cells that lacked mortalin-p53 interaction (Fig. 17.1d, e).

Mortalin-p53 interaction that causes inactivation of p53 function was also targeted by a small molecule (MKT-077, a cationic rhodacyanine dye) and peptide ($p53^{312-352}$, carboxyl-terminal amino acid resides 312–352 of p53), previously shown to bind to mortalin and reactivate wild-type p53 leading to growth arrest (Kaul et al. 2005; Wadhwa et al. 2000). When subjected to MKT-077 or $p53^{312-352}$ peptide

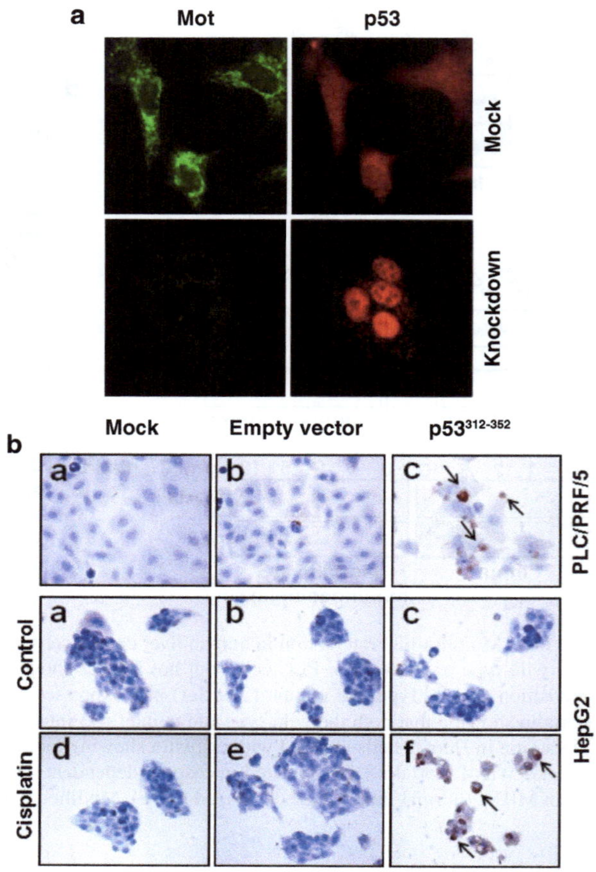

Fig. 17.2 Mortalin knockdown resulted in the nuclear translocation and accumulation of p53 (**a**). Induction of apoptosis in PLC cells by p53(312–352) peptide that abrogates mortalin-p53 interaction. HepG2 cells that lack mortalin-p53 interaction do not undergo apoptosis unless they are treated with cisplatin (a stress condition that induced mortalin-p53 interaction) (**b**). Modified from Lu et al. (2011b)

expression, PLC cells (high level of stress as indicated by mutant and phosphorylated p53) underwent apoptosis (Lu et al. 2011b). Consistent with the occurrence of apoptosis, nuclear translocation and accumulation of p53 was observed (Fig. 17.2a). In contrast to PLC cells, HepG2 cells that lacked mortalin-p53 interactions did not show any apoptosis either with $p53^{312-352}$ or with MKT-077 (Fig. 17.2b). However, cells treated with cisplatin condition that induced mortalin-p53 interaction responded both to the $p53^{312-352}$ and MKT-077 by undergoing apoptosis (Fig. 17.2b). These findings led us to propose a model on the regulation of mortalin-p53 interaction by stress. In normal, immortalized and non-malignant cancer cells that are unstressed and have low p53 level, mortalin and p53 do not interact (a). In these cells, p53 freely translocates from cytoplasm to nucleus. Mortalin knockdown does not cause apoptosis in this category of cells. Interestingly, mortalin-p53 interaction can be induced by exposure to the genotoxic stress that also results in p53 phosphorylation. In this category of cells, apoptotic ability of p53 is compromised by its binding to mortalin (b). Knockdown of mortalin in this category hence induces reactivation

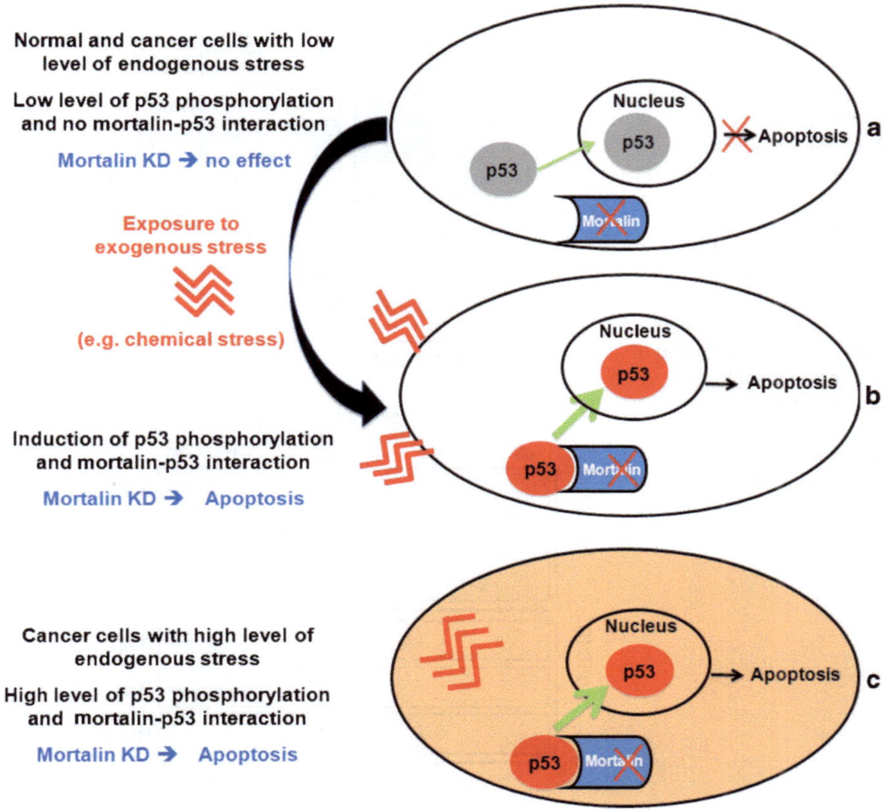

Fig. 17.3 A model showing the effect of mortalin knockdown and nuclear translocation of p53 in cells with different levels of stress. (**a**) In normal, immortalized and non-malignant cancer cells with low p53 phosphorylation, mortalin does not interact with p53 and hence p53 freely translocates from cytoplasm to nucleus. However, unphosphorylated p53 is inactive, unstable and insufficient for apoptotic activity. Mortalin knockdown has no effect in this category of cells. (**b**) Exposure of cells to stress (genotoxic or malignancy) causes p53 phosphorylation and accumulation that induces mortalin-p53 interaction. Mortalin silencing by shRNA in this scenario causes nuclear translocation of the phosphorylated p53 that is stable and active resulting in apoptotic death of cells. (**c**) In malignant cancer cells with high stress, p53 is heavily phosphorylated and stable. In these cells, mortalin captures p53 in the cytoplasm and blocks its transcriptional activation, growth arrest and apoptotic functions. In this scenario, mortalin knockdown causes nuclear translocation of p53 leading to apoptotic death of cells. According to the proposed model, mortalin-p53 interaction is selective for stressed cancer cells and hence could serve as a safe target for cancer therapeutics. Modified from Lu et al. (2011a)

of p53-mediated apoptosis. Malignant cancer cells are physiologically stressed and accumulate p53 (mutant) that is highly phosphorylated and have mortalin-p53 interaction (c). In these cells, mortalin knockdown causes nuclear translocation of p53 leading to apoptotic death of cells. According to this model, mortalin-p53 interaction is selective for stressed cancer cells (normal cells lack this interaction) and hence could serve as a safe target for cancer therapeutics (Fig. 17.3).

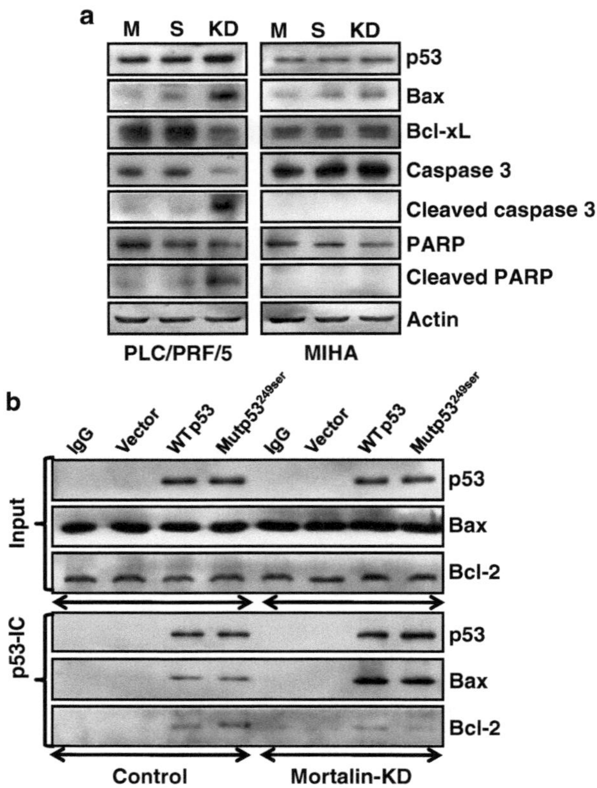

Fig. 17.4 Mortalin knockdown caused caspase-mediated apoptosis in PLC (cancer), but not in MIHA (normal) cells. In PLC, but not in MIHA, proapoptotic protein Bax was increased and antiapoptotic protein, Bcl-xL was decreased. PARP cleavage occurred in PLC cells only (**a**). Mortalin-knockdown cells that underwent apoptosis showed increased amount of Bax in p53 complexes suggesting that mortalin-shRNA induced apoptosis in cancer cells involved p53-Bax pathway (**b**). Modified from Lu et al. (2011b)

17.4 Mortalin Knockdown Causes Reactivation of Mutant p53 Dependent Apoptosis in Liver Cancer Cells

Investigations on the molecular markers of apoptosis in mortalin knockdown cancer cells revealed that the mortalin-compromised PLC cells that harbor mutant p53 have increase in pro-apoptotic protein-Bax, cleavage of caspase 3 and PARP and decrease in anti-apoptotic protein Bcl-xL (Fig. 17.4a). Of note, consistent with the absence of apoptosis, these molecular events were not detected in MIHA cells subsequent to mortalin knockdown (Fig. 17.4a). p53 inhibitors, Pifithrin-α (PFT-α) and Pifithrin-μ (PFT-μ) when used to competitively block its transcriptional (as indicated by nuclear accumulation) and apoptotic (by mitochondrial accumulation) activities revealed that mortalin-silencing induced p53-mediated apoptosis involve its nuclear as well as mitochondrial functions. Normal immortalized hepatocytes-MIHA that lacked mortalin-mutant p53 interaction did not exhibit apoptosis, changes in the apoptotic proteins and subcellular localization of p53 with/without mortalin shRNA (Figs. 17.2a and 17.4a).

Tumor suppressor protein p53 is regarded as a central player in tumor suppression. In vast majority of tumors, loss of p53 function attained by mutations is required for tumor maintenance. It controls apoptosis and cellular senescence directly or indirectly at multiple levels of the tumor suppression network by invoking a myriad of mechanisms that involve its transcription-dependent and -independent functions. Since deregulated apoptosis (tumor clearance mechanism) is a major cancer-sustaining mechanism, therapeutic strategies aimed at reactivation of apoptotic activities of mutant p53 in tumors are emerging as a promising cancer therapeutics approach. The majority of p53 mutations observed in human cancers are missense mutations in its DNA binding domain that abrogate its sequence-specific binding to the wild-type p53 responsive element making it transcriptionally inactive. However, mutant p53 has been reported to have functions (growth arrest and induction of apoptosis) independent to its transcriptional activation function. It was also shown that the p53-mediated apoptosis could occur in the presence of p53-DNA binding inhibitors, like PIASγ, suggesting that sequence specific transactivation is not essential for p53-dependent apoptosis. Our findings suggested that the mutant p53 binds to mortalin in tumor cells and causes inactivation of its apoptotic function (Lu et al. 2011b). In order to determine whether such p53-dependent apoptosis involved restoration of the wild-type p53 trans-activation in the mutant p53, we performed wild-type p53 dependent luciferase reporter assays in Hep3B (p53-null) cells with exogenous expression of either the wild type or the mutant p53. We found that mortalin knockdown enhanced transactivation ability of the wild-type p53. In the same assays, mutant p53 did not show any transcriptional activation function in control or mortalin-shRNA transfected cells suggesting that the apoptosis caused by mortalin knockdown involved activation of mutant p53 functions, independent of the wild-type p53 transactivation. Nevertheless, the apoptotic ability of mutant p53 was stronger than the wild-type p53 suggesting that mortalin-shRNA is a good candidate for HCC therapy as they are frequently associated with p53 mutations. Transcription-independent apoptotic ability of mutant p53 has been reported in several studies (Haupt et al. 1995; Caelles et al. 1994; Nelson et al. 2001). It was shown that the p53 mediated apoptosis involves (i) activation of Bax dimer formation and changes in the mitochondrial membrane permeabilization (Chipuket al. 2004) and (ii) formation of nuclear Bax/p53 complexes (Raffo et al. 2000) and their crosstalk with nuclear chaperone, nucleophosmin (Geng et al. 2010). We found that in mutant p53 harboring cells, mortalin knockdown resulted in the nuclear accumulation of p53, increased levels of Bax and increased level of mutant p53-Bax interaction (Fig. 17.4b) suggesting that mortalin silencing caused activation of p53-Bax apoptosis pathway.

The identification of a non-toxic chemicals and small molecules capable of restoring the tumor suppression and apoptotic function of mutant p53 is an exciting prospective for future cancer therapy. Our findings that mortalin silencing reactivated p53 activities selectively in cancer cells proposed that mortalin-p53 interaction could serve as a target for selective cancer cell killing. Abrogation of this interaction by cationic dye (MKT-077), a natural phytochemical withanone and p53 carboxyl-terminus peptide ($p53^{312-352}$) trigger apoptosis in cancer cells with mutant p53. Since large majority of cancers possess mutant p53, such assignment of apoptotic function to mutant p53 is expected to be of great advantage in cancer therapeutics. Normal cells lack mortalin-p53 interaction and hence such therapeutic strategy will be highly selective for cancer cells.

17.5 Summary

Development of novel and effective cancer treatment is a current challenge to clinicians and medical researchers. In the past years, antagonists, small molecule inhibitors, compounds targeted to cancer-related proteins have shown their potentials in counteracting cancer progression and development. Recently, the focus is drifted towards exploration of cancer-driven protein-protein interactions in malignant cells as compared to their nonmalignant neighbors within the tumor microenvironment. These approaches are expected to yield outcomes that are superior, sensitive and selective in cancer treatment. Mortalin-p53 interaction is one such attractive target for cancer therapy.

Acknowledgments This work was done by WJ Lu in partial fulfillment of her PhD degree and based on her thesis submitted to The University of Hong Kong.

References

Arakawa H (2005) p53, apoptosis and axon-guidance molecules. Cell Death Differ 12:1057–1065

Bode AM, Dong Z (2004) Post-translational modification of p53 in tumorigenesis. Nat Rev Cancer 4:793–805

Bosari S, Viale G, Roncalli M, Graziani D, Borsani G, Lee AKC, Coggi G (1995) P53 Gene-Mutations, P53 Protein Accumulation and Compartmentalization in Colorectal Adenocarcinoma. Am J Pathol 147:790–798

Bourdon JC, Renzing J, Robertson PL, Fernandes KN, Lane DP (2002) Scotin, a novel p53-inducible proapoptotic protein located in the ER and the nuclear membrane. J Cell Biol 158:235–246

Caelles C, Helmberg A, Karin M (1994) p53-dependent apoptosis in the absence of transcriptional activation of p53-target genes. Nature 370:220–223

Chipuk JE, Kuwana T, Bouchier-Hayes L, Droin NM, Newmeyer DD, Schuler M, Green DR (2004) Direct activation of Bax by p53 mediates mitochondrial membrane permeabilization and apoptosis. Science 303:1010–1014

Dundas SR, Lawrie LC, Rooney PH, Murray GI (2005) Mortalin is over-expressed by colorectal adenocarcinomas and correlates with poor survival. J Pathol 205:74–81

Fuchs SY, Adler V, Buschmann, T, Wu X, Ronai Z (1998) Mdm2 association with p53 targets its ubiquitination. Oncogene 17:2543–2547

Geng Y, Walls KC, Ghosh AP, Akhtar RS, Klocke BJ, Roth KA (2010) Cytoplasmic p53 and activated Bax regulate p53-dependent, transcription-independent neural precursor cell apoptosis. J Histochem Cytochem 58:265–275

Haupt Y, Rowan S, Shaulian E, Vousden KH, Oren M (1995) Induction of apoptosis in HeLa cells by trans-activation-deficient p53. Genes Dev 9:2170–2183

Helmbrecht K, Zeise E, Rensing L (2000) Chaperones in cell cycle regulation and mitogenic signal transduction: a review. Cell Prolif 33:341–365

Hoffman WH, Biade S, Zilfou JT, Chen J, Murphy M (2002) Transcriptional repression of the anti-apoptotic survivin gene by wild type p53. J Biol Chem 277:3247–3257

Kaul SC, Aida S, Yaguchi T, Kaur K, Wadhwa R (2005) Activation of wild type p53 function by its mortalin-binding, cytoplasmically localizing carboxyl terminus peptides. J Biol Chem 280:39373–39379

Kaul SC, Wadhwa R, Matsuda Y, Hensler PJ, Pereira-Smith OM, Komatsu Y, Mitsui Y (1995) Mouse and human chromosomal assignments of mortalin, a novel member of the murine hsp70 family of proteins. FEBS Lett 361:269–272

King FW, Wawrzynow A, Hohfeld J, Zylicz M (2001) Co-chaperones Bag-1, Hop and Hsp40 regulate Hsc70 and Hsp90 interactions with wild-type or mutant p53. EMBO J 20:6297–6305

Lindquist S, Craig EA (1988) The heat-shock proteins. Annu Rev Genet 22:631–677

Liu S, Li J, Tao Y, Xiao X (2007) Small heat shock protein alphaB-crystallin binds to p53 to sequester its translocation to mitochondria during hydrogen peroxide-induced apoptosis. Biochem Biophys Res Commun 354:109–114

Lu WJ, Lee NP, Kaul SC, Lan F, Poon RT, Wadhwa R, Luk JM (2011a) Mortalin-p53 interaction in cancer cells is stress dependent and constitutes a selective target for cancer therapy. Cell Death Differ 18:1046–1056

Lu WJ, Lee NP, Kaul SC, Lan F, Poon RT, Wadhwa R, Luk JM (2011b) Induction of mutant p53-dependent apoptosis in human hepatocellular carcinoma by targeting stress protein mortalin. Int J Cancer 129:1806–1814

Luo J, Su F, Chen D, Shiloh A, Gu W (2000) Deacetylation of p53 modulates its effect on cell growth and apoptosis. Nature 408:377–381

Ma Z, Izumi H, Kanai M, Kabuyama Y, Ahn NG, Fukasawa K (2006) Mortalin controls centrosome duplication via modulating centrosomal localization of p53. Oncogene 25:5377–5390

Melchior F, Hengst L (2002) SUMO-1 and p53. Cell Cycle 1:245–249

Miyashita T, Krajewski S, Krajewska M, Wang HG, Lin HK, Liebermann DA, Hoffman B, Reed JC (1994) Tumor suppressor p53 is a regulator of bcl-2 and bax gene expression in vitro and in vivo. Oncogene 9:1799–1805

Moll UM, Ostermeyer AG, Haladay R, Winkfield B, Frazier M, Zambetti G (1996) Cytoplasmic sequestration of wild-type p53 protein impairs the G1 checkpoint after DNA damage. Mol Cell Biol 16:1126–1137

Muller L, Schaupp A, Walerych D, Wegele H, Buchner J (2004) Hsp90 regulates the activity of wild type p53 under physiological and elevated temperatures. J Biol Chem 279:48846–48854

Murray-Zmijewski F, Slee EA, Lu X (2008) A complex barcode underlies the heterogeneous response of p53 to stress. Nat Rev Mol Cell Biol 9:702–712

Nakano K, Vousden KH (2001) PUMA, a novel proapoptotic gene, is induced by p53. Mol Cell 7:683–694

Nelson V, Davis GE, Maxwell SA (2001) A putative protein inhibitor of activated STAT (PIASy) interacts with p53 and inhibits p53-mediated transactivation but not apoptosis. Apoptosis 6:221–234

Nikolaev AY, Li M, Puskas N, Qin J, Gu W (2003) Parc: a cytoplasmic anchor for p53. Cell 112:29–40

Oda E, Ohki R, Murasawa H, Nemoto J, Shibue T, Yamashita T, Tokino T, Taniguchi T, Tanaka N (2000) Noxa, a BH3-only member of the Bcl-2 family and candidate mediator of p53-induced apoptosis. Science 288:1053–1058

Oki E, Ando K, Saeki H, Ohga T, Kakeji Y, Tsujitani S, Kusumoto T, Maehara Y (2011) Prognostic significance of mortalin expression in gastric adenocarcinoma with wild-type p53. 2011 ASCO Annual Meeting citation. J Clin Oncol 29(suppl abstr):e14662)

Petros AM, Gunasekera A, Xu N, Olejniczak ET, Fesik SW (2004) Defining the p53 DNA-binding domain/Bcl-x(L)-binding interface using NMR. FEBS Lett 559:171–174

Pinhasi-Kimhi O, Michalovitz D, Ben-Zeev A, Oren M (1986) Specific interaction between the p53 cellular tumour antigen and major heat shock proteins. Nature 320:182–184

Powers MV, Workman P (2007) Inhibitors of the heat shock response: biology and pharmacology. FEBS Lett 581:3758–3769

Raffo AJ, Kim AL, Fine RL (2000) Formation of nuclear Bax/p53 complexes is associated with chemotherapy induced apoptosis. Oncogene 19:6216–6228
Sax JK, Fei P, Murphy ME, Bernhard E, Korsmeyer SJ, El-Deiry WS (2002) BID regulation by p53 contributes to chemosensitivity. Nat Cell Biol 4:842–849
Selkirk JK, Merrick BA, Stackhouse BL, He C (1994) Multiple p53 protein isoforms and formation of oligomeric complexes with heat shock proteins Hsp70 and Hsp90 in the human mammary tumor, T47D, cell line. Appl Theor Electrophor 4:11–18
Shaw P, Freeman J, Bovey R, Iggo R (1996) Regulation of specific DNA binding by p53: evidence for a role for O-glycosylation and charged residues at the carboxy-terminus. Oncogene 12:921–930
Sherman M, Multhoff G (2007) Heat shock proteins in cancer. Ann N Y Acad Sci 1113:192–201
Sugars KL, Budhram-Mahadeo V, Packham G, Latchman DS (2001) A minimal Bcl-x promoter is activated by Brn-3a and repressed by p53. Nucleic Acids Res 29:4530–4540
Sugito K, Yamane M, Hattori H, Hayashi Y, Tohnai I, Ueda M, Tsuchida N, Ohtsuka K (1995) Interaction between hsp70 and hsp40, eukaryotic homologues of DnaK and DnaJ, in human cells expressing mutant-type p53. FEBS Lett 358:161–164
Takano S, Wadhwa R, Yoshii Y, Nose T, Kaul SC, Mitsui Y (1997) Elevated levels of mortalin expression in human brain tumors. Exp Cell Res 237:38–45
Takimoto R, El-Deiry WS (2000) Wild-type p53 transactivates the KILLER/DR5 gene through an intronic sequence-specific DNA-binding site. Oncogene 19:1735–1743
Vogelstein B, Lane D, Levine AJ (2000) Surfing the p53 network. Nature 408:307–310
Vousden KH, Lu X (2002) Live or let die: the cell's response to p53. Nat Rev Cancer 2:594–604
Wadhwa R, Sugihara T, Yoshida A, Nomura H, Reddel RR, Simpson R, Maruta H, Kaul SC (2000). Selective toxicity of MKT-077 to cancer cells is mediated by its binding to the hsp70 family protein mot-2 and reactivation of p53 function. Cancer Res 60:6818–6821
Wadhwa R, Takano S, Kaur K, Deocaris CC, Pereira-Smith OM, Reddel RR, Kaul SC (2006). Upregulation of mortalin/mthsp70/Grp75 contributes to human carcinogenesis. Int J Cancer 118:2973–2980
Wadhwa R, Takano S, Robert M, Yoshida A, Nomura H, Reddel RR, Mitsui Y, Kaul SC (1998) Inactivation of tumor suppressor p53 by mot-2, a hsp70 family member. J Biol Chem 273:29586–29591
Wesierska-Gadek J, Bugajska-Schretter A, Cerni C (1996) ADP-ribosylation of p53 tumor suppressor protein: mutant but not wild-type p53 is modified. J Cell Biochem 62:90–101
Whitesell L, Sutphin P, An WG, Schulte T, Blagosklonny MV, Neckers L (1997) Geldanamycin-stimulated destabilization of mutated p53 is mediated by the proteasome in vivo. Oncogene 14:2809–2816
Yi X, Luk JM, Lee NP, Peng J, Leng X, Guan XY, Lau GK, Beretta L, Fan ST (2008) Association of mortalin (HSPA9) with liver cancer metastasis and prediction for early tumor recurrence. Mol Cell Proteomics 7:315–325
Yonish-Rouach E, Resnitzky D, Lotem J, Sachs L, Kimchi A, Oren M (1991) Wild-type p53 induces apoptosis of myeloid leukaemic cells that is inhibited by interleukin-6. Nature 352:345–347
Zaika A, Marchenko N, Moll UM (2000) Cytoplasmically "sequestered" wild type p53 protein is resistant to Mdm2-mediated degradation. J Biol Chem 275:11538–11538
Zhao LY, Liao DQ (2003) Sequestration of p53 in the cytoplasm by adenovirus type 12 E1B 55-kilodalton oncoprotein is required for inhibition of p53-mediated apoptosis. J Virol 77:13171–13181

Chapter 18
Mortalin Targeting Gadgets for Cancer Therapy

Chae-Ok Yun and Renu Wadhwa

Abstract Mortalin, also known as GRP75/mtHSP70/PBP74, is a member of the hsp70 family of proteins. It is frequently enriched in cancers and shows distinct subcellular distribution in normal and immortal/cancer cells. Mortalin-overexpressing immortalized human cells show enhanced malignant properties, including anchorage-independent growth, formation of tumors in nude mice and chemotaxis. Targeting mortalin with antisense, ribozyme or siRNA specific for mortalin induced tumor cell-specific growth arrest and apoptosis suggesting that the development of various modalities for targeting mortalin may be useful in clinical applications. In this chapter, we discuss the various strategies that have been explored to generate mortalin-targeted cancer gene therapeutics.

Keywords Mortalin · Targeting · Antisense · Ribozyme · siRNA · Oncolytic adenoviruses · Apoptosis · Therapy

18.1 Mortalin as a Target for Anti-Cancer Treatment

Mortalin is frequently enriched in human cancers and has been functionally related to human carcinogenesis (Cussac et al. 2006; Dundas et al. 2005; Ma et al. 2006; Shin et al. 2003; Wadhwa et al. 2006; Lu et al. 2011). Comparison of the level of mortalin expression in normal, immortalized and tumor derived cell lines revealed that it is upregulated in all the immortalized and tumor-derived cells. The data suggested that the upregulation of mortalin is tightly associated with human carcinogenesis and could be an early event in the development of tumors. It was shown to interact with tumor suppressor protein p53 causing inactivation of its transcriptional activation, apoptotic and control of centrosome duplication functions; all of these promote uncontrolled proliferation of cancer cells (Kaul et al. 2001, 2007a;

C.-O. Yun (✉)
Department of Bioengineering, College of Engineering, Hanyang University,
17 Haengdang-Dong, Seongdong-Gu, Seoul 133-791, Korea
e-mail: chaeok@hanyang.ac.kr

R. Wadhwa
National Institute of Advanced Industrial Science & Technology (AIST), Central 4,
1-1-1 Higashi, Tsukuba, Ibaraki 305-8562, Japan

S. C. Kaul, R. Wadhwa (eds.), *Mortalin Biology: Life, Stress and Death*,
DOI 10.1007/978-94-007-3027-4_18,

Wadhwa et al. 1998, 2002c, Ma et al. 2006; Lu et al. 2011). Forced overexpression of mortalin in normal human fibroblasts resulted in their extended lifespan that was also shown to be mediated, at least in part, by its interactions with p53 (Kaul et al. 2001; Wadhwa et al. 1998, 2002c). A proteomic study investigating protein markers associated with post-operative early recurrence (ER) of hepatocarcinoma demonstrated that mortalin was significantly associated with liver cancer metastasis and predicted early tumor recurrence (Yi et al. 2008). Quantitative estimation of protein expression revealed that the tumor cells that express high levels of mortalin have more aggressive tumor phenotypes, including tumor volume and metastasis (Dundas et al. 2005; Lu et al. 2011). Furthermore, level of mortalin expression in tumor tissues positively correlated with poor patient survival (Dundas et al. 2005) and tumor reoccurrence after the surgery (Yi et al. 2008). In agreement with these studies, forced overexpression of mortalin in non-malignant breast and bone cancer cells converted them to the tumorigenic and metastatic cells (Wadhwa et al. 2006). Taken together, these data suggested that the clinical use of molecules to specifically target mortalin might serve as effective therapeutic strategy for treating a variety of cancers. Importantly, mortalin-specific targeting molecules might also be potentially effective in preventing cancer recurrence.

18.2 Mortalin-specific Antisense Oligonucleotides for Cancer Therapy

Many researchers have used antisense oligonucleotides to specifically and selectively downregulate gene expression at the level of messenger RNA (Aharinejad et al. 2004; Cerutti et al. 1996; Dias and Stein 2002; Lee and Roth 2003). Oligonucleotides, which are short sequences of DNA, have been designed to hybridize via the specificity of Watson-Crick base-pair interactions to a defined target messenger RNA and can inhibit its translation into protein. The specificity of oligonucleotide hybridization makes the treatment with antisense oligonucleotides an attractive approach to selectively modulate the expression of genes involved in diseases such as viral infection, cancers, or inflammatory disorders (Dias and Stein 2002; Forte et al. 2005; Kurreck 2003). Zamecnik and Stephenson first investigated the use of antisense oligonucleotides in 1978 by inhibiting generation of the Rous sarcoma virus in chicken fibroblasts (Zamecnik and Stephenson 1978). The use of these oligonucleotides have since been greatly elaborated upon. Numerous oligonucleotides are currently being tested in clinical trials (Cossu and Sampaolesi 2007; Gibbs 2000; Jain et al. 2006; Kurreck 2003).

Transfection of an expression plasmid including mortalin-specific antisense oligonucleotides suppresses the growth of transformed cells (Wadhwa et al. 2004a). WI-38 human embryonic fibroblasts immortalized by stable transfections of expression plasmids encoding E6E7 and catalytic subunit of hTERT protein were transfected with an antisense mortalin expression plasmid. In this study, authors

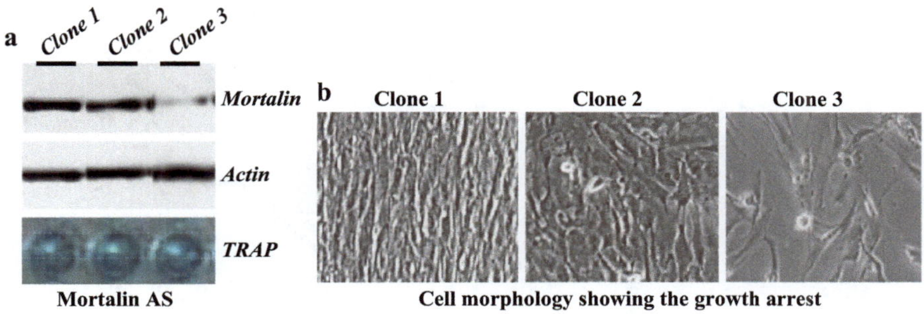

Fig. 18.1 Mortalin silencing by antisense expression plasmid in human embryonic fibroblasts immortalized with E6E7 and telomerase. Cells compromised for mortalin showed growth arrest, in spite of the active telomerase, compare clone 1, 2 and 3 for mortalin, TRAP (**a**) and growth arrest (**b**) as seen by cell morphology. Modified from Wadhwa et al. (2004a)

first selected the pool cultures and examined the expression of mortalin by RT-PCR and Western blotting assays. However, they found that none of the ten-pooled cultures showed any change in the expression as compared to the untransfected and vector transfected cells. There were two possible interpretations for this: either the antisense expression of mortalin was not effective or the cells with effective antisense mortalin expression do not survive and thus the cultures were overtaken by the unaffected fast dividing cell population. In order to resolve these issues, attempts to isolate individual colonies after transfection with the antisense plasmid were undertaken. Fifteen colonies isolated after the selection of transfected cells were individually passaged. Seven colonies showed growth comparable to the vector and untransfected cells (representative clone 1, Fig. 18.1), four colonies showed retarded growth as compared to the controls (representative clone 2, Fig. 18.1) and the remaining four showed severe growth arrest (representative clone 3, Fig. 18.1). Interestingly, these four clones that were severely retarded in their growth showed maximum reduction in mortalin level (Fig. 18.1). These were, however, positive for the telomerase activity (Wadhwa et al. 2004a). Results of this study suggested that the targeting mortalin might induce growth arrest in telomerase positive cancer cells.

18.3 Mortalin Ribozyme for Cancer Therapy

Ribozymes are also called molecular scissors. They are the small RNAs that have enzymatic properties (Cech 1986) and fine-tune gene expression cascades such as those that induce replication, transcription, and translation. These naturally occurring RNA molecules catalyze essential steps in gene expression, including cleavage and ligation (Cech 2000; Doudna and Cech 2002; Zhang and Cech 1997). Hammerhead ribozymes are the smallest ribozymes, and are used as so-called ‘molecular scissors’ in molecular biology and biotechnology to eliminate gene function.

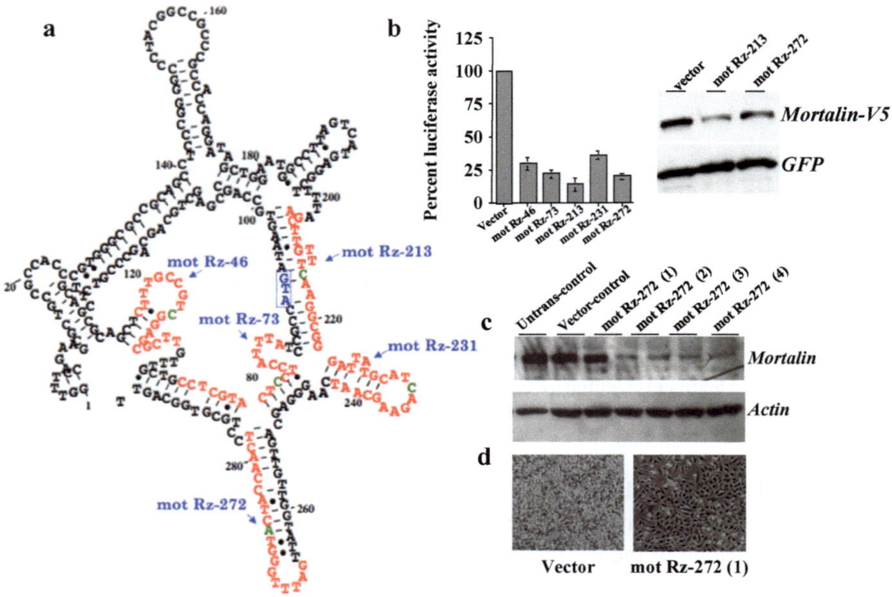

Fig. 18.2 Mortalin silencing by ribozymes in human cancer cells. Sites for the mortalin targeting ribozymes (**a**), their effect in mortalin-dependent reporter assays (**b**), endogenous mortalin (**c**) and growth of human cancer is shown (**d**). Modified from Wadhwa et al. (2003)

RNA-helicase-linked hybrid ribozymes that target the 5′ region of mortalin RNA have been developed to specifically suppress the expression of mortalin Fig. 18.2a (Wadhwa et al. 2003). All of these when tested by the luciferase reporter, linked to either the N-terminal region of mortalin or an exogenously transfected V5-tagged mortalin, showed significant suppression of the reporters (Fig. 18.2b). However, the majority of the pooled cell cultures selected after the transfections of ribozymes neither showed reduction in mortalin nor the growth arrest (Wadhwa et al. 2003). Therefore, approach to isolate the individual cell clones after transfection of ribozymes similar to the antisense expression plasmid as described above was adopted. As shown in Fig. 18.2c, d, clones that were compromised for mortalin were growth arrested and showed decreased colony-forming efficiency. Moreover, strong suppression of colony-forming efficiency was evident in cells that expressed wild-type p53 (HT1080, U2OS, and MCF7), whereas milder suppression of colony-forming efficiency was observed in p53-null cells (Saos-2). The data suggested that the mechanism inducing growth arrest involves both p53 inactivation as well as signaling pathways that are independent of p53 and may involve mitochondrial import and chaperone functions of mortalin, as suggested by other studies (Liu et al. 2001; Wadhwa et al. 2002a, b). The results of these studies suggested that the effective targeting of mortalin causes growth arrest of tumor cells in both p53-dependent and -independent pathways.

18.4 Mortalin siRNA for Cancer Therapy

RNA interference (RNAi) is a sequence-specific, post-transcriptional gene silencing mechanism present in a wide range of organisms, including plants, *Caenorhabditis elegans*, and Drosophila (Elbashir et al. 2001; Shin et al. 2003). Triggered by double-stranded RNA (dsRNA), RNAi causes degradation of mRNA, homologous in sequence to the dsRNA. In 1998, Fire et al. reported that dsRNA molecules in *C. elegans* system inhibit translation of targeted mRNA more efficiently than individual single-stranded RNAs, and the resulting phenotype was inheritable by its progeny (Fire et al. 1998). Exogenous dsRNA is processed into 19- to 23-nucleotide small interfering RNAs (siRNAs) by the worm enzyme Dicer. Incorporation of these siRNAs into an RNA-induced silencing complex (RISC) is required for the post-transcriptional recognition and subsequent degradation of homologous mRNAs.

The initial studies investigating RNAi in mammalian tissues relied on transfection of cells with synthetic RNA oligonucleotides (Lou et al. 2003; Wacheck et al. 2003). The cost of chemically synthesized siRNAs is high and their effect is transient. An alternative strategy is to use the endogenous expression of siRNAs by various promoter expression cassettes that allow transcription of functional siRNAs. Many reports in recent past have described successful expression and effect of vector-based siRNA in mammalian cells providing a low-cost alternative to the chemically synthesized siRNAs. Furthermore, DNA-vector-based strategy is capable of mediating stable target gene inhibition, thus allowing gene function analysis over an extended period of time. Pol III promoters such as, U6, H1 and 7SK are often employed for the expression of siRNA in mammalian cells for the reasons such as, (i) Pol III promoters generate relatively small RNAs such as tRNA and snRNA, (ii) their RNA products are neither 5′-capped nor 3′-polyadenylated, (iii) RNA polymerase III recognizes poly (T) sequence as a termination codon, thus it is possible to design small RNAs that carry 3′ overhangs of one or two uridines (a structural feature for siRNA to be effective *in vitro*), (iv) Pol III starts transcription at G or A, thus the start site of the siRNA can be predetermined and offers a system in which any sequences can be inserted downstream of the transcription start site, (v) controlled RNA polymerase III promoters such as regulated by tetracycline or Cre-loxP can be generated for the analysis of temporally expressed genes or lethal genes (Wadhwa et al. 2004b; Azkur et al. 2005; Spankuch et al. 2004; Wannenes et al. 2005). Recently, Pol II (CMV) and Pol III (tRNA Val) specific promoters have been used to suppress the gene expression in cultured mammalian cells, although their utility is limited by the short half-life of the siRNAs.

A plasmid driven expression of siRNA *in vivo* is based on two designs. In the first case, Pol III promoter is employed to direct the synthesis of small inverted repeats separated by a spacer region of a varying length. Since U6 and H1 promoters start transcription at G or A, the resulting RNA form hairpins containing stems that have specific target sequence, loop sequence (3–9 nu) and 3′ U overhangs. These hairpin RNAs are processed *in vivo* by Dicer (Fig. 18.3). In the second case, two promoters are placed in tandem or in two separate vectors to direct the transcription of a sense

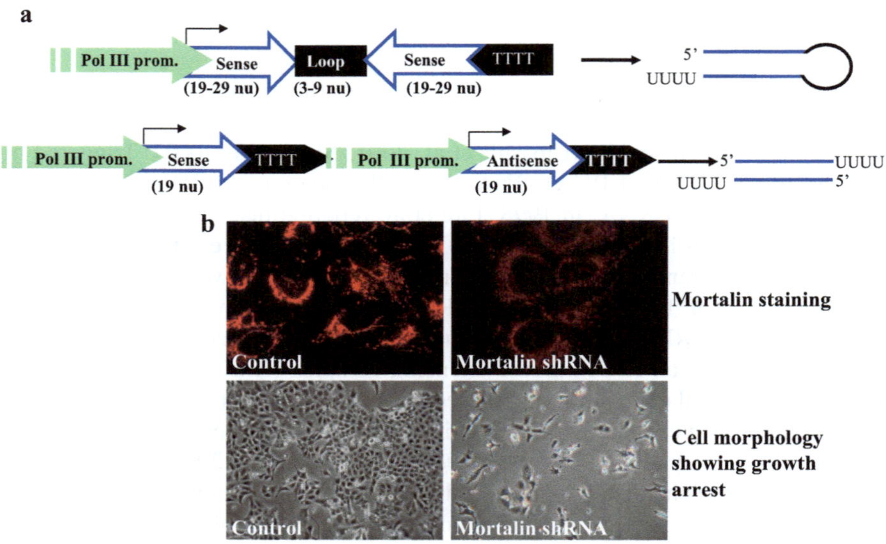

Fig. 18.3 Mortalin silencing by siRNA in human cancer cells. Schematic representation of Pol III promoter-driven shRNA expression plasmid (**a**). Effect of mortalin shRNA as examined by mortalin immuno-cytostaining and cell growth as examined by phase contrast imaging of cells (**b**). Modified from Yoo et al. (2010)

and antisense strand of a small RNA with 19 nt specific gene targeting sequence and 3′-UUUU overhangs. Sense and antisense strands form duplexes *in vivo* and function similar to the *in vitro* siRNA (Fig. 18.3). In most of the studies, both tandem and hairpin loop vectors driven siRNA suppressed the target gene expression; the hairpin-type siRNA-expression vector had significantly higher suppressive activity than the tandem-type siRNA-expression vector at a low concentration. Mortalin siRNA expression driven by Pol III promoter in human cancer cells caused their growth arrest or apoptosis (Fig. 18.3) (Wadhwa et al. 2004b; Lu et al. 2011).

18.5 Mortalin Targeting Oncolytic Adenoviruses for Cancer Therapy

As discussed above, effective mortalin silencing in cancer cells by antisense, ribozymes and siRNA was able to cause their growth arrest or apoptosis. However, there were still the technical issues on obtaining the effective silencing and a pool culture showing sufficient silencing for the growth arrest phenotype. Furthermore, achieving therapeutic *in vivo* gene silencing in mammalian tissues requires highly robust and long-lasting intracellular expression. One strategy to achieve such stable expression of siRNA is the use of a vector-based delivery system such as recombinant viral vectors. Adenovirus (Ad) vectors have been extensively used because of their unique ability to efficiently deliver DNA for the expression of genes in many different cell types (Russell 2000). To-date, over 1,600 gene therapy clinical protocols have been conducted. Ad vectors have been used in more than 300 protocols since 1993, and have become the most common gene delivery method for clinical gene therapy (http://www.wiley.co.uk/genmed/clinical). E1/E3-deleted replication-defective Ad derived from human Ad type 5 is commonly used for cancer gene therapy. In contrast to other vectors, this Ad offers a much higher transduction efficiency and transgene expression levels in a broad spectrum of cell types (Jolly 1994). However, replication-deficient viral vectors have thus far been used with limited success in cancer gene therapy, mainly due to the limited transduction efficiency and short duration of therapeutic gene expression (Carette et al. 2004). It is thus expected that similar technical difficulties are inherent in using non-replicating vectors for the delivery of shRNAs that interfere with the expression of genes involved in tumor cell survival.

In order to overcome the technical difficulties inherent in non-replicating vectors, oncolytic Ads are being developed to serve as selectively-replicating anti-tumoral agents. A number of clinical trials are currently ongoing to test the efficacy of such viruses in treating a variety of cancer types (Ganly et al. 2000; Heise et al. 2000; Khuri et al. 2000; Kirn 2001a, b). One clear benefit of a selectively-replicating system is its potential for amplification: the replicating vector can infect adjacent cancer cells and deliver the therapeutic gene, ultimately enhancing the potential for a viral-based therapy to treat complex human tumors. Applying shRNA technology in the context of oncolytic viruses may be particularly effective in overcoming the short inhibitory effect of siRNA, as well as avoiding non-specific shRNA expression in normal cells (Kirn 2007). The duration of gene suppression is largely dependent on the rate of cell growth, as well as the turnover rate of the targeted protein. Because cancer cells are actively dividing, the duration of siRNA-mediated gene silencing *in vivo* is likely to be governed by the efficiency of the siRNA-expressing vector and the stability of the functional siRNA. Oncolytic Ad-mediated siRNA expression improves efficacy over non-replicating vectors because of the self-perpetuating nature of viral multiplication and secondary infection of adjacent cells. This process of multiplication and infection enables the delivery of therapeutic siRNA molecules that are replenished over time (Yoo et al. 2008).

An effort to combine the shRNA therapeutic platform with an oncolytic viral platform was first undertaken using Ad-ΔB7-shVEGF, a VEGF-specific shRNA-expressing oncolytic Ad in human xenograft tumor models (Yoo et al. 2007). Ad-ΔB7-shVEGF elicited an enhanced antitumor response and a survival advantage compared to Ad-ΔB7, its cognate control oncolytic Ad suggesting that the oncolytic Ad-mediated siRNA silencing is a powerful approach to treat human cancers. Tumor histological analysis revealed that Ad-ΔB7-shVEGF induced a significant reduction in tumor vasculature, verifying that the mechanism was anti-angiogenic in nature. Furthermore, the duration and magnitude of the gene silencing effect following infection with Ad-ΔB7-shVEGF was longer in duration and more effective compared to the non-replicating Ad. Interestingly, despite the fact that the initial viral load of the Ad-ΔB7-shVEGF was 1000-fold less than the non-replicating Ad expressing

VEGF-specific shRNA, the overall effect of Ad-ΔB7-shVEGF was much greater (Yoo et al. 2007).

Various advantages and disadvantages are inherent in the type of RNA polymerase encoded in the Ad. Most shRNA-expressing vectors have used RNA polymerase III promoters, such as U6, H1 or 7SK, because polymerase III promoters efficiently direct the synthesis of small noncoding transcripts (Makinen et al. 2006; Robb et al. 2005; Yoo et al. 2008). RNA polymerase II promoters allow inducible, tissue- or cell type-specific RNA expression, which is not possible with polymerase III promoters. In one study, the efficacy of a RNA polymerase II promoter-driven vector was compared to polymerase III using an Ad expressing shRNA specific for IL-8. The Ad expressing IL-8-specific shRNA under the control of the polymerase III promoter was more effective in reducing IL-8 expression compared to the polymerase II promoter-driven vector. In *in vivo*, intra-tumoral injection of oncolytic Ad expressing IL-8-specific shRNA under the control of U6 promoter significantly inhibited the growth of several human tumor xenografts as well as disseminated breast cancer metastases (Yoo et al. 2008). These studies indicated that the use of cancer cell-specific, replicating oncolytic Ad to deliver tumor-targeted shRNA offers multiple potential benefits including restricted and renewable siRNA expression within the tumor microenvironment, an additive anti-tumor outcome through viral oncolysis, and siRNA-mediated silencing of genes involved in tumorigenesis.

Mortalin-targeting oncolytic Ads have been generated to selectively and effectively silence the expression of mortalin in cancer cells (Fig. 18.4a) (Yoo et al. 2010). Ad-ΔB7-shMot, a mortalin-specific shRNA-expressing oncolytic Ad, induced a strong cytotoxic effect in cancer cells, whereas minimal cytotoxicity was observed in normal cells (Fig. 18.4b). Moreover, the amount of cancer cell death was positively correlated with the degree of mortalin silencing. *In vivo* intra-tumoral injection of Ad-ΔB7-shMot significantly inhibited the growth of mortalin overexpressing MCF7/mot breast cancer xenografts in mice (Fig. 18.4c). The survival advantage was also significantly enhanced in mice treated with Ad-ΔB7-shMot compared to those treated with a control oncolytic Ad (Yoo et al. 2010). Analysis of Ad-ΔB7-shMot oncolytic Ad treated tumors showed that they undergo high rate of apoptosis and have poor blood vessel formation, suggesting that mortalin may also have significant role in angiogenesis and its silencing in tumors will affect tumor growth as well as invasion (Yoo et al. 2010).

18.6 Mortalin Antibody-based Cancer Therapy

Gene therapy is a potential solution to both hereditary and non-hereditary diseases, but its success is dependent on safe and specific gene delivery (Shiota et al. 2007). To generate cancer cell-specific gene delivery vectors, antibodies/ligands that specifically bind proteins/receptors are conjugated with cationic carriers, although it is necessary that the expression of proteins/receptors is either enriched in or exclusive to cancer cells (Aoki et al. 2001; Chiu et al. 2004; Densmore et al. 2001; Gautam et al. 2000; Guo and Lee 1999; Hattori et al. 2004; Hwa Kim et al. 2005; Zanta et al. 1997).

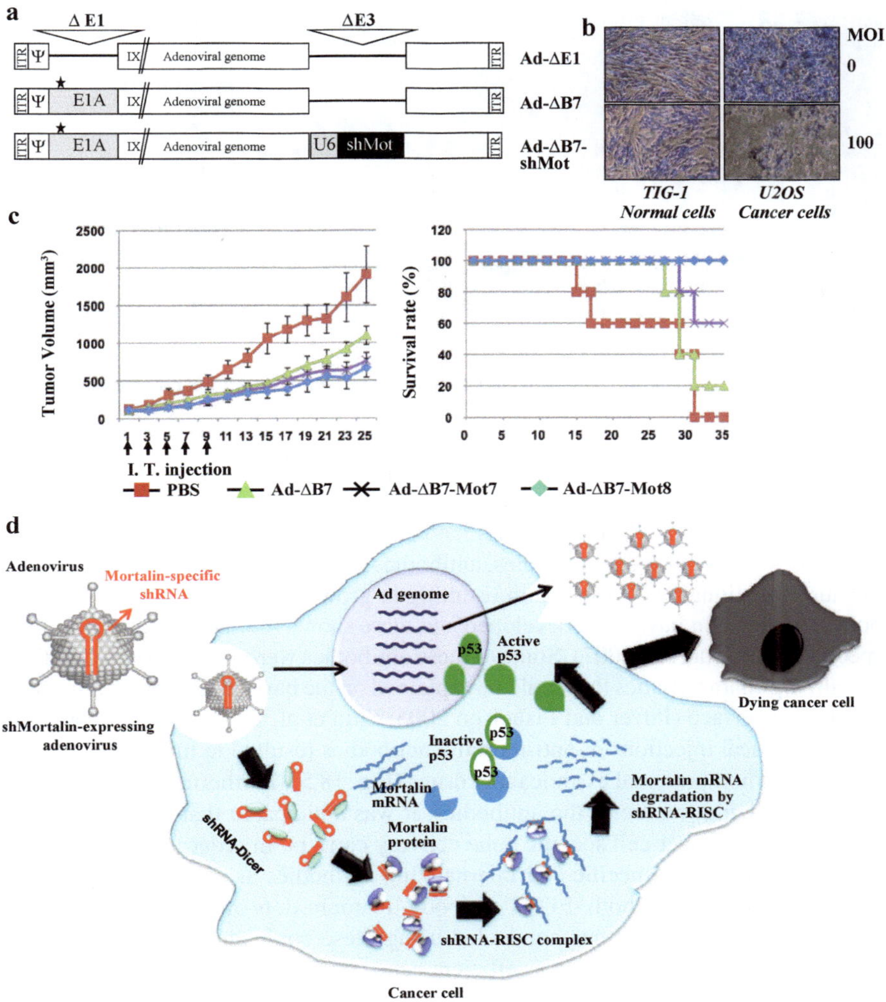

Fig. 18.4 Mortalin silencing by mortalin shRNA-armed adeno-oncolytic virus in human cancer cells. Schematic representation of adeno-oncolytic virus for targeting mortalin (**a**). Effect of mortalin-targeting adeno-oncolytic virus on human normal and cancer cells (**b**). Enhanced tumor suppression and survival advantage of mortalin-targeting adeno-oncolytic viruses (**c**). A model showing cancer cell killing by mortalin shRNA-armed adeno-oncolytic virus (**d**). Modified from Yoo et al. (2010)

Antibodies are preferred for targeted delivery because of the high affinity for their specific antigens. The antibodies used so far for targeted delivery include anti-human epidermal growth factor receptor-2 (HER2) antibody (trastuzumab) to target HER2 protein expressed in breast cancer cells (Chiu et al. 2004; Strehblow et al. 2005), lymphoma cell-specific antibodies (Guillem et al. 2002), ovarian carcinoma-associated

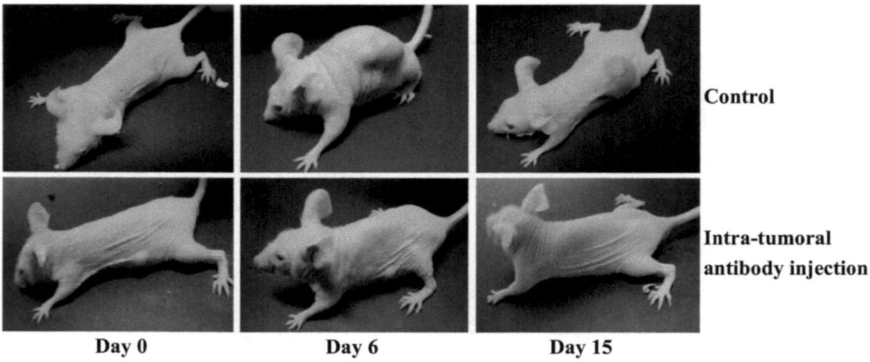

Fig. 18.5 Tumor suppression with polyclonal mortalin antibody. Intra-tumoral injections of the antibody were given every alternate day

antigen (OA3)-specific antibody to target ovarian carcinoma cells (Merdan et al. 2003) and prostate-specific membrane antigen (PSMA)-specific antibody to target prostate cancer cells (Moffatt and Cristiano 2006).

Besides the use of small molecules, antibodies specific for mortalin are anticipated to cause functional loss of mortalin resulting in growth arrest of cancer cells. Antibodies to mortalin have been developed and were shown to bind mortalin with high specificity (Shiota et al. 2007). Some of these antibodies were shown to have a cell internalizing characteristics that could be explained on the basis of presence of mortalin on the cell surface (Pilzer and Fishelson 2005; Shin et al. 2003). Intra-tumoral and intra-peritoneal injections of anti-mortalin antibodies resulted in tumor suppression confirming their potential as anticancer drugs (Fig. 18.5). Furthermore, based on the cell-internalizing feature of the antibodies, it was anticipated that these antibodies could serve as cancer cell specific gene delivery carriers. In order to investigate the potential of mortalin-specific cell internalizing antibodies as nanocarriers, various conjugates such as antibody-FITC, antibody-quantum dots and antibody-plasmids were generated. Experiments conducted using these conjugates showed that anti-mortalin antibodies could serve as efficient nanocarriers. Furthermore, they showed cancer cell-specific delivery characteristics suggesting their use in therapeutics (Kaul et al. 2007b; Shiota et al. 2007).

A variety of mortalin targeting gadgets have been developed and have proven to cause suppression of the cancer cell growth. Although the mechanism of cancer cell death is a continuing theme of research, it is evident that mortalin targeting gadgets could be useful cancer therapeutic tools.

References

Aharinejad S, Paulus P, Sioud M, Hofmann M, Zins K, Schafer R, Stanley ER, Abraham D (2004) Colony-stimulating factor-1 blockade by antisense oligonucleotides and small interfering RNAs suppresses growth of human mammary tumor xenografts in mice. Cancer Res 64:5378–5384

Aoki K, Furuhata S, Hatanaka K, Maeda M, Remy JS, Behr JP, Terada M, Yoshida T (2001) Polyethylenimine-mediated gene transfer into pancreatic tumor dissemination in the murine peritoneal cavity. Gene Ther 8:508–514

Azkur AK, Kim B, Suvas S, Lee Y, Kumaraguru U, Rouse BT (2005) Blocking mouse MMP-9 production in tumor cells and mouse cornea by short hairpin (sh) RNA encoding plasmids. Oligonucleotides 15:72–84

Carette JE, Overmeer RM, Schagen FH, Alemany R, Barski OA, Gerritsen WR, Van Beusechem VW (2004) Conditionally replicating adenoviruses expressing short hairpin RNAs silence the expression of a target gene in cancer cells. Cancer Res 64:2663–2667

Cech TR (1986) Biologic catalysis by RNA. Harvey Lect 82:123–144

Cech TR (2000) Structural biology. The ribosome is a ribozyme. Science 289:878–879

Cerutti J, Trapasso F, Battaglia C, Zhang L, Martelli ML, Visconti R, Berlingieri MT, Fagin JA, Santoro M, Fusco A (1996) Block of c-myc expression by antisense oligonucleotides inhibits proliferation of human thyroid carcinoma cell lines. Clin Cancer Res 2:119–126

Chiu SJ, Ueno NT, Lee RJ (2004) Tumor-targeted gene delivery via anti-HER2 antibody (trastuzumab, Herceptin) conjugated polyethylenimine. J Control Release 97:357–369

Cossu G, Sampaolesi M (2007) New therapies for Duchenne muscular dystrophy: challenges, prospects and clinical trials. Trends Mol Med 13:520–526

Cussac D, Pichereaux C, Colomba A, Capilla F, Pont F, Gaits-Iacovoni F, Lamant L, Espinos E, Burlet-Schiltz O, Monsarrat B et al (2006) Proteomic analysis of anaplastic lymphoma cell lines: identification of potential tumour markers. Proteomics 6:3210–3222

Densmore CL, Kleinerman ES, Gautam A, Jia SF, Xu B, Worth LL, Waldrep JC, Fung YK, T'Ang A, Knight V (2001) Growth suppression of established human osteosarcoma lung metastases in mice by aerosol gene therapy with PEI-p53 complexes. Cancer Gene Ther 8:619–627

Dias N, Stein CA (2002) Potential roles of antisense oligonucleotides in cancer therapy. The example of Bcl-2 antisense oligonucleotides. Eur J Pharm Biopharm 54:263–269

Doudna JA, Cech TR (2002) The chemical repertoire of natural ribozymes. Nature 418:222–228

Dundas SR, Lawrie LC, Rooney PH, Murray GI (2005) Mortalin is over-expressed by colorectal adenocarcinomas and correlates with poor survival. J Pathol 205:74–81

Elbashir SM, Harborth J, Lendeckel W, Yalcin A, Weber K, Tuschl T (2001) Duplexes of 21-nucleotide RNAs mediate RNA interference in cultured mammalian cells. Nature 411:494–498

Fire A, Xu S, Montgomery MK, Kostas SA, Driver SE, Mello CC (1998) Potent and specific genetic interference by double-stranded RNA in Caenorhabditis elegans. Nature 391:806–811

Forte A, Cipollaro M, Cascino A, Galderisi U (2005) Small interfering RNAs and antisense oligonucleotides for treatment of neurological diseases. Curr Drug Targets 6:21–29

Ganly I, Kirn D, Eckhardt G, Rodriguez GI, Soutar DS, Otto R, Robertson AG, Park O, Gulley ML, Heise C et al (2000) A phase I study of Onyx-015, an E1B attenuated adenovirus, administered intratumorally to patients with recurrent head and neck cancer. Clin Cancer Res 6:798–806

Gautam A, Densmore CL, Waldrep JC (2000) Inhibition of experimental lung metastasis by aerosol delivery of PEI-p53 complexes. Mol Ther 2:318–323

Gibbs JB (2000) Mechanism-based target identification and drug discovery in cancer research. Science 287:1969–1973

Guillem VM, Tormo M, Revert F, Benet I, Garcia-Conde J, Crespo A, Alino SF (2002) Polyethyleneimine-based immunopolyplex for targeted gene transfer in human lymphoma cell lines. J Gene Med 4:170–182

Guo W, Lee RL (1999) Receptor-targeted gene delivery via folate-conjugated polyethylenimine. AAPS PharmSci 1:E19

Hattori Y, Kawakami S, Suzuki S, Yamashita F, Hashida M (2004) Enhancement of immune responses by DNA vaccination through targeted gene delivery using mannosylated cationic liposome formulations following intravenous administration in mice. Biochem Biophys Res Commun 317:992–999

Heise C, Ganly I, Kim YT, Sampson-Johannes A, Brown R, Kirn D (2000) Efficacy of a replication-selective adenovirus against ovarian carcinomatosis is dependent on tumor burden, viral replication and p53 status. Gene Ther 7:1925–1929

Hwa Kim S, Hoon Jeong J, Chul Cho K, Wan Kim S, Gwan Park T (2005) Target-specific gene silencing by siRNA plasmid DNA complexed with folate-modified poly(ethylenimine). J Control Release 104:223–232
Jain RK, Duda DG, Clark JW, Loeffler JS (2006) Lessons from phase III clinical trials on anti-VEGF therapy for cancer. Nat Clin Pract Oncol 3:24–40
Jolly D (1994) Viral vector systems for gene therapy. Cancer Gene Ther 1:51–64
Kaul SC, Reddel RR, Mitsui Y, Wadhwa R (2001) An N-terminal region of mot-2 binds to p53 in vitro. Neoplasia 3:110–114
Kaul SC, Deocaris CC, Wadhwa R (2007a) Three faces of mortalin: a housekeeper, guardian and killer. Exp Gerontol 42:263–274
Kaul Z, Yaguchi T, Harada JI, Ikeda Y, Hirano T, Chiura HX, Kaul SC, Wadhwa R (2007b) An antibody-conjugated internalizing quantum dot suitable for long-term live imaging of cells. Biochem Cell Biol 85:133–140
Khuri FR, Nemunaitis J, Ganly I, Arseneau J, Tannock IF, Romel L, Gore M, Ironside J, MacDougall RH, Heise C et al (2000) a controlled trial of intratumoral ONYX-015, a selectively-replicating adenovirus, in combination with cisplatin and 5-fluorouracil in patients with recurrent head and neck cancer. Nat Med 6:879–885
Kirn D (2001a) Clinical research results with dl1520 (Onyx-015), a replication-selective adenovirus for the treatment of cancer: what have we learned? Gene Ther 8:89–98
Kirn D (2001b) Oncolytic virotherapy for cancer with the adenovirus dl1520 (Onyx-015): results of phase I and II trials. Expert Opin Biol Ther 1:525–538
Kirn D (2007) Armed interference: oncolytic viruses engineered to carry antitumor shRNAs. Mol Ther 15:227–228
Kurreck J (2003) Antisense technologies. Improvement through novel chemical modifications. Eur J Biochem 270:1628–1644
Lee LK, Roth CM (2003) Antisense technology in molecular and cellular bioengineering. Curr Opin Biotechnol 14:505–511
Liu Q, Krzewska J, Liberek K, Craig EA (2001) Mitochondrial Hsp70 Ssc1: role in protein folding. J Biol Chem 276:6112–6118
Lou TF, Gray CW, Gray DM (2003) The reduction of Raf-1 protein by phosphorothioate ODNs and siRNAs targeted to the same two mRNA sequences. Oligonucleotides 13:313–324
Lu WJ, Lee NP, Kaul SC, Lan F, Poon RT, Wadhwa R, Luk JM (2011) Mortalin–p53 interaction in cancer cells is stress dependent and constitutes a selective target for cancer therapy. Cell Death Differ 18:1046–1056
Ma Z, Izumi H, Kanai M, Kabuyama Y, Ahn NG, Fukasawa K (2006) Mortalin controls centrosome duplication via modulating centrosomal localization of p53. Oncogene 25:5377–5390
Makinen PI, Koponen JK, Karkkainen AM, Malm TM, Pulkkinen KH, Koistinaho J, Turunen MP, Yla-Herttuala S (2006) Stable RNA interference: comparison of U6 and H1 promoters in endothelial cells and in mouse brain. J Gene Med 8:433–441
Merdan T, Callahan J, Petersen H, Kunath K, Bakowsky U, Kopeckova P, Kissel T, Kopecek J (2003) Pegylated polyethylenimine-Fab' antibody fragment conjugates for targeted gene delivery to human ovarian carcinoma cells. Bioconjug Chem 14:989–996
Moffatt S, Cristiano RJ (2006) PEGylated J591 mAb loaded in PLGA-PEG-PLGA tri-block copolymer for targeted delivery: in vitro evaluation in human prostate cancer cells. Int J Pharm 317:10–13
Pilzer D, Fishelson Z (2005) Mortalin/GRP75 promotes release of membrane vesicles from immune attacked cells and protection from complement-mediated lysis. Int Immunol 17:1239–1248
Robb GB, Brown KM, Khurana J, Rana TM (2005) Specific and potent RNAi in the nucleus of human cells. Nat Struct Mol Biol 12:133–137
Russell WC (2000) Update on adenovirus and its vectors. J Gen Virol 81:2573–2604
Shin BK, Wang H, Yim AM, Le Naour F, Brichory F, Jang JH, Zhao R, Puravs E, Tra J, Michael CW et al (2003) Global profiling of the cell surface proteome of cancer cells uncovers an abundance of proteins with chaperone function. J Biol Chem 278:7607–7616

Shiota M, Ikeda Y, Kaul Z, Itadani J, Kaul SC, Wadhwa R (2007) Internalizing antibody-based targeted gene delivery for human cancer cells. Hum Gene Ther 18:1153–1160

Spankuch B, Matthess Y, Knecht R, Zimmer B, Kaufmann M, Strebhardt K (2004) Cancer inhibition in nude mice after systemic application of U6 promoter-driven short hairpin RNAs against PLK1. J Natl Cancer Inst 96:862–872

Strehblow C, Schuster M, Moritz T, Kirch HC, Opalka B, Petri JB (2005) Monoclonal antibody-polyethyleneimine conjugates targeting Her-2/neu or CD90 allow cell type-specific nonviral gene delivery. J Control Release 102:737–747

Wacheck V, Losert D, Gunsberg P, Vornlocher HP, Hadwiger P, Geick A, Pehamberger H, Muller M, Jansen B (2003) Small interfering RNA targeting bcl-2 sensitizes malignant melanoma. Oligonucleotides 13:393–400

Wadhwa R, Takano S, Robert M, Yoshida A, Nomura H, Reddel RR, Mitsui Y, Kaul SC (1998) Inactivation of tumor suppressor p53 by mot-2, a hsp70 family member. J Biol Chem 273:29586–29591

Wadhwa R, Taira K, Kaul SC (2002a) An Hsp70 family chaperone, mortalin/mthsp70/PBP74/Grp75: what, when, and where? Cell Stress Chaperones 7:309–316

Wadhwa R, Taira K, Kaul SC (2002b) Mortalin: a potential candidate for biotechnology and biomedicine. Histol Histopathol 17:1173–1177

Wadhwa R, Yaguchi T, Hasan MK, Mitsui Y, Reddel RR, Kaul SC (2002c) Hsp70 family member, mot-2/mthsp70/GRP75, binds to the cytoplasmic sequestration domain of the p53 protein. Exp Cell Res 274:246–253

Wadhwa R, Ando H, Kawasaki H, Taira K, Kaul SC (2003) Targeting mortalin using conventional and RNA-helicase-coupled hammerhead ribozymes. EMBO Rep 4:595–601

Wadhwa R, Takano S, Taira K, Kaul SC (2004a) Reduction in mortalin level by its antisense expression causes senescence-like growth arrest in human immortalized cells. J Gene Med 6:439–444

Wadhwa R, Kaul SC, Miyagishi M, Taira K (2004b) Know-how of RNA interference and its applications in research and therapy. Mutat Res 567:71–84

Wadhwa R, Takano S, Kaur K, Deocaris CC, Pereira-Smith OM, Reddel RR, Kaul SC (2006) Upregulation of mortalin/mthsp70/Grp75 contributes to human carcinogenesis. Int J Cancer 118:2973–2980

Wannenes F, Ciafre SA, Niola F, Frajese G, Farace MG (2005) Vector-based RNA interference against vascular endothelial growth factor-A significantly limits vascularization and growth of prostate cancer in vivo. Cancer Gene Ther 12:926–934

Yi X, Luk JM, Lee NP, Peng J, Leng X, Guan XY, Lau GK, Beretta L, Fan ST (2008) Association of mortalin (HSPA9) with liver cancer metastasis and prediction for early tumor recurrence. Mol Cell Proteomics 7:315–325

Yoo JY, Kim JH, Kwon YG, Kim EC, Kim NK, Choi HJ, Yun CO (2007) VEGF-specific short hairpin RNA-expressing oncolytic adenovirus elicits potent inhibition of angiogenesis and tumor growth. Mol Ther 15:295–302

Yoo JY, Kim JH, Kim J, Huang JH, Zhang SN, Kang YA, Kim H, Yun CO (2008) Short hairpin RNA-expressing oncolytic adenovirus-mediated inhibition of IL-8: effects on antiangiogenesis and tumor growth inhibition. Gene Ther 15:635–651

Yoo JY, Ryu J, Gao R, Yaguchi T, Kaul SC, Wadhwa R, Yun CO (2010) Tumor suppression by apoptotic and anti-angiogenic effects of mortalin-targeting adeno-oncolytic virus. J Gene Med 12:586–595

Zamecnik PC, Stephenson ML (1978) Inhibition of Rous sarcoma virus replication and cell transformation by a specific oligodeoxynucleotide. Proc Natl Acad Sci U S A 75:280–284

Zanta MA, Boussif O, Adib A, Behr JP (1997) In vitro gene delivery to hepatocytes with galactosylated polyethylenimine. Bioconjug Chem 8:839–844

Zhang B, Cech TR (1997) Peptide bond formation by in vitro selected ribozymes. Nature 390:96–100

Part V
Mortalin Based Technologies

Chapter 19
Cell Internalizing Anti-Mortalin Antibody for Generation of Illuminating MSCs for Long-Term *In vitro and In vivo* Tracking

Toshimasa Uemura, Masanori Nishi, Sunil C. Kaul and Renu Wadhwa

Abstract Visualize the invisible is the most recent challenge of modern biotechnology in which the advancement of molecular and cell imaging has been prioritized. Fluorescence microscopy has become an essential tool to study biological molecules, pathways and events *in vitro* and *in vivo* for diagnostics and therapeutics. The field not only involves the development of new generation of molecules that are nontoxic and noninvasive, but also concerns their long-term stability, sensitivity and resolution in visual assays. This chapter reviews the use of quantum dots (small, light-emitting semiconductor nanocrystals) in conjugation with the anti-mortalin cell-internalizing antibody for bioimaging. The conjugate was able to enter the cells and illuminate them. The use of such illuminated cells for long-term tracking and imaging in mouse and rabbit models of cartilage and bone differentiation *in vitro* and *in vivo* is described.

Keywords Anti-mortalin antibody · Cell internalization · Quantum dots · Conjugation · Cell tracking · Imaging

19.1 Introduction

Recent progress in stem cell research has transformed the conventional view of multipotent mesenchymal stem cells (MSCs) differentiating into a variety of mesodermal lineage cell types (myocytes, adipocytes, chondrocytes and osteocytes) and tissues (muscle, fat, cartilage and bone, respectively) to their reprogramming and trans-differentiation into other germ layers and tissues with more specialized functions (referred to as stem cell plasticity) (Phinney and Darwin 2007; Thowfeequ et al. 2007). Breakthrough in iPS (induced pluripotent stem cells) field has further established that the pluripotent stem cells could be artificially derived from a non-pluripotent adult somatic cells by inserting certain genes (Okita et al. 2007; Yu et al. 2007; Nakagawa et al. 2008). While holding great potential for basic, clinical, regenerative and therapeutic medicine, stem cell research is yet to cross barriers of safety, efficacy and functionality in *in vitro* and *in vivo* settings. Mesenchymal stem

T. Uemura (✉) · M. Nishi · S. C. Kaul · R. Wadhwa
National Institute of Advanced Industrial Science and Technology (AIST),
1-1-1 Higashi, Tsukuba, Ibaraki 305-8562, Japan
e-mail: t.uemura@aist.go.jp

S. C. Kaul, R. Wadhwa (eds.), *Mortalin Biology: Life, Stress and Death*,
DOI 10.1007/978-94-007-3027-4_19, © Springer Science+Business Media B.V. 2012

cells (MSCs) are being tested widely as a cell resource for disease modeling and drug discoveries, tissue engineering and therapeutics, and clinical trials (Ohgushi et al. 2004; Miyahara et al. 2006; Scott and Baker 2007; Swieszkowski et al. 2007; Kagiwada et al. 2008). Core requisites to such basic and clinical research are the cell labeling technologies that can track the implanted cells from the host tissues. In contrast to the rapidly developing stem cell technologies, there are only limited methods to label cells in a stable and non-invasive way, precluding the long-term functional studies. Current methods for labeling MSCs include either exogenously introducing radioactive probes (Stodilka et al. 2006), ultra small iron particles (Flexman et al. 2006), organic fluorescent dyes (Templin et al. 2006), semiconductor nanoparticles (Quantum dots) (Gao et al. 2002; Kaji et al. 2007; Muller-Borer et al. 2007) or the fluorescent proteins expressed by the cells (Rubart 2004; Stadtfeld et al. 2005). However, evidence to their genetic non-invasiveness and functional inertness is still limiting.

19.2 Quantum Dots for Bio-Imaging

Genetically encoded fluorescent proteins such as green fluorescent protein (GFP) have been widely used for cell labeling. GFPs are spontaneous fluorescent proteins isolated from a jellyfish, *Aequorea victoria* (Lippincott-Schwartz and Smith 1997; Stahl et al. 2005). GFPs are transfected into the cells via either viral (retrovirus, lentivirus) or nonviral approaches. In comparison with the organic dyes such as FITC, GFPs have a number of advantages such as better stability and pH tolerance in addition to longer luminescence time. However, GFPs suffer from a number of intrinsic deficiencies such as sensitivity to proteolytic enzymes and overlap with autofluorescence signal, thus making it difficult for *in vivo* cell tracking (Jaiswal et al. 2004; Gao and Nie 2005).

Quantum dots (QDs) are small, light-emitting semiconductor nanocrystals, typically in the size of 15–20 nm (Frangioni 2003, 2006). QDs are generally composed of atoms from groups II–VI (e.g., CdSe, CdTe, CdS, and ZnSe) or III–V (e.g., InP and InAs) of the periodic table, and are nanoparticles with physical dimensions smaller than the excitation Bohr radius. The diameter of QDs determines their emission and excitation spectra and can be fine-tuned (Jaiswal and Simon 2004). For biological application, highly luminescent QDs were made water soluble and tethered to biomolecules for cell labeling (Bruchez et al. 1998; Chan and Nie 1998). Typical QDs with multiple emission spectra are shown in Fig. 19.1. QDs have several marked advantages for long-term cell labeling and *in vivo* cell tracking. They are generally stable and maintain fluorescent intensity in a biological setting for a long time (Marks and Nolan 2006; Alivisatos 2004).

Based on the advantages of QDs, they have been applied to several biological studies since some pioneering works had been published (Bruchez et al. 1998; Chan and Nie 1998; Han et al. 2001; Chan et al. 2002; Jaiswal et al. 2003). Many of the studies relate to cancer cells and a long-term tracking of their metastatic behaviors *in vivo* (Voura et al. 2004; Cai et al. 2006). Several methods have been designed to

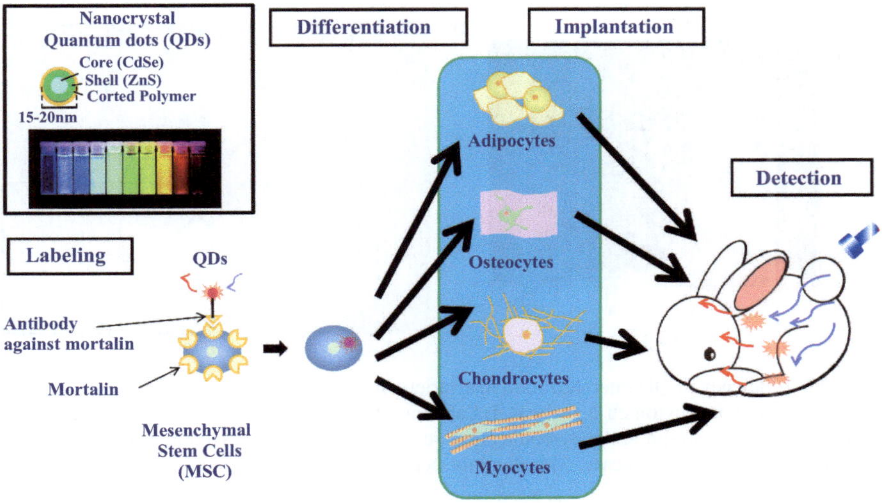

Fig. 19.1 Stable and non-disruptive *in vivo* labeling of MSCs by internalizing QDs. Typical quantum dots (QDs) for biological application: Core: cadmium selenide. Shell: zinc sulfide. These quantum dots are polymer coated for their biocompatibility; The evaluation protocol of the regenerative process by internalization of QDs to MSCs. Antibodies against mortalin were conjugated to QDs, internalized into MSCs. After differentiation to each mesenchymal tissue, MSCs derived cells were implanted and optically detected for their tracking

label live cells by QDs. These include their endocytotic uptake, selective labeling of cell surface proteins with QDs conjugated to either the antibodies (Jaiswal et al. 2003) or peptides (Cai et al. 2006; Yukawa et al. 2010).

19.3 Labeling Stem Cells with Quantum Dots

Stem cells give rise to tissue progenitor cells, which differentiate into tissue-forming cells. For example, mesenchymal stem cells (MSCs) give rise to osteoprogenitor cells, which differentiate into bone-forming osteoblasts (Friedenstein et al. 1970; Caplan 1991; Colter et al. 2001). In addition to osteogenic differentiation, MSCs are capable of differentiating into chondrocytes, myocytes, adipocytes, fibroblasts, etc. (Pittenger et al. 1999; Alhadlaq and Mao 2004; Marion and Mao 2006). In spite of the recent development of iPS (induced pluripotent stem cells) and ES (embryonic stem cells) cells, MSCs are still expected to be a major cell source for tissue engineering, especially for mesenchymal tissue repair because of their easy extraction from bone marrow and treatment. Recently, it is becoming widely recognized that the tracking of the implanted cells *in vivo* is clinically important after implantation of stem cells or their derived cells to the patients for confirming their contribution to tissue regeneration and safety. However, appropriate methods have not yet been developed. In the present chapter, a labeling method of MSCs by anti-mortalin antibody-QD conjugate

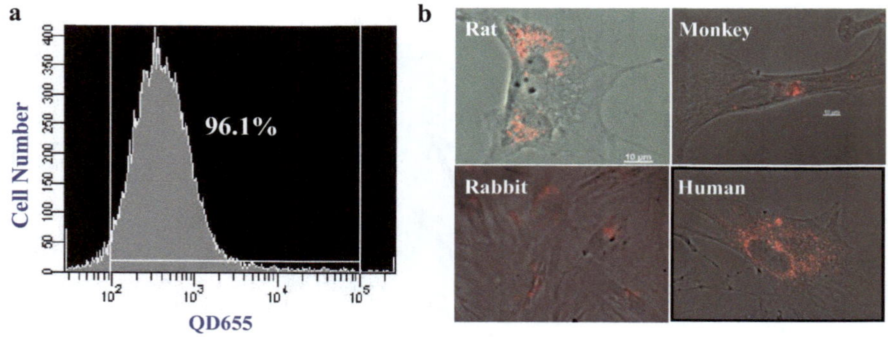

Fig. 19.2 Efficiency of QD internalization in a variety of cells. **a** High internalization efficiency of QDs to MSCs by flowcytometric analysis. **b** Labeling of rat, rabbit, monkey and human MSCs by internalizing anti-mortalin antibody-conjugated quantum dots (i-QDs). Highly efficient internalization was observed in each case. Almost all the cells in culture showed uptake of i-QDs (Ohyabu et al. 2009)

as a candidate for labeling techniques in stem cell therapy (Fig. 19.1) including tissue engineering is described.

19.4 *In Vitro/In Vivo* Labeling of MSCs by Internalizing QDs

For labeling of MSCs, internalizing quantum dots (i-QDs) were synthesized by conjugating them to a unique internalizing antibody against a heat shock protein 70 (hsp70) family stress chaperone, mortalin, that is upregulated and expressed on the surface of dividing cells. Previously, it was reported that the i-QDs could be used for long-term imaging of cancer cells in culture (Kaul et al. 2007). In this review, the use of i-QDs is reported for stable, genetically noninvasive, and functionally inert labeling of MSCs for long-term *in vivo* tracking.

As mortalin is a highly conserved protein, anti-mortalin antibody based i-QDs were expected to be functional in multiple species. i-QDs were prepared by conjugation of Qdot 655 (Invitrogen, 655 nm emission) with an internalizing anti-mortalin monoclonal antibody using an antibody conjugation kit as described (Kaul et al. 2007; Shiota et al. 2007). For generating i-QD-labeled mesenchymal cells, QD-antibody conjugates were added to cell culture medium for 24 h. Flow cytometric analysis showed that the bone marrow derived MSCs were labeled by i-QDs with high internalization efficiency of 96% as shown in Fig. 19.2. Moreover, MSCs from rat, rabbit, monkey and human were all labeled efficiently by i-QDs (Ohyabu et al. 2009). In each case, more than 90% of cells took up i-QDs and were detected on serial sectioning of cells (Fig. 19.2). This internalization efficiency was higher than a peptide conjugation method that gives about 60% efficiency.

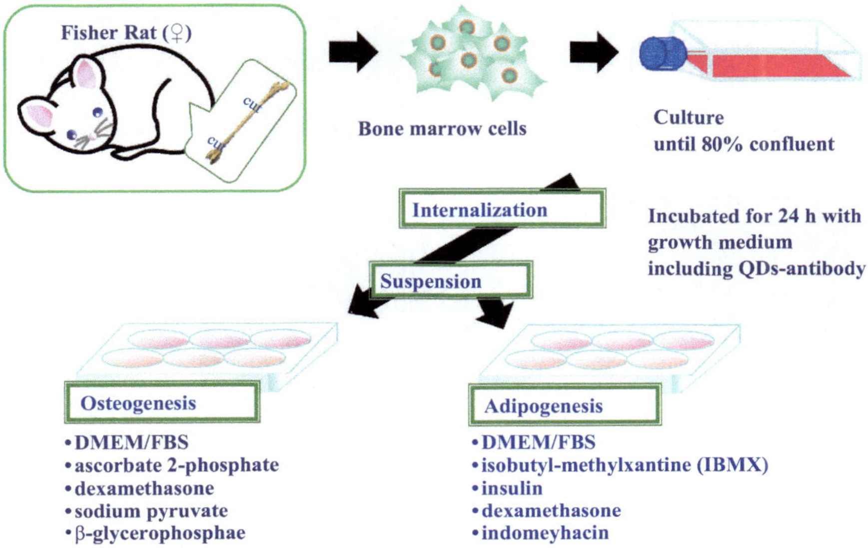

Fig. 19.3 Schematic diagram showing the *in vitro* experiment of the QDs internalization to rat MSCs and differentiation into osteoblasts and adipocytes. Rat bone marrow cells harvested from the femora of male were cultured until 80% confluent. After internalization of QD into the MSCs, the cells were differentiated into osteoblasts and adipocytes in osteogenesis and adipogenesis induction medium, respectively

19.5 *In vitro* Differentiation of i-QDs Labeled MSCs

Whether i-QD-labeled MSCs (i-MSCs) were fully functional and undergo the expected level of *in vitro* differentiation was examined by comparing with untreated MSCs (c-MSCs). The i-QD-labeled cells were subcultured for three kinds of differentiation: (1) adipogenic differentiation, (2) osteogenic differentiation and (3) chondrogenic differentiation (Fig. 19.3), by two representative culture systems: pellet (static) culture and RWV (rotating wall vessel) bioreactor 3D culture (Ohyabu et al. 2009; Yoshioka et al. 2011). The cells subjected to osteogenic and adipogenic differentiation were stained with tissue specific markers: alizarin red for osteoblasts and oil red O for adipocytes (Fig. 19.4). The data suggested that rat i-MSCs underwent adipogenesis and osteogenesis with the same efficiency and time course as c-MSCs (Fig. 19.4). There were no differences in alizarin red staining intensities at 1 or 2 weeks of differentiation. Similarly, oil red O staining, an established marker of adipocytes, revealed that the i-MSCs were not affected in their adipogenic potential. Similar results were obtained for monkey and human MSCs. For chondrogenic differentiation, rabbit i-MSCs were cultured in chondrogenic differentiation medium by RWV (rotating wall vessel) bioreactor following the method described earlier

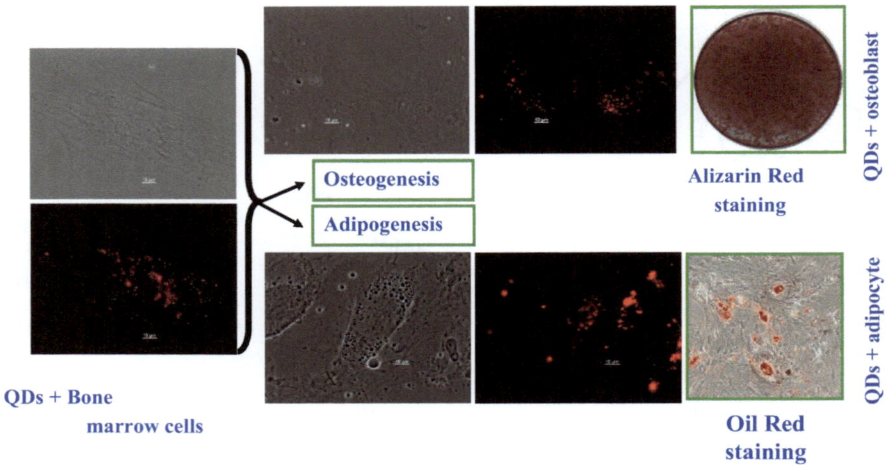

Fig. 19.4 Differentiation of i-QDs labeled rat bone marrow derived MSCs into adipocytes and osteoblasts confirmed by Oil Red and Alizarin Red staining, respectively. The cells underwent osteogenesis and adipogenesis with the same efficiency and time course as the control MSCs (Ohyabu et al. 2009)

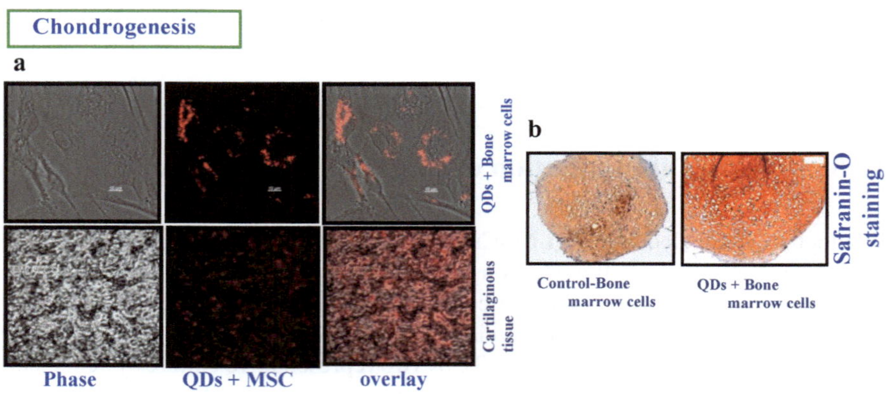

Fig. 19.5 Differentiation of i-QD-labeled rabbit MSCs to chondrocytes and cartilage tissues. **a** i-QDs were detected both in chondrocytes and cartilage. The efficiency of chondrogenesis among c-MSCs and i-MSCs was examined by **b** Safranin O and Toluidine blue staining (Ohyabu et al. 2009)

(Ohyabu et al. 2006; Yoshioka et al. 2007; Sakai et al. 2009). Chondrogenic differentiation of the cultured i-MSCs as examined by Safranin O and Toluidine blue staining (Fig. 19.5), and measurement of GAG content revealed no difference between the differentiation potential of c-MSCs and i-MSCs. Differentiated cells and tissues retained strong QD signals.

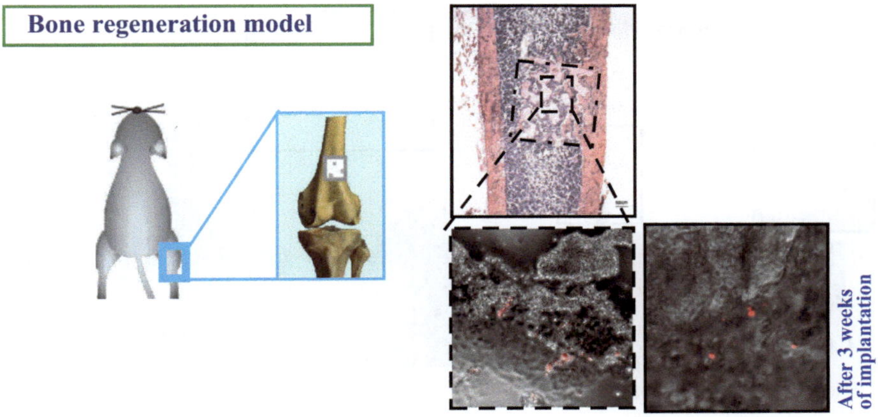

Fig. 19.6 *In vivo* osteogenic differentiation of i-QD-labeled rat MSCs (cultured bone transplantation model). Normal bone structures were formed as seen by histochemistry. Substantial QD fluorescence was retained even after differentiation (Ohyabu et al. 2009)

19.6 *In vivo* Osteogenic Differentiation of i-QDs Labeled MSCs

Due to their stable and long term fluorescent properties, QDs were anticipated, so far, most useful for *in vivo* imaging of MSCs. Here, its application to cultured bone transplantation model is introduced. This bone regeneration model was first developed by Caplan (Caplan 1991) using bone marrow derived MSCs (Maniatopoulos et al. 1988). Beta-tricalcium phosphate (β-TCP) cubic blocks (2 × 2 × 2 mm; average pore diameter, 200 mm) were used as scaffolds for rat bone marrow derived i-QD-labeled MSCs (i-MSCs) and control MSCs (c-MSCs). β-TCP/MSCs composites were incubated in osteogenic medium and transplanted into rat femur orthotopic defects (Dong et al. 2002; Uemura et al. 2003). Differentiation of the implanted MSCs was examined on Day 26 and at 8 weeks by histologic analysis as shown in Fig. 19.6. The c-MSCs and i-MSCs differentiated to form bone matrix with the same efficiency and time course. Hematoxylin and eosin staining of rat bone sections, three weeks after implantation, revealed extensive bone formation and the presence of i-MSCs in the newly formed bone structures. Furthermore, the QD signals remained strong enough to be detected by microscopy. The results revealed that the regenerated bone in the porous area of the scaffold was derived mostly from implanted donor cells and their derivative populations. The results are consistent with the previous reports (Kojima and Uemura 2005).

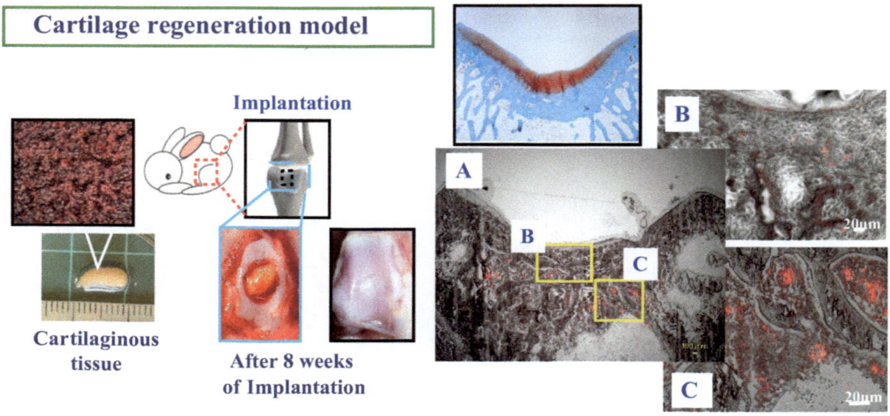

Fig. 19.7 *In vivo* chondrogenic differentiation of i-QD-labeled rabbit MSCs (osteochondral defect model). Three-dimensional cartilaginous tissues regenerated from rabbit i-MSCs by RWV bioreactor were transplanted into osteochondral defects of rabbit knee joints, the fate of transplanted i-QDs labeled cells were tracked both phenotypically and spatially for up to 26 weeks (Ohyabu et al. 2009; Yoshioka et al. 2007)

19.7 *In vivo* Chondrogenic Differentiation of i-QDs Labeled MSCs

Currently the transplantation of MSCs is investigated as a potential treatment for cartilage repair in animal and clinical studies (Wakitani et al. 2008). In this study, three dimensional cartilaginous tissues regenerated from rabbit i-MSCs by RWV bioreactor were transplanted into osteochondral defects of rabbit knee joints, the fate of transplanted i-QDs labeled cells was tracked both phenotypically and spatially for a long time as shown in Fig. 19.7. Cartilaginous aggregates of about 7-mm along the major axis labeled or non-labeled were transplanted into the cylindrical defects 5×5 mm in area and 4 mm in depth created on the patellar grooves. After the operation, the animals were then caged and allowed to move freely without any splinting after recovery, before finally being sacrificed at 4, 8 and 26 weeks. After the animals were sacrificed, the defects of the patellar groove were examined macroscopically. At 4 weeks, the defects were covered with a white, glossy, smooth membrane, which resembled articular cartilage and remained for the entire 26 weeks. At 4 weeks, a QD signal was observed mainly in the middle and deep zones of the regenerated tissue. At 8 weeks, signal was observed throughout the regenerated tissue in the cartilage layer, subchondral bone layer and bone marrow stroma, with some labeling observed at the surface of the trabecular bone. Signal was still observed at 26 weeks after transplantation, mainly in bone marrow stroma but also on the surface of the trabecular bone. This study showed that the labeling of rabbit MSCs with anti-mortalin antibody-conjugated i-QDs is a well-tolerable procedure and provides a stable long-term fluorescence signal during the cartilage repair process for up to 26 weeks.

19.8 Summary and Future Prospects

Internalizing quantum dots (i-QDs) generated by their conjugation with an internalizing antibody against mortalin, offered an efficient, genetically noninvasive, nontoxic, and functionally inert way to label mesenchymal stem cells (MSCs). The i-QD-labeled MSCs underwent normal adipogenic, osteogenic and chondrogenic differentiation *in vitro* and *in vivo*. Based on these data, the potential application of i-QDs in *in vivo* diagnostics, regenerative and therapeutic medicine is suggested. Clinical application of i-QDs warrants further studies on their risk assessment and generation of modified bio-compatible QDs and their detection in deeply seated tissue sites by multiphoton and near-field optical imaging. Combining these different technologies will promote an interdisciplinary research for novel bioimaging methodology for clinical medicine.

Acknowledgments The authors thank Z. Kaul, Y. Ohyabu, T. Yoshioka, S. Sakai, H. Mishima, N. Ochiai and K. Inoue for their collaboration, technical assistance and encouragement through this work.

References

Alhadlaq A, Mao JJ (2004) Mesenchymal stem cells: isolation and therapeutics. Stem Cells Dev 13:436–48
Alivisatos P (2004) The use of nanocrystals in biological detection. Nat Biotechnol 22:47–52
Bruchez M Jr, Moronne M, Gin P, Weiss S, Alivisatos AP (1998) Semiconductor nanocrystals as fluorescent biological labels. Science 281:2013–2016
Cai W, Shin DW, Chen K, Gheysens O, Cao Q, Wang SX, Gambhir SS, Chen X (2006) Peptide-labeled near-infrared quantum dots for imaging tumor vasculature in living subjects. Nano Lett 6:669–676
Caplan AI (1991) Mesenchymal stem cells. J Orthop Res 9:641–650
Chan WC, Nie S (1998) Quantum dot bioconjugates for ultrasensitive nonisotopic detection. Science 281:2016–2018
Chan WC, Maxwell DJ, Gao X, Bailey RE, Han M, Nie S (2002) Luminescent quantum dots for multiplexed biological detection and imaging. Curr Opin Biotechnol 13:40–46
Colter DC, Sekiya I, Prockop DJ (2001) Identification of a subpopulation of rapidly self-renewing and multipotential adult stem cells in colonies of human marrow stromal cells. Proc Natl Acad Sci U S A 98:7841–7845
Dong J, Uemura T, Shirasaki Y, Tateishi T (2002) Promotion of bone formation using highly pure porous beta-TCP combined with bone marrow-derived osteoprogenitor cells. Biomaterials 23:4493–4502
Frangioni JV (2003) In vivo near-infrared fluorescence imaging. Curr Opin Chem Biol 7:626–634
Frangioni JV (2006) Self-illuminating quantum dots light the way. Nat Biotechnol 24:326–328
Flexman JA, Minoshima S, Kim Y, Cross DJ (2006) Magneto-optical labeling of fetal neural stem cells for in vivo MRI tracking. Conf Proc IEEE Eng Med Biol Soc 1:5631–5634
Friedenstein AJ, Chailakhjan RK, Lalykina KS (1970) The development of fibroblast colonies in monolayer cultures of guinea-pig bone marrow and spleen cells. Cell Tissue Kinet 3:393–403
Gao X, Chan WC, Nie S (2002) Quantum-dot nanocrystals for ultrasensitive biological labeling and multicolor optical encoding. J Biomed Opt 7:532–537
Gao X, Nie S (2005) Quantum dot-encoded beads. Methods Mol Biol 303:61–71

Han M, Gao X, Su JZ, Nie S (2001) Quantum-dot-tagged microbeads for multiplexed optical coding of biomolecules. Nat Biotechnol 19:631–635

Jaiswal JK, Mattoussi H, Mauro JM, Simon SM (2003) Long-term multiple color imaging of live cells using quantum dot bioconjugates. Nat Biotechnol 21:47–51

Jaiswal JK, Goldman ER, Mattoussi H, Simon SM (2004) Use of quantum dots for live cell imaging. Nat Methods 1:73–78

Jaiswal JK, Simon SM (2004) Potentials and pitfalls of fluorescent quantum dots for biological imaging. Trends Cell Biol 14:497–504

Kagiwada H, Yashiki T, Ohshima A, Tadokoro M, Nagaya N, Ohgushi H (2008) Human mesenchymal stem cells as a stable source of VEGF-producing cells. J Tissue Eng Regen Med 2:184–189

Kaji N, Tokeshi M, Baba Y (2007) Single molecule measurement with a single quantum dot. Chem Rec 7:295–304

Kaul Z, Yaguchi T, Harada JI, Ikeda Y, Hirano T, Chiura HX, Kaul SC, Wadhwa R (2007) An antibody-conjugated internalizing quantum dot suitable for long-term live imaging of cells. Biochem Cell Biol 85:133–140

Kojima H, Uemura T (2005) Strong and rapid induction of osteoblast differentiation by Cbfa1/Til-1 overexpression for bone regeneration. J Biol Chem 280:2944–2953

Lippincott-Schwartz J, Smith CL (1997) Insights into secretory and endocytic membrane traffic using green fluorescent protein chimeras. Curr Opin Neurobiol 7:631–639

Maniatopoulos C, Sodek J, Melcher AH (1988) Bone formation in vitro by stromal cells obtained from bone marrow of young adult rats. Cell Tissue Res 254:317–3130

Marks KM, Nolan GP (2006) Chemical labeling strategies for cell biology. Nat Methods 3:591–596

Marion NW, Mao JJ (2006) Mesenchymal stem cells and tissue engineering. Methods Enzymol 420:339–361

Miyahara Y, Nagaya N, Kataoka M, Yanagawa B, Tanaka K, Hao H, Ishino K, Ishida H, Shimizu T, Kangawa K, Sano S, Okano T, Kitamura S, Mori H (2006) Monolayered mesenchymal stem cells repair scarred myocardium after myocardial infarction. Nat Med 12:459–465

Muller-Borer BJ, Collins MC, Gunst PR, Cascio WE, Kypson AP (2007) Quantum dot labeling of mesenchymal stem cells. J Nanobiotechnol 5:9

Nakagawa M, Koyanagi M, Tanabe K, Takahashi K, Ichisaka T, Aoi T, Okita K, Mochiduki Y, Takizawa N, Yamanaka S (2008) Generation of induced pluripotent stem cells without Myc from mouse and human fibroblasts. Nat Biotechnol 26:101–106

Ohgushi H, Kitamura S, Kotobuki N, Hirose M, Machida H, Muraki K, Takakura Y (2004) Clinical application of marrow mesenchymal stem cells for hard tissue repair. Yonsei Med J 45(Suppl):61–67

Ohyabu Y, Kida N, Kojima H, Taguchi T, Tanaka J, Uemura T (2006) Cartilaginous tissue formation from bone marrow cells using rotating wall vessel (RWV) bioreactor. Biotechnol Bioeng 95:1003–1008

Ohyabu Y, Kaul Z, Yoshioka T, Inoue K, Sakai S, Mishima H, Uemura T, Kaul SC, Wadhwa R (2009) Stable and nondisruptive in vitro/in vivo labeling of mesenchymal stem cells by internalizing quantum dots. Hum Gene Ther 20:217–224

Okita K, Ichisaka T, Yamanaka S (2007) Generation of germline-competent induced pluripotent stem cells. Nature 448:313–317

Phinney DG, Darwin J (2007) Concise review: mesenchymal stem/multipotent stromal cells: the state of transdifferentiation and modes of tissue repair. Stem cells 35:2896–2902

Pittenger MF, Mackay AM, Beck SC, Jaiswal RK, Douglas R, Mosca JD, Moorman MA, Simonetti DW, Craig S, Marshak DR (1999) Multilineage potential of adult human mesenchymal stem cells. Science 1284:143–147

Rubart M (2004) Two-photon microscopy of cells and tissue. Circ Res 95:1154–1166

Sakai S, Mishima H, Ishii T, Akaogi H, Yoshioka T, Ohyabu Y, Chang F, Ochiai N, Uemura T (2009) Rotating three-dimensional dynamic culture of adult human bone marrow-derived cells for tissue engineering of hyaline cartilage. J Orthop Res 27:517–521

Scott CT, Baker M (2007) Overhauling clinical trials. Nat Biotechnol 25:287–292

Shiota M, Ikeda Y, Kaul Z, Itadani J, Kaul SC, Wadhwa R (2007) Internalizing antibody-based targeted gene delivery for human cancer cells. Hum Gene Ther 18:1153–1160

Stadtfeld M, Varas F, Graf T (2005) Fluorescent protein-cell labeling and its application in time-lapse analysis of hematopoietic differentiation. Methods Mol Med 105:395–412

Stahl A, Wu X, Wenger A, Klagsbrun M, Kurschat P (2005) Endothelial progenitor cell sprouting in spheroid cultures is resistant to inhibition by osteoblasts: a model for bone replacement grafts. FEBS Lett 579:5338–5342

Stodilka RZ, Blackwood KJ, Kong H, Prato FS (2006) A method for quantitative cell tracking using SPECT for the evaluation of myocardial stem cell therapy. Nucl Med Commun 27:807–813

Swieszkowski W, Tuan BH, Kurzydlowski KJ, Hutmacher DW (2007) Repair and regeneration of osteochondral defects in the articular joints. Biomol Eng 24:489–495

Templin C, Kotlarz D, Marquart F, Faulhaber J, Brendecke V, Schaefer A, Tsikas D, Bonda T, Hilfiker-Kleiner D, Ohl L, Naim HY, Foerster R, Drexler H, Limbourg FP (2006) Transcoronary delivery of bone marrow cells to the infarcted murine myocardium: feasibility, cellular kinetics, and improvement in cardiac function. Basic Res Cardiol 101:301–310

Thowfeequ S, Myatt E, Tosh D (2007) Tansdifferentiation in developmental biology, disease, and in therapy. Dev Dyn 236:3208–3217

Uemura T, Dong J, Wang Y, Kojima H, Saito T, Iejima D, Kikuchi M, Tanaka J, Tateishi T (2003) Transplantation of cultured bone cells using combinations of scaffolds and culture techniques. Biomaterials 24:2277–2286

Voura EB, Jaiswal JK, Mattoussi H, Simon SM (2004) Tracking metastatic tumor cell extravasation with quantum dot nanocrystals and fluorescence emission-scanning microscopy. Nat Med 10:993–998

Wakitani S, Kawaguchi A, Tokuhara Y, Takaoka K (2008) Present status of and future direction for articular cartilage repair. J Bone Miner Metab 26:115–122

Yoshioka T, Mishima H, Ohyabu Y, Sakai S, Akaogi H, Ishii T, Kojima H, Tanaka J, Ochiai N, Uemura T (2007) Repair of large osteochondral defects with allogeneic cartilaginous aggregates formed from bone marrow-derived cells using RWV bioreactor. J Orthop Res 25:1291–1298

Yoshioka T, Mishima H, Kaul Z, Ohyabu Y, Sakai S, Ochiai N, Kaul SC, Wadhwa R, Uemura T (2011) Fate of bone marrow mesenchymal stem cells following the allogeneic transplantation of cartilaginous aggregates into osteochondral defects of rabbits. J Tissue Eng Regen Med 5:437–443

Yu J, Vodyanik MA, Smuga-Otto K, Antosiewicz-Bourget J, Frane JL, Tian S, Nie J, Jonsdottir GA, Ruotti V, Stewart R, Slukvin II, Thomson JA (2007) Induced pluripotent stem cell lines derived from human somatic cells. Science 318:1917–1920

Yukawa H, Kagami Y, Watanabe M, Oishi K, Miyamoto Y, Okamoto Y (2010) Quantum dots labeling using octa-arginine peptides for imaging of adipose tissue-derived stem cells. Biomaterials 31:4094–4103

Chapter 20
Mortalin Staining Pattern as a Reporter for Cell Based Anti-Cancer Drug Screening

Ran Gao, Zeenia Kaul, Tomoko Yaguchi and Renu Wadhwa

Abstract Normal cells can divide only for a limited number of times *in vitro* (Hayflick's limit) and reach an irreversible non-proliferative state referred to as replicative cellular senescence. Since many tumor suppressor genes that are inactivated in cancers are highly expressed in senescing cells, cellular senescence is widely regarded as a tumor suppressor mechanism that acts as a safeguard for cancer. Indeed, forced expression of tumor suppressor proteins have been shown to induce growth arrest of cancer cells that often mimics replicative senescence. Mortalin is a member of the Hsp70 family of proteins and exhibits pancytoplasmic distribution pattern in normal and perinuclear in human cancer cells. Cancer cells when induced to senesce by a variety of chemicals, such as BrdU, 5AZA-dC, MKT-077 and phytochemicals, showed shift in mortalin staining pattern from the perinuclear to the pancytoplasmic type, associated with nuclear translocation and activation of p53 function. Using mortalin staining as a model reporter, human shRNA library was screened for anticancer candidate. Bioinformatics analysis of the gene targets of the selected shRNAs revealed the involvement of DNA damage signaling pathway as a candidate target for anticancer therapy. In this chapter, use of mortalin staining as a reporter for anticancer drug screening and its outcome will be discussed.

Keywords Mortalin · Immunostaining · Perinuclear · Pancytoplasmic · Drug screening

20.1 Introduction

Mortalin is a member of Hsp70 family of proteins, first identified in the cytoplasmic fractions of normal and hybrid (generated from the fusion of normal and immortal cells) mouse cells with mortal phenotype (Wadhwa et al. 1993a). A specific antibody

R. Wadhwa (✉) · R. Gao · T. Yaguchi
National Institute of Advanced Industrial Science and Technology (AIST), Central 4, 1-1-1 Higashi, Tsukuba, Ibaraki 305-8562, Japan
e-mail: renu-wadhwa@aist.go.jp

Z. Kaul
Center for Childhood Cancer, The Research Institute at Nationwide Children's Hospital, Columbus, Ohio, USA

S. C. Kaul, R. Wadhwa (eds.), *Mortalin Biology: Life, Stress and Death*,
DOI 10.1007/978-94-007-3027-4_20, © Springer Science+Business Media B.V. 2012

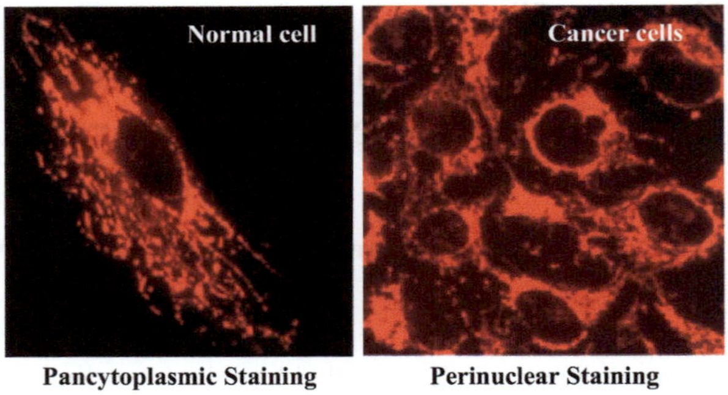

Fig. 20.1 Differential staining pattern of mortalin in normal (pancytoplasmic) and cancer (perinuclear) cells. Modified from Gao et al. (2010)

raised against the protein did not crossreact to any other member of hsp70 family proteins and revealed that in normal cells mortalin exists widely in the cytoplasm, hence called pancytoplasmic, and in immortal and cancer cells it concentrate mostly around the nucleus, hence called perinuclear (Fig. 20.1) (Wadhwa et al. 1993b). Whereas the latter was observed in a large variety of human immortal, transformed and malignant cells, all normal cells showed pancytoplasmic staining for mortalin (Wadhwa et al. 1995) suggesting that mortalin staining pattern could be a reliable marker for senescence phenotype. The data was further supported by inducing senescence in cancer cells either by genes or drugs and visualizing their mortalin staining pattern.

20.2 Induction of Senescence in Human Transformed Cells by Single Chromosomes and Chromosome Fragments was Associated with Change in Mortalin Staining Pattern

Replicative cellular senescence of normal human cells is an established phenomenon. Several studies have demonstrated a genetic basis for senescence. The hybrids yielded by the fusion of normal with immortal human cells ceased proliferating, indicating that the senescence is dominant over immortalization and had led to the idea of senescence playing a role in tumor suppression (Pereira-Smith and Smith 1983; Smith and Pereira-Smith 1996). Fusion of different immortal human cell lines resulted in the identification of four complementation groups for indefinite division (Group A–D), suggesting the involvement of, at least, four genetic pathways in causing senescence. Human chromosomes carrying senescence genes have been identified as chromosomes 1 (for Group C), 4 (for Group B), and 7 (for Group D). The identification was based on the ablility of the chromosome to induce senescence in multiple immortal cell lines assigned to one of the complementation groups, with no effect on the proliferation of cell lines assigned to the other groups (Hensler et al. 1994; Ning et al. 1991; Ogata et al. 1995). Nakabayashi et al. (1997) have shown that the introduction of chromosome 7 by microcell-mediated chromosome transfer induced growth arrest in SUSM-1 that involves senescence, tumor suppression, terminal differentiation, and apoptosis. Also, in the microcell hybrids that stopped dividing, mortalin distributed uniformly in the cytoplasm, similar to the cytoplasmic type seen in mortal cells (Nakabayashi et al. 1997). Bertram et al. (1999) have cloned mortality factor 4 (MORF 4) on chromosome 4 that induces a senescent-like phenotype in immortal cell lines assigned to the complementation group B with concomitant changes in two markers for senescence, β-galactosidase and mortalin (Bertram et al. 1999). Cells that stopped dividing showed increased level of expression of senescence-associated β-galactosidase and shift in staining pattern of mortalin from the perinuclear to the pancytoplasmic type.

20.3 Drug Induced Senescence in Cancer Cells Involves Change in Mortalin Staining Pattern

5-Bromodeoxyuridine was found to induce flat and enlarged cell shape, characteristics of senescent cells, and senescence-associated β-galactosidase in mammalian cells regardless of cell type or species. In immortal human cells, fibronectin, collagenase I, and $p21^{WAF1}$ mRNAs were immediately and very strongly induced. Furthermore, the staining pattern of mortalin also changed to the pancytoplasmic (mortal) type from the perinucler (immortal) type (Michishita et al. 1999).

5′-upstream regulatory regions of genes often contain 5-methylcytosine that impose an important control on regulation of gene expression, DNA replication and chromatin-organization. Methylated DNA recruits histone deacetylase (HDAC) resulting in deacetylated histones and condensed chromatin structure in which access of transcriptional machinery to the target gene is blocked leading to transcriptional repression (Sakajiri et al. 2005). Hyper-methylation and hence the silencing of tumor suppressor genes are viewed as important regulatory events in immortalization, tumor initiation and progression of cancer (Enokida et al. 2004; Huang et al. 2003; Lichtenstein and Kisseljova 2001; Sugimoto et al. 2004). On the other hand, global hypo-methylation (reduction in genomic 5-methylcytosine content) has been recognized as a cause of oncogenesis that correlates with transformation and tumor progression (Das and Singal 2004; Hiltunen et al. 2002; Lin et al. 2001). The DNA methyltransferase (DNMT) is a major enzyme that determines the genomic methylation patterns. Consistent with the increased levels of many tumor suppressors in senescent cells, it was shown that the levels of DNMT enzyme activity decreases as normal fibroblasts were cultured to senescence. In contrast, SV40-infected transformed cells show high levels and activity of DNMT implying that an ability to maintain DNMT level is acquired with SV40-induced escape from senescence (Casillas et al. 2003; Vertino et al. 1994). An increased level of DNMT has also been associated with the initiation and promotion of a variety of cancers (Eads et al. 1999; Kaneda et al. 2002; Shen et al. 2002).

5-aza-2′ deoxycytidine (5AZA-dC) is a DNA demethylating agent. It has been shown to induce cellular senescence-like phenotype in a variety of human caner cells by demethylation of CpG-rich promoters resulting in transcription of genes that are normally silenced by methylation (Fairweather et al. 1987; Holliday 1987; Kulaeva et al. 2003). p16^{INK4a}, a tumor suppressor gene and critical player in replicative and premature senescence in human cells, has a CpG-rich promoter and is repressed by methylation in a large variety of *in vitro* transformed and tumor-derived cell lines (Swafford et al. 1997). Reactivation of p16^{INK4a} gene promoter by demethylation or by its exogenous expression causes a senescence-like growth arrest (Hara et al. 1996; Serrano et al. 1997; Uhrbom et al. 1997). An alternative mechanism involving covalent binding and trapping of DNA methyltransferase by 5AZA-dC has been proposed as its primary effect; demethylation of genomic DNA was demonstrated as a secondary event (Juttermann et al. 1994). Although the molecular mechanism of 5AZA-dC induced senescence in cancer cells is still a subject of investigation, it has been shown that 5AZA-dC kills cancer cells selectively sparing the normal human cells.

Human osteosarcoma (U2OS) cells in which p16^{INK4A} is silenced by hypermethylation provides a good model to study molecular events associated with 5AZA-dC induced cellular senescence. 5AZA-dC treated cells showed the expression of p16^{INK4A}, enlarged cell size, growth arrest and senescence associated β-gal staining (Fig. 20.2a, b). In this model of induced senescence, we examined the intracellular distribution of mortalin. As shown in Fig. 20.2c, the perinuclear mortalin staining pattern in U2OS cells shifted to the pancytoplasmic type (typical of normal cells) in 5AZA-dC-treated cells. It was associated with induction of p53 and p21^{WAF1} (Fig. 20.2d). It has been established that the perinuclear mortalin binds to p53 and inactivates its function in transformed cells (Mihara et al. 2003; Wadhwa et al. 1998). Shift in perinuclear mortalin to pancytoplasmic type in 5AZA-dC treated cells, therefore, suggested that the p53 activation seen in these cells might be the result of abrogation of mortalin-p53 complexes. Indeed, we found that the 5AZA-dC treated cells showed nuclear translocation of p53 (Fig. 20.2e). A time course study was undertaken to reveal the molecular events associated with 5AZA-dC induced senescence. As shown in Fig. 20.2e, whereas shift in mortalin staining pattern occurred as early as 4 h of treatment, p53 translocation to the nucleus occurred at about 8 h. The data suggested that shift in mortalin distribution is an early event and might be independent to that of the demethylation effects as suggested by induction of p16^{INK4A} that took about 24–48 h of treatment (Fig. 20.2d). Involvement of mortalin in 5AZA-dC induced senescence was further tested by generating mortalin overexpressing derivative cells. The latter were found to be significantly more sensitive to 5AZA-dC (Widodo et al. 2007a). The data matched with the fact that mortalin is enriched in majority of cancer cells and are sensitive to 5AZA-dC (Dundas et al. 2005; Srokowski et al. 2004; Wadhwa et al. 2006) suggesting that the overexpression of mortalin sensitizes the cells to 5AZA-dC.

Effect of 5AZA-dC on mortalin was studied by physicochemical assays. The ANS dye can bind to accessible hydrophobic regions in protein molecules and becomes highly fluorescent, a property that makes it a reliable reporter assay for studying structural changes in a protein. Purified recombinant mortalin protein mixed with 5AZA-dC showed loss in fluorescence in a dose-dependent manner (Widodo et al. 2007a). Structural changes associated with 5AZA-dC-mortalin interactions were investigated by the circular dichroism (CD). With addition of 5AZA-dC, the spectra

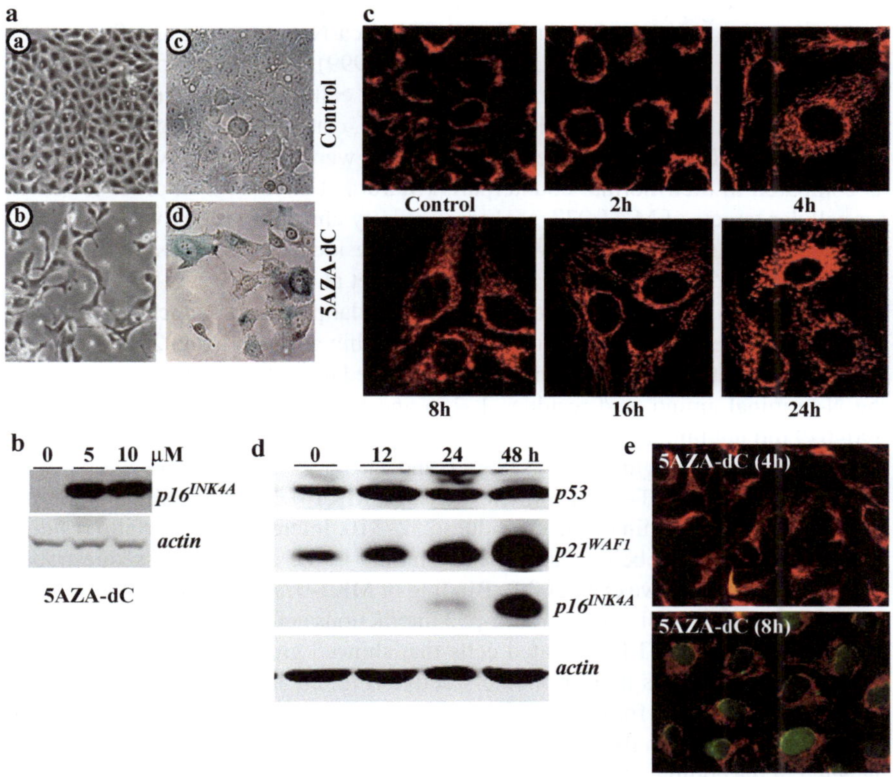

Fig. 20.2 Involvement of mortalin in 5AZA-dC induced senescence. Detection of 5AZA-dC induced senescence by **a** β-gal staining, **b** induction of p16^{INK4A} expression as seen in Western blot. 5AZA-dC induced senescence was associated with shift in mortalin staining pattern **c** and activation of p53 function seen by increased level of its expression and of the downstream effector p21^{WAF1} **d** Shift in mortalin staining pattern and translocation of p53 occurred as early as 4 h and 8 h (**b** and **e**). Modified from Widodo et al. (2007a)

of mortalin manifested lower α-helical content depicting an increased unordered structure suggesting that 5AZA-dC causes perturbations in mortalin structure by binding to its hydrophobic regions. Such interaction might also be responsible for abrogation of mortalin-p53 complexes and translocation of p53 to the nucleus as seen in Fig. 20.2e and account for DNA demethylation-independent effect of 5AZA-dC for induction of senescence in cancer cells.

MKT-077 (1-ethyl-2-[[3-ethyl-5-(3-methylbenzothiazolin-2-yliden)]-4-oxothiazolidin-2-ylidene]-methyl] pyridinium chloride also known as FJ-776) is a water soluble mitochondrion-seeking delocalized cationic rhodacyanine dye that causes selective death of cancer cells. Its accumulation in the mitochondria induces gross ultrastructural alterations (such as loss of cristae, presence of dense matrix inclusions and organellar swelling)(Modica-Napolitano et al. 1996). It is preferentially retained in the cancer cell mitochondria due to their higher mitochondrial membrane potential as compared to the normal cells (Modica-Napolitano et al. 1996; Chiba et al. 1998a, b; Petit et al. 1999; Tatsuta et al. 1999). In addition to its mitochondrial toxicity, MKT-077 has been shown to target cellular proteins including RAS, mortalin and actin. Human bladder carcinoma (EJ, carrying an oncogenic c-Ha-Ras mutant) and v-Ha-ras-transformed NIH 3T3 cells were more sensitive to MKT-077 as compared to their normal counterparts (Maruta et al. 1999). In order to search for cellular targets of MKT-077, sepharose affinity chromatography was performed and two cellular proteins, actin and Hsc70 were identified as possible targets in ras-transformed cells (Maruta et al. 1999; Tikoo et al. 1999). MKT-077-conjugated sepharose beads were then used to pull dowm cellular proteins in which mortalin was identified by Western blotting with mortalin specific antibody (Fig. 20.3a). Similar assay using lysates from cells expressing V5-tagged mortalins revealed that whereas the N-terminal amino acid residues 1–256 and C-terminal residues 310–679 and 390–679 did not bind to MKT-077 precipitated, the full-length protein, its carboxy-terminal truncated fragments (1–538 a. a. and 1–435 a. a.) and 252–679 a. a. were pulled down with MKT-077. From these assays, MKT-077 binding site on mortalin was deduced to be within the a. a. residues 252–310. Interestingly, mortalin residues 252–283 are found to be essential for its binding to the cytoplasmic sequestration domain of p53 (Wadhwa et al. 2000). Binding of MKT-077 to this region of mortalin was hence thought to disrupt mortalin-p53 interactions and cause reactivation of p53 function. Indeed, MKT-077 treated cells that showed growth arrest exhibited p53 in the nucleus and was transcriptionally active (Fig. 20.3b). In order to investigate mortalin-p53 complex formation in control and MKT-077 treated cells, p53 was immunoprecipitated from the control and MKT-077 treated cells wherein mortalin was detected in the complex by anti-mortalin antibody. As seen in Fig. 20.3c, expression level of mortalin was enhanced in response to MKT-077 treatment. However, it was not co-immunoprecipitated with p53 (Fig. 20.3d) confirming that MKT-077 abrogated mot-p53 interactions. Notably, cellular distribution of mortalin in MKT-077 treated transformed cells reverted back to normal pancytoplasmic type (Fig. 20.3e). Normal cells (MRC-5) treated with equivalent concentrations of MKT-077 exhibited increase in mortalin similar to the transformed cells, but there was no change in the subcellular distribution of either mortalin or p53 (Fig. 20.3b). A similar alteration in mortalin staining pattern in cancer cells treated with MKT-077 analogues (FJ5002, FJ-5744, FJ-5847, FJ-5826 and FJ-5850 that caused selective killing) was observed. Since pancytosolic mortalin and p53 do not interact in normal cells, the action of MKT-077 by this pathway accounts, at least in part, for its selectivity for cancer cells. Furthermore, it was shown that MKT-077 binding to the nucleotide-binding domain of mortalin causes tertiary structural changes in the protein resulting in an inactivation of its chaperone function and senescence in human tumor cell lines (Deocaris et al. 2007). Interestingly, in tumor cells with elevated level of mortalin expression, fairly low drug doses were sufficient to induce senescence that was correlated with the shift in mortalin staining pattern from perinuclear to pancytoplasmic type (Fig. 20.3e). Since MKT-077 was found to bind to the nucleotide-binding domain (NBD) (a. a. 252–310) of mortalin, homology model based on the structure template coordinates

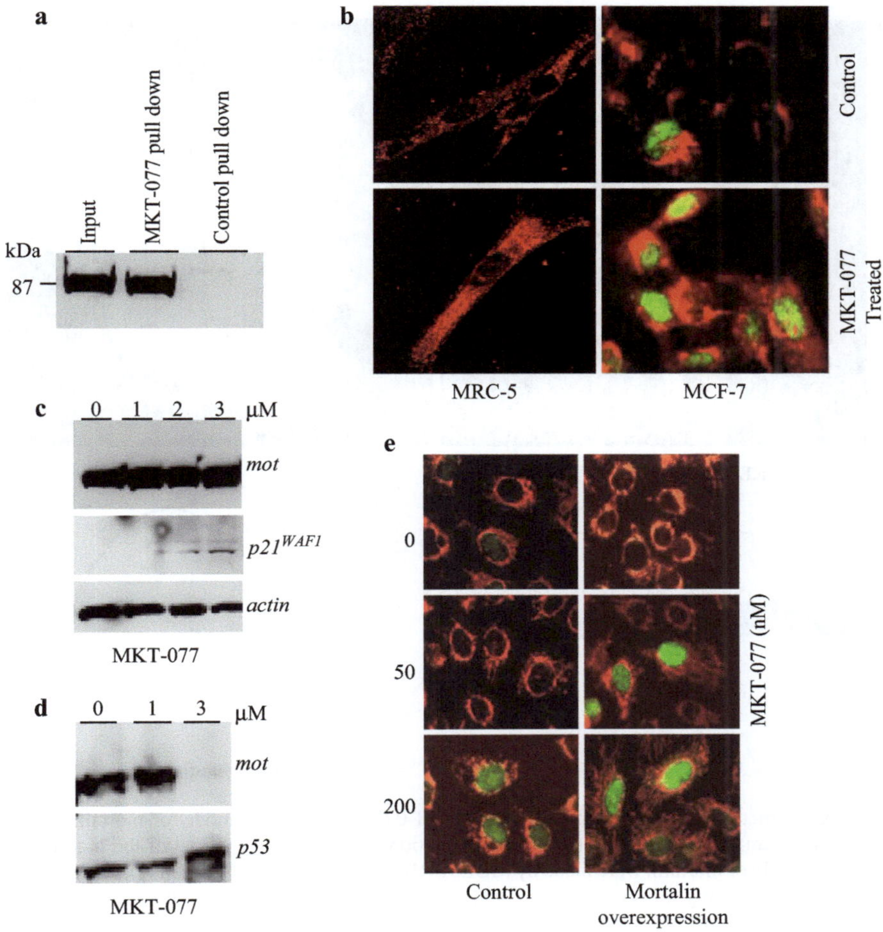

Fig. 20.3 Mortalin is the cellular target of MKT-077. MKT-077 binds to mortalin protein (**a**). Accumulation of p53 in the nucleus as a result of MKT-077 treatment (**b**). MKT-077 dissociates mot (mortalin) from p53, resulting in an increased $p21^{WAF1}$ level. **c** Western blotting of MKT-077 treated cells with anti-mortalin and anti-$p21^{WAF1}$ antibodies. Increase in mot (mortalin) and $p21^{WAF1}$ levels were detected subsequent to MKT-077 treatment. Cells treated with indicated concentrations of MKT-077 were used for immunoprecipitation of p53 protein with polyclonal anti-p53 antibody. p53-immunocomplexes were analyzed for the presence of mot (mortalin) by Western blotting with monoclonal anti-mortalin antibody. Immunoprecipitated p53 was detected by a monoclonal anti-p53 antibody. Mortalin was not detected in p53 immunocomplexes from cells treated with high concentrations of MKT-077 (**d**). MKT-077 treatment at the low dose induced shift in mortalin staining pattern from perinuclear to pancytoplasmic type (**e**). Modified from Deocaris et al. (2007) and Wadhwa et al. (2000)

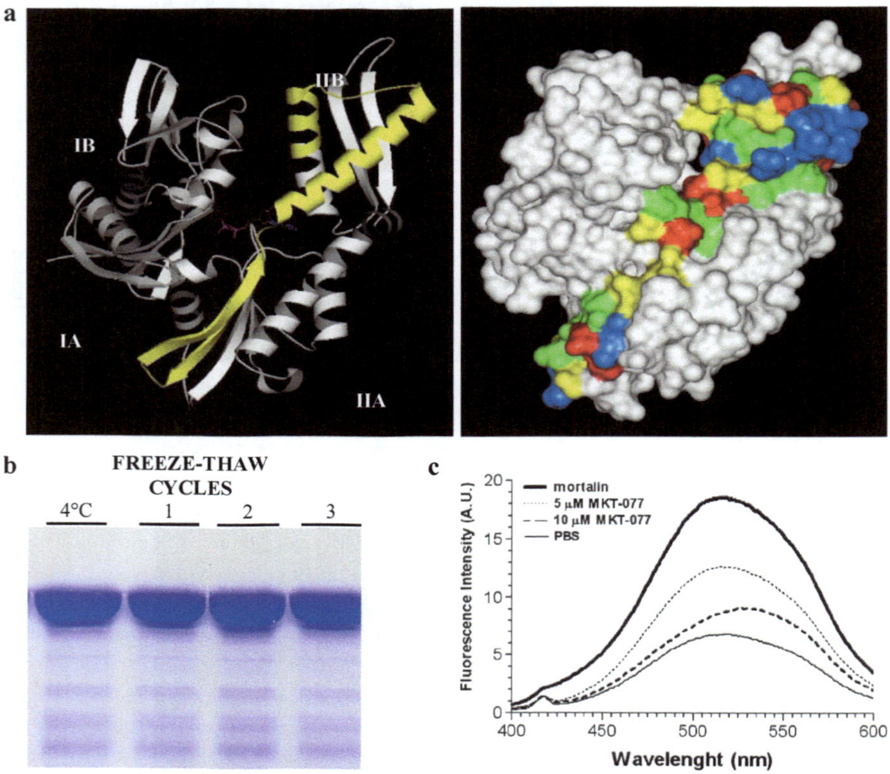

Fig. 20.4 Structural changes in mortalin with MKT-077. **a** Homology model of mortalin NBD. Ribbon (*left*) and contact surface (*right*) representations of a homology model of the N-terminal ATPase domain of mortalin were made by MOLSCRIPT and INSIGHT II (Accelrys), respectively. The MKT-077-binding region is shown in yellow (in the ribbon diagram). For the contact surface model of mortalin, residues belonging to the MKT-077-binding regions are colored as follows: hydrophobic (Ala, Ile, His, Leu, Met, Phe, Trp, Tyr, and Val: *green*), negatively charged (Asp and Glu: *red*), positively charged (Arg and Lys: *blue*), and remainder (*yellow*). **b** Purity of His-tagged recombinant mortalin protein is shown in fresh and after freeze-thaw cycles (lanes 1–3, from −20°C to RT). **c** Fluorescence spectra of bis-ANS bound to mortalin-MKT-077 complex. MKT-077 was incubated with 0.5 mM mortalin for 15–20 min. Bis-ANS was finally added to a final concentration of 2.5 mM and fluorescence intensity was recorded. Modified from Deocaris et al. (2007)

of human Hsp70 (1S3X), using the Swiss-Model ver. 3.5, a comparative protein-modeling server (http://swissmodel.expasy.org) was generated. It showed two main lobes separated by a wide cleft that binds to ATP. Each lobe contains distinct sub-domains IA-B and IIA-B, with sub-domain IA containing both N- and C-terminus and links to the remaining part of mortalin i.e. the substrate-binding domain. The MKT-077 binding region, as delineated by pull-down experiments (Fig. 20.4a) is shown (yellow, Fig. 20.4a, *left*; yellow, red, green and blue, Fig. 20.4a, *right*) spanning a pair of α-helices (a. a. 274–295; a. a. 301–319) in sub-domain IIB, and an opposite anti-parallel βsheets (a. a. 252–258; a. a. 261–270) located in sub-domain IIA.

In order to characterize the direct interaction of MKT-077 with mortalin, recombinant His-tagged mortalin protein prepared in *E. coli*, and confirmed to be stable through the several freeze-thaw cycles (Fig. 20.4b) was used for examination of the hydrophobic topography of mortalin in the presence of MKT-077. ANS, a reporter dye that preferentially binds to exposed hydrophobic patches in proteins and becomes appreciably fluorescent at 480 nm showed good binding to mortalin. However, the addition MKT-077 to the protein resulted in a dose-dependent loss in ANS fluorescence (Fig. 20.4c) suggesting that MKT-077 induced changes in the tertiary structure of mortalin that would also affect it chaperone function (Deocaris et al. 2007). The latter was investigated by *in vitro* (insulin light scattering) and *in vivo* (luciferase denaturation and refolding) chaperone assays. The data demonstrated that MKT-077 affected the biophysical characteristics of mortalin, its chaperoning activity and induced senescence in cancer cells that was tightly associated with shift in mortalin staining pattern in cells.

20.4 Alcoholic Extract of Ashwagandha Leaves and Withanone

Ashwagandha (*Withania somnifera*; Solanaceae) is one of the most commonly used plants in Ayurveda (world's oldest system of home medicine system). Extracts from different parts of the plant are known to possess alkaloids and steroidal lactones (commonly known as withanolides) and have been in use for a variety of health promoting effects. Alcoholic extract of Ashwagandha leaves (i-Extract) was shown to possess strong anti-cancer activity. By chemical fractionation, the activity was assigned to its constituent withanone. It was shown that the i-Extract and withanone kill cancer cells selectively by activation of p53 tumor suppressor gene and induction of oxidative stress, also marked by shift in mortalin staining pattern from the perinuclear to the pancytoplasm type (Fig. 20.5a) (Widodo et al. 2007b). Similar to MKT-077, withanone was found to bind to mortalin at its N-terminal peptide-binding domain and release p53 from the mortalin-p53 complexes (Grover et al. unpublished observations) accounting for p53 activation and induction of senescence function. Of note, hydrogen peroxide and epoximicin, established senescence-inducing drugs, caused similar shift in mortalin staining pattern (and was associated with induction of p53 activity and senescence-associated β-gal staining, marker for senescence).

In order to establish mortalin staining as a marker for induction of senescence, cancer cells were treated with fifteen well known anticancer drugs. IC50 of each of the drug was determined by cell viablity assays (Table 20.1). Mortalin staining pattern was examined in response to the drug treatments and it was found to shift from perinulear (control) (typical of cancer cells) to pancytoplasmic (typical of normal cells) in response to all the drugs examined (Fig. 20.5b). Of note, the concentration of the drug that resulted in shift in mortalin staining pattern was 10–100 fold lower than their IC50 suggesting that the low doses of drugs induced senescence in cancer cells in

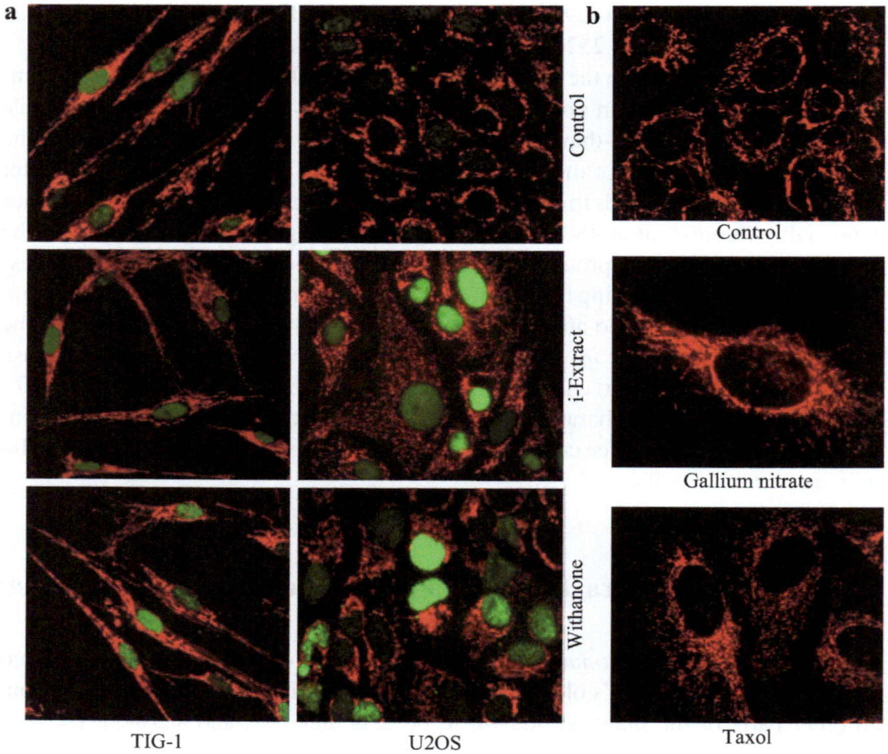

Fig. 20.5 i-Extract and withanone from *Withania somnifera* plant activate wild type tumor suppressor p53 in cancer cells only. **a** i-Extract caused shift in staining pattern of mortalin (*red* staining) in cancer (U2OS) cells. Consistent with the shift in mortalin staining pattern (from perinuclear to pancytoplasmic), p53 was translocated into the nucleus of cancer cells (*green* nuclear staining). **b** Shift in mortalin staining from perinuclear to pancytoplasmic in response to low doses of drug treatment. Modified from Widodo et al. (2007b) and Kaul et al. (2007)

contrast to apoptosis caused at high doses. These studies had strongly suggested that mortalin staining pattern could serve as induced-Senescence (i-Senescence) detection assay of screening for drugs, peptides and siRNAs for safe anti-cancer therapeutics.

20.5 Use of Mortalin Staining as Anti-Cancer siRNA Screening

The recent discovery of RNA interference (RNAi) triggered by small interfering RNA (siRNA) or short hairpin RNA (shRNA) in eukaryote cells has revolutionized the biology, especially the study of gene function. RNAi is already a valuable and widely used tool to silence the genes with high specificity and efficacy to analyze of molecular mechanisms and pathways for a variety of human diseases, including

Table 20.1 IC50 and senescence-inducing concentrations of drugs obtained on the basis of shift of mortalin staining from the perinuclear to the pancytoplasmic. Modified from Kaul et al. (2007)

	Drug name	IC 50	Senescence inducing concentration
1	Adramycin	0.25 μg	0.025 μM
2	Curcumin	10 μM	1 μM
3	Campotothecin	10 μM	1 μM
4	Doxorubicin	10 μM	1 μM
5	ECG	50 μM	0.5 μM
6	EGCG	50 μM	0.05 μM
7	EGC	50 μM	0.5 μM
8	Etoposide	10 μM	0.01 μM
9	Ga-nitrate	1 mM	0.1 mM
10	Mitoxantrone	2.5 μM	0.0025 μM
11	Nocodazole	10 μM	0.01 μM
12	Retinoic acid	20 μM	1 μM
13	Taxol	10 μM	0.01 μM
14	Trichostatin	2 μM	0.1 μM
15	Tunicamycin	2 μM	0.01 μg/ml

cancers. Prolonged gene silencing has been achieved by expressing siRNAs as a short hairpin (shRNA) precursor structure of around 70 nucleotides from plasmid vectors that contain specific promoters. These shRNA precursors expressed from RNA polymerase promoters II or III are processed into a fully functional siRNA by an enzyme called Drosha (Paddison et al. 2002; Walton et al. 2010).

20.6 Outcome of Anti-Cancer shRNA Screening Using Mortalin Staining as a Reporter

Human breast carcinoma (MCF7) and osteocarcinoma (U2OS) cells were stably transfected with a library of shRNA expression plasmids designed to express a short hairpin RNA driven by human U6 promoter. Cells were fixed and stained with anti-mortalin antibody and scanned under the automated scanning system attached to Axiovert 200 M microscope (Zeiss). shRNAs that caused shift in mortalin staining from perinuclear to pancytoplasmic were selected for the second round of screening. The cycle was repeated four times and resulted in the identification of 22 shRNAs (Table 20.2) with potential to induce shift in mortalin staining pattern in cancer cells. The 22 shRNAs were then examined for i-Senescence phenotype in six other cancer (four breast carcinomas-SKBR3, T47D, MDA-MB231, MDA-MB157 and two gall bladder carcinomas-TGBC-2, SKCH) cells. By this step, 9 of the 22 were found to cause shift in mortalin staining pattern in all the six cell types. Interestingly, the nine selected gene targets (BCL2A1, MAP3K7, MAPK1, TERT, Gp96, TNKS1BP1, IL1A, KPNA2, CALR and USP10) were found to be the cancer promoting genes as evidenced by number of other studies (Gao et al. 2010).

Table 20.2 List of 22 anti-cancer candidate shRNAs selected by using mortalin staining as a reporter. Modified from Gao et al. (2010)

Sr. No.	Gene name	Accession no.	Evidence for role in carcinogenesis/Function
1	BRCA1	NM007294-1	DNA damage response
2	MRE11	NM005590-1	DNA damage and telomere maintenance
3	NBS1	NM002485-2	DNA damage and telomere maintenance
4	TNPO1	NM002270-1	Nuclear import of proteins
5	TNPO2	NM013433-2	Nuclear import of proteins
6	TNPO3	NM012470-2	Nuclear import of proteins
7	MDC1	XM376479-1	Maintain genome integrity
8	TNF2	NM012461-2	Telomere length regulation
9	IRAK1	NM001569	Systemic lupus erythematosus
10	RIPK3	NM006871	Immune response
11	TP53BP1	NM005657-1	Associated with squamous cell carcinoma
12	BIN1	NM004305-1	Downregulated in cancer
13	BCL2A1	NM004049	Upregulated in cancers
14	MAP3K7	NM145333	Upregulated in cancers
15	MAPK1	NM002745	Upregulated in cancers
16	TERT	NM003219-2	Upregulated in cancers
17	Gp96	NM003299-1	Upregulated in cancers
18	TNKS1BP1	NM033396-1	Upregulated in cancers
19	IL1A	NM000575-1	Upregulated in cancers
20	KPNA2	NM002266-2	Upregulated in cancers
21	CALR	NM004343-1	Upregulated in cancers
22	USP10	NM005153-1	Upregulated in cancers

Bioinformatics analysis using MetaCore Pathway and Data Mining (http://www.genego.com/genego_lp.php) revealed that the selected genes were involved in a wide range of cellular functions including regulation of apoptosis, telomere maintenance, DNA damage regulation, immune response, protein degradation and protein transport. Whereas the most complicated gene target identified was MAPK1 with signaling network that involved hundreds of other genes, the simplest was TIN-2 that showed crosstalk with only seven proteins studies (Gao et al. 2010). We also examined to see whether i-Senescence induced by various shRNAs involved activation of p53 and/or p16 pathways by performing their immunostaining assays. p53 was activated in some, but not all, cases of i-Senescence; cells with mutant p53 also showed induction of senescence by different shRNAs. Similarly, $p16^{INK4A}$ was induced in some, but not all, cases. Interestingly, the expression of p16 was found to be elevated in cells when i-Senescence occurred by NBS1, MRE11 and BRCA1 shRNAs; the three genes merged into one signaling pathway that regulates DNA damage response of cells. They have also been reported to regulate each other and are involved in telomere length maintenance mechanisms (Zhong et al. 2007). The data suggested that the DNA damage response and senescence pathways have tight link. When DNA damage response machinery is disturbed, cells may enter senescence. Whereas the significance of each of the selected gene target in carcinogenesis and its validation as anticancer gene target remains to be clarified, the data suggested that mortalin staining pattern could be used as a potent reporter to detect the senescence induced by multiple and divergent pathways.

Mortalin is expressed in all human cell types and tissues. Multiple binding partners of mortalin have been identified. These include p53, FGF-1, IL-1 receptor type 1, glucose regulated ER chaperone (GRP94), voltage-dependent anion-selective channel (VDAC), an inner mitochondrial membrane protein (NADH dehydrogenase), mevalonate pyruvate decarboxylase, MPD (a peroxisomal protein) and Tim23 (mitochondrial inner membrane translocase) (Kaul et al. 2002; Mizukoshi et al. 1999, 2001; Takano et al. 2001; Wadhwa et al. 1998, 2002a, 2003 unpublished observations). The functional role of mortalin and its binding partners in biological processes including cellular senescence, immortalization and stress response is not yet completely understood. The interaction of mortalin with p53 is the best characterized so far. These two proteins interact in the cytoplasm of transformed cells resulting in cytoplasmic retention and transcriptional inactivation of p53 (Wadhwa et al. 1998, 2002b) thus offering a mechanism of inactivation of wild type p53 in tumors. Abrogation of mot-p53 interactions by some drugs indeed restored nuclear translocation and activation of p53 function followed by growth arrest of tumor cells. Besides this, inhibition of chaperoning activity of mortalin by drugs was shown to be responsible for growth arrest of cancer cells. In each case, it was tightly associated with shift in mortalin staining from perinuclear to pancytoplasmic region endorsing its use as a reporter for anticancer drug screenings.

References

Bertram MJ, Berube NG, Hang-Swanson X, Ran Q, Leung JK, Bryce S, Spurgers K, Bick RJ, Baldini A, Ning Y et al (1999) Identification of a gene that reverses the immortal phenotype of a subset of cells and is a member of a novel family of transcription factor-like genes. Mol Cell Biol 19:1479–1485

Casillas MA, Jr Lopatina N, Andrews LG, Tollefsbol TO (2003) Transcriptional control of the DNA methyltransferases is altered in aging and neoplastically-transformed human fibroblasts. Mol Cell Biochem 252:33–43

Chiba Y, Kubota T, Watanabe M, Matsuzaki SW, Otani Y, Teramoto T, Matsumoto Y, Koya K, Kitajima M (1998a) MKT-077, localized lipophilic cation: antitumor activity against human tumor xenografts serially transplanted into nude mice. Anticancer Res 18:1047–1052

Chiba Y, Kubota T, Watanabe M, Otani Y, Teramoto T, Matsumoto Y, Koya K, Kitajima M (1998b) Selective antitumor activity of MKT-077, a delocalized lipophilic cation, on normal cells and cancer cells in vitro. J Surg Oncol 69:105–110

Das PM, Singal R (2004) DNA methylation and cancer. J Clin Oncol 22:4632–4642

Deocaris CC, Widodo N, Shrestha BG, Kaur K, Ohtaka M, Yamasaki K, Kaul SC, Wadhwa R (2007) Mortalin sensitizes human cancer cells to MKT-077-induced senescence. Cancer Lett 252:259–269

Dundas SR, Lawrie LC, Rooney PH, Murray GI (2005) Mortalin is over-expressed by colorectal adenocarcinomas and correlates with poor survival. J Pathol 205:74–81

Eads CA, Danenberg KD, Kawakami K, Saltz LB, Danenberg PV, Laird PW (1999) CpG island hypermethylation in human colorectal tumors is not associated with DNA methyltransferase overexpression. Cancer Res 59:2302–2306

Enokida H, Shiina H, Igawa M, Ogishima T, Kawakami T, Bassett WW, Anast JW, Li LC, Urakami S, Terashima M et al (2004) CpG hypermethylation of MDR1 gene contributes to the pathogenesis and progression of human prostate cancer. Cancer Res 64:5956–5962

Fairweather DS, Fox M, Margison GP (1987) The in vitro lifespan of MRC-5 cells is shortened by 5-azacytidine-induced demethylation. Exp Cell Res 168:153–159

Gao R, Yaguchi T, Horimoto K, Ishii T, Kaul SC, Wadhwa R (2010) Identification of anti-cancer shRNAs based on the staining pattern of mortalin. Tiss Cult Res Communn 29:147–153

Hara E, Smith R, Parry D, Tahara H, Stone S, Peters G (1996) Regulation of p16CDKN2 expression and its implications for cell immortalization and senescence. Mol Cell Biol 16:859–867

Hensler PJ, Annab LA, Barrett JC, Pereira-Smith OM (1994) A gene involved in control of human cellular senescence on human chromosome 1q. Mol Cell Biol 14:2291–2297

Hiltunen MO, Turunen MP, Hakkinen TP, Rutanen J, Hedman M, Makinen K, Turunen AM, Aalto-Setala K, Yla-Herttuala S (2002) DNA hypomethylation and methyltransferase expression in atherosclerotic lesions. Vasc Med 7:5–11

Holliday R (1987) The inheritance of epigenetic defects. Science 238:163–170

Huang Y, de la Chapelle A, Pellegata NS (2003) Hypermethylation, but not LOH, is associated with the low expression of MT1G and CRABP1 in papillary thyroid carcinoma. Int J Cancer 104:735–744

Juttermann R, Li E, Jaenisch R (1994) Toxicity of 5-aza-2′-deoxycytidine to mammalian cells is mediated primarily by covalent trapping of DNA methyltransferase rather than DNA demethylation. Proc Natl Acad Sci U S A 91:11797–11801

Kaneda A, Kaminishi M, Yanagihara K, Sugimura T, Ushijima T (2002) Identification of silencing of nine genes in human gastric cancers. Cancer Res 62:6645–6650

Kaul SC, Taira K, Pereira-Smith OM, Wadhwa R (2002) Mortalin: present and prospective. Exp Gerontol 37:1157–1164

Kaul Z, Yaguchi T, Chiura HX, Kaul SC, Wadhwa R (2007) Quantum dot-based mortalin staining as a visual assay for detection of induced senescence in cancer cells. Ann N Y Acad Sci 1100:368–372

Kulaeva OI, Draghici S, Tang L, Kraniak JM, Land SJ, Tainsky MA (2003) Epigenetic silencing of multiple interferon pathway genes after cellular immortalization. Oncogene 22:4118–4127

Lichtenstein AV, Kisseljova NP (2001) DNA methylation and carcinogenesis. Biochemistry (Mosc) 66:235–255

Lin CH, Hsieh SY, Sheen IS, Lee WC, Chen TC, Shyu WC, Liaw YF (2001) Genome-wide hypomethylation in hepatocellular carcinogenesis. Cancer Res 61:4238–4243

Maruta H, Tikoo A, Shakri R, Shishido T (1999) The anti-RAS cancer drug MKT-077 is an F-actin cross-linker. Ann N Y Acad Sci 886:283–284

Michishita E, Nakabayashi K, Suzuki T, Kaul SC, Ogino H, Fujii M, Mitsui Y, Ayusawa D (1999) 5-Bromodeoxyuridine induces senescence-like phenomena in mammalian cells regardless of cell type or species. J Biochem 126:1052–1059

Mihara M, Erster S, Zaika A, Petrenko O, Chittenden T, Pancoska P, Moll UM (2003) p53 has a direct apoptogenic role at the mitochondria. Mol Cell 11:577–590

Mizukoshi E, Suzuki M, Loupatov A, Uruno T, Hayashi H, Misono T, Kaul SC, Wadhwa R, Imamura T (1999) Fibroblast growth factor-1 interacts with the glucose-regulated protein GRP75/mortalin. Biochem J 343:461–466

Mizukoshi E, Suzuki M, Misono T, Loupatov A, Munekata E, Kaul SC, Wadhwa R, Imamura T (2001) Cell-cycle dependent tyrosine phosphorylation on mortalin regulates its interaction with fibroblast growth factor-1. Biochem Biophys Res Commun 280:1203–1209

Modica-Napolitano JS, Koya K, Weisberg E, Brunelli BT, Li Y, Chen LB (1996) Selective damage to carcinoma mitochondria by the rhodacyanine MKT-077. Cancer Res 56:544–550

Nakabayashi K, Ogata T, Fujii M, Tahara H, Ide T, Wadhwa R, Kaul SC, Mitsui Y, Ayusawa D (1997) Decrease in amplified telomeric sequences and induction of senescence markers by introduction of human chromosome 7 or its segments in SUSM-1. Exp Cell Res 235:345–353

Ning Y, Weber JL, Killary AM, Ledbetter DH, Smith JR, Pereira-Smith OM (1991) Genetic analysis of indefinite division in human cells: evidence for a cell senescence-related gene(s) on human chromosome 4. Proc Natl Acad Sci U S A 88:5635–5639

Ogata T, Oshimura M, Namba M, Fujii M, Oishi M, Ayusawa D (1995) Genetic complementation of the immortal phenotype in group D cell lines by introduction of chromosome 7. Jpn J Cancer Res 86:35–40

Paddison PJ, Caudy AA, Bernstein E, Hannon GJ, Conklin DS (2002) Short hairpin RNAs (shRNAs) induce sequence-specific silencing in mammalian cells. Genes Dev 16:948–958

Pereira-Smith OM, Smith JR (1983) Evidence for the recessive nature of cellular immortality. Science 221:964–966

Petit T, Izbicka E, Lawrence RA, Nalin C, Weitman SD, Von Hoff DD (1999) Activity of MKT 077, a rhodacyanine dye, against human tumor colony- forming units. Anticancer Drugs 10:309–315

Sakajiri S, Kumagai T, Kawamata N, Saitoh T, Said JW, Koeffler HP (2005) Histone deacetylase inhibitors profoundly decrease proliferation of human lymphoid cancer cell lines. Exp Hematol 33:53–61

Serrano M, Lin AW, McCurrach ME, Beach D, Lowe SW (1997) Oncogenic ras provokes premature cell senescence associated with accumulation of p53 and p16INK4a. Cell 88:593–602

Shen H, Wang L, Spitz MR, Hong WK, Mao L, Wei Q (2002) A novel polymorphism in human cytosine DNA-methyltransferase-3B promoter is associated with an increased risk of lung cancer. Cancer Res 62:4992–4995

Smith JR, Pereira-Smith OM (1996) Replicative senescence: implications for in vivo aging and tumor suppression. Science 273:63–67

Srokowski T, Pfeifer JD, Li J, Olson LM, Rader JS (2004) Expression and localization of GRP75 in human epithelial tumors and normal tissues. Appl Immunohistochem Mol Morphol 12:132–138

Sugimoto S, Maass N, Takimoto Y, Sato K, Minei S, Zhang M, Hoshikawa Y, Junemann KP, Jonat W, Nagasaki K (2004) Expression and regulation of tumor suppressor gene maspin in human bladder cancer. Cancer Lett 203:209–215

Swafford DS, Middleton SK, Palmisano WA, Nikula KJ, Tesfaigzi J, Baylin SB, Herman JG, Belinsky SA (1997) Frequent aberrant methylation of p16INK4a in primary rat lung tumors. Mol Cell Biol 17:1366–1374

Takano S, Wadhwa R, Mitsui Y, Kaul SC (2001) Identification and characterization of molecular interactions between glucose-regulated proteins (GRPs) mortalin/GRP75/peptide-binding protein 74 (PBP74) and GRP94. Biochem J 357:393–398

Tatsuta N, Suzuki N, Mochizuki T, Koya K, Kawakami M, Shishido T, Motoji N, Kuroiwa H, Shigematsu A, Chen LB (1999) Pharmacokinetic analysis and antitumor efficacy of MKT-077, a novel antitumor agent. Cancer Chemother Pharmacol 43:295–301

Tikoo A, Cutler H, Lo SH, Chen LB, Maruta H (1999) Treatment of Ras-induced cancers by the F-actin cappers tensin and chaetoglobosin K, in combination with the caspase-1 inhibitor N1445. Cancer J Sci Am 5:293–300

Uhrbom L, Nister M, Westermark B (1997) Induction of senescence in human malignant glioma cells by p16INK4A. Oncogene 15:505–514

Vertino PM, Issa JP, Pereira-Smith OM, Baylin SB (1994) Stabilization of DNA methyltransferase levels and CpG island hypermethylation precede SV40-induced immortalization of human fibroblasts. Cell Growth Differ 5:1395–1402

Wadhwa R, Kaul SC, Ikawa Y, Sugimoto Y (1993a) Identification of a novel member of mouse hsp70 family: its association with cellular mortal phenotype. J Biol Chem 268:6615–6621

Wadhwa R, Kaul SC, Mitsui Y, Sugimoto Y (1993b) Differential subcellular distribution of mortalin in mortal and immortal mouse and human fibroblasts. Exp Cell Res 207:442–448

Wadhwa R, Pereira-Smith OM, Reddel RR, Sugimoto Y, Mitsui Y, Kaul SC (1995) Correlation between complementation group for immortality and the cellular distribution of mortalin. Exp Cell Res 216:101–106

Wadhwa R, Takano S, Robert M, Yoshida A, Nomura H, Reddel RR, Mitsui Y, Kaul SC (1998) Inactivation of tumor suppressor p53 by mot-2, a hsp70 family member. J Biol Chem 273:29586–29591

Wadhwa R, Sugihara T, Yoshida A, Nomura H, Reddel RR, Simpson R, Maruta H, Kaul SC (2000) Selective toxicity of MKT-077 to cancer cells is mediated by its binding to hsp70 family protein, mot-2, and reactivation of p53 function. Cancer Res 60:6818–6821

Wadhwa R, Taira K, Kaul SC (2002a) An hsp70 family chaperone, mortalin/mthsp70/PBP74/Grp75: what, when and where? Cell Stress & Chaperones 7:309–316

Wadhwa R, Yaguchi T, Hasan MK, Mitsui Y, Reddel RR, Kaul SC (2002b) Hsp70 family member, mot-2/mthsp70/GRP75, binds to the cytoplasmic sequestration domain of the p53 protein. Exp Cell Res 274:246–253

Wadhwa R, Yaguchi T, Hasan MK, Taira K, Kaul SC (2003) Mortalin-MPD (mevalonate pyrophosphate decarboxylase) interactions and their role in control of cellular proliferation. Biochem Biophys Res Commun 302:735–742

Wadhwa R, Takano S, Kaur K, Deocaris CC, Pereira-Smith OM, Reddel RR, Kaul SC (2006) Upregulation of mortalin/mthsp70/Grp75 contributes to human carcinogenesis. Int J Cancer 118:2973–2980

Walton SP, Wu M, Gredell JA, Chan C (2010) Designing highly active siRNAs for therapeutic applications. FEBS J 277:4806–4813

Widodo N, Deocaris CC, Kaur K, Hasan K, Yaguchi T, Yamasaki K, Sugihara T, Ishii T, Wadhwa R, Kaul SC (2007a) Stress chaperones, mortalin, and pex19p mediate 5-aza-2′ deoxycytidine-induced senescence of cancer cells by DNA methylation-independent pathway. J Gerontol A Biol Sci Med Sci 62:246–255

Widodo N, Kaur K, Shrestha BG, Takagi Y, Ishii T, Wadhwa R, Kaul SC (2007b) Selective killing of cancer cells by leaf extract of Ashwagandha: Identification of a tumor inhibitory factor and the first molecular insights to its effect. Clinical Cancer Res 13:2298–2306

Zhong ZH, Jiang WQ, Cesare AJ, Neumann AA, Wadhwa R, Reddel RR (2007) Disruption of telomere maintenance by depletion of the MRE11/RAD50/NBS1 complex in cells that use alternative lengthening of telomeres. J Biol Chem 282:29314–29322

Chapter 21
Cell Internalizing Anti-mortalin Antibody as a Nanocarrier

Zeenia Kaul, Tomoko Yaguchi, Renu Wadhwa and Sunil C. Kaul

Abstract Nanocarriers are the small molecules with capability of adhering to or pass through the cell membrane and deliver the genes, peptides, growth factors or toxins to living cells. They form an extremely important part of gene delivery and molecular therapeutics. Besides the viral modes of gene delivery, cationic carrier molecules (e.g., liposome and polymers) that form non-covalent complexes with DNA constitute a major class of nanocarriers. Although relatively less efficient than viral systems, cationic carrier molecules have inherent advantage of flexibility and safety. Their derivatives in conjugation with functional molecules such as, peptides, proteins, growth factors and antibodies have been in focus to generate nanocarriers with low toxicity, high stability, high efficiency and cell-specific targeting features. In this chapter, we discuss the use of cell internalizing antibodies against mortalin protein as nanocarriers. Internalizing anti-mortalin antibodies was employed for (i) internalization of nanoparticles (quantum dots, QD) to generate illuminating cells and (ii) gene delivery. Furthermore, cationic polymer polyethylenimine (PEI) and internalizing anti-mortalin antibody complex was shown to enhance the gene delivery specifically to cancer cells and thus enabling the latter to serve as a novel cancer-targeting nanocarrier.

Keywords Anti-mortalin antibody · Internalization · Nanocarrier · Conjugation · Quantum dots · Imaging

21.1 Introduction

In order to generate cancer cell specific non-viral vectors, antibodies/ligands to protein/receptors that are either enriched or are exclusive to cancer cells are conjugated with cationic carriers (Zanta et al. 1997; Densmore et al. 2001; Chiu et al. 2004;

S. C. Kaul (✉) · T. Yaguchi · R. Wadhwa
National Institute of Advanced Industrial Science and Technology (AIST), Central 4,
1-1-1 Higashi, Tsukuba, Ibaraki 305-8562, Japan
e-mail: s-kaul@aist.go.jp

Z. Kaul
Center for Childhood Cancer, The Research Institute at Nationwide Children's Hospital,
Columbus, Ohio, USA

S. C. Kaul, R. Wadhwa (eds.), *Mortalin Biology: Life, Stress and Death*,
DOI 10.1007/978-94-007-3027-4_21, © Springer Science+Business Media B.V. 2012

Hanifi et al. 2010). Some examples of such polyplexes generated for targeted delivery include the use of folic acid as a high affinity ligand to target folate receptor in ovarian cancers (Guo and Lee 1999), transferrin as a targeting ligand for delivery to transferrin receptor-expressing tumor cells (Hu-Lieskovan et al. 2005) and RGD (arginine-glycine-aspartic) peptides for targeting integrin receptor of endothelial cells of tumor metastases (Kunath et al. 2003). Antibodies have been preferred for targeted delivery due to their high affinity to the antigens. Some of the antibodies used for targeted delivery so far include anti-HER2 antibody (Trastuzumab) for targeting human epidermal growth factor receptor-2 (HER2) in breast cancers (Chiu et al. 2004; Strehblow et al. 2005; Gao et al. 2009; Wang et al. 2010), lymphoma cell specific antibodies (Guillem et al. 2002; Peer et al. 2007), OA3 antibody for ovarian carcinoma (Merdan et al. 2003) and PSMA antibody for prostate cancer cells (Moffatt and Cristiano 2006). In the recent past, many studies have investigated antibody-mediated targeting of liposomal anticancer drugs to antigen epitopes expressed at the surface of cancer cells (Maruyama et al. 1997; Lopes de Menezes et al. 1998; Sugano et al. 2000; Park et al. 2002). Upon binding of the antibody to the cell surface antigen, liposome contents are released over time at or near the cell surface, and the released drug enters the cell by passive diffusion or normal transport mechanisms. In this case, although increased concentrations of drug is achieved at the cell surface, the rate of diffusion and redistribution of the released drug away from the cell might exceed the rate at which the drug enters the cell. Therefore, internalizing antibodies are preferred for targeting liposomal drugs. They trigger receptor-mediated uptake of the immunoliposomal drug package into the cell interior, where the drug contents are released subsequent to degradation of liposomes by lysosmal and endosomal enzymes, and higher concentrations of drug are achieved directly into the cell interior. Liposomes targeted to internalizing receptors have demonstrated increased therapeutic activity in tumor models (Lopes de Menezes et al. 1998; Sugano et al. 2000; Park et al. 2002; Allen 2002; Sapra and Allen 2002).

21.2 Cell-Internalizing Polyclonal Anti-Mortalin Antibody

Mortalin, a dynamic and multifunctional stress chaperone, has differential subcellular distribution and function in normal and cancer cells (Wadhwa et al. 1993a, b; Kaul et al. 2005). Some of its so far characterized functions include mitochondrial biogenesis, intracellular trafficking, cell proliferation and signaling, stress response and inactivation of tumor suppressor protein p53 in cancer cells (Kaul et al. 2005, 2007a; Wadhwa et al. 2006). It is upregulated and is functionally relevant in human carcinogenesis (Shin et al. 2003; Dundas et al. 2004; Cussac et al. 2006; Ma et al. 2006; Wadhwa et al. 2006; Lu et al. 2011). Mortalin overexpressing human immortalized cells showed malignant properties including anchorage-independent growth, tumor formation in nude mice and chemotaxis (Wadhwa et al. 2006). On the other hand, anti-mortalin molecules, such as mortalin-specific ribozymes, siRNA, small chemical compounds were found to suppress the growth of tumor cells (Wadhwa et al. 2000, 2003, 2004; Pilzer et al. 2010) and hence antibodies to mortalin were anticipated to be useful for cancer therapy.

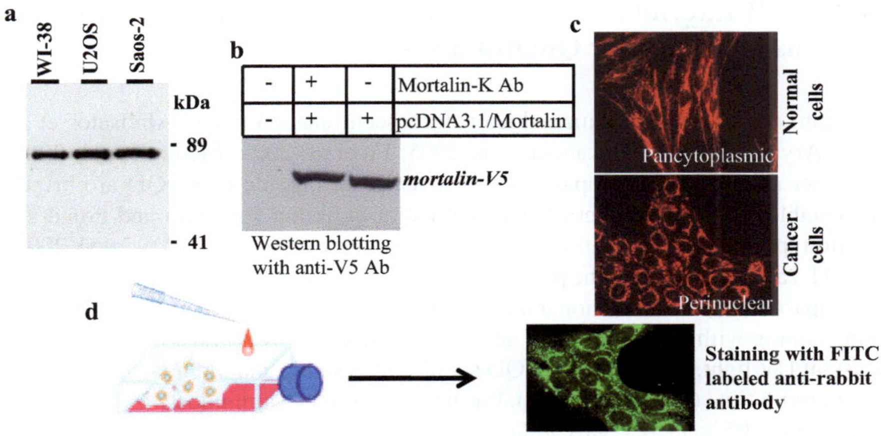

Fig. 21.1 Mortalin-specific polyclonal antibody with cell-internalizing characteristics. **a** Western blotting of human cell lysates with anti-mortalin antibody showing its reactivity to a single protein. **b** Immunoprecipitation of V5-tagged mortalin protein by anti-mortalin antibody showing specificity of antibody to mortalin. **c** Immunostaining of mortalin in normal and cancer cells with K antibody showing pancytoplasmic distribution in normal and perinuclear in cancer cells. **d** Internalization of anti-mortalin antibody in human cancer cells. Antibody was added to the culture medium and was detected inside the cells by staining with FITC-conjugated secondary antibody. Modified from Shiota et al. (2007)

We first generated polyclonal anti-mortalin antibody (K-antibody) in rabbit using full-length recombinant mortalin expressed in bacteria as an antigen. As shown in Fig. 21.1a, b, K-antibody was examined for its reactivity by Western blotting and immunoprecipitation assays. It recognized a single band from the protein lysates prepared from human normal and cancer cells. Its specificity to mortalin was confirmed by immunoprecipitation assays. V5-tagged mortalin protein was expressed in human cells from an exogenous promoter and was immunoprecipitated with the K-antibody. Immunoprecipitated protein was detected by anti-V5 (Fig. 21.1b) antibody by Western blotting demonstrating that the K-antibody was extremely specific for mortalin. Furthermore, the antibody detected the differential staining pattern of mortalin in normal and transformed human cells as reported earlier (Fig. 21.1c) (Wadhwa et al. 1993b). Based on the global profiling of cell surface proteins, mortalin was detected on the surface of cancer cells (Shin et al. 2003). We attempted to examine this by incubating the cells with antibody-supplemented medium followed by their staining with fluorescence tagged secondary antibody (Fig. 21.1d). By this protocol, we discovered that the K-antibody got internalized into cells. Number of cell lines including (osteosarcoma, U2OS and Saos-2), breast carcinoma (MCF7, SKBR3, T47D), fibrosarcoma (HT1080), cervical carcinoma (HeLa), lung carcinoma (A549) and glioblastoma (A172), urothelial carcinoma (J82), hepatocarcinoma (HepG2) and embryonic carcinoma (NEC8) tested so far confirmed the cell internalizing property of anti-mortalin (K) antibody (Shiota et al. 2007).

21.3 Cell-Internalizing Polyclonal Anti-Mortalin Antibody as a Carrier for Quantum Dots

Quantum dots (QD) are semiconductor fluorescent nanoparticles (Alivisatos et al. 2005; Arya et al. 2005; Bakalova et al. 2005; Fu et al. 2005; Medintz et al. 2005; Santra et al. 2005). In comparison to conventional organic dyes, QD are bright, photostable, posses size-controllable multicolor emission spectrum and broad absorption spectra allowing their visualization with single light source (Arya et al. 2005) (Fig. 21.2a). We confirmed the photostability of QD by using them as staining probes in comparison to the conventional fluorochrome, Alexa. As shown in Fig. 21.2b, mortalin staining with Alexa bleached and became invisible by 8 min when exposed to constant UV light. In comparison, QD staining was stable and did not show any difference with time (Kaul et al. 2003). Furthermore, double-immunostaining with two kinds of QD (655 and 525) gave a clear high-resolution overlay (Fig. 21.2c) (Wadhwa et al. 2005) suggesting that the QDs could be recruited for multicolor protein imaging.

Since QD are highly resistant to chemical and metabolic degradation, they are considered suitable for long-term bioimaging. Some bio-applications of QD include QD-based cell motility assay and *in vivo* imaging of cancer cells (Ballou et al. 2005; Gu et al. 2005; Santra et al. 2005; Pinaud et al. 2006; Jaiswal and Simon 2007). However, in spite of such several advantages over conventional fluorescent dyes including inertness, greater photostability and a wider range of excitation/emission wavelengths, their toxicity has been an issue for its wider use as analytic, diagnostic and therapeutic tools. Hence the generation of modified QD with low toxicity, high hydrophilicity, high resolution and photostability have been the subject of research at molecular, cellular and animal levels (Gao et al. 2004; Braydich-Stolle et al. 2005; Fan et al. 2005; Hasegawa et al. 2005). One of the hurdles in the use of QD is their delivery to cells. Several modifications of QD have been formulated for their efficient internalization into the cells (Hoshino et al. 2004; Alivisatos et al. 2005; Bentzen et al. 2005; Mason et al. 2005; Silver and Ou 2005; Tomlinson et al. 2005; Young and Rozengurt 2006). Many studies have shown that coating QDs with cationic peptides resulted in their rapid endocytosis (Voura et al. 2004; Silver and Ou 2005; Tomlinson et al. 2005; Vu et al. 2005). Hasegawa et al. (2005) reported that CHPNH2-QD nanoparticles, monodisperse hybrid nanoparticles (38 nm in diameter) of QD mixed with nanogels of cholesterol-bearing pullulan (CHP) modified with amino groups (CHPNH2), get internalized into various human cells with higher efficiency than conventional, cationic liposome carrier (Hoshino et al. 2004). In spite of the continuous developments in QD modifications, there toxicity has been an issue of debate. A concentration-dependent toxicity of Ag nanoparticles was reported in a mouse spermatogonial stem cell line as a model *in vitro* (Fan et al. 2005). CdSe-core QDs were found to be acutely toxic to primary hepatocytes due to the liberation of free Cd2 + ions resulting from deterioration of the CdSe lattice. When appropriately coated, CdSe-core QDs could be rendered nontoxic and used to track cell migration and reorganization *in vitro* (Derfus et al. 2004).

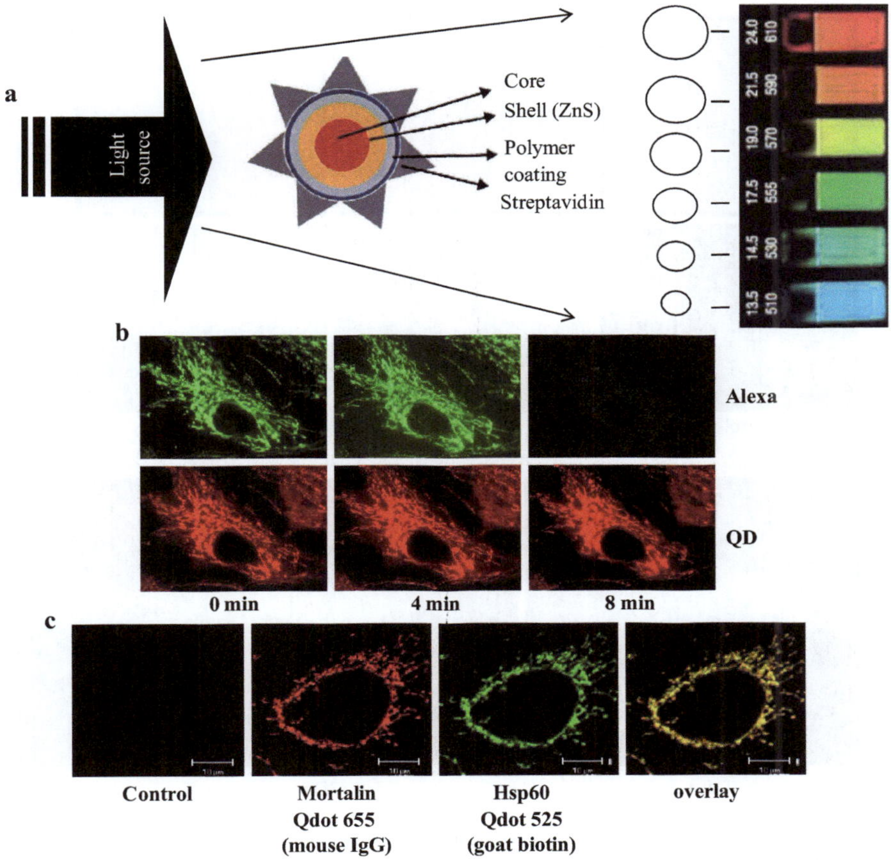

Fig. 21.2 Use of mortalin antibody conjugated with quantum dots for immunostaining. **a** Diagrammatic presentation of size-tunable fluorescence properties of quantum dots; spectral range of the six QDs excited at single light source (365 nm UV) is shown. **b** Stability of QD staining as compared to Alexa was determined by time-course experiment. **c** Double immunolocalization of mortalin and HSP60 as obtained by Qdot-655 and Qdot-525 labeled secondary antibodies, respectively, is shown. Modified from Kaul et al. (2003) and Wadhwa et al. (2005)

We prepared a conjugate of QD with an internalizing K-antibody. The QD-K antibody conjugate was added into the cell culture medium followed by their examination in the cells by microscopy. As shown in Fig. 21.3a, we found that QD was internalized into the cells along with the K-Antibody. Furthermore (i) optical sections revealed the presence of QD inside the cells in cytoplasm and (ii) QDs could be observed in cells even after a number of divisions (Kaul et al. 2007b). These data demonstrated that the QD-K antibody conjugate efficiently gets internalized and are nontoxic to cells. It thus provides a sensitive tool for long-term molecular imaging.

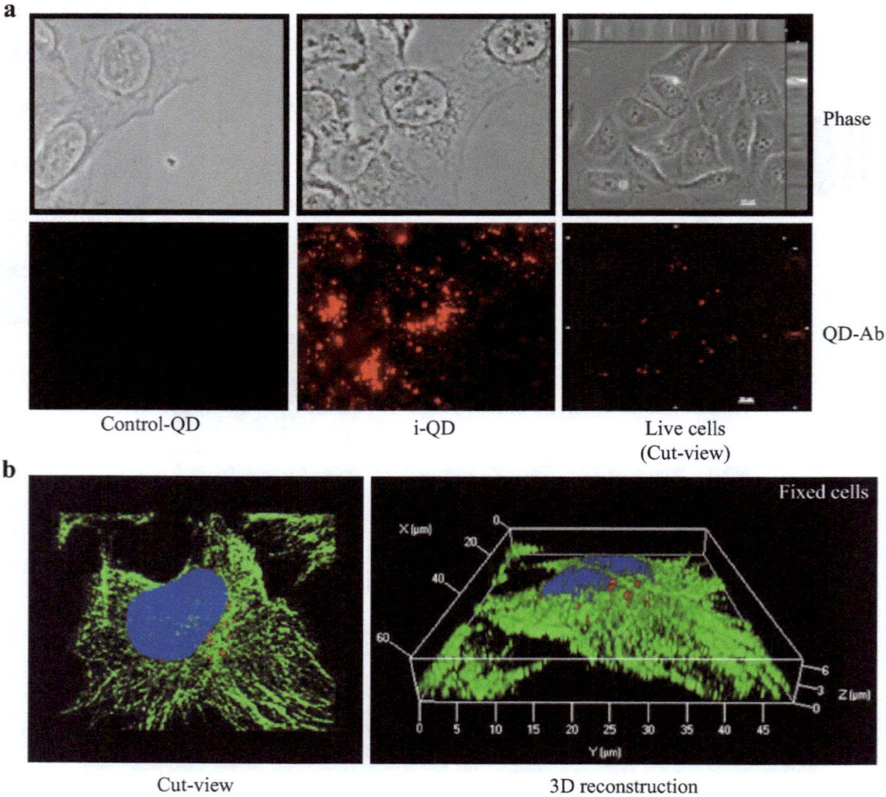

Fig. 21.3 Use of anti-mortalin antibody conjugated with quantum dots for live imaging. **a** Imaging of cells labeled with QD-Ab conjugates. For internalization, QD-Ab conjugates ($\sim 0.2-0.4$ mg/ml) were added in the cell culture medium for 24 h followed by direct examination under the microscope for live imaging or the cells were fixed and observed by confocal microscope to examine the presence of QD inside the cells (**b**). Modified from Kaul et al. (2007b)

21.4 Cell-internalizing Monoclonal Anti-mortalin Antibody as a Nano-carrier

Since only a few polyclonal antibodies raised against mortalin have cell internalizing feature, we anticipated that the internalization of mortalin antibody might depend on the amount of cell surface mortalin as well as epitope and paratopes of the antibody. To characterize this phenomenon, we raised monoclonal antibodies against the full-length mortalin in mouse. A large number of hybridoma clones were screened for their specificity, reactivity and internalizing and non-internalizing features. We found that although many clones showed excellent specificity to mortalin in Western blotting and immunostaining assays, only a fraction of them (Fig. 21.4b) possessed

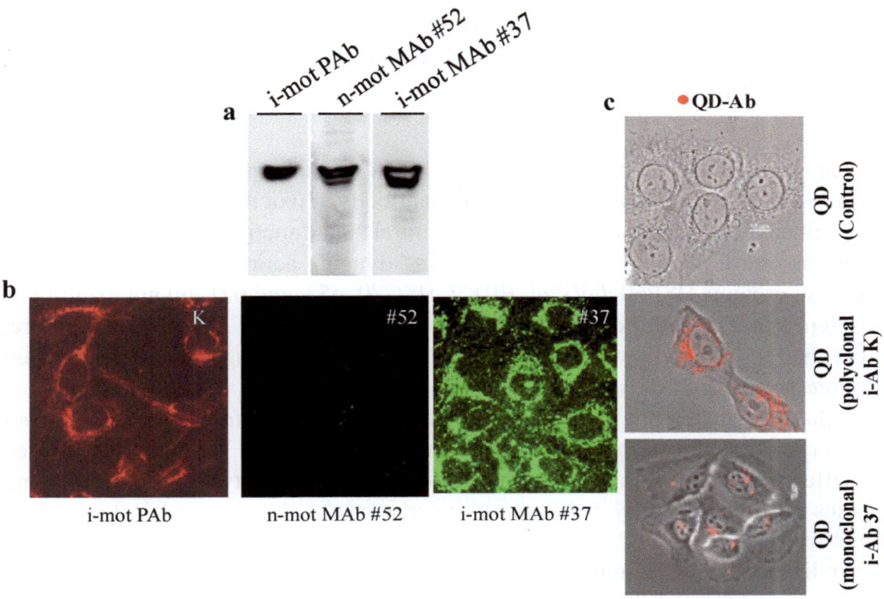

Fig. 21.4 Mortalin-specific monoclonal antibody with cell-internalizing characteristics. **a** Western blotting of U2OS cell lysate with polyclonal (i-mot PAb) and monoclomal internalizing (i-mot MAb) and non-internalizing (n-mot MAb) anti-mortalin antibody clones. **b** Immunostaining of cells with secondary antibodies after incubation with primary non-internalizing (#52) and internalizing (K and #37) antibodies. **c** Imaging of live cells labeled with QD conjugated with either the polyclonal (i-Ab K) or monoclonal (i-Ab 37) antibody. Modified from Kaul et al. (2007b) and Shiota et al. (2007)

internalization feature. Clone #52 was equally reactive in the Western blotting and immunocytochemistry, but did not show internalization (Fig. 21.4a, b). These data confirmed that the internalization is a specific phenomenon. To further rule out the possibility of internalization as a staining artifact, internalization of monoclonal anti-mortalin antibody (i-mot MAb) was examined in cells grown in low and high FBS-medium, subjected to acid treatment, and extensive washings with PBS/Triton X-100. We found that whether the cells are in lag phase of growth or arrested by serum starvation, QD-Ab conjugate got internalized as efficiently as in the growing cells. Time course experiment revealed that the internalization becomes visible at about 1–2 h of incubation of cells and increases progressively in next 24 h. As shown in Fig. 21.4c, both polyclonal K and monoclonal (#37) anti-mortalin antibodies conjugated to Qdot 655 were able to internalize. Antibodies, QD and QD-Ab conjugate were added to the culture medium of cells and their internalization was examined under the microscope after 24 h. QD alone did not show any internalization during the same time. The presence of internalized QD (iQD) inside the cells was also confirmed by cross-section analysis as shown in Fig. 21.3b. The data demonstrate that the specific anti-mortalin antibodies with cell-internalizing properties could serve as a carrier to deliver QDs into the cells.

21.5 Generation of Illuminating Cells by Incorporation of Quantum Dots Conjugated to Internalizing Antibodies to Mortalin

We next generated illuminating cells (i-cells) by incorporating quantum dots conjugated to internalizing anti-mortalin antibody. Toxicity of QD to cells was determined by investigations on the structural and functional proteins and cell proliferation. Cytoskeleton proteins, ezrin, dynein, actin and vimentin; stress response proteins HSP60, GRP94, ERK2, HSC70, p53, and p21 did not exhibit any difference in iQD cells compared to control cells (Kaul et al. 2007b). Furthermore, more than 95% of cells with iQD also incorporated BrdU suggesting that iQD did not interfere with DNA synthesis (Fig. 21.5a). All the stages of cell division (from interphase to late telophase) were identified in i-cells demonstrating that QD did not interfere with cell division. Cell viability when examined by WST-1 assay showed no difference in control and QD-internalized cells. Sensitivity of iQD for imaging studies was examined by pulse-chase experiments as diagrammed in Fig. 21.5b. Cells were exposed to iQD for 24 h following which the cells were split onto three coverslips and cultured in normal medium during the next three days. The medium with un-incorporated/leftover iQD conjugate was transferred to fresh cells and processed for next 24 h. The process was repeated for at least 5–7 days. We found that the iQD was visible in cells even after five times recycling of the medium (Fig. 21.5b) indicating that only a minute quantities of QD are required for imaging analysis. The data was further supported by the cell-splitting regime followed in our pulse chase experiment in which cells were cultured for 3–4 days during which they underwent about five-seven population doublings as indicated by increase in cell number. QDs were visible even in cells that had divided five-eight times (Fig. 21.5b) suggesting that i-cells could be tracked for long time for imaging.

21.6 Cell Internalizing Antibodies as Nano-carriers for DNA

Polethylenimine (PEI)-based cationic polyplexes have been widely adopted as non-viral transfection reagents (Boussif et al. 1995; Duan et al. 2008; Park et al. 2009). However, there are yet unsolved issues on their toxicity and targeted gene delivery. We generated PEI-mot (anti-mortalin) antibody complex by covalently coupling it with N-(6-Maleimidocaproyloxy)succinimide (EMCS) linker as shown in Fig. 21.6a. The complex was confirmed to bind to mortalin by ELISA assay. i-motMAb-PEI/DNA polyplexes were then generated and analyzed by gel retardation assays on a 0.8% agarose gel (Shiota et al. 2007). Both PEI-DNA and i-motMAb-PEI-DNA polyplexes were evaluated for *in vitro* transfection efficiency in U2OS cells using expression plasmid for DsRed2 protein. Expression of the plasmid was examined by direct observation under the microscope (Fig. 21.6b) and flow cytometry (Shiota et al. 2007). These data showed that i-motMAb-PEI was highly efficient. To further verify that the

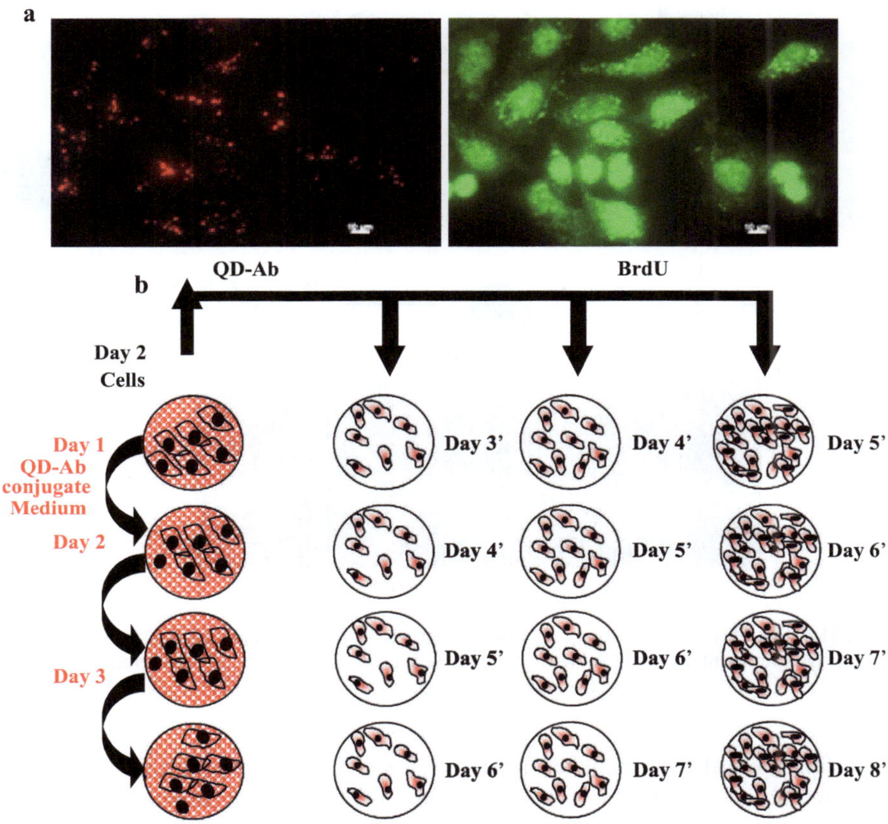

Fig. 21.5 Imaging of cells illuminated with quantum dots and anti-mortalin antibody conjugate. **a** Cells labeled with QDs (left panel) showed normal DNA synthesis as seen by bromodeoxyuridine (BrdU) labeling (right panel). There were no differences in the BrdU labeling in control and QD-internalized cells. **b** Pulse chase experiment scheme for tracking the iQD in proliferating cells. QD-Ab conjugate (0.2–0.4 mg/ml) was added to the cell culture medium on day 0. On day 1, medium with remaining QD-Ab conjugate was transferred to fresh cells. The cells were split and divided into three and examined for the presence of internalized QD for next three days as shown. Internalized QDs were visible in cells even at 8 days (Day 8') of the experiment. Modified from Kaul et al. (2007b)

delivery of DNA into the cells was indeed enhanced by internalizing anti-mortalin monoclonal antibody (i-motMAb), we used various controls including (i) control antibodies and (ii) mortalin-overexpressing isogenic derivatives. We found that the gene delivery was mediated by i-motMAb and was dependent on the expression level of mortalin protein. Since mortalin is frequently upregulated during carcinogenesis (Wadhwa et al. 2006), we anticipated that i-motMAb-PEI nanocarrier could be useful to selectively enhance delivery to cancer cells. We tested this hypothesis using a variety of human cancer and normal cells and luciferase as a reporter gene

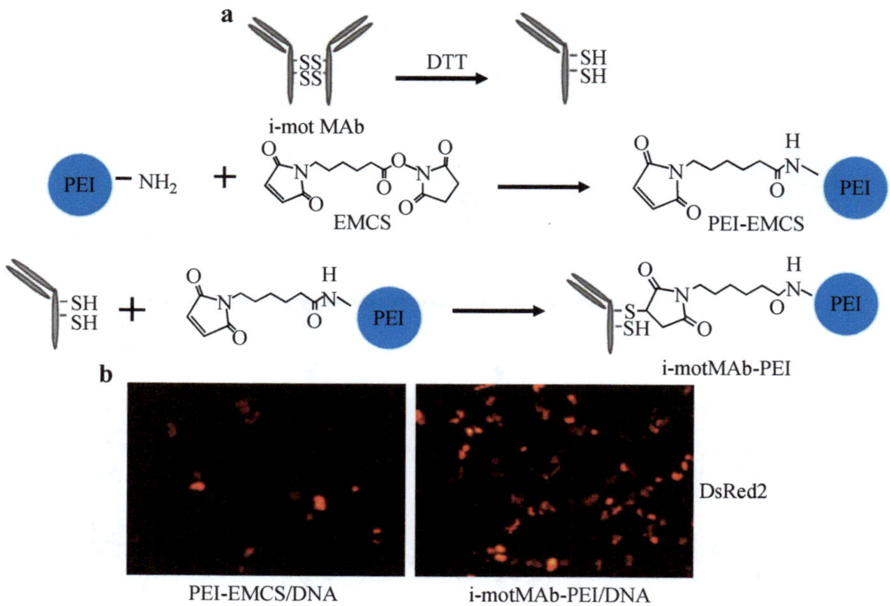

Fig. 21.6 Use of mortalin antibody for gene delivery. **a** Schematic presentation of the synthesis of polyethylenimin-anti mortalin antibody conjugates. **b** Expression of DsRed2 delivered by PEI-mot (mortalin) antibody complex to the human cancer cells. Modified from Shiota et al. (2007)

transfer. i-motAb-PEI mediated luciferase transfer and expression was enhanced in osteosarcoma (U2OS), cervical carcinoma (HeLa), breast carcinoma (SKBR3, T47D) cell lines, but not in normal cells (TIG-1, HMES3 and HMES4) confirming that i-motMAb could serve as a cancer cell specific nanocarrier and provides a useful system to target chemical toxins, small functional molecules and peptides to cancer cells.

21.7 Conclusions

As described in this chapter the cell internalizing anti-mortalin polyclonal and monoclonal antibodies serve as nanocarrier for imaging and gene delivery. In particular, we discussed the use of internalizing anti-mortalin antibody conjugates with the quantum dots to generate illuminated cells. The quantum dots thus internalized inside the cells with the help of these specific antibodies are nontoxic and could remain inside the cells for a long time thus enabling their long-term imaging *in vitro* and *in vivo*. These unique anti-mortalin antibodies have strong potential for its use in biomedical application including drug/toxin delivery to specific cell type/disease.

References

Alivisatos AP, Gu W, Larabell C (2005) Quantum dots as cellular probes. Annu Rev Biomed Eng 7:55–76

Allen TM (2002) Ligand-targeted therapeutics in anticancer therapy. Nat Rev Cancer 2:750–763

Arya H, Kaul Z, Wadhwa R, Taira K, Hirano T, Kaul SC (2005) Quantum dots in bio-imaging: Revolution by the small. Biochem Biophys Res Commun 329:1173–1177

Bakalova R, Zhelev Z, Ohba H, Baba Y (2005) Quantum dot-based western blot technology for ultrasensitive detection of tracer proteins. J Am Chem Soc 127:9328–9329

Ballou B, Ernst LA, Waggoner AS (2005) Fluorescence imaging of tumors in vivo. Curr Med Chem 12:795–805

Bentzen EL, Tomlinson ID, Mason J, Gresch P, Warnement MR, Wright D, Sanders-Bush E, Blakely R, Rosenthal SJ (2005) Surface modification to reduce nonspecific binding of quantum dots in live cell assays. Bioconjug Chem 16:1488–1494

Boussif O, Lezoualc'h F, Zanta MA, Mergny MD, Scherman D, Demeneix B, Behr JP (1995) A versatile vector for gene and oligonucleotide transfer into cells in culture and in vivo: polyethylenimine. Proc Natl Acad Sci U S A 92:7297–7301

Braydich-Stolle L, Hussain S, Schlager JJ, Hofmann MC (2005) In vitro cytotoxicity of nanoparticles in mammalian germline stem cells. Toxicol Sci 88:412–419

Chiu SJ, Ueno NT, Lee RJ (2004) Tumor-targeted gene delivery via anti-HER2 antibody (trastuzumab, Herceptin) conjugated polyethylenimine. J Control Release 97:357–369

Cussac D, Pichereaux C, Colomba A, Capilla F, Pont F, Gaits-Iacovoni F, Lamant L, Espinos E, Burlet-Schiltz O, Monsarrat B et al (2006) Proteomic analysis of anaplastic lymphoma cell lines: identification of potential tumour markers. Proteomics 6:3210–3222

Densmore CL, Kleinerman ES, Gautam A, Jia SF, Xu B, Worth LL, Waldrep JC, Fung YK, T'Ang A, Knight V (2001) Growth suppression of established human osteosarcoma lung metastases in mice by aerosol gene therapy with PEI-p53 complexes. Cancer Gene Ther 8:619–627

Derfus AM, Chan WCW, Bhatia SN (2004) Probing the cytotoxicity of semiconductor quantum dots. Nano Lett 4:11–18

Duan Y, Guan X, Ge J, Quan D, Zhuo Y, Ye H, Shao T (2008) Cationic nano-copolymers mediated IKKbeta targeting siRNA inhibit the proliferation of human Tenon's capsule fibroblasts in vitro. Mol Vis 14:2616–2628

Dundas SR, Lawrie LC, Rooney PH, Murray GI (2004) Mortalin is over-expressed by colorectal adenocarcinomas and correlates with poor survival. J Pathol 205:74–81

Fan H, Leve EW, Scullin C, Gabaldon J, Tallant D, Bunge S, Boyle T, Wilson MC, Brinker CJ (2005) Surfactant-assisted synthesis of water-soluble and biocompatible semiconductor quantum dot micelles. Nano Lett 5:645–648

Fu A, Gu W, Larabell C, Alivisatos AP (2005) Semiconductor nanocrystals for biological imaging. Curr Opin Neurobiol 15:568–575

Gao X, Cui Y, Levenson RM, Chung LW, Nie S (2004) In vivo cancer targeting and imaging with semiconductor quantum dots. Nat Biotechnol 22:969–976

Gao J, Li B, Li H, Zhang X, Zhang D, Zhao L, Wang C, Fang C, Qian W, Hou S et al (2009) Development and characterization of a fully functional small anti-HER2 antibody. BMB Rep 42:636–641

Gu W, Pellegrino T, Parak WJ, Boudreau R, Le Gros MA, Gerion D, Alivisatos AP, Larabell CA (2005) Quantum-dot-based cell motility assay. Sci STKE 2005:l5

Guillem VM, Tormo M, Revert F, Benet I, Garcia-Conde J, Crespo A, Alino SF (2002) Polyethyleneimine-based immunopolyplex for targeted gene transfer in human lymphoma cell lines. J Gene Med 4:170–182

Guo W, Lee RL (1999) Receptor-targeted gene delivery via folate-conjugated polyethylenimine. AAPS PharmSci 1:E19

Hanifi A, Fathi MH, Sadeghi HM, Varshosaz J (2010) Mg2 +substituted calcium phosphate nano particles synthesis for non viral gene delivery application. J Mater Sci Mater Med 21:2393–2401

Hasegawa U, Nomura SM, Kaul SC, Hirano T, Akiyoshi K (2005) Nanogel-quantum dot hybrid nanoparticles for live cell imaging. Biochem Biophys Res Commun 331:917–921
Hoshino A, Hanaki K, Suzuki K, Yamamoto K (2004) Applications of T-lymphoma labeled with fluorescent quantum dots to cell tracing markers in mouse body. Biochem Biophys Res Commun 314:46–53
Hu-Lieskovan S, Heidel JD, Bartlett DW, Davis ME, Triche TJ (2005) Sequence-specific knockdown of EWS-FLI1 by targeted, nonviral delivery of small interfering RNA inhibits tumor growth in a murine model of metastatic Ewing's sarcoma. Cancer Res 65:8984–8992
Jaiswal JK, Simon SM (2007) Optical monitoring of single cells using quantum dots. Methods Mol Biol 374:93–104
Kaul Z, Yaguchi T, Kaul SC, Hirano T, Wadhwa R, Taira K (2003) Mortalin imaging in normal and cancer cells with quantum dot immuno-conjugates. Cell Res 13:503–507
Kaul SC, Aida S, Yaguchi T, Kaur K, Wadhwa R (2005) Activation of wild type p53 function by its mortalin-binding, cytoplasmically localizing carboxyl terminus peptides. J Biol Chem. 280:39373–39379
Kaul SC, Deocaris CC, Wadhwa R (2007a) Three faces of mortalin: A housekeeper, guardian and killer. Exp Gerontol 42:263–274
Kaul Z, Yaguchi T, Harada JI, Ikeda Y, Hirano T, Chiura HX, Kaul SC, Wadhwa R (2007b) An antibody-conjugated internalizing quantum dot suitable for long-term live imaging of cells. Biochem Cell Biol 85:133–140
Kunath K, Merdan T, Hegener O, Haberlein H, Kissel T (2003) Integrin targeting using RGD-PEI conjugates for in vitro gene transfer. J Gene Med 5:588–599
Lopes de Menezes DE, Pilarski LM, Allen TM (1998) In vitro and in vivo targeting of immunoliposomal doxorubicin to human B-cell lymphoma. Cancer Res 58:3320–3330
Lu WJ, Lee NP, Kaul SC, Lan F, Poon RT, Wadhwa R, Luk JM (2011) Mortalin-p53 interaction in cancer cells is stress dependent and constitutes a selective target for cancer therapy. Cell Death Differ 18:1046–1056
Ma Z, Izumi H, Kanai M, Kabuyama Y, Ahn NG, Fukasawa K (2006) Mortalin controls centrosome duplication via modulating centrosomal localization of p53. Oncogene. 25:5377–5390
Maruyama K, Takahashi N, Tagawa T, Nagaike K, Iwatsuru M (1997) Immunoliposomes bearing polyethyleneglycol-coupled Fab' fragment show prolonged circulation time and high extravasation into targeted solid tumors in vivo. FEBS Lett 413:177–180
Mason JN, Tomlinson ID, Rosenthal SJ, Blakely RD (2005) Labeling cell-surface proteins via antibody quantum dot streptavidin conjugates. Methods Mol Biol 303:35–50
Medintz IL, Uyeda HT, Goldman ER, Mattoussi H (2005) Quantum dot bioconjugates for imaging, labelling and sensing. Nat Mater 4:435–446
Merdan T, Callahan J, Petersen H, Kunath K, Bakowsky U, Kopeckova P, Kissel T, Kopecek J (2003) Pegylated polyethylenimine-Fab' antibody fragment conjugates for targeted gene delivery to human ovarian carcinoma cells. Bioconjug Chem 14:989–996
Moffatt S, Cristiano RJ (2006) PEGylated J591 mAb loaded in PLGA-PEG-PLGA tri-block copolymer for targeted delivery: in vitro evaluation in human prostate cancer cells. Int J Pharm 317:10–13
Park JW, Hong K, Kirpotin DB, Colbern G, Shalaby R, Baselga J, Shao Y, Nielsen UB, Marks JD, Moore D, et al (2002) Anti-HER2 immunoliposomes: enhanced efficacy attributable to targeted delivery. Clin Cancer Res 8:1172–1181
Park JS, Na K, Woo DG, Yang HN, Kim JM, Kim JH, Chung HM, Park KH (2009) Non-viral gene delivery of DNA polyplexed with nanoparticles transfected into human mesenchymal stem cells. Biomaterials 31:124–132
Peer D, Karp JM, Hong S, Farokhzad OC, Margalit R, Langer R (2007) Nanocarriers as an emerging platform for cancer therapy. Nat Nanotechnol 2:751–760
Pilzer D, Saar M, Koya K, Fishelson Z (2010) Mortalin inhibitors sensitize K562 leukemia cells to complement-dependent cytotoxicity. Int J Cancer 126:1428–1435

Pinaud F, Michalet X, Bentolila LA, Tsay JM, Doose S, Li JJ, Iyer G, Weiss S (2006) Advances in fluorescence imaging with quantum dot bio-probes. Biomaterials 27:1679–1687

Santra S, Dutta D, Walter GA, Moudgil BM (2005) Fluorescent nanoparticle probes for cancer imaging. Technol Cancer Res Treat 4:593–602

Sapra P, Allen TM (2002) Internalizing antibodies are necessary for improved therapeutic efficacy of antibody-targeted liposomal drugs. Cancer Res 62:7190–7194

Shin BK, Wang H, Yim AM, Le Naour F, Brichory F, Jang JH, Zhao R, Puravs E, Tra J, Michael CW et al (2003) Global profiling of the cell surface proteome of cancer cells uncovers an abundance of proteins with chaperone function. J Biol Chem 278:7607–7616

Shiota M, Ikeda Y, Kaul Z, Itadani J, Kaul SC, Wadhwa R (2007) Internalizing antibody-based targeted gene delivery for human cancer cells. Hum Gene Ther 18:1153–1160

Silver J, Ou W (2005) Photoactivation of quantum dot fluorescence following endocytosis. Nano Lett 5:1445–1449

Strehblow C, Schuster M, Moritz T, Kirch HC, Opalka B, Petri JB (2005) Monoclonal antibody-polyethyleneimine conjugates targeting Her-2/neu or CD90 allow cell type-specific nonviral gene delivery. J Control Release 102:737–747

Sugano M, Egilmez NK, Yokota SJ, Chen FA, Harding J, Huang SK, Bankert RB (2000) Antibody targeting of doxorubicin-loaded liposomes suppresses the growth and metastatic spread of established human lung tumor xenografts in severe combined immunodeficient mice. Cancer Res 60:6942–6949

Tomlinson ID, Mason JN, Blakely RD, Rosenthal SJ (2005) Peptide-conjugated quantum dots: imaging the angiotensin type 1 receptor in living cells. Methods Mol Biol 303:51–60

Voura EB, Jaiswal JK, Mattoussi H, Simon SM (2004) Tracking metastatic tumor cell extravasation with quantum dot nanocrystals and fluorescence emission-scanning microscopy. Nat Med 10:993–998

Vu TQ, Maddipati R, Blute TA, Nehilla BJ, Nusblat L, Desai TA (2005) Peptide-conjugated quantum dots activate neuronal receptors and initiate downstream signaling of neurite growth. Nano Lett 5:603–607

Wadhwa R, Kaul SC, Ikawa Y, Sugimoto Y (1993a) Identification of a novel member of mouse hsp70 family. Its association with cellular mortal phenotype. J Biol Chem 268:6615–6621

Wadhwa R, Kaul SC, Mitsui Y, Sugimoto Y (1993b) Differential subcellular distribution of mortalin in mortal and immortal mouse and human fibroblasts. Exp Cell Res 207:442–448

Wadhwa R, Sugihara T, Yoshida A, Nomura H, Reddel RR, Simpson R, Maruta H, Kaul SC (2000) Selective toxicity of MKT-077 to cancer cells is mediated by its binding to the hsp70 family protein mot-2 and reactivation of p53 function. Cancer Res 60:6818–6821

Wadhwa R, Ando H, Kawasaki H, Taira K, Kaul SC (2003) Targeting mortalin using conventional and RNA-helicase-coupled hammerhead ribozymes. EMBO Rep 4:595–601

Wadhwa R, Takano S, Taira K, Kaul SC (2004) Reduction in mortalin level by its antisense expression causes senescence-like growth arrest in human immortalized cells. J Gene Med 6:439–444

Wadhwa R, Takano S, Kaur K, Aida S, Yaguchi T, Kaul Z, Hirano T, Taira K, Kaul SC (2005) Identification and characterization of molecular interactions between mortalin/mtHsp70 and HSP60. Biochem J 391:185–190

Wadhwa R, Takano S, Kaur K, Deocaris CC, Pereira-Smith OM, Reddel RR, Kaul SC (2006) Upregulation of mortalin/mthsp70/Grp75 contributes to human carcinogenesis. Int J Cancer 118:2973–2980

Wang G, Qiu J, Wang R, Krause A, Boyer JL, Hackett NR, Crystal RG (2010) Persistent expression of biologically active anti-HER2 antibody by AAVrh.10-mediated gene transfer. Cancer Gene Ther 17:559–570

Young SH, Rozengurt E (2006) Qdot nanocrystal conjugates conjugated to bombesin or ANG II label the cognate G protein-coupled receptor in living cells. Am J Physiol Cell Physiol 290:C728–732

Zanta MA, Boussif O, Adib A, Behr JP (1997) In vitro gene delivery to hepatocytes with galactosylated polyethylenimine. Bioconjug Chem 8:839–844

Index

S. C. Kaul, R. Wadhwa (eds.), *Mortalin Biology: Life, Stress and Death*,
DOI 10.1007/978-94-007-3027-4, © Springer Science+Business Media B.V. 2012

MIX
Papier aus verantwortungsvollen Quellen
Paper from responsible sources
FSC® C105338

If you have any concerns about our products,
you can contact us on
ProductSafety@springernature.com

In case Publisher is established outside the EU,
the EU authorized representative is:
Springer Nature Customer Service Center GmbH
Europaplatz 3, 69115 Heidelberg, Germany

Printed by Libri Plureos GmbH
in Hamburg, Germany